Human Virology

Human Virology

Edited by Janice Campbell

◻SYRAWOOD
PUBLISHING HOUSE

New York

Published by Syrawood Publishing House,
750 Third Avenue, 9th Floor,
New York, NY 10017, USA
www.syrawoodpublishinghouse.com

Human Virology
Edited by Janice Campbell

International Standard Book Number: 978-1-68286-847-8 (Hardback)

Cataloging-in-Publication Data

Human virology / edited by Janice Campbell.
    p. cm.
Includes bibliographical references and index.
ISBN 978-1-68286-847-8
1. Medical virology. 2. Virus diseases. 3. Virology. I. Campbell, Janice.
QR201.V55 H86 2020
616.910 1--dc23

Safety and Efficacy of Hepatitis B Vaccination in Cirrhosis of Liver

D. Ajith Roni,[1] Rama Mohan Pathapati,[2] A. Sathish Kumar,[1] Lalit Nihal,[1] K. Sridhar,[1] and Sujith Tumkur Rajashekar[2]

[1] *Medical Gastroenterology, Narayana Medical College Hospital, Nellore, Andhra Pradesh 524002, India*
[2] *Clinical Pharmacology, Narayana Medical College Hospital, Nellore, Andhra Pradesh 524002, India*

Correspondence should be addressed to Rama Mohan Pathapati; pill4ill@yahoo.co.in

Academic Editor: Masao Matsuoka

Introduction. Patients with chronic liver disease (CLD) are more likely to have severe morbidity and fatality rate due to superimposed acute or chronic hepatitis B (HBV) infection. The literature has shown that hepatitis B vaccines are safe and effective in patients with CLD, but the data in cirrhosis liver is lacking. We assessed the safety and immunogenicity of HBV vaccine in patients with cirrhosis liver. *Methods*. CTP classes A and B CLD patients negative for hepatitis B surface antigen and antibody to hepatitis B core antigen were included. All patients received three doses of hepatitis B vaccine 20 mcg intramuscularly at 0, 30, and 60 days. Anti-HBs antibody was measured after 120 days. *Results*. 52 patients with mean age 47.48 ± 9.37 years were studied. Response rates in CTP classes A and B were 88% and 33.3%. We observed that the alcoholic chronic liver disease had less antibody response (44%) than other causes of chronic liver disease such as cryptogenic 69% and HCV 75%. *Conclusions*. Patients with cirrhosis liver will have low antibody hepatitis B titers compared to general population. As the age and liver disease progress, the response rate for hepatitis B vaccination will still remain to be weaker.

1. Introduction

Globally, chronic HBV infection affects over 350 million people, and up to 40% of these cases may progress to cirrhosis, liver failure, or hepatocellular carcinoma [1]. Chronic liver disease [CLD] contributes to approximately 400000 hospitalizations and nearly 30,000 deaths annually worldwide [2, 3]. When compared with patients without liver disease, patients with CLD are more likely to have severe complications and also severe fatality rate due to superimposed acute or chronic HBV infection. Both acute and chronic coinfections with HBV can be prevented by HBV vaccination [4, 5]. Strong epidemiological evidence suggests an increased occurrence of fulminant liver failure, cirrhosis and hepatocellular carcinoma in patients with HBV, and HCV coinfection [6, 7]. HBV vaccination is safe and well tolerated and has high seroconversion rates in patients with mild to moderate CLD but has reduced efficacy in advanced liver disease and after liver transplantation [8–17]. To minimize the occurrence of HBV infection in CLD, a variety of organizations have recommended HBV vaccination for these patients [18, 19]. The immune response to HBV vaccines among patients with CLD varies from 70% to 90%. Hence in evaluating HBV vaccination in patients with cirrhosis liver, three questions need to be answered: (1) who needs vaccination? And (2) is the vaccination safe? and (3) is it effective? Answering these questions will provide an effective strategy for applying HBV vaccines in CLD. To this purpose we studied the safety and immunogenicity of 3 doses of $20 \mu g$ of HBV vaccine in patients with cirrhosis liver.

2. Methods

This prospective open label study was conducted in the department of medical gastroenterology. Institutional ethics committee approved the study protocol. Informed consent was obtained from study participants. Patients with CLD of CTP class A and B were enrolled in the study. Both male and female patients between 18 and 60 years who were serologically negative for hepatitis B surface antigen, antibody to hepatitis B core antigen and have no history of hepatitis B vaccination were included. Patients with CTP class-C, having malignancy, acute liver disease, HIV, receiving immunosuppressive medications and life expectancy less than 120 days were excluded. All patients received three doses of HBV vaccine (Shanvac, M/s Shantha Biotech) 20 mcg intramuscularly over the deltoid region during all the three visits at 0, 30, and 60 days. A 5 mL of blood was collected in a plain vacutainer tube, the serum was separated, and the analysis for anti-Hbc total and titer for anti-Hbs was done on the same day of the collection of samples. Blood analysis for anti-Hbc total and titer for anti-HBs was done with chemiluminescence analyzer-Access 2 Immunoassay analyzer, and Beckman coulter. Anti-HBs antibody was measured after 120 days, and according to antibody titers patients were classified into good responders, poor responders, and nonresponders. Good responders were defined as those having the anti-HBs titer were > or =100 mUI/mL, poor responders having anti-HBs titer between 10 and 99 mUI/mL, and nonresponders having anti-HBs titer <10 mUI/mL. The secondary outcome was to assess the safety of HBV vaccination in CLD. Patients reported adverse events; infusion site reactions and routine laboratory parameters were considered safety markers of the study.

2.1. Statistical Analysis. Data was entered into excel spreadsheet 2007 and analyzed by using the GraphPad Prism Software version 4. All the continuous data will be expressed as mean ± SD. Categorical data was expressed as numbers and percentages. Chi-square test was used to detect differences between groups. A two-tailed P value < 0.05 was considered statistically significant.

3. Results

52 patients received 3 doses of HBV vaccine; the mean age of patients was 47.48 ± 9.37 years. There were 37 males and 15 females. 30 patients were less than 50 years of age and 22 were more than 50 years. 25 patients were in CTP class A and the rest 27 were in CTP class B. The reasons for CLD were alcohol 27/52 (52%), cryptogenic 13/52 (25%), hepatitis C 8/52 (15%), and others 4/52 (8%). (Figure 1). In this study 31/52 (60%) were good responders, 10/52 (19%) were of poor responders and 11/52 (21%) were nonresponders. (Figure 2). Patient characteristics and clinical profiles were shown in Table 1.

Among the patients who were less than 50 years of age, 21/30 (70%) of them were good responders, 6/30 (20%) were poor responders, and 3/30 (10%) were nonresponders. In patients more than 50 years, 9/22 (40%) were good responders, 5/22 (23%) were poor responders, and 8/22 (36%) were nonresponders. On comparing the responders rate between the age groups, patients in age group less than 50 years had significant responder rates (70%) than patients with age more than 50 years. (40%) ($P = 0.03$).

We observed that 20/37 (54%) of the males and 11/15 (74%) of the females were good responders, 8/37 (22%) males and 2/15 (13%) females were poor responderss and

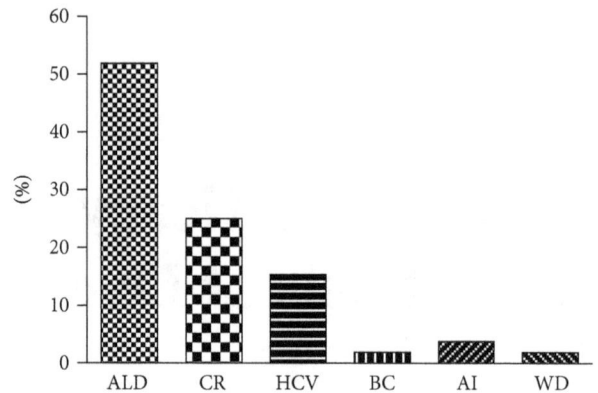

FIGURE 1: ALD: alcoholic liver disease; CR: cryptogenic hepatitis; HCV: hepatitis C virus; BC: budd chiarri; AI: autoimmune hepatitis; and WD: wilson's Disease.

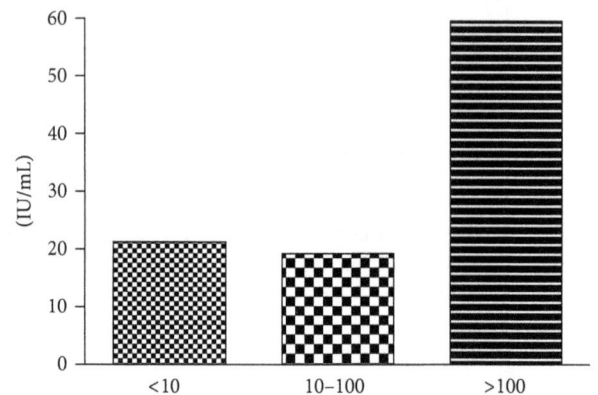

FIGURE 2: Antibody response shown in percentages after administration of HBV vaccine in chronic liver disease patients.

nonresponders were 9/37 (24%) in males and 2/15 (13%) in females (P value 0.43).

When we were individually analyzing the vaccine response across different etiologies, we found that in 27 of the patients with alcohol induced CLD, 12/27 (44%) had good response, 7/27 (26%) had poor response, and 8/27 (30%) were non responders and in 8 patients with HCV related CLD 6/8 (76%) were good responders, 1/8 (12%) poor responders, and 1/8 (12%) were nonresponders. Overall response rates in 52 patients were as follows: 31/52 (60%) had good response, 10/52 (19%) had poor response, and 11/52 (21%) were nonresponders. ($P = 0.36$).

We also compared response rates with the child scores; out of 25 patients who were child A score, 22/25 (88%) were good responders, 1/25 (40%) were poor responders, and 2/25 (8%) were nonresponders and in child B class only 9/27 (33%) had good antibody response and 9/27 (33%) had poor response, 9/27 (33%) were nonresponders ($P < 0.0001$). None of our patients had suffered significant systemic or local adverse reactions. All patients complained of pain and redness during vaccine administration. No other adverse events were observed.

TABLE 1: Patient characteristics, clinical profiles, and immunological outcomes.

Antibody titers 226.88 ± 164 [IU/mL]	Total	Nonresponders [<10 IU/mL]	Partial responders [10–100 IU/mL]	Responders [>100 IU/mL]	P value
Age group 47.48 ± 9.37 [years]					
<50	30	3 [10%]	6 [20%]	21 [70%]	
>50	22	8 [36%]	5 [23%]	9 [41%]	0.03
Total	**52**	**11 [21%]**	**11 [21%]**	**30 [58%]**	
Gender					
Female	15	2 [13%]	2 [13%]	11 [74%]	
Male	37	9 [24%]	8 [22%]	20 [54%]	0.43
Total	**52**	**11 [21%]**	**10 [19%]**	**31 [60%]**	
Etiology					
Alcohol	27	8 [30%]	7 [26%]	12 [44%]	
Hepatitis C	8	1 [12%]	1 [12%]	6 [76%]	
Cryptogenic	13	2 [15%]	2 [15%]	9 [70%]	0.36
Others	4	0	0	4 [100%]	
Total	**52**	**11 [21%]**	**10 [19%]**	**31 [60%]**	
Child's class					
A	25	2 [8%]	1 [4%]	22 [88%]	
B	27	9 [33.3%]	9 [33.3%]	9 [33.3%]	<0.0001
Total	**52**	**11 [21%]**	**10 [19%]**	**31 [60%]**	

4. Discussion

Vaccination with HBV vaccine is extremely safe in general population and in patients with chronic liver disease. The immunogenicity rates of vaccination in general population are >90% whereas in CLD it varied from 18% to 100%. In the present study, the response to standard HBV vaccination with a dose of 20 μg at 0, 1, and 2 months in cirrhosis of liver with various etiologies was evaluated and compared. We found that good responders were only 60%, poor responders 19% percent, and nonresponders 21%. When we analyzed the various aetiologies of CLD and the vaccine response rate, we observed that the patients with alcoholic chronic liver disease (ALD) had poor antibody response (44%) as compared to other aetiologies of chronic liver disease such as cryptogenic (69%) and HCV (75%) related liver disease. Severity of the chronic liver disease predicts the response rate; the good responders were in CTP class A (88%) as compared to CTP class B (33.3%). It was observed that apart from severity of liver disease, the age of the patients also had contributed to antibody response; patients less than 50 years had a higher rate of response with hepatitis B vaccination than patients above 50 years.

Studies conducted by Lee et al., Wiedmann et al. and Keeffe and Krause [20–22] in CLD patients with HCV who had received 20 ug showed response rates of 100%, 89%, and 69%, respectively. However these studies did not include any patients with cirrhosis liver. In our study the good responders in patients with HCV related cirrhosis liver were 76%. However a fewer number of patients with HCV had participated

in our study. In chronic ALD patients the response rates observed by Mendenhall et al., Bronowicki et al., and Rosman et al. [23–25] were 18%, 69%, and 46%, respectively. The previous studies had included only fewer cirrhotic patients. In our study the good responders in patients with alcoholic induced cirrhosis were only 44%, and all the patients were with child A or child B cirrhosis when compared to previous studies in the literature.

These observations suggest that if we are vaccinating at an earlier age (<50 years) and also at an early stage of chronic liver disease (child A), the immunogenicity of hepatitis B vaccination is superior as compared to patients with age more than 50 years and with child B cirrhosis liver. These observations suggest that if the patient is having advanced cirrhosis liver or in the age group of more than 50 years, it is always better to try different regimens like 40 ug or 80 ug of hepatitis vaccination or other routes of vaccine administrations like multiple intradermal dose. And the literature search had shown that the safety profile of these higher doses was comparable with that of normal dose. There were only few studies having data on multiple intradermal dose of hepatitis B vaccination in chronic liver disease, but it has not been yet approved in vaccination schedule.

5. Study Limitations

We had evaluated only using standard dose of hepatitis B vaccine (Shanvac, M/s Shantha biotech) in patients with cirrhosis liver of CTP classes A and B only and had excluded CTP class C patients. However we have not compared the efficacy of Shanvac HBV vaccine with other available vaccines in the market. Our study populations were diversified and the subjects included in the different etiologic groups were small. We should have attempted with higher dose and/or weekly intradermal doses as per the evidence from the literature of patients with chronic kidney disease. Additionally long-term persistence of antibody titers, the frequency of estimating the postvaccination anti-Hbs titer, and the need for booster dose of hepatitis B vaccine were not evaluated due to technical and financial constraints. Another limitation of our study is that all the patients received the same quantity of HBV vaccine (20 μg), and thus comparisons between high and low doses cannot be anticipated.

6. Conclusions

Patients with cirrhosis liver when compared to general population will have low postimmunization antibody titres against hepatitis B. As the age and the stage of liver disease progress, the immunogenicity of standard dose of hepatitis B vaccination against hepatitis B infection will still remain weak. Hence all the cirrhotics with non-HBV etiologies should be initiated on hepatitis B vaccination protocol at the time of diagnosis to achieve better protection against HBV.

If the patient had not achieved seroconversion with standard universal dose of hepatitis B vaccine, reimmunization with a higher dose of hepatitis B vaccine schedule or with multiple intradermal route of vaccination may be considered. Further studies are needed to assess the antibody titres by considering type of vaccine, the dose to be administered, the route of administration, the frequency of antibody testing, and the requirement for booster dose.

References

[1] A. S. Lok, "Chronic hepatitis B," *The New England Journal of Medicine*, vol. 346, no. 22, pp. 1682–1683, 2002.

[2] L. J. Kozak, M. F. Owings, and M. J. Hall, "National Hospital Discharge Survey: 2001 annual summary with detailed diagnosis and procedure data," *Vital and Health Statistics*, vol. 13, no. 156, pp. 1–198, 2004.

[3] E. Arias, R. N. Anderson, H. C. Kung, S. L. Murphy, and K. D. Kochanek, "Deaths: final data for 2001," *National Vital Statistics Reports*, vol. 52, no. 3, pp. 1–115, 2003.

[4] R. S. Koff, "Risks associated with hepatitis A and hepatitis B in patients with hepatitis C," *Journal of Clinical Gastroenterology*, vol. 33, no. 1, pp. 20–26, 2001.

[5] G. Reiss and E. B. Keeffe, "Review article: hepatitis vaccination in patients with chronic liver disease," *Alimentary Pharmacology & Therapeutics*, vol. 19, pp. 715–727, 2004.

[6] A. Alberti, P. Pontisso, L. Chemello et al., "The interaction between hepatitis B virus and hepatitis C virus in acute and chronic liver disease," *Journal of Hepatology, Supplement*, vol. 22, no. 1, pp. 38–41, 1995.

[7] L. Benvegnu, G. Fattovich, F. Noventa et al., "Concurrent hepatitis B and C virus infection and risk of hepatocellular carcinoma in cirrhosis. A Prospective Study," *Cancer*, vol. 74, pp. 2442–2448, 1994.

[8] S. Chlabicz and A. Grzeszczuk, "Hepatitis B virus vaccine for patients with hepatitis C virus infection," *Infection*, vol. 28, no. 6, pp. 341–345, 2000.

[9] S. D. Lee, C. Y. Chan, M. I. Yu, R. H. Lu, F. Y. Chang, and K. J. Lo, "Hepatitis B vaccination in patients with chronic hepatitis C," *Journal of Medical Virology*, vol. 59, pp. 463–468, 1999.

[10] M. Wiedmann, U. G. Liebert, U. Oesen et al., "Decreased immunogenicity of recombinant hepatitis B vaccine in chronic hepatitis C," *Hepatology*, vol. 31, no. 1, pp. 230–234, 2000.

[11] A. S. Rosman, P. Basu, K. Galvin, and C. S. Lieber, "Efficacy of a high and accelerated dose of hepatitis B vaccine in alcoholic patients: a randomized clinical trial," *American Journal of Medicine*, vol. 103, no. 3, pp. 217–222, 1997.

[12] E. B. Keeffe and D. S. Krause, "Hepatitis B vaccination of patients with chronic liver disease," *Liver Transplantation*, vol. 4, pp. 437–439, 1998.

[13] C. Mendenhall, G. A. Roselle, L. A. Lybecker et al., "Hepatitis B vaccination: response of alcoholic with and without liver injury," *Digestive Diseases and Sciences*, vol. 33, pp. 263–269, 1998.

[14] J. P. Bronowicki, F. Weber-Larivaille, J. P. Gut, M. Doffoël, and D. Vetter, "Comparison of immunogenicity of anti-HBV vaccination end serovaccination in alcoholic patients with cirrhosis," *Gastroenterologie Clinique et Biologique*, vol. 21, no. 11, pp. 848–853, 1997.

[15] E. Villeneuve, J. Vincelette, and J. P. Villeneuve, "Ineffectiveness of hepatitis B vaccination in cirrhotic patients waiting for liver transplantation," *Canadian Journal of Gastroenterology*, vol. 14, pp. 59B–62B, 2000.

[16] M. Dominguez, R. Barcena, M. Garcia, A. Lopez-Sanroman, and J. Nuno, "Vaccination against hepatitis B virus in cirrhotic patients on liver transplant waiting list," *Liver Transplantation*, vol. 6, pp. 440–442, 2000.

[17] M. Arslan, R. H. Wiesner, C. Sievers, K. Egan, and N. N. Zein, "Double-dose accelerated hepatitis B vaccine in patients with end-stage liver disease," *Liver Transplantation*, vol. 7, no. 4, pp. 314–320, 2001.

[18] Centers for Disease Control and Prevention, "Recommendations and reports: hepatitis A and B vaccines," *Morbidity and Mortality Weekly Report*, vol. 52, pp. 34–36, 2003.

[19] National Institutes of Health, "National Institutes of Health Consensus Development Conference Statement: management of hepatitis C: 2002—June 10-12, 2002," *Hepatology*, vol. 36, supplement 1, pp. S3–S20, 2002.

[20] S. D. Lee, C. Y. Chan, M. I. Yu, R. H. Lu, F. Y. Chang, and K. J. Lo, "Hepatitis B vaccination in patients with chronic hepatitis C," *Journal of Medical Virology*, vol. 59, pp. 463–468, 1999.

[21] M. Wiedmann, U. G. Liebert, U. Oesen et al., "Decreased immunogenicity of recombinant hepatitis B vaccine in chronic hepatitis C," *Hepatology*, vol. 31, no. 1, pp. 230–234, 2000.

[22] E. B. Keeffe and D. S. Krause, "Hepatitis B vaccination of patients with chronic liver disease," *Liver Transplantation*, vol. 4, pp. 437–439, 1998.

[23] A. S. Rosman, P. Basu, K. Galvin, and C. S. Lieber, "Efficacy of a high and accelerated dose of hepatitis B vaccine in alcoholic patients: a randomized clinical trial," *American Journal of Medicine*, vol. 103, no. 3, pp. 217–222, 1997.

[24] C. Mendenhall, G. A. Roselle, L. A. Lybecker et al., "Hepatitis B vaccination. Response of alcoholic with and without liver injury," *Digestive Diseases and Sciences*, vol. 33, pp. 263–269, 1998.

Clinical Symptoms of Human Rotavirus Infection Observed in Children in Sokoto, Nigeria

B. R. Alkali,[1] A. I. Daneji,[1] A. A. Magaji,[1] and L. S. Bilbis[2]

[1]Faculty of Veterinary Medicine, Usmanu Danfodiyo University, PMB 2346, Sokoto, Sokoto State, Nigeria
[2]Faculty of Science, Usmanu Danfodiyo University, PMB 2346, Sokoto, Sokoto State, Nigeria

Correspondence should be addressed to B. R. Alkali; balkali@yahoo.co.uk

Academic Editor: Jay C. Brown

Rotavirus has been identified among the most important causes of infantile diarrhoea, especially in developing countries. The present study was undertaken to determine the occurrence and clinical symptoms of human rotavirus disease among children presenting with varying degree of diarrhoea in selected urban hospitals in Sokoto metropolis, Nigeria. Diarrhoea samples were collected from 200 diarrheic children younger than 5 years of age and tested using a commercially available DAKO Rotavirus ELISA kit which detects the presence of human group A rotaviruses. A questionnaire, based on WHO generic protocol, was completed for each child to generate the primary data. Of the total number of samples collected, 51 were found to be positive for human group A rotavirus indicating 25.5% prevalence of the disease in Sokoto state. The symptoms associated with the disease were analyzed and discussed.

1. Introduction

Diarrhoea illnesses were reported to consistently rank as one of the top six causes of all deaths, one of the top three causes of death from infectious disease, and one of the top two causes of death when considering years of life lost [1–3]. Rotavirus was identified to be responsible for up to 20% of these deaths [4]. Also reports have shown that 39% of diarrhoea episodes seen at health centers were rotavirus positive [5, 6].

Rotavirus is a genus in the Family of Reoviridae with the characteristic wheel-like (i.e., Rota is Latin for wheel) appearance. The inner capsid contains the viral genome of 11 segments of double stranded RNA that encode six structural and six nonstructural proteins [7]. The structural proteins of the virion are depicted as three concentric circles, forming an equal number of layers around the dsRNA genome (triple layered particle) [8]. It is a nonenveloped triple layered icosahedral virus consisting of an inner core containing proteins VPl, VP2, and VP3, encoded by segments 1–3, a middle capsid made up of protein VP6, encoded by gene segment 6 and an outer capsid made up of a VP7 shell and a VP4 spike protein encoded by segments 7, 8 or 9, and 4, respectively [7]. The external layer of the virus is discontinuous and looks like a sponge, because of the multiple small extensions of the VP4 spike [9].

Rotavirus strains had been classified into eight main (A–H) serotype groups (or serogroups) on the basis of antigenic sites located on the VP6 protein [10]. The most virulent and commonly isolated strains belong to serogroup A (GARVs) as the group constitute an important cause of acute infectious diarrhoea in children and various domestic mammalian and avian species.

Indeed group A rotaviruses were reported to constitute the major cause of severe gastroenteritis in young children and animals worldwide affecting nearly all animals from whales and snakes to cows and pigs [11, 12]. Studies have also shown that by the age of two years almost all children are infected by rotavirus with children in industrialized countries experiencing their first infection at comparatively older age compared to those in developing countries [5, 13].

In Nigeria, a high incidence of childhood diarrhoea is estimated to account for over 160 000 of all deaths in children less than 5 years of age annually and of this number approximately 20% had been associated with rotavirus infection [14]. Although diarrhoea, vomiting, and dehydration are frequently associated with the disease, there is need to

Recent Advances in Diagnosis, Prevention, and Treatment of Human Respiratory Syncytial Virus

Swapnil Subhash Bawage, Pooja Munnilal Tiwari, Shreekumar Pillai, Vida Dennis, and Shree Ram Singh

Center for NanoBiotechnology Research, Alabama State University, Montgomery, AL 36104, USA

Correspondence should be addressed to Shree Ram Singh; ssingh@alasu.edu

Academic Editor: Subhash Verma

Human respiratory syncytial virus (RSV) is a common cause of respiratory infection in infants and the elderly, leading to significant morbidity and mortality. The interdisciplinary fields, especially biotechnology and nanotechnology, have facilitated the development of modern detection systems for RSV. Many anti-RSV compounds like fusion inhibitors and RNAi molecules have been successful in laboratory and clinical trials. But, currently, there are no effective drugs for RSV infection even after decades of research. Effective diagnosis can result in effective treatment, but the progress in both of these facets must be concurrent. The development in prevention and treatment measures for RSV is at appreciable pace, but the implementation into clinical practice still seems a challenge. This review attempts to present the promising diverse research approaches and advancements in the area of diagnosis, prevention, and treatment that contribute to RSV management.

1. Introduction

Worldwide, there are reportedly about 12 million severe and 3 million very severe cases of lower respiratory tract infection (LRTI) in children [1]. Respiratory syncytial virus (RSV) is a common contributor of respiratory infections causing bronchiolitis, pneumonia, and chronic obstructive pulmonary infections in people of all ages but affects mainly children and elderly along with other viral infections leading to high mortality and morbidity [2–4]. A recent global survey suggests that RSV is not prevalent throughout the year in the tropical regions of the globe, but the incidence peaks in winter with a wide ranging persistence depending on the geographical topology [5]. RSV has been reported to be a prevalent lower respiratory tract pathogen distributed worldwide including countries from both, the developed and developing world. The major countries with RSV seasonal outbreaks include USA, Canada, Cambodia, Mexico, Uruguay, Brazil, Peru, France, Finland, Norway, Sweden, Latvia, Denmark, Germany, Netherlands, Ireland, Italy, Turkey, Iran, Saudi Arabia, Australia, New Zealand, China, Korea, Hong Kong, Japan, India, Pakistan, Bangladesh, Nepal, Taiwan, Vietnam,

Myanmar, Thailand, Madagascar, Kenya, Zambia, Nigeria, and Columbia. The data about human RSV described in literature over the years seem to have been unchanged significantly, indicating the severity of RSV and the urgent concern to address this issue. An estimate of more than 2.4 billion US dollars per year is the economic cost of viral lower respiratory tract infection in children [6].

RSV is a *Paramyxovirus* belonging to the genus *Pneumovirus*. RSV is an enveloped, nonsegmented, negative, single stranded linear RNA genome virus (Figure 1). RSV genome (~15 kb) has 10 genes encoding 11 proteins with two open reading frames of gene M2 [7, 8]. Other genes include nonstructural proteins NS1 and NS2 (type I interferon inhibitors), L (RNA polymerase), N (nucleoprotein), P (Phosphoprotein cofactor for L), M (Matrix protein), M2.1 and M2.2 (required for transcription) SH (small hydrophobic protein) G (glycoprotein), and F (fusion protein). Being a negative strand RNA genome virus, RSV packages its own polymerase into the nucleocapsid. Of these proteins, fusion protein (F) is indispensable for viral attachment to the host and entry into the host cell. Although the G protein is responsible for the preliminary attachment, the F protein is

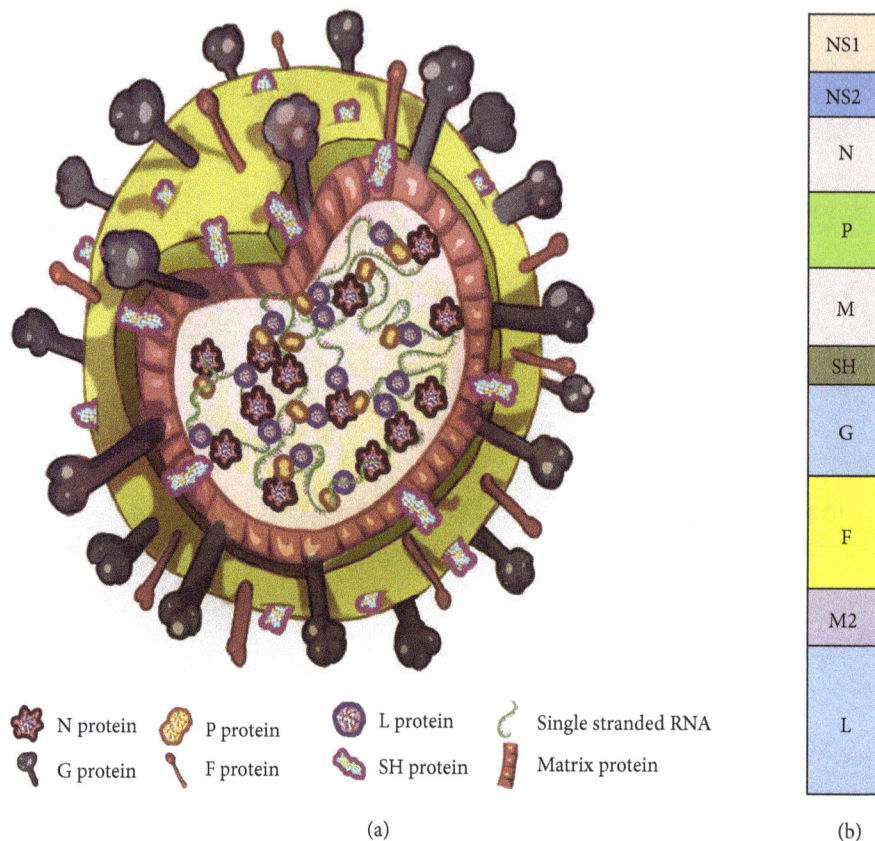

FIGURE 1: Structure and genome organization of respiratory syncytial virus. (a) Approximately 200 nm RSV virion particle and (b) single stranded negative RNA genome consisting of 10 genes.

necessary for the fusion, budding, and spread of the virus [9, 10]. After attachment to the host cell, RSV fuses with the host cell membrane using the F protein through the 6 helix coiled-coil bundle of the F protein, a mechanism characteristically found in *paramyxoviridae* members [11]. Although the detailed mechanism of RSV infection is not fully understood, the most accepted mechanism is the entry of the nucleocapsid into the host cell mediated by the F protein through clathrin mediated endocytosis [12]. The RNA is first converted into a plus strand, which serves as the template for replication; whereas for transcription, the RNA genome itself transcribes mRNA for protein synthesis without any intermediate.

Almost all children of 2 years of age will have had an RSV infection and leading to 160,000–600,000 deaths per year [4]. Approximately, 25% to 40% of infants and children at the first exposure to RSV have signs or symptoms of bronchiolitis or pneumonia. These symptoms include rhinorrhea, low-grade fever, cough, and wheezing. The symptoms in adults may include common cold, with rhinorrhea, sore throat, cough, malaise, headache, and fever. It can also lead to exacerbated symptoms such as severe pneumonia in the elderly, especially residing in nursing homes [13]. Usually, children show symptoms within 4 to 6 days of infection and most of them recover in 1 to 2 weeks while serving

as carriers of the virus for 1 to 3 weeks. RSV infection in children of nosocomial origin is associated with higher mortality than community-acquired illness because of the pre-existing morbidity [14, 15]. Severe RSV disease risk hovers for the elderly and adults with chronic heart or lung disease or with weakened immune system [16]. RSV infection does not provoke lasting immunity [17] therefore, reinfection is very common [18]. Recently, RSV infection was reported to account for hospitalizations and mortality in elderly people [19]. RSV accounted for severe lower respiratory tract infections including chronic lung disease, systemic comorbidities, and even death. At present, there is no specific treatment for RSV infection ever since its first discovery in 1956 [20]. Currently, Food and Drug Administration (FDA) approved prophylactic drug for RSV that includes palivizumab and ribavirin; administered along with symptomatic treatment drugs and supportive care. Currently, techniques used for diagnosis of RSV include ELISA, direct immunofluorescence, western blot, PCR, and real-time PCR. The diagnosis and treatment scenario has significantly changed with the advent of advanced techniques and in-depth understanding of RSV biology, but the execution of these clinical developments in practice requires extensive study and time. This review presents the recent advances in the diagnosis, prevention, and treatment of RSV (Figure 2).

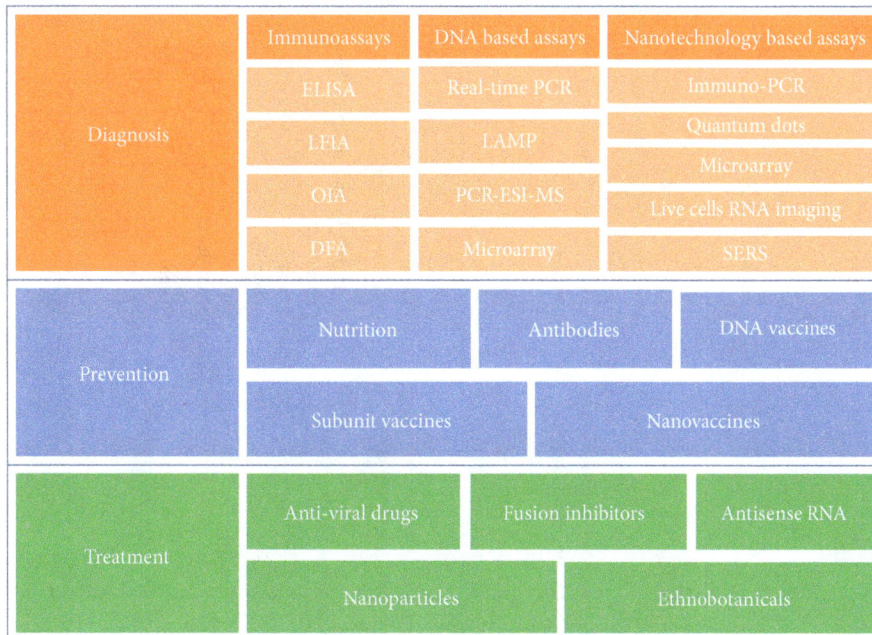

FIGURE 2: A schematic representation of RSV management through coordinated diagnosis, prevention, and treatment.

2. Diagnosis

There are ten known genotypes of RSV based on the sequence analysis of the RSV genome, which suggests that the pathogen changes with time and the resulting genotypes pose a threat to the public health. Information from comparative genome analysis has been utilized for evolutionary studies and most importantly for the development of detection and treatment studies [21, 22]. Diagnostic assays for detection of pathogens have a pivotal role in public-health monitoring [23]. However, diagnosis becomes difficult when the causative agent exhibits overlapping symptoms with other disease(s) or remains nonsymptomatic. Hence the correct diagnosis has to be empirically derived for the treatment. Some of the promising molecular and biophysical techniques for RSV diagnosis are discussed below (Table 1).

2.1. Immunoassays

2.1.1. Enzyme-Linked Immunosorbent Assay. Enzyme-linked immunosorbent assay (ELISA) is the enzyme facilitated colorimetric detection of specific protein-antibody complexes. ELISA is extensively used for detection of various proteins at very low concentration in different sample types, thus making it clinically significant in routine diagnosis of pathogens. ELISA for RSV detection is mainly based on targeting RSV F protein (antigen). Recently, several modifications of the classical ELISA technique have been efficiently developed and employed for the detection of RSV. Sensitivity of ELISA was increased by using the high affinity anti-RSV F antibody peptides derived from the motavizumab [85]. Motavizumab is a high affinity antibody based therapeutic against RSV which binds to RSV F protein; however, it was disapproved by the FDA due to higher hypersensitivity in patients receiving motavizumab as compared to palivizumab. Motavizumab

also caused urticaria [34]. It is presumed to be a better binding target than the conventional F protein ELISA. This method could prove more effective than the F protein ELISA due to its higher sensitivity to the degradation of motavizumab. In a simple thin layer amperometric enzyme immunoassay, RSV was detected as early as 25 minutes, at low cost with comparative sensitivity to that of real-time PCR and immunofluorescence assays [35]. The assay is based on the development of a double layer sandwich method similar to ELISA. It involves a polystyrene microarray slide coated with monoclonal antibody which captures the antigen (RSV). The antigen-antibody complex is detected by horse radish peroxidase conjugated secondary antibody on a screen printed electrochemical cell coated with the substrate.

2.1.2. Immunofluorescence Assay. Presently, immunofluorescence is one of the most common and rapid RSV detection techniques used, where the antigen is detected by a fluorescently tagged antibody. Direct fluorescent antibody assay (DFA) is a standard detection technique used for decades and other RSV detection techniques are often compared to DFA for evaluating their efficiency [24]. An indirect assay which uses secondary fluorescently tagged antibody is another option to DFA. Currently, there are many molecular techniques that are more sensitive and reliable than DFA, but DFA is widely used for RSV detection in clinical samples due to the ease and rapidity. The sensitivity and specificity of DFA are a subject of variance as the success of the technique is dependent on numerous factors, mainly the skills of the technician and nature of the sample. A study showed that DFA could detect RSV with a sensitivity and specificity of 77.8% and 99.6%, respectively (positive predictive value was 98.6% and the negative predictive value was 94%). DFA is a reliable technique for RSV detection for patients tested

TABLE 1: Comparison of RSV detection techniques.

Technique	Reference	Principle	Advantages	Drawbacks	Current usage status
(A) Fluorescence based methods					
(1) DFA	[24, 25]	Microscopic detection of RSV with specific antibody conjugated with fluorophore.	Easy procedure	Human error, fading of dyes	Research intent, Hospital based procedure, commercial diagnostic assays
(2) QDs	[26–31]	Detection of signals from fluorescent nanoparticles upon encounter with RSV either through microscopy or flow cytometry	Photostable, inorganic in nature, resistant to metabolic degradation	Toxicity, insolubility	Research intent
(3) Molecular beacon based imaging	[32, 33]	Hairpin DNA functionalized gold nanoparticle with fluorophore hybridization with target mRNA	Live cell imaging with real-time detection	Probable gene silencing, metabolic degradation	Research intent
(B) Immunoassays					
(1) ELISA	[34, 35]	Specific binding and colorimetric detection of antigen-antibody complex	Easy protocol, high specificity and sensitivity	Cumbersome, prone to human errors	Hospital based procedure, commercial diagnostic assays
(2) OIA	[36–38]	Presence of specific antigen-antibody complex formed alters the reflective surfaces properties which is visually detected	Easy, rapid, specificity, cost effective	Needs confirmation by other tests for negative samples	Research intent, not commercialized
(3) LFIA	[39–42]	Immuno-complexes detected chromatographically	Easy, rapid, handy, cost effective, FDA approved	Nonquantitative, limit of sample volume limits detection	Hospital based procedure, commercial diagnostic assays
(C) Molecular methods					
(1) LAMP	[41, 43–45]	Colorimetric/turbidimetric detection of isothermal amplification of DNA using specific primer	Sensitivity and specificity	Semiquantitative, designing compatible primer set	Research intent, not commercialized
(2) PCR	[46, 47]	Amplification of viral cDNA and visualization of PCR product	Rapid and sensitive than conventional culture methods	High limits of detection	Research intent, hospital based procedure
(3) Real-Time PCR	[48–53]	Real-time amplification of target DNA or cDNA	Rapid (3–5 hours), highly sensitive and very low limits of detection	Expensive	Research intent, hospital based procedure, commercial assay
(4) Multiplex PCR	[54–57]	Use of multiple primer and/or probe sets	Simultaneous detection of multiple pathogenic species or strains	Less sensitive	Research intent, hospital based procedure
(5) Immuno-PCR	[58, 59]	A combination of immunoassay and real-time PCR	Very low limits of detection, improved limits of detection over individual ELISA, and PCR (4000 and 4 fold. respectively)	Complex experimental design	Research intent, not commercialized
(6) Microarray	[60–74]	Hybridization of sample biomolecules to immobilized target DNA or protein on a chip	Highly sensitive, large scale identification of multiple pathogens; protein and nucleic acid targets	Cost-ineffective	Research intent, hospital based procedure, commercial assay
(D) Biophysical method					
(1) PCR-ESI-MS	[75, 76]	Mass spectroscopy of PCR-amplicons through electron spray dispersion	Highly sensitive and specific even at strain level and efficient multiple pathogens detection.	Expensive	Research intent, not commercialized
(2) SERS	[77–84]	Inelastic scattering of monochromatic radiation upon interaction with an analyte with low-frequency vibrational and/or rotational energy	Rapid and nondestructive detection of analytes with high sensitivity	Sample preparation	Research intent, not commercialized

within first 3 days after onset of symptoms, but the sensitivity decreases if tested 4–7 days after the onset of the symptoms [24]. DFA can detect RSV alone or as a multivalent test for other respiratory viruses (influenza A and B viruses, parainfluenza virus types 1 to 3, and adenovirus) using SimulFluor Respiratory Screen assay [25].

2.1.3. Optical Immunoassay.

2.1.3. Optical Immunoassay. The direct visualization of an antigen-antibody complex often referred to as optical immu-noassay (OIA) is used for qualitative detection of *Streptococ-cus* and influenza virus. Visualization of immunocomplexes can be enhanced by immobilizing them on a special reflecting surface. The antigen-antibody complex forms a thin layer, which changes the reflective properties of the surface [36]. The technique is simple, rapid, and sensitive and as low as 1 ng of antibody per mL can also be detected [37]. OIA is now used for detection of RSV with a sensitivity, specificity, positive, and negative predictive values of 87.9%, 99.6%, 98.9%, and 94.5%, respectively. Though OIA offers rapid and cost-effective RSV detection, it is recommended that negative results of OIA should be confirmed by other tests [38].

2.1.4. Lateral Flow Immunoassay. Lateral flow immunoassay (LFIA) is an immunochromatographic technique known for the rapid detection of RSV from nasal washes or nasal aspirates. The lateral flow of antigen-antibody complex on the substrate matrix reaches the reaction area and results in formation of colored band indicating the presence of antigen in the specimen. There are many LFIA kits like Remel Xpect, Binax Now RSV, BD Directigen EZ RSV, QuickLab RSV Test, and RSV Respi-Strip [39–42]. The sensitivity and specificity are normally above 90% and 95%, respectively, but vary as per the manufacturer. Modifications of LFIA may include colloidal gold conjugated with antibody specific to RSV in the matrix for trapping the antigen and assisting the gold mediated reaction for color band development.

2.2. Loop-Mediated Isothermal Amplification. Loop-mediated isothermal amplification (LAMP) is a nucleic acid based detection method used for bacterial and viral pathogens. It can be used for RNA viruses where the additional step of cDNA synthesis is required and commonly designated as reverse transcription LAMP (RT-LAMP). LAMP consists of pairs of primers specific to cDNA, which is then amplified by autocycling strand displacement activity of DNA polymerase generally at 60°C. The reaction time of 1 to 1.5 hours is sufficient and can be monitored by real-time turbidimeter and the resulting product can also be viewed by agarose gel electrophoresis. This method can distinguish RSV strain A and B upon restriction digestion of the product [41]. Alternatively, RSV strain A and B can be detected designing specific primer sets for them. The efficiency of RT-LAMP was tested for RSV detection from nasopharyngeal aspirates and compared with viral isolation, enzyme immunoassay, immuno-chromatographic assay, and real-time PCR. RT-LAMP was the most sensitive among all the methods tested, with the exception of real-time PCR. Also, the RT-LAMP was specific for RSV and did not react with any other respiratory

pathogens like RNA viruses, DNA virus, or bacteria [43]. Although this method is rapid, the requirement of the turbidimeter makes it slightly inconvenient as a cost-effective detection technique. Development in the chemistry of the LAMP has made it possible to get rid of the turbidimeter as it is now possible to visually detect the presence of pathogen in the test sample. LAMP is now referred to as colorimetric detection of loop-mediated isothermal amplification reaction due to incorporation of the dye in the chemistry to aid visual detection. Wang et al. (2012) have developed such an RT-LAMP assay for the detection of human *metapneumoviruses* (which include rhinovirus, RSV, influenza virus A/PR/8/34 (H1N1)), which can detect as low as ten viral RNA copies with better efficiency than RT-PCR [44]. It is now possible to detect RSV A and B strains in nasopharyngeal specimen using multiplex LAMP (M-LAMP) in just 30 minutes. M-LAMP had a sensitivity and specificity of 100% when compared with PCR [45].

2.3. Polymerase Chain Reaction (PCR) Based Detection

2.3.1. Conventional PCR. The applicability of PCR is ubiquitous and has made the diagnosis of pathogens tremendously rapid and sensitive [46]. RSV is usually challenging to detect due to poor viral titer and sensitivity to antigen based detection methods. Thus, a PCR based method was developed and compared to serological and culture based detection methods in adults having respiratory infections [47]. The PCR method was based on the reverse transcription-nested PCR technique involving the outer and inner primers designed from the F gene of RSV strain A, over a two day procedure with a sensitivity of 73%. This method is faster as compared to the conventional culture method which usually takes 3–5 days, resulting in faster treatments.

2.3.2. Real-Time PCR. Although conventional reverse transcriptase-PCR (RT-PCR) is sensitive as compared to the culture methods, it suffers from the lower sensitivity. This problem has been obviated by the development of real-time PCR based methods. Real-time PCR is by far the most sensitive method for the detection and diagnosis of a wide array of pathogens, including RSV. Several studies have been conducted for the development of real-time PCR assays for RSV detection, especially the RT-PCR. In one such study, a rapid, sensitive, and specific method was devised based on TaqMan real-time PCR for the detection of both RSV A and RSV B strain [48]. The primer and the probe sets were designed from the nucleocapsid gene (N). The sensitivity of this method was found to be 0.023 PFU (plaque forming unit) or two copies of mRNA for RSV A, whereas the sensitivity for RSV B was 0.018 PFU or nine mRNA copies. This method is fast and efficient as the diagnosis was performed within 6 hours of the procurement of samples.

A similar assay based on real-time PCR was developed and employed in the detection of RSV in the bronchoalveolar lavage (BAL) of lung transplant patients or patients with respiratory infections [49]. The assay was designed based on the RSV N gene and first involved a screening step of the RSV positive samples using a SYBR green based assay, followed by quantitative real-time PCR using TaqMan based assay to be more cost effective. Also, the assay was developed for the detection of both RSV A and B subgroups separately. The assay was found positive in 16% of transplant patients, thus indicating some possible association between RSV infection and lung transplant tolerance.

Real-time PCR based on RSV N gene has also been used to quantify RSV from the nasal aspirates of children [50]. The method was one log more sensitive than the conventional culture method. The real-time PCR assay resulted in 56% of positives, whereas the immunofluorescence assay had 41.3% and culture method had 45.3% of positives. Surprisingly, the RSV to GAPDH ratio did not differ in children with severe or nonsevere infection, raising confusion about the correlation of viral load to the severity of infection due to the intensity of viral replication, genetic susceptibility of the host and immune responses, thus, making it necessary to consider other factors besides the viral load.

A similar assay was developed for the detection of RSV using primer-probe sets from the RSV F gene [51]. A comparative analysis was made between the nested RT-PCR, real-time PCR, and ELISA, wherein real-time PCR was 25% more sensitive than the conventional nested PCR, thus making it more applicable for clinical samples. Additionally, the real-time PCR was performed in two steps to increase the sensitivity of the samples. This method was more rapid as the results were obtained in 3.5 to 4 hours upon receiving the samples. Several other reports also emphasize the applicability of real-time PCR in detection of RSV in immunocompromised patients [52]. Besides the RSV F and N gene, the RSV matrix gene and polymerase gene have also been used to detect RSV in children [53]. The latter method had the ability to quantify and classify RSV with high efficiency and rapidity. However, the sensitivity of the technique relies on the age of the patients.

2.3.3. Multiplex PCR.
The multiplex PCR approach has been used for the detection and subtyping of RSV and human influenza virus simultaneously [54] and as many as 18 respiratory viruses could be detected [55]. The method was based on the primer sets of hemagglutinin and nucleoprotein gene of influenza virus and nucleocapsid protein gene of RSV and found to be rapid, specific, and sensitive. However, the method suffered from the drawback that it could not distinguish between RSV A and B. This method could prove significant during a respiratory outbreak for surveillance. In such situations, multiplex real-time PCR assay, such as the commercial "Simplexa Flu A/B & RSV kit" that differentiates influenza A virus, influenza B virus, and RSV, is very useful [56]. Likewise, recently, a rapid and sensitive detection assay for RSV and other respiratory viruses was developed by Idaho Technologies christened as FilmArray (Idaho Technologies, Salt Lake City, UT). This is an automatic real-time molecular station with capability of nucleic acid extraction, initial reverse transcription, and multiplex PCR followed by singleplex second-stage PCR reactions for specific virus detection [57].

2.4. Microarray.
Correlating gene expression signatures with disease progression of the patient's individual genetic profile is possible by comprehensive understanding and interpretation of the dynamics of biome-interactions in the lungs. This would result in more efficient therapy for respiratory diseases via the concept of personalized medicine, wherein the microarray finds its applications. Microarrays are the miniaturized assay platforms with a high-density array of immobilized DNA or protein. Hybridization of sample biomolecules to corresponding DNA or protein on the chip is detected, which allows determination of a variety of analytes present in the samples in a single experiment [60]. Microarray has proved to be a robust and reliable tool to understand the echelon of genomics, transcriptomics, and proteomics. Implementation of microarray with the metagenomic approach serves for rapid virus identification. Microarray based identification and characterization of viruses in clinical diagnostics are possible by designing oligonucleotide probes using the sequence data available in the public database. This approach is capable of discovering novel viruses, even though there are no conserved genes that can be targeted by sequencing. In the case of respiratory diseases, it can be executed by collecting the bronchoalveolar lavage enriched by nuclease treatment followed by filtration and extraction of total nucleic acids. Further, the nucleic acids are amplified by a random priming-based method, referred to as sequence independent single primer amplification, giving near-full-length reads of genomes of RNA or DNA viruses, which could then be compared to known viruses and used for designing the oligomers for the microarray [61]. Thus, sequencing coupled with microarray is a powerful high throughput diagnostic tool.

The utility of a metagenomics based strategy for broad-spectrum diagnostic assay using microarrays was demonstrated for screening viruses like Rhino virus, Parainfluenza virus, Sendai virus, Poliovirus, Adenovirus, and RSV from clinical samples [62]. A well-known example is "Virochip" which is a pan-viral microarray, designed to simultaneously detect all known viruses, and has comparable or superior sensitivity and specificity to conventional diagnostics [63, 64]. Similarly, an influenza microarray, "FluChip-55 microarray", for the rapid identification of influenza A virus subtypes H1N1, H3N2, and H5N1 was developed [65]. The procedure is simple and follows few steps including RNA extraction from clinical samples, subsequently their reverse transcription, second-strand cDNA synthesis, and then PCR amplification of randomly primed cDNA. The hybridization of nucleic acids with probes is done by fluorescent dyes or by an alternative method of electrochemical detection. The latter method relies on a redox reaction to generate electrical current on the array for measurement. Incorporation of Cy3 fluorescent dye and hybridization to the microarray [66] has been empowered by specific algorithms to match the diagnostic needs, leading to the final and critical step of scanning and analysis [67].

The genes involved in the pathways of neuroactive ligand-receptor interaction, p53 signaling, ubiquitin mediated proteolysis, Jak-STAT signaling, cytokine-cytokine receptor interaction, hematopoietic cell lineage, cell cycle, apoptosis, and

TABLE OF CONTENTS

PREFACE

The main aim of this book is to educate learners and enhance their research focus by presenting diverse topics covering this vast field. This is an advanced book which compiles significant studies by distinguished experts in the area of analysis. This book addresses successive solutions to the challenges arising in the area of application, along with it; the book provides scope for future developments.

Virology is a sub-field of microbiology that deals with the study of viruses, their structure, properties, classification and evolution. It also studies the interactions of viruses with host cells, their modes of infecting hosts for reproduction, the diseases that they cause and their use in therapy and research, among others. Viral classification is an important area of virology. Plant viruses, bacteriophages, fungal viruses and animal viruses are the classifications of viruses based on the host cells that they infect. Humans can only be infected by viruses from other vertebrates. These viruses cause many important infectious diseases such as dengue fever, AIDS, polio, hepatitis, influenza, common cold, rabies, etc. Viruses such as human papillomavirus, and hepatitis B and C viruses contribute to the development of cervical cancer and liver cancer respectively. Therefore, the study of virology is of immense importance in medical science. Our understanding of viruses has undergone rapid development over the past few decades. This book presents the complex subject of human virology in the most comprehensible and easy to understand language. It includes contributions of experts and scientists, which will provide innovative insights into this field.

It was a great honour to edit this book, though there were challenges, as it involved a lot of communication and networking between me and the editorial team. However, the end result was this all-inclusive book covering diverse themes in the field.

Finally, it is important to acknowledge the efforts of the contributors for their excellent chapters, through which a wide variety of issues have been addressed. I would also like to thank my colleagues for their valuable feedback during the making of this book.

Editor

comprehensively evaluate the symptoms and signs associated with rotavirus disease especially because various pathogens have been identified to cause severe diarrhoeal diseases including viruses and bacteria. Thus, the study was designed to provide baseline information and insight into the general symptoms of rotavirus disease and identify the symptoms that may be significantly associated with the disease among children in Sokoto, Nigeria.

2. Study Area

The study was conducted in three urban hospitals located in Sokoto state, namely, Usmanu Danfodiyo University Teaching Hospital, Sokoto (UDUTH), Specialist Hospital, Sokoto, and Women and Children Hospital, Sokoto. These urban hospitals also service rural communities from all parts of the state, including neighboring states. Sokoto state lies between longitude $11°$ $30'$ to $13°$ $50'$ East and latitude $4°$ to $6'$ North. The state falls within the savannah zone and is located in northwestern Nigeria where life expectancy for men and women is 51 years and 52 years, respectively. The GNP per capita is 320 dollars.

2.1. Sampling Method. Simple random sampling method was adopted in the study where each child in the population had equal chance of being selected. This sampling technique provided opportunity in the realistic generalization of the research population. A questionnaire based on WHO generic protocol was administered to generate the primary data along with sample bottle where adequate information on every child was obtained. Patient information such as identification number, address, and admission diagnosis, date of admission, and presenting symptoms were collected. In order to enhance the validity of the research questionnaire, the instrument was validated by both validity and reliability tests. The validity of the questionnaire was determined by the critique of the research experts of the questionnaire. The modification of the questionnaire was based on the experts' comments and advice. The reliability of the questionnaire was determined through the administration of the modified copy to some nurses and matrons of the hospitals selected for the study. The results provided the basis for the final modification of the questionnaire.

3. Data Analysis

3.1. Samples Collection. Statistical Programme for Social Sciences (SPSS17.0) was used to analyze the data. Data was analyzed by simple inferential statistics. The frequencies of findings and the percentages they represent were highlighted on tables, graphs, and charts. Also Chi-square analysis was used for significance testing in drawing inferences.

Diarrhoea samples were collected from all diarrheic children under 5 years of age that were presented at the identified hospitals after obtaining parental consent. Diarrhoea in the study was defined as the passage of more than 3 looser than normal stools within 24 hours. The stool samples were collected aseptically in sterile commercial bijou bottles, adequately labeled (patient ID and date of collection), and

transported on ice to the Veterinary Microbiology Laboratory of Usmanu Danfodiyo University, Sokoto, where they were stored at $-20°C$ until they were transported on ice to Noguchi Memorial Institute for Medical Research (NMIMR) in Accra, Ghana, where they were stored at $-20°C$ until they were tested. A stool specimen logbook was kept in the laboratory where information on all diarrhoeal children was checked regularly and matched with the information in the questionnaire to ensure proper entry of information. Also, data form for analysis of rotavirus diarrhoea was adapted from the WHO generic protocol with some modifications.

4. Determination of Rotavirus Antigen by ELISA

A commercial DAKO Rotavirus ELISA kit was used to detect the presence of human group A rotaviruses in stool samples according to the manufacturer's instructions. Briefly, 2 drops ($100\,\mu L$) of each of the prepared 10% stool suspension were added into each well of the provided 96-well microtiter plate precoated with rotavirus specific rabbit polyclonal antibody except the first three wells designated as blank, negative, and positive controls, respectively. Two drops of the conjugate contained in the kit were then added into each microwell and mixed gently by swirling on table's top. The plates were then incubated at room temperature for 1 hour. The contents were then discarded and the plates were tapped upside down against paper towel to remove all liquid from the wells. The wells were then overflowed with freshly prepared washing buffer and contents were discarded. The plates were tapped upside down against paper towel to remove excess wash buffer. The washing was repeated 5 times. Two drops of the substrate contained in the kit were then added to each microwell and the plate was incubated at room temperature for 10 minutes. Results were then observed visually within 10–20 minutes after the incubation. Finally the reaction was stopped by the addition of stopping solution (H_2SO_4) to each microwell and the results were finally read spectrophotometrically within 30 minutes of stopping the reaction on Multiskan ELISA reader (Multiskan Plus, Labsystems Oy, Pulttitie 8, P.O. Box 8, 00881 Helsinki, Finland) at a wavelength of 450 nm.

5. Interpretation of the Results

5.1. Visual Observation. All negative controls were colourless or faintly blue while samples with a more intense blue colour than negative control were observed as positive. Samples that showed equal or less colour than the negative control were observed as negative.

5.2. Photometric Determination/Readings. The negative control or mean of the negative controls should be less than 0.15 absorbance units. The cutoff value was calculated by adding 0.100 absorbance units to the negative control value. All samples with absorbance value above the cutoff value were read as positive while all samples with absorbance value below the cutoff point were read as negative.

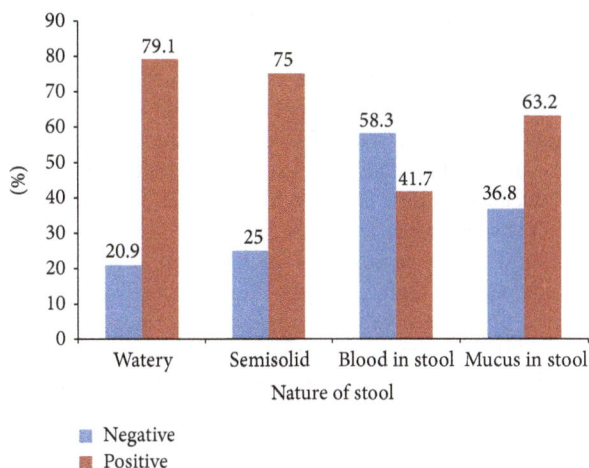

FIGURE 1: Distribution of rotavirus diarrhoea in children presenting with different types of stool in Sokoto.

TABLE 1: Duration of rotavirus diarrhoea in children in Sokoto.

Duration of diarrhoea in days	Number of positive cases	% positive	Cumulative %
0–2	22	43.1	43.1
3-4	12	23.5	66.7
5–7	14	27.5	94.1
8–10	2	3.9	98
>10 days	1	2	100
Total	51	100	

TABLE 2: Frequency of vomiting in rotavirus diarrhoea in children in Sokoto.

Vomiting	Number of positive cases	Percentage positive
Yes	40	78.4
No	11	21.6
Total	51	100.0

TABLE 3: Duration of vomiting in rotavirus diarrhoea in children in Sokoto.

Duration of vomiting in days	Number of positive cases	% positive	Cumulative %
No response	11	0	0
0–2	36	90	90
3-4	1	2.5	92.5
5–7	3	7.5	100
Total	51	100	

6. Results

6.1. Rate of Rotavirus Detection among Children in Sokoto, Nigeria. Out of the 200 human diarrhoea stools examined by ELISA, rotavirus was detected in 51 of the samples, indicating a prevalence of 25.5%.

6.2. Stool Analysis of Rotavirus Diarrhoea in Children in Sokoto. Figure 1 showed the summary of data on the frequency of rotavirus detection according to the nature of stools. The data showed a high frequency of detection in watery stool tinged with blood (58.3%) indicating possible mixed infection with other parasites. The detection of the virus in stool mixed with mucus was 36.8% which further supports the possibility of mixed infection.

6.3. Analysis of Duration of Rotavirus Diarrhoea in Children in Sokoto. The results showed that, for the 51 rotavirus positive children, diarrhoea lasted for 2 days in majority of cases (43.1%). However, the diarrhoea could last for up to 7 days as observed in 27.5% of rotavirus positive children. Only in few cases (2%) did the duration of the diarrhoea reach 10 days (Table 1).

6.4. Analysis of Vomiting in Rotavirus Diarrhoea in Children in Sokoto. The results showed that vomiting was present in over 78.4% of all rotavirus diarrhoea while vomiting was absent in 22.6% of the cases (Table 2). Chi-square analysis indicated significant association between rotavirus diarrhoea and vomiting ($P < 0.05$). The duration of vomiting in days observed in 51 rotavirus positive children showed that majority of cases occurred within 1-2 days (90%) with very few cases occurring up to seven days (7.5%) (Table 3).

6.5. Analysis of Dehydration in Rotavirus Diarrhoea in Children in Sokoto. The data on the level of dehydration in rotavirus diarrhoea positive children in Sokoto showed that none, mild, or severe dehydration was present in 7.8%, 37.3%, and 45.1%, respectively, as summarized in Figure 2. The

result showed that the level of dehydration in the majority of children suffering from rotavirus diarrhoea was severe. Chi-square analysis also indicated statistically significant association between rotavirus diarrhoea and dehydration ($P < 0.05$).

6.6. Analysis of Other Symptoms Present in Rotavirus Diarrhoea in Children in Sokoto. The data indicated that majority of the children suffering from rotavirus diarrhoea had either fever (72.5%) or fever and respiratory symptoms (11.8%). The prevalence of rotavirus diarrhoea in children showing respiratory symptoms without fever was 3.9% (Table 4). Chi-square analysis did not indicate any significant association between rotavirus diarrhoea and these symptoms ($P > 0.05$).

7. Discussion

World Health Organization (WHO) estimated that 42 percent of the total 10.6 million deaths among children younger than five years of age worldwide occur in the African region [15]. Although mortality rates among these children had declined globally, the situation in Africa was considered strikingly different [16]. This was because the mortality rate of children younger than 5 years of age in the African region was said to be seven times higher than that in the European region

TABLE 4: Presence of other symptoms in rotavirus diarrhoea in children in Sokoto.

Other symptoms present	Frequency	Percent	Cumulative percent
Fever	37	72.5	82.2
Respiratory symptoms	2	3.9	86.7
Respiratory symptoms and fever	6	11.8	100.0
Total	45	88.2	
No response	6	11.8	
Total	51	100.0	

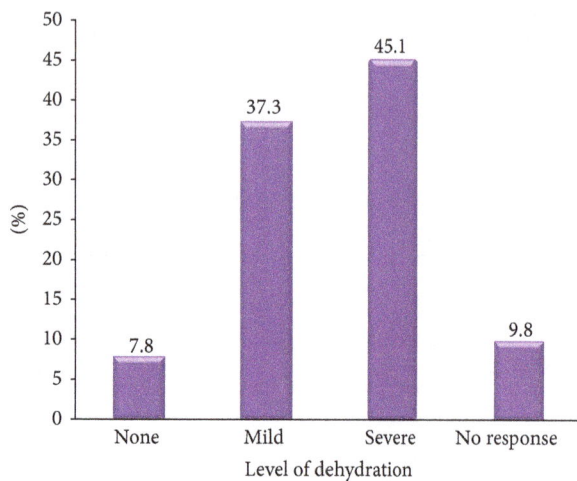

FIGURE 2: Dehydration status of rotavirus diarrhoea positive children in Sokoto.

[16]. Furthermore, earlier report by Cunliffe et al. [5] showed that, of the 25 million children born each year in sub-Saharan Africa, 4.3 million (about 1 in 6) would die by the age of 5 years and about 1/5 of these deaths (850,000) would be from diarrhoea. Interestingly, rotavirus was identified to be the single most important pathogen associated with diarrhoea cases in both hospital patients and outpatients [5].

In this study, 51 (25.5%) out of the 200 diarrhoeic children tested were found to be positive for rotavirus while 149/200 (74.5%) tested negative for rotavirus. Thus, the prevalence of rotavirus diarrhoea accounted for 25.5% of diarrhoea cases among children younger than five years of age presented to hospitals in Sokoto metropolis.

The result of this study is consistent with the sentinel based rotavirus surveillance system and hospital based study results within the African region [17].

Interestingly, however, earlier studies carried out in different parts of northern Nigeria reported low prevalence. Pennap and Umoh [18] reported rotavirus infection prevalence of 15.6% among children (0–60 months old) that presented with diarrhoea in northeastern Nigeria. Aminu et al. [19] similarly reported rotavirus prevalence of 18% among diarrheic children and 7.2% among nondiarrheic children in

a hospital setting in northern Nigeria and prevalence of 9% among children younger than five years of age in a community based study in the same region. Similarly, other investigators reported lower prevalence of the infection in the northern region [20]. The low prevalence reported in the community based study is expected as higher prevalence of rotavirus infection is more likely to be encountered in hospital based studies since rotavirus positive cases are often severe and likely represented in hospitals [21]. However, generally, studies from southern Nigeria had shown higher rotavirus prevalence values than those from northern Nigeria [22–25]. The differences in the prevalence recorded by different investigators had been attributed to differences in time of sample collection, method of screening samples, geographical location of the study, or changing trends of the burden of the rotavirus disease over the years [26].

Earlier studies indicated that stools in rotavirus diarrhoea were nonbloody and generally lack faecal leukocytes and mucus may be found in about 20% of cases [27, 28]. But surprisingly the result in this study showed a high frequency of rotavirus detection in watery stool tinged with blood (58.3%). This is also in contrast with the recent observation that blood tinged diarrhoea was rare in rotavirus infection [18]. However, the observation of high prevalence of rotavirus in blood watery stool may likely be a result of mixed infection with other pathogens such as Shigella because, in developing areas like Sokoto, transmission of enteric pathogens and coinfection are high as a result of poor sanitation, low immunity, lack of access to treatment, imbalanced diet, and poor nutrition. The detection rate of the virus in stool mixed with mucus in this study was 36.8% which further supports the possibility of mixed infection even though stool in rotavirus infection had been reported to often contain large amounts of mucus [29].

The result on the occurrence of vomiting in children with rotavirus diarrhoea showed that vomiting was present in over 33% of all rotavirus positive children while vomiting was absent in 13.8% of the cases. There was significant association between vomiting and rotavirus diarrhoea ($P < 0.05$). Indeed, vomiting had always been a common occurrence in rotavirus diarrhoea and had been reported to precede the diarrhoea in approximately half of all rotavirus diarrhoea cases [30]. The duration of vomiting in days observed in the rotavirus positive children showed that majority of cases occurred within 1-2 days (90%) with very few cases occurring up to seven days (7.5%). This is in agreement with the observation of Pennap and Umoh [18]. But, generally rotavirus disease is usually self-limiting, lasting for four to eight days, and the overall duration of symptoms was reported to be between 2 and 22 days [31]. Recent report showed that, in severe rotavirus cases, children may suffer from symptoms of gastroenteritis for up to 9 days and then recover [32].

Rotavirus had often been associated with severe dehydration which is actually responsible for death associated with the infection [33]. In addition, children with dehydration had been found to be about two times more likely to have rotavirus diarrhoea [6]. In this study, the prevalence of rotavirus diarrhoea in children with none, mild, or severe dehydration was found to be 15.9%, 17.8%, and 42.4%, respectively. The result showed that the level of dehydration in the majority of children suffering from rotavirus diarrhoea was severe. Chi-square analysis also indicated significant association between rotavirus diarrhoea and dehydration ($P < 0.05$). The result is in conformity with the report of Pennap and Umoh [18]. Indeed, rotavirus infection had been associated with severe diarrhoea episodes and vomiting which often led to severe dehydration in babies and young children [33].

The analysis of other symptoms observed with rotavirus diarrhoea in children in Sokoto showed that the majority of the children suffering from rotavirus diarrhoea had either fever (26.8%) or fever and respiratory symptoms (25%). The prevalence of rotavirus diarrhoea in children showing respiratory symptoms without fever was 21.1%. Chi-square analysis did not indicate any significant association between rotavirus diarrhoea and these symptoms ($P > 0.05$). When the frequency of occurrence of fever was considered alone or in combination with respiratory symptoms, the result showed that fever was present in 51.8% of the cases. This is in consonance with many reports that indicated presence of fever in about 45%–84% of patients suffering from rotavirus diarrhoea [34–37]. The observation of the presence of respiratory symptoms in 25% of the cases is also in agreement with earlier reports that indicated presence of various upper and lower respiratory infections, including otitis media, laryngitis, pharyngitis, and pneumonia during rotavirus illness [38–40].

8. Conclusion

Rotavirus detection was the greatest in children with blood tinged watery stool indicating high possibility of mixed infections occurring in this environment. The symptoms of vomiting and dehydration were significantly associated with rotavirus diarrhoea while other symptoms such as fever and/or respiratory symptoms singly or in combination occur in rotavirus diarrhoea but are not significantly associated with the disease.

Acknowledgment

The authors wish to acknowledge The Noguchi Memorial Institute for Medical Research (NMIMR), University of Ghana, Legon, Ghana, for providing space to carry out the laboratory analysis.

References

[1] C. J. Murray and A. D. Lopez, "Global mortality, disability, and the contribution of risk factors: global Burden of Disease Study," *The Lancet*, vol. 349, no. 9063, pp. 1436–1442, 1997.

[2] World Health Organization, "The world health report life in the 21st century. A vision for all," Tech. Rep., World Health Organization, Geneva, Switzerland, 1998.

[3] World Health Organization, *Department of Vaccines and Biologicals: Report of the Meeting on Future Directions for Rotavirus Vaccine Research in Developing Countries*, World Health Organization, Geneva, Switzerland, 2000.

[4] I. de Zoysa and R. G. Feachem, "Interventions for the control of diarrhoeal diseases among young children: rotavirus and cholera immunization," *Bulletin of the World Health Organization*, vol. 63, no. 3, pp. 569–583, 1985.

[5] N. A. Cunliffe, P. E. Kilgore, J. S. Bresee et al., "Epidemiology of rotavirus diarrhoea in Africa: a review to assess the need for rotavirus immunization," *Bulletin of the World Health Organization*, vol. 76, no. 5, pp. 525–537, 1998.

[6] F. N. Binka, F. K. Anto, A. R. Oduro et al., "Incidence and risk factors of paediatric rotavirus diarrhoea in northern Ghana," *Journal of Tropical Medicine & International Health*, vol. 8, no. 9, pp. 840–846, 2003.

[7] M. K. Estes, "Rotaviruses and their replication," in *Fields Virology*, D. M. Knipe, P. M. Howley, D. E. Griffin et al., Eds., vol. 2, pp. 1747–1785, Lippincott Williams & Wilkins, Philadelphia, Pa, USA, 4th edition, 2001.

[8] B. McClain, E. Settembre, B. R. S. Temple, A. R. Bellamy, and S. C. Harrison, "X-ray crystal structure of the rotavirus inner capsid particle at 3.8 A resolution," *Journal of Molecular Biology*, vol. 397, no. 2, pp. 587–599, 2010.

[9] E. C. Settembre, J. Z. Chen, P. R. Dormitzer, N. Grigorieff, and S. C. Harrison, "Atomic model of an infectious rotavirus particle," *The EMBO Journal*, vol. 30, no. 2, pp. 408–416, 2011.

[10] J. Matthijnssens, P. H. Otto, M. Ciarlet, U. Desselberger, M. van Ranst, and R. Johne, "VP6-sequence-based cut-off values as a criterion for rotavirus species demarcation," *Archives of Virology*, vol. 157, no. 6, pp. 1177–1182, 2012.

[11] A. Z. Kapikian and R. M. Chanock, "Rotaviruses," in *Fields Virology*, B. N. Fields, D. M. Knipe, P. M. Howley et al., Eds., vol. 2, pp. 1657–1708, Lippincostt-Raven, Philadelphia, Pa, USA, 3rd edition, 1996.

[12] V. Martella, N. Decaro, A. Pratelli, M. Tempesta, and C. Buonavoglia, "Variation of rotavirus antigenic specificity in a dairy herd over a long-term survey," in *Genomic Diversity and Molecular Epidemiology of Rotaviruses*, N. Kobayashi, Ed., Research Signpost, Trivandrum, India, 2003.

[13] J. S. Bresee, R. I. Glass, B. Ivanoff, and J. R. Gentsch, "Current status and future priorities for rotavirus vaccine development, evaluation and implementation in developing countries," *Vaccine*, vol. 17, no. 18, pp. 2207–2222, 1999.

[14] U. D. Parashar, E. G. Hummelman, J. S. Bresee, M. A. Miller, and R. I. Glass, "Global illness and deaths caused by rotavirus disease in children," *Emerging Infectious Diseases*, vol. 9, no. 5, pp. 565–572, 2003.

[15] J. Bryce, C. Boschi-Pinto, K. Shibuya, R. E. Black, and World Child Health Epidemiologic Reference Group, "WHO estimates of the causes of death in children," *The Lancet*, vol. 365, no. 9465, pp. 1147–1152, 2005.

[16] World Health Organization, *Child Health*, WHO, Regional Office for Africa, 2005, http://www.afro.who.int/en/clusters-a-programmes/frh/child-and-adolescent-health/programme-components/child-health.html.

[17] World Health Organization, "Global networks for surveillance of rotavirus gastroenteritis, 2001–2008," *Weekly Epidemiological Record*, vol. 83, no. 47, pp. 421–425, 2008.

[18] G. Pennap and J. Umoh, "The prevalence of group A rotavirus infection and some risk factors in pediatric diarrhea in

Zaria, North central Nigeria," *African Journal of Microbiology Research*, vol. 4, no. 14, pp. 1532–1536, 2010.

[19] M. Aminu, A. A. Ahmad, J. U. Umoh, J. Dewar, M. D. Esona, and A. D. Steele, "Epidemiology of rotavirus infection in North-Western Nigeria," *Journal of Tropical Pediatrics*, vol. 54, no. 5, pp. 340–342, 2008.

[20] M. I. Adah, A. Rohwedder, O. D. Olaleye, O. A. Durojaiye, and H. Werchau, "Further characterization of field strains of rotavirus from Nigeria VP4 genotype P6 most frequently identified among symptomatically infected children," *Journal of Tropical Pediatrics*, vol. 43, no. 5, pp. 267–274, 1997.

[21] I. Banerjee, S. Ramani, B. Primrose et al., "Comparative study of the epidemiology of rotavirus in children from a community-based birth cohort and a hospital in South India," *Journal of Clinical Microbiology*, vol. 44, no. 7, pp. 2468–2474, 2006.

[22] P. O. Abiodun and H. Omoigberale, "Prevalence of nosocomial rotavirus infection in hospitalized children in Benin City, Nigeria," *Annals of Tropical Paediatrics*, vol. 14, no. 1, pp. 85–88, 1994.

[23] O. O. Omotade, O. D. Olayele, C. O. Oyejide, R. M. Avery, A. Pawley, and A. P. Shelton, "Rotavirus serotypes and subgroups in gastroenteritis," *Nigerian Journal of Paediatrics*, vol. 22, pp. 11–17, 1995.

[24] R. Audu, S. A. Omilabu, I. Peenze, and D. Steele, "Viral diarrhoea in young children in two districts in Nigeria," *Central African Journal of Medicine*, vol. 48, no. 5-6, pp. 59–63, 2002.

[25] M. S. Odimayo, W. I. Olanrewaju, S. A. Omilabu, and B. Adegboro, "Prevalence of rotavirus-induced diarrhoea among children under 5 years in Ilorin, Nigeria," *Journal of Tropical Pediatrics*, vol. 54, no. 5, pp. 343–346, 2008.

[26] CDC, "Rotavirus surveillance—worldwide, 2001–2008," *Morbidity and Mortality Weekly Report*, vol. 57, no. 46, pp. 1255–1257, 2008, http://www.cdc.gov/mmwr/preview/mmwrhtml/mm5746a3.htm.

[27] L. K. Pickering, H. L. DuPont, J. Olarte, R. Conklin, and C. Ericsson, "Fecal leukocytes in enteric infections," *American Journal of Clinical Pathology*, vol. 68, no. 5, pp. 562–565, 1977.

[28] L. Huicho, D. Sanchez, M. Contreras et al., "Occult blood and fecal leukocytes as screening tests in childhood infectious diarrhea: an old problem revisited," *The Pediatric Infectious Disease Journal*, vol. 12, no. 6, pp. 474–477, 1993.

[29] A. Ferdrick, E. Murphy, P. J. Gibbs, M. C. Horzineck, and M. J. Studdert, "Gasteroenteritis," *Nigerian Journal of Paediatrics*, vol. 22, pp. 11–17, 2002.

[30] I. E. Haffejee, "The pathophysiology, clinical features and management of rotavirus diarrhoea," *The Quarterly Journal of Medicine*, vol. 79, no. 288, pp. 289–299, 1991.

[31] R. G. Wyatt, R. H. Yolken, J. J. Urrutia et al., "Diarrhea associated with rotavirus in rural Guatemala: a longitudinal study of 24 infants and young children," *The American Journal of Tropical Medicine and Hygiene*, vol. 28, no. 2, pp. 325–328, 1979.

[32] C. W. Bass and K. N. Dorsey, "Rotavirus and other agents of viral gastroenteritis," in *Nelson Textbook of Pediatrics*, E. Richard and F. Behrman, Eds., pp. 107–110, Raven Press, Philadelphia, Pa, USA, 2004.

[33] P. A. Offit and M. F. Clark, "Reoviruses," in *Principles and Practice of Infectious Diseases*, G. L. Mandell, J. E. Bennett, and R. Dolin, Eds., pp. 1696–1703, Churchill Livingstone, Philadelphia, Pa, USA, 5th edition, 2000.

[34] W. J. Rodriguez, H. W. Kim, J. O. Arrobio et al., "Clinical features of acute gastroenteritis associated with human reovirus-like agent in infants and young children," *The Journal of Pediatrics*, vol. 91, no. 2, pp. 188–193, 1977.

[35] M. C. Steinhoff, "Rotavirus: the first five years," *The Journal of Pediatrics*, vol. 96, no. 4, pp. 611–622, 1980.

[36] I. Uhnoo, E. Olding-Stenkvist, and A. Kreuger, "Clinical features of acute gastroenteritis associated with rotavirus, enteric adenoviruses, and bacteria," *Archives of Disease in Childhood*, vol. 61, no. 8, pp. 732–738, 1986.

[37] A. Kovacs, L. Chan, C. Hotrakitya, G. Overturf, and B. Portnoy, "Rotavirus gastroenteritis," *American Journal of Diseases of Children*, vol. 141, no. 2, pp. 161–166, 1987.

[38] H. M. Lewis, J. V. Parry, H. A. Davies et al., "A year's experience of the rotavirus syndrome and its association with respiratory illness," *Archives of Disease in Childhood*, vol. 54, no. 5, pp. 339–346, 1979.

[39] M. Santosham, R. H. Yolken, E. Quiroz et al., "Detection of rotavirus in respiratory secretions of children with pneumonia," *The Journal of Pediatrics*, vol. 103, no. 4, pp. 583–585, 1983.

[40] B. J. Zheng, R. X. Chang, G. Z. Ma et al., "Rotavirus infection of the oropharynx and respiratory tract in young children," *Journal of Medical Virology*, vol. 34, no. 1, pp. 29–37, 1991.

cancer were upregulated in RSV-infected BEAS-2B cells. RSV infection up-regulated 947 and 3047 genes at 4 h and 24 h, respectively, and 124 genes were common at both instances. Moreover, 1682 and 3771 genes were downregulated at 4 h and 24 h, respectively, and only 192 genes were same. Respiratory disease biomarkers like ARG2, SCNN1G, EPB41L4B, CSF1, PTEN, TUBB1, and ESR2 were also detected. RSV infection signs and symptoms render a partial consequence of host pathogen interaction, but the transcription profile is better exemplified by microarray. The transcription profiles of RSV infected mice lungs and lymph nodes showed gene expression of antigen processing and inflammation. The response is higher in lungs on the day 1 after RSV infection than day 3. The gene expression profile shortly after RSV infection can be accounted as a biomarker and can be scaled up for diagnostics with *in vitro* and *in vivo* profiles [68, 69].

Sometimes DNA microarrays do not confer their utility in very specific investigations in the case of personalized medicine wherein the manifestation of disease occurs at the transcription level. This situation may arise in complications of RSV with other associated disorders like asthma [70, 71]. As protein is the abundant functional biomolecule, reflecting the physiological or pathological state of the organ, protein microarray profile is an option to access under this situation [72]. Protein microarray has analytical and functional applications to study the protein-protein, protein-DNA, and protein-ligand interactions. These features enable the profiling of immune responses and are thus important for diagnostics and biomarker discovery. Protein microarray can serve as a rapid, sensitive, and simple tool for large-scale identification of viral-specific antibodies in sera [73, 74]. An extensive study was done during the 2002 SARS pandemic using coronavirus protein microarray to screen antibodies in human sera (>600 samples) with >90% accuracy and at least as sensitive as, and more specific than the available ELISA tests [73]. Thus, this system has enormous potential to be used as an epidemiological tool to screen viral infections.

2.5. Mass Spectroscopy. Mass spectroscopy (MS) has become a method of choice for molecular investigation of pathogens as the reliability is reinforced due the well-characterized sequence information of nucleic acids or proteins and even for intact viruses [86, 87]. But pragmatic usage of MS is possible when coupled with various chromatography and affinity-based techniques. The combination of affinity based viral detention and nucleic acid based MS serves as a solution for low detection limits. Affinity-based methods employing nanotechnology can be used to find traces of target pathogen to improve detection limits. PCR amplification of pathogen nucleic acid combined with MS can be used as substitutes [88–90]. MS has the advantage of rapid identification of multiple viruses at the same time and even identifies the protein modification status [86, 91]. Now, MS is no longer confined to proteomics based analysis, and MS based genomics have become common practice. There are several variations of MS, like Matrix-assisted laser desorption/ionization mass spectroscopy (MALDI-MS), Surface-enhanced laser desorption/ionization mass spectroscopy (SELDI-MS), Bioaerosol mass spectrometry (BAMS), Pyrolysis gas chromatography mass spectrometry (Py/GC/MS), Capillary electrophoresis-MS, and Liquid chromatography mass spectrometry (LC-MS) are the few examples that are used for the identification of pathogens [89]. But the MS technique that promises robust practical application for pathogen detection is the electrospray ionization mass spectrometry (ESI-MS) [92, 93] Figure 3(a).

A notable ESI-MS for global surveillance of influenza virus was accomplished by Sampath et al. [75] by coupling MS with reverse-transcription PCR (RT-PCR). The study correctly identified 92 mammalian and avian influenza isolates (which represented 30 different H and N types, including 29 avian H5N1 isolates). The analysis showed more than 97% sensitivity and specificity in the identification of 656 clinical human respiratory specimens collected over a seven-year period (1999–2006) at species and subtypes level. The surveillance of samples from 2005-2006 influenza virus incidence showed evidence of new genotypes of the H3N2 strains. The study also suggested approximately 1% mixed viral quasi-species in the 2005-2006 samples providing insight into viral evolution. This study led to a number of RT-PCR/ESI-MS based detection methods for RSV and related respiratory viruses like influenza A and B, parainfluenza types 1–4, *adenoviridae* types A–F, *coronaviridae*, human bocavirus, and human metapneumovirus screening [76]. These assays had 87.9% accuracy, compared to conventional clinical virology assays and pathogens undetected by traditional clinical virology methods could be successfully detected. The advantage of RT-PCR/ESI-MS platform is the ability to determine the quantity of pathogens, detailed pathogen characterization, and the detection of multiple pathogens with high sensitivity and efficiency.

2.6. Nanotechnology Based Detection. Recent advancements in nanotechnology have changed the perception and perspective of research. When scaled down to nanometers, the properties of matter change and nanotechnology exploits these new properties and harnesses them with the existing technologies to exceptional capabilities. Techniques discussed below are examples of the nanotechnology based detection approaches which utilize basic traditional but indispensable detection techniques.

2.6.1. Nanoparticle Amplified Immuno-PCR. Immuno-PCR, a combination of ELISA and PCR, is used widely for the detection of various bacterial and viral antigens with lower titers as meagre as zepto moles [58]. Perez et al. [59] reported a modification of immuno-PCR using gold nanoparticles for RSV detection. Target extraction was enhanced by using magnetic microparticles (MMPs) functionalized with anti-RSV antibody to capture the antigen (RSV). The MMP-RSV complex is then countered with gold nanoparticles functionalized with both, palivizumab (Synagis), an anti-RSV F protein antibody, and DNA sequence partially hybridized with a tag DNA sequence (fAuNP). The MMP-RSV-fAuNP complex is then heated to release the partially hybridized tag DNA sequence, which is then quantified from the supernatant by real-time PCR. These modifications enabled detection of RSV even at 4.1 PFU/mL. This assay offers

(a) PCR-Electrospray Ionization/Mass Spectroscopy

(b) Surface Enhanced Raman Spectroscopy

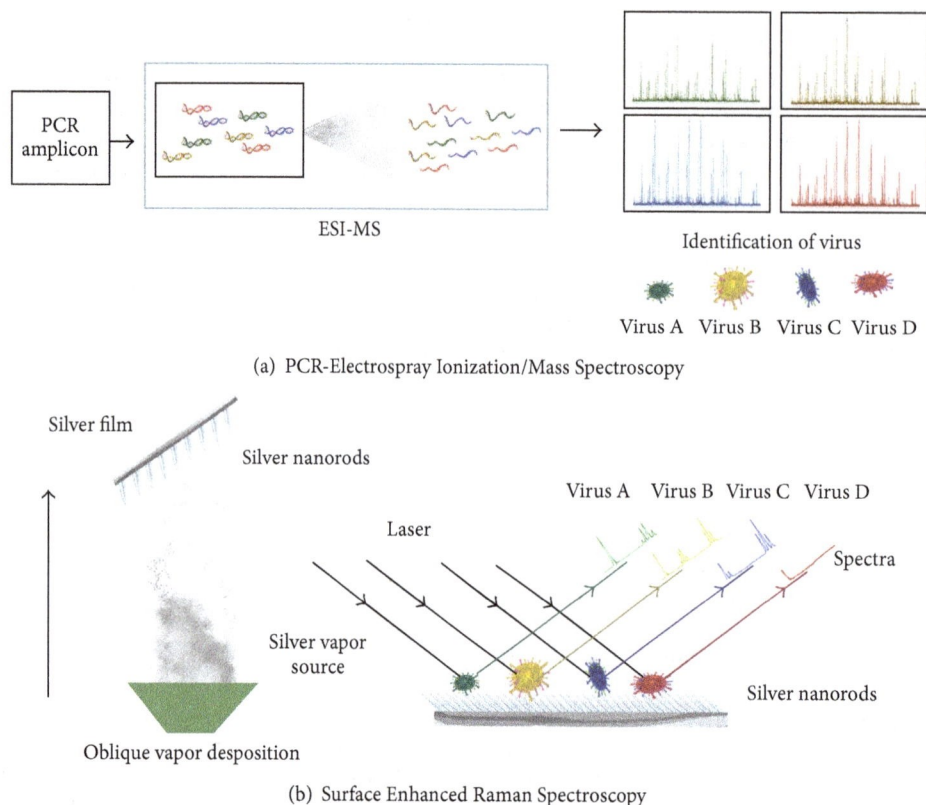

FIGURE 3: A schematic representation of biophysical method of RSV detection. (a) PCR-electrospray ionization mass spectroscopy and (b) Surface enhanced Raman spectroscopy.

a 4000-fold improvement in the limit of detection over ELISA and a 4-fold improvement over detection when compared with real-time RT-PCR [59].

2.6.2. *Live Cell RNA Imaging.* Live cell imaging can provide the benefit of effective diagnosis and treatment, if the capability of identifying, monitoring, and quantifying biomolecules is developed. However, development of such systems for detection of viral agents is difficult, especially in the early stages of infection. With the advent of molecular beacon technology, it is now possible to track mRNA of host and RSV [32]. Based on this framework, a modification by oligonucleotide-functionalized gold nanoparticulate probe was suggested as an improvement, wherein a gold nanoparticle is functionalized with DNA hairpin structure by thiol linkage. The DNA is so designed that the loop portion has a complimentary sequence to RNA to be detected and the 5' stem is linked to gold nanoparticle by thiol group and the 3' end is linked to a fluorophore. On hybridization to the target RNA with loop, the fluorophore goes away from the quenching gold and the emission is tracked. Thus, live imaging of mRNA is possible. This mechanism of hairpin DNA functionalized gold nanoparticles (hAuNP) was executed by Jayagopal et al. in detecting RSV mRNA in HEp-2 cells [33]. This technique offers the advantage of real-time detection of multiple mRNA at the same time, including the mRNA of RSV and glyceraldehyde 3-phosphate dehydrogenase of the HEp-2 cell. Quantitative assay using this approach was

possible using DNA hairpin structures functionalized to a gold filament, which is immersed in a capillary tube containing viral RNA and scanned for fluorescence. The set-up was able to detect as low as 11.9 PFUs, which was ~200 times better than a standard comparison ELISA.

2.6.3. *Quantum Dots.* Immunofluorescence microscopy based detection of RSV, that is, direct fluorescent-antibody assay (DFA), is considered as gold standard [39], but the comparative efficacy of DFA with other assays does not seem to have reached a consensus [39, 40, 94–97]. The prime possible disadvantages and inconsistencies of DFA can be attributed to the fading of the dyes, conjugating antibodies with dyes, limited sensitivity due to background staining, and excitation at two different wavelengths [26–28]. To address these issues, fluorescent nanoparticles, that is, quantum dots (QDs), appear as promising candidates for field clinical diagnostics. Due to their inorganic nature, they are less susceptible to metabolic degradation. QDs are photostable; that is, they do not lose fluorescence on long exposure to light and can be excited at the same wavelength while radiating at different wavelengths and hence can be used for multiplexing [27]. Various successful attempts were made to ameliorate RSV detection using QDs. The progression of RSV infection in the HEp-2 cell line was studied using confocal microscopy by QDs probing F and G proteins and it was found that this method was more sensitive than real-time, quantitative RT-PCR, particularly at early infection [29]. This approach was used by Tripp et al. *in vitro*, on Vero cell lines and was extrapolated by an *in vivo* BALB/c mice study, which concluded the approach beyond diagnostics as it can be used for multiplexed virus and/or host cell antigen detection and intracellular tracking studies [28].

Flow cytometry is now widely used in diagnostics and the reliability is improved because millions of cells are analyzed at a time and comprehensive data is produced [30]. Tracking and targeting cellular proteins and the ease of multiple parameters correlation allow flow cytometry to be used for various qualitative and quantitative assays. Flow cytometer could detect RSV with sensitivity and reproducibility [31]. Agrawal et al. [26] showed that antibody conjugated QDs could detect RSV rapidly and sensitively using the principles of microcapillary flow cytometry (integrated with a fixed-point confocal microscope) and single-molecule detection. A 40-nm carboxylate-modified fluorescent nanoparticles and streptavidin-coated QDs were used in their study, which could estimate relative levels of surface protein expression.

2.6.4. Gold Nanoparticle Facilitated Microarray. The FDA approved microarray systems, semiautomated respiratory virus nucleic acid test (VRNAT), and the fully automated respiratory virus nucleic acid test *SP* (RVNAT*SP*) (Nanosphere, Northbrook, IL) are examples of the microarray based detection systems for influenza A virus, influenza B virus, RSV A, and RSV B [56, 98]. These systems are based on the efficient detection of microarray based hybridization using gold nanoparticles. The hybridization between the oligonucleotide probes and target DNA/RNA is detected specifically by hybridizing them again to gold nanoparticles functionalized with oligonucleotide and the signal is generated by gold facilitated reduction of silver in the presence of a reducing agent [99].

2.6.5. Surface Enhanced Raman Spectroscopy. Raman spectroscopy works on the basis of the inelastic scattering of monochromatic radiation like near infrared, visible, or near ultraviolet, interacting with an analyte with low-frequency vibrational and/or rotational energy. But the signal generated is low and hence the signal is enhanced using silver or gold matrix substrates. There are many modifications of Raman spectroscopy, but the most widely used one is the surface enhanced Raman spectroscopy (SERS) (Figure 3(b)) [77]. Direct intrinsic, indirect intrinsic and extrinsic detection are three major SERS detection configurations [78]. In contrast to IR spectroscopy, the Raman spectra can be obtained without the interference of water molecules and thus biological analytes can be studied in their native conformation. SERS is routinely used for various bioanalytical purposes due to the rapid and nondestructive detection of analytes with sensitivity, specificity, and precision even for a single molecule or live cells [79, 80]. SERS can be targeted to various analytes that may constitute DNA, RNA, proteins, or other organic compounds. Nucleic acids are the preferred candidates in biological SERS investigations, as the influence of base composition and sequence, conformation (local and/or global) of nucleic acids, or intermolecular dynamics with protein or ligand is correspondingly expressed as typical

spectral signature [81]. Bacteria and viruses from various biological samples can be identified, characterized, and classified from clinical samples [77, 82]. SERS can distinguish between DNA or RNA viruses like Adenovirus, Rhinovirus, Rotavirus [83], and RSV [84]. Rotavirus is the common cause of gastroenteritis in children and the rapid and sensitive detection was demonstrated with SERS [83]. SERS is an established powerful tool for sensitive, expedited, and specific detection of various respiratory pathogens, like *Mycoplasma pneumoniae* and RSV. Using silver nanorod arrays (NA) platform for SERS, differentiation of *M. pneumoniae* in culture and in spiked and true clinical throat swab samples was achieved [82]. The notable sensitivity of SERS to resolve strain level differences for RSV strains A/Long, A2, ΔG, and B1 can be exploited for clinical diagnosis instrumentation [84].

3. Prevention

Prevention is the most important aspect of healthcare and has prime contribution to the culmination of the disease than the treatment measures. Effective preventive measures reduce the mortality and economic burden of the disease. Until now there is no effective vaccine for prevention for RSV. Direct or indirect contact with the nasopharyngeal secretions or droplets (sneezing, coughing and kissing), fomites, and food from infected patients can potentially transmit RSV. Live virus can survive on surfaces for several hours [100], but at the cellular level, the viral spread is a series of systematic events of invasion including viral attachment and fusion followed by viral replication and protein synthesis. The F protein accumulates in the host membrane and then surrounds the budding progeny viruses, thus spreading the infection to adjacent cells and exacerbating the infection (Figure 4). Thus these proteins are considered as the potential candidate for the development of prevention measures such as antibodies, DNA vaccines, and subunit vaccines.

3.1. Antibodies. RSV has three envelope proteins F, G, and SH. Both F and G are glycosylated and represent the targets of neutralizing antibodies. The RSV F protein emerged as a good vaccine candidate due to its conserved and vital role in cell attachment. Passive immunization is a direct approach to counter RSV (Figure 5). Initially, polyclonal antibodies from healthy human individuals resistant to RSV were successful in preventing RSV infection in high risk infants and these pooled and purified immunoglobins were popular as RespiGam. The monoclonal antibody specifically neutralizing F protein conferred effective protection against RSV as compared to RespiGam, and this licensed monoclonal antibody, palivizumab (Synagis), is now used to passively protect high risk infants from severe disease, thus replacing the RespiGam. The efficacy of the recombinant monoclonal antibody, palivizumab, has been tested for prophylaxis and therapy in immunocompromised cotton rats [101]. Repeated doses of palivizumab were required to prevent rebound RSV replication. Palivizumab is administered alone or in combination with aerosolized ribavirin. Palivizumab cannot

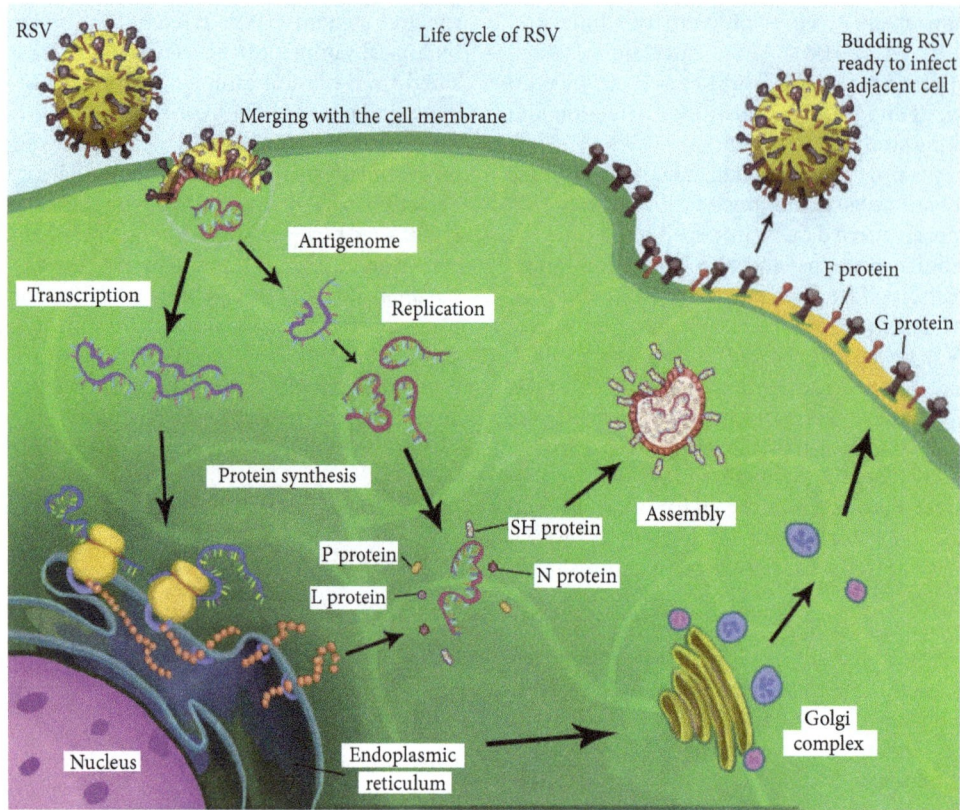

FIGURE 4: A schematic representation of RSV life cycle.

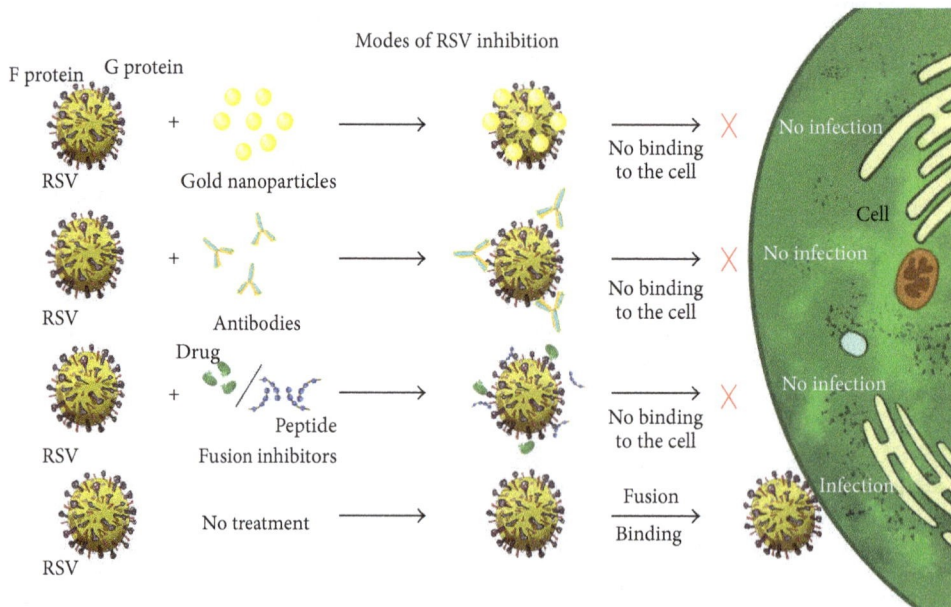

FIGURE 5: A schematic representation of various compounds inhibiting RSV binding to the cell.

cure or treat serious RSV disease but neutralization of RSV can help in preventing serious RSV infections. Motavizumab has been found to neutralize RSV by binding the RSV fusion protein F after attachment to the host, but before the viral transcription [102]. Viral entry was not inhibited by palivizumab or motavizumab when pretreated with RSV, but there was a reduction in viral transcription, thus inhibiting both cell-cell and virus-cell fusion most likely by preventing the conformational changes in the F protein needed for viral fusion.

The effective use of palivizumab is limited due to the cost and its use in infants with high risk of bronchiolitis based on the coverage by different healthcare systems [103]. In spite of these restrictions on palivizumab, it has a wide societal impact on use in infants with chronic lung disease due to premature birth or those with haemodynamically significant cardiac disease. According to the modified recommendations of the Committee on Infectious Diseases of the Centers for Disease Control and Prevention of RSV, palivizumab is recommended for infants with congenital heart disease (CHD), chronic lung disease (CLD), and birth before 32 weeks [104]. Minimum 5 doses are recommended irrespective of the month of the first dose for all geographical locations for infants with a gestational age of 32 weeks 0 days to 32 weeks 6 days without hemodynamically significant CHD or CLD. The new recommendations of the committee were aimed at the high risk groups including infants attending child care or one or more siblings or other children younger than 5 years living with the child. Also, the infants were qualified for receiving prophylaxis only until they reached 90 days of age. Palivizumab, although effective, is costly and thus is not beneficial to the recipients especially during the periods when RSV is not circulating. A cost effective means of producing RSV F neutralizing antibodies was experimented in phages and plants. Much success in this regard of palivizumab production was observed in the *Nicotiana benthamiana* plant system which offered glycosylation and high production at lower upstream and equivalent downstream cost, when compared to mammalian derived palivizumab. The efficacy of the plant derived palivizumab was more than the mammalian derived palivizumab or the plant derived human monoclonal antibodies in cotton rats [105].

3.2. DNA Vaccines. RSV genome codes structural and functional proteins that are immunogenic and referred for vaccine development against RSV. DNA based vaccines were developed based on these proteins because the concept was simple, as it involved a DNA fragment coding part or whole protein of RSV that was inserted into an appropriate expression plasmid vector under a constitutive promoter control (Figure 6). The initial work with this approach was successful in the expression in cells and *in vivo* murine models to eliminate the RSV infection, but the problem of RSV associated Th2 type immune response was persistent. This problem was attempted to be resolved by manipulating the parameters of choice of the protein to be expressed, the expression vector, adjuvants, formulations, and intracellular stability of the plasmid.

Mice challenged with the RSV-G construct had balanced systemic and pulmonary Th1/Th2 cytokines and RSV neutralizing antibody responses [106]. However, the RSV F protein gene is considered as a widely used and prospective target for the development of vaccines and is often a favourite model for DNA vaccines against RSV [107]. But the wild type RSV F protein expressed from DNA plasmid was poorly expressed, so a codon optimized DNA vaccine was designed for better *in vivo* expression and hence was more immunogenic and reduced the RSV titer [108]. To address the low immunogenicity of F protein, the antagonistic activity

of RSV for IFN and immunopathology, a construct was designed by inserting the F gene into Newcastle disease virus (NDV) vector (NDV-F). This modification served the purpose of higher elicitation of IFN by NDV-F than RSV, protection against RSV infection without immunopathology and enhanced adaptive immunity in BALB/c mice [109]. Wu et al. [110] developed a DNA vaccination strategy against RSV using a mucosal adjuvant. Two DNA vaccine vectors, namely, DRF-412 and DRF-412-P containing residues 412–524 of the RSV F gene, were cloned into the phCMV1 DNA vaccine vector. The DNA vaccine vectors DRF-412 contained the cholera toxin gene region called ctxA2B acting as a mucosal adjuvant. The DNA vaccine was successfully expressed in mouse muscle tissue, which was confirmed by immunohistological analysis and RT-PCR. The immunized mice induced neutralization antibody, systemic Ab (IgG, IgG1, IgG2a, and IgG2b) responses, and mucosal antibody responses (Ig A) which mimicked the challenge with live RSV. The mice immunized with the DRF-412 vector contained less RSV RNA in lung tissue and induced a higher mixed Th1/Th2 cytokine immune response and had better protection than those immunized with the DRF-412-P vector, which was confirmed by lung immunohistology studies [110]. A rational approach to confer protection against various pathogens through single vaccination or more often known as combined or composite vaccines was employed for RSV, influenza A virus (INF-A), and herpes simplex virus type-1 [111]. Mice were immunized either by injection or by gene gun (gold beads as carrier of the plasmid DNA) with mixture of four plasmids: INF-A haemagglutinin (HA), INF-A nucleoprotein (NP), HSV-1 glycoprotein D (gD) and RSV glycoprotein F. This led to protection of mice from the respective pathogens of challenge; in addition, it also offered protection from *Mycoplasma pulmonis* challenge as well [111]. Recently several developments have been made in the field of recombinant vaccines. In one such approach, *Mycobacterium bovis* Bacillus Calmette-Gue'rin (BCG) vaccine was modified to carry RSV N or M2 and was tested and found to establish the Th1 type immunity in RSV challenged mice [112]. The recombinant vaccine also elicited the activation of RSV specific T cells producing IFN-γ and IL-2, along with reduction in weight loss, and lung viral protein load, thus establishing a Th1-polarized immune response.

Bactofection is bacteria mediated transfer of plasmid DNA into mammalian cells. Xie et al. employed this interesting way of delivering and expressing the DNA vaccine in mice against RSV [113]. This approach serves the purpose of naturally activating the immunostimulatory response of the host and also delivery of the DNA vaccine. Attenuated *Salmonella typhimurium* strain SL7207 expressing vector pcDNA3.1/F containing the RSV F gene was orally administered to BALB/c mice, triggering efficient antigen-specific humoral, cellular, and mucosal immunity [113, 114].

A novel method of developing RSV DNA vaccine was devised by replacing the structural genes with RSV genes in an attenuated strain of Venezuelan equine encephalitis virus (VEEV). VEEV has an ssRNA (+) genome and contains a strong subgenomic promoter. The replicon particles were prepared by providing helper RNAs encoding VEEV

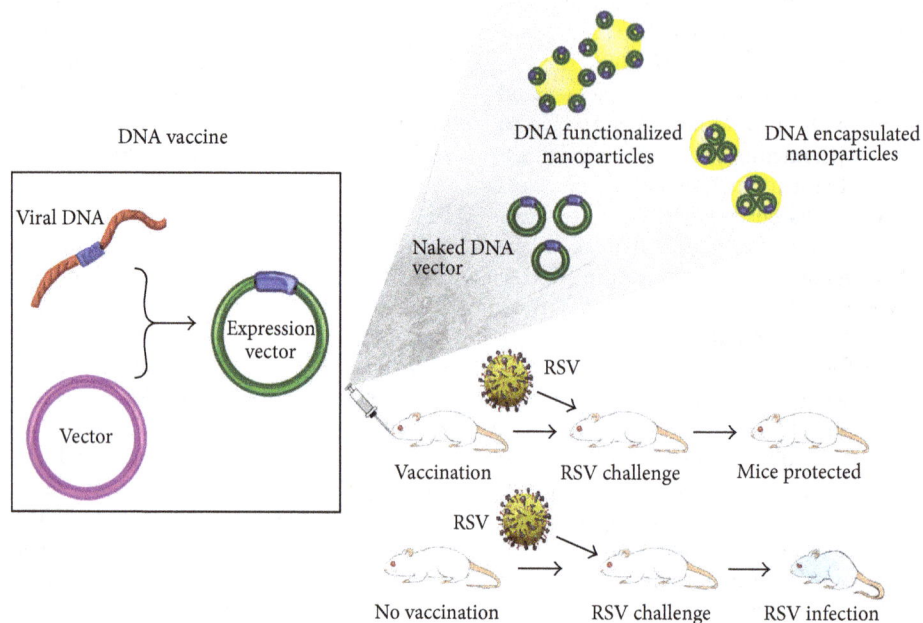

FIGURE 6: A schematic representation of a simple DNA vaccine administered as a naked DNA vector or functionalized nanoparticle or as an encapsulation for controlled delivery.

capsid and envelope glycoprotein which comprise structural proteins, and all of these, when transfected into Vero cells, resulted into a replicon particle that could independently synthesize the RSV protein, thereby activating the immune response and protection. The system could be modulated by the administration of helper RNAs. This strategy was applied to the mice and rhesus monkey models, conferring protection against RSV and a desirable extent of a balanced Th1/Th2 type immune response was received [115].

3.3. *Subunit Vaccines.* Several approaches have been considered for developing an effective vaccine against RSV including the formalin inactivated RSV vaccine developed in late 1960s, which instead resulted in an enhanced infection [116]. An efficient RSV vaccine would be the one with proper balance between immunogenicity and protection without any allergic response [8].

RSV F has widely been accepted as the vaccine candidate due to its conserved nature among various strains as well as among the other paramyxoviruses [8, 117–119]. An antigenic region corresponding to RSV F protein (region 255–278) was cloned into a vector with ctxA2 B gene of cholera toxin and named as rF255, which elicited a helper T cell type 1 immune response in mice. It also resulted in higher expression of serum neutralizing antibody in immunized mice [120]. Similarly, a multivalent recombinant protein was developed by cloning RSV F, M2, and G protein into a bacterial pET32a (+) vector (called rFM2G), which resulted in enhanced serum IgG titers [121]. rFM2G was also used in conjunction with flagellin as an adjuvant which did not increase the IgG titers. In an effort to address the issues of immunogenicity, several studies have used the approach of combining other viruses with RSV genes. One such study uses

virus like particles (VLP) developed from NP and M proteins of Newcastle disease virus (NDV) and a chimeric protein consisting of cytoplasmic and transmembrane domains of NDV HN protein and ectodomain of the RSV G protein (H/G) [122]. These VLPs resulted in an immune response better than UV-inactivated RSV and provided complete protection in mice from RSV even at a single dose, with elicited neutralizing antibodies. The VLP-H/G-immunized mice did not show enhanced pathology as compared to FI-RSV. In another approach to solve the immunogenicity related problems in devising an RSV vaccine, a recombinant vector, based on either murine PIV type 1 or Sendai virus was used to deliver RSV G protein through reverse genetics [123]. This provided effective protection against RSV in cotton rats. Another similar study using Sendai virus (SeV) as a vaccine carrying RSV F gene provided protection against four pathogens including hPIV-1, mouse PIV-1, hPIV-3, and RSV [124]. This approach has been used to compare two vaccine models, one with SeV backbone and the other with PIV-3 backbone. The SeV based vaccine showed a decrease in RSV load in African green monkeys lung titers.

Another study with recombinant vaccine utilizes recombinant simian Varicella viruses (rSVVs) which express RSV G and M2 protein genes and was evaluated in the Vero cell line [125]. Such a recombinant vaccine approach could be very useful in elderly people with risk of infection with *Varicella* and RSV, as well as children with chickenpox and RSV. rSVV based vaccines resulted in enhanced immune responses to RSV antigens serving as suitable vaccines in rhesus monkeys. A very similar approach was adopted for developing a recombinant *alphavirus* or immune stimulating complex (ISCOM) antigen against RSV [126]. The recombinant vector was designed by using a self-abortive alphavirus called Semiliki forest virus carrying RSV F and G genes. The advantages of using such recombinant abortive viruses are that they can undergo viral infection just once, eliminating chances of adverse effects resulting from further infection, and infecting the host cell resulting in expression of the inserted RSV gene followed by humoral and cell mediated immune response. The ISCOM used was composed of *Quillaja* saponin, lipids, and RSV antigens, having adjuvant properties as well. ISCOM has been shown to enhance antibody production, T cell proliferation, and MHC class-I responses as well as RSV specific neutralizing antibodies, IgG and IgA. The recombinant vaccine SFV/F or SFV/FG resulted in IFN gamma response along with resistance to RSV infection without worsened RSV disease, whereas the results were opposite in ISCOM/FG with enhanced goblet cell hyperplasia post-RSV challenge.

3.4. Nanovaccines. Recently, several applications of nanotechnology have appeared in the development of vaccines popularly known as Nanovaccines. DNA vaccine is prone to rapid degradation when introduced into an animal system; so to increase the retention and increase the efficacy of the DNA vaccines, they can be encapsulated into a polymer that will protect and facilitate controlled release (Figure 5). Various synthetic or natural polymers are now experimented for targeted delivery and controlled release of the carrier. Chitosan is the polymer of great interest in respiratory disease treatment, because of its mucoadhesive property and biodegradability, which balances the purpose of longer retention and controlled release of carrier molecules encapsulated. Thus, chitosan nanoparticles are being developed against RSV. A DNA vaccine (DR-FM2G) constitutive expression vector consisting of antigenic regions of RSV F, M2, and G genes driven by human cytomegalovirus promoter was encapsulated using chitosan nanoparticles (DCNPs). The advantages of DCNPs are that it was more stable than naked DNA or chitosan and so ideal for protection against DNA degradation by nucleases. The sustainability of the DCNPs in mice was more than naked DNA which was correspondingly indicated by higher RSV protein expression evident by immunohistochemical and real-time PCR studies [127, 128]. A similar study was conducted by Eroglu et al. [129], where the highly conserved RSV F gene was cloned into pHCMV1 expression vector and encapsulated into poly-hydroxyethyl-methacrylate nanospheres coated by chitosan and transfected into Cos-7 cells. The transfection efficiency of the system was at par with commercially available transfecting agents. *In vivo* studies with the BALB/C mice also indicated F protein expression and reduced RSV infection.

A vaccine based on the recombinant RSV N gene rings, called N-SRS, enclosing a bacterial RNA has been developed and assessed intranasally in BALB/c mice [130]. N-SRS was adjuvanted with *E. coli* enterotoxin LT (R192G) with efficient protection against RSV and high titers of IgG1, IgG2a and IgA, and so forth. Although this nanovaccine elicited a mild inflammatory response in airways of the mice, it enhanced the expression of antigen specific CD4$^+$ and CD8$^+$ responses. In another novel approach, nanoemulsion (NE) was used as an adjuvant to mucosal RSV vaccine in a mouse model

and found to provide significant protection against RSV, along with RSV specific humoral responses and an enhanced Th1/Th2 response [131]. The NE rendered the RSV ineffective when treated for 2 or 3 hours, with a reduction in viral titer up to 10-fold as compared to media controls. NE-RSV also led to higher expression of RSV specific antibody, with significant decrease in viral load, no hyperreactivity, and no Th2 cytokine induction. Thus, this novel approach seems to be a promising and safe vaccine option against RSV.

The bitter episode of formalin inactivated RSV vaccine has impeded the vaccine development and in fact has raised serious concern in the use of native RSV or its components. This approach was revived with a novelty, wherein the engineered RSV F protein aggregates formed nanoparticles and were used as vaccine, and these nanoparticles induced protective immunity in cotton rats [132]. To combat RSV, host neutralizing antibodies are always more preferred than the therapeutic antibodies. But the epitopes of the neutralizing antibodies are larger than those of the therapeutic neutralizing antibodies (palivizumab and motavizumab). Thus, the use of these epitopes for neutralizing antibodies as a vaccine requires the retention of immuno functional conformation while getting rid of the undesired protein. RSV F oligomeric protein nanoparticle was synthesized by inserting recombinant RSV F gene into Baculovirus and expressed in Sf9 insect cell lines. This resulted in high recombinant protein expression compared to native protein. This also resulted in the conformation of rosette nanoparticles which was the aggregate of multiple RSV F oligomers found to be immunogenic. The study was extrapolated to phase 1 human clinical trials, which showed its safety and efficacy against RSV [133].

3.5. Nutrition. Nutrition is not considered as treatment but is a prerequisite for homeostasis, and any nutritional imbalance attracts disorders and diseases. Supplementary nutrients could reverse the adverse effects of the disorders and diseases, so *in sensu* nutrition can serve both as treatment and prevention. Carbohydrate rich diets and diet lacking antioxidants like fruits and vegetables during pregnancy are suspected for RSV susceptibility [134]. Resveratrol (trans-3,4,5-trihydroxystilbene), a polyphenol from grapes has demonstrated reduction in RSV replication and inflammation [135, 136]. A study demonstrated that the cord blood deficient in vitamin D was associated with RSV bronchiolitis; the neonates were at higher risk of RSV in the first year of life [137]. Vitamin D inhibits NF-κB signalling which is responsible for RSV inflammation without affecting the antiviral activity of the host [138]. Lower levels of micronutrients like zinc, copper, selenium and retinol (Vitamin A), and alpha-tocopherol (Vitamin E) were observed in the children affected by RSV and human metapneumovirus [139]. In one study, RSV infected children were administered vitamin A to compensate the lower vitamin A serum level which was found beneficial for children with severe RSV infection [140]. Also, probiotic diet has shown to boost resistance against pathogen by modulating immune response, as in case of RSV, *Lactobacillus rhamnosus* (a probiotic bacteria) treated BALB/c mice showed significantly reduced lung viral loads and pathology after the RSV challenge [141]. These studies indicate the importance and association of nutrition with RSV susceptibility.

4. Treatment

There are very limited treatment options available for RSV. However, there are many drugs for the symptoms associated with RSV infection. The target genes and proteins vital for RSV infection (discussed in Section 3) are important for developing preventative and treatment measures. The mode of action and potency of a drug determines the approach of prophylactic or curative application. Considering the proposed life cycle of RSV, theoretically, there are numerous modes to interfere with RSV infection, but these options may not be practical. Replication, transcription and fusion are the few target processes for drug development against RSV. A focus is therefore on development of potent drug which holds conformity in the human trials. Some of the approaches are described below that promise to be a potential treatment for RSV (Table 2).

4.1. Antiviral Drugs. Ribavirin or 1-[(2R, 3R, 4S, 5R)-3,4-dihydroxy-5-(hydroxymethyl)oxolan-2-yl]-1H-1,2,4-triazole-3-carboxamide is a widely used broad spectrum synthetic anti-viral drug for both DNA and RNA virus treatment. Oral and nasal administration of ribavirin for treatment of severe lower respiratory tract infections caused by RSV and influenza virus are an option [104]. Ribavirin is phosphorylated in the cells and competes with adenosine-5′-triphosphate and guanosine-5′-triphosphate for viral RNA-dependent RNA polymerases in RNA viruses. However, the mechanism of ribavirin differs for DNA viruses, as it is a competitive inhibitor of inosine monophosphate dehydrogenase (IMPDH), causing deletion of GTP and messenger RNA (mRNA) guanylyl transferase (mRNA capping enzyme) and adversely affecting protein synthesis [142, 143]. The exact mechanism of anti-viral activity of ribavirin against RNA and DNA viruses is still not clear. The usefulness of ribavirin against viruses is not only due to its anti-viral activity but also due to its capability to modulate the immune system. Ribavirin is suggested to have immuno-stimulatory effects on Th cells [144]. The derivatives of ribavirin such as viramidine, merimepodib, and other IMPDH inhibitory molecules like mycophenolate and mizoribine have shown antiviral activity against the hepatitis C virus, and hence, there is scope for investigating them as potential anti-RSV drugs [143, 145]. There are many other compounds that can inhibit RSV replication and a well-known compound RSV 604 ((S)-1-(2-fluorophenyl)-3-(2-oxo-5-phenyl-2,3-dihydro-1H-benzo[e][1,4]diazepin-3-yl)-urea) showed promising results against RSV [146].

A derivative of antibiotic geldanamycin 17-ally-lamino-17-demethoxygeldanamycin (17AAG) and 17-dimethylaminoethylamino-17-demethoxygeldanamycin (17DMAG) targeted against cancer has now attracted researchers due to its antiviral property. These compounds are HSP90 inhibitors and thus helpful against RSV, as RSV is dependent on heat shock protein (HSP90) for its replication [165]. The anti-RSV activity of these compounds was seen in human airway epithelial cells (HAEC) and is considered as drug resistant therapeutics, due to the highly conserved target chaperon protein. These compounds are also known to inhibit replication of HPIV, influenza virus, and rhinovirus. The anti-viral activity dose is not toxic to the cells and inhalation mode of treatment can increase the local efficacy and avoid unnecessary exposure to other organs [166].

4.2. Fusion Inhibitors. Recent advances in the development of anti-viral drugs include the fusion inhibitors. The fusion inhibitors are usually synthetic compounds or molecules interrupting the fusion of virus with the host cell usually by binding the fusion proteins (Figure 5). The fusion inhibitors have been widely studied as anti-viral agents in several viruses including HIV, RSV, Henipavirus, Hendra virus, Nipah virus, Paramyxovirus, metapneumoviruses, HIV, and RSV [11, 167–173]. The first reports of the use of peptide(s) as fusion inhibitors include the development of DP-178, a synthetic peptide based on the leucine zipper region of the HIV fusion glycoprotein gp41 [167], which showed an IC50 at 0.38 nM against HIV-1. Fusion inhibitors for the paramyxoviruses have also been developed based on the conserved region of the fusion protein F. The F protein is widely known for its conserved nature among the *Paramyxoviridae* family [7]. Lambert et al. [174] developed the fusion inhibitors belonging to the conserved heptad repeat (HR) domains of F1 region of F protein which is analogous to the peptides DP-107 and DP-178 of HIV gp41. These fusion inhibitors were tested against RSV, human parainfluenza virus 3, and measles virus, which showed antiviral activity specific to the species of origin. DP-178 is an FDA approved anti-HIV drug with International Nonproprietary Name (INN) Enfuvirtide and trade name Fuzeon. Out of the peptides tested, the peptide T-118 developed from RSV was the most effective, with an EC50 of 0.050 μM. These fusion inhibitors were then further characterized and tested. It was shown that different fusion inhibitors derived from same HR region differ in their anti-viral activity [147]. The HR121 and HR212 peptides showed an IC50 of 3.3 and 7.95 μM, respectively, against RSV. Similarly, another peptide inhibitor was developed from Rho-A which showed inhibition of syncytia formation induced by RSV [148]. RhoA (a small GTPase) is involved in many biological processes and was shown to bind the RSV-F protein at amino acids 146–155. A peptide derived from the RSV F binding domain of RhoA (RhoA77-95) was shown to inhibit RSV and PIV-3 infection and syncytium formation, block cell-to-cell fusion, and reduce viral titers and illness in mice.

The fusion inhibitors are not only limited to peptide inhibitors; but a range of chemical inhibitors have also been tested against RSV and benzimidazoles are well-known fusion inhibitors [175]. A lead compound was identified to subsequently synthesize an analogue JNJ2408068, a low molecular weight benzimidazole, which showed high antiviral activity. It had an EC50 of 0.16 nM, 100,000 times better than that of ribavirin. This compound showed anti-fusion activity against RSV in a dual mode of action including prevention of cell-virus fusion activity as well as cell-cell

Recent Advances in Diagnosis, Prevention, and Treatment of Human Respiratory Syncytial Virus

TABLE 2: Comparison of different treatment approaches for RSV.

Treatment	Mechanism	Example	Remark	References
Antiviral drugs	Replication inhibition	RSV 604		[104]
	Mutation	Ribavirin, viramidine, merimepodib	Effective against RSV, but adverse effect on the host	[142–146]
	Inhibitor of inosine monophosphate dehydrogenase	Ribavirin, mycophenolate, mizoribine		
	Immunostimulatory effects	Ribavirin		
Fusion inhibitors	Inhibiting fusion protein attachment to cell	Peptide—HR121, HR212, RhoA Chemical—BMS-433771, RFI-641	Peptide fusion inhibitors promising anti RSV drug; chemical fusion inhibitors have side effects	[147–150]
Nanoparticles	Inhibiting attachment to cell	Silver nanoparticles, gold nanoparticles siRNA-ALN-RSV01	Emerging field, conclusive studies required Effective and safe; ALN-RSV01 completed phase IIb clinical trials	[151, 152]
Antisense therapy	RNA interference	Phosphorodiamidate morpholino oligomers Plant extracts—*Cinnamomum cassia*, *Cimicifuga foetida*,		[153–157]
Ethnobotanicals	Probably fusion inhibitors, anti-inflammatory	Sheng-Ma-Ge-Gen-Tang, Ginger, etc. Decoctions-modified Dingchuan, Liu-He-Tang, water extract of Licorice	Promising but conclusive studies required	[158–164]

fusion activity. However, it was found ineffective against other viruses of the family including HPIV-3 and measles virus. A vast screening of as many as 16,671 compounds (source ChemBioNet library) was conducted for anti-RSV activity *in vitro* and two novel compounds, N-(2-hydroxyethyl)-4-methoxy-N-methyl-3-(6-methyl[1,2,4]triazolo[3,4-a]phthal-azin-3-yl)benzenesulfonamide (named as P13) and the 1,4-bis(3-methyl-4-pyridinyl)-1,4-diazepane (named as C15) were mined, which reduced the virus infectivity with IC50 values of 0.11 and 0.13 μM, respectively [176].

Recently, several groups have reported synthetic fusion inhibitors of RSV, especially the benzotriazole derivatives [177]. After evaluating the structure-activity relationship (SAR) of these compounds, named as series **1** compounds, it was observed that the topology of the side chains of these compounds is important and facilitates the modification of their physical properties, as many of these compounds showed poor therapeutic indices (cytotoxic effects) to the host cells tested. In order to address these issues, a second series of derivatives of the compound **1** were developed and evaluated for SAR and their functionality as fusion inhibitors [178]. These compounds were developed from **1**, which had a tolerant diethylaminoethyl side chain with both polar and nonpolar functionality against RSV and had a replacement of the benzotriazole with benzimidazole-2-one. These were potent inhibitors of RSV *in vitro*. These compounds were named **2** and had an additional structural vector absent in **1**, which accounted for enhanced potency as fusion inhibitors and served as the base for further development of fusion inhibitors. Further, these group **2** compounds were modified by introducing acidic and basic functional groups into the side chains [179]. The oxadiazolone had anti-RSV activity comparable to that of ribavirin, whereas the ester modified group **2** compounds were suitable for oral admin-istration. These studies further led to the identification of a benzimidazole-2-one derivative called BMS-433771 which was an orally active RSV inhibitor [149]. Another compound studied was the 5-aminomethyl analogue **10aa** with potent anti-RSV activity towards BMS-433771 resistant RSV. The compound BMS-433771 was further modified at side chains and with the introduction of an aminomethyl substituent at the 5-position of the core benzimidazole moiety [180]. The aminomethyl substitution in the benzimidazole ring was found to enhance the antiviral activity.

Furthermore, the consecutive modification of benzimi-dazole resulted in benzimidazole-isatin oximes which were evaluated for anti-RSV activity [181]. The compound was ana-lyzed for its antiviral activity, cell permeability, and metabolic stability in human live microsomes. Several other derivatives with modification such as O-alkylation and addition of nitrogen atoms to isatin phenyl ring were implemented also to enhance antiviral activity. Three compounds **18j**, **18i**, and **18n** were shown to have anti-viral activity against RSV in the BALB/c mice. Further, a compound RFI-641 was identified which was found to be the most potent anti-RSV agent inhibiting RSV both *in vitro* and *in vivo* and is in phase I clinical trials. RFI-641 is a biphenyl triazine synthesized by coupling diaminobiphenyl to two chlorotriazine molecules under microwave conditions [150]. RFI-641 was found to be

effective against 6 laboratory and 18 clinical viruses at con-centrations between 0.008 and 0.11 mM (0.013–0.18 mg/mL). The compound reduced the viral load to an extent of 1.7 logs in the African green monkey model and also in mice and cotton rats. In order to further enhance the antiviral activity of RFI-641, it was modified by replacing its triazine linkers with pyrimidine [182]. However, this modification did not have much difference in the anti-viral activity, thus rendering this modification not much of practical use. There are several novel nonbenzimidazole based compounds, showing anti-RSV activity *in vitro*, but a more polar compound thiazole-imidazole 13 was selected on the compound potency, mod-erate permeability, and low metabolic rate in rats, and more detailed *in vivo* studies are further anticipated [183].

4.3. Nanoparticles. It has been established that metals like silver [184] and gold [185] have anti-microbial activity, but cytotoxic effects of these reactive metals make them unsuit-able for their use in humans. The reactivity and behaviour of metals can be modulated by reducing their size to nanoscale. Carbon nanotubes (CNTs) are emerging nanomaterials for biomedical application [186]. Polyvinylpyrrolidone (PVP) conjugated silver nanoparticles showed low toxicity to HEp-2 cells at low concentrations and exhibited 44% RSV inhi-bition [151] (Figure 6). Singh et al. used fusion inhibitor peptide functionalized gold nanoparticles and carboxylated gold nanoparticles of size 13 nm against RSV, which showed 83% and 88% inhibition of RSV, respectively [152]. Similar approach was employed by recombinant RSV F protein functionalized on gold nanorods [187]. The emergence of nanotechnology has opened new avenues for RSV treatment.

4.4. Antisense Treatment. RNA interference (RNAi) which is a normal cellular event has become a powerful means of controlling gene regulation. The interference mediated by siRNA was used against human immunodeficiency virus, poliovirus, hepatitis C, and parainfluenza virus (PIV) in cell culture [188–190]. The concept of inhibiting RSV infection using targeted antisense mechanism was applied by Jairath et al. by silencing the RSV-NS2 gene [191]. Following the RNAi approach, Bitko et al. designed siRNA against the P gene of RSV and PIV which protected mice against individual and mixed infections upon intranasal administration [192]. The effectiveness of siRNA action was observed with and without the use of transfection reagents. This approach was also effective when targeting the RSV-F gene [193]. Similar work on HEp-2 cell lines was replicated using four siRNA, designed to silence RSV F gene, which showed inhibitory action against RSV at various concentrations [194]. Silencing different RSV genes too had an inhibitory action on the RSV, a plasmid encoding siRNA which was complexed with chitosan targeting RSV-NS1 gene decreased RSV infection in BALB/c mice and Fischer 344 rats and also reduced the associated inflammation [195, 196]. Zhang et al. showed that siRNA nanoparticle targeting RSV NS1 gene resulted in increased IFN-β and IFN-inducible genes in A549 cells and in human dendritic cells, elevated type-1 IFN, and increased differentiation of CD4$^+$ T cells to Th1 cells [196]. Also mice treated with siNS1 nanoparticles exhibited significant decrease in lung viral titers and inflammation.

An interferon-inducible enzyme, 2-5A-dependent RNase, present in higher vertebrates requires $5'$-phosphorylated, $2',5'$-linked oligoadenylate (2-5A) for its endoribonuclease activity against single-stranded RNAs. This feature of 2-5A-dependent RNase is looked upon as an effective RSV treatment [197]. RSV upon infection elicits immune response of the host and particularly the interferon levels [68], and this phenomenon is exploited for the anti-RSV activity of 2-5A-dependent RNaseL. The endoribonuclease activity of 2-5A-dependent RNaseL was used for targeting RSV M2 gene specifically by covalent $2'-5'$ oligoadenylate target antisense, which resulted in the reduction of RSV replication. The endoribonuclease activity had negligble effect with an inactive dimeric form of 2-5A linked to antisense, 2-5A linked to a randomized sequence of nucleotides, and antisense molecules lacking 2-5A and did not affect the other RSVs or cellular RNAs [198]. Additionally, to widen the range of the approach, the effects of modification of oligonucleotides and RNA target sites were studied [199]. This model was improved with respect to the specificity and activity by a chimera of 2-5A-antisense, christened as NIH351. The sequence information of RSV genome was used to develop NIH351 and was 50- to 90-fold more potent against RSV strain A2 than ribavirin [200]. Administration of siRNA in combination with ribavirin was recommended for effective treatment [201]. The parent molecule was chemically modified to further increase the in vivo stability and specificity and potency of NIH351. The resulting new version RBI034 was ~50% more effective than the parent molecule against RSV (strain A and B) and was not cytotoxic in the effective dose ranges. RBI034 treatment of African green monkeys shows promising results [202].

Alvarez and coworkers [153] came up with a new RSV-NS1 gene specific siRNA (ALN-RSV01) having a broad spectrum of antiviral activity that targeted the nucleocapsid gene of RSV. In vivo BALB/c murine studies demonstrated that intranasal dosing of ALN-RSV01 resulted in a 2.5- to 3.0-log-unit reduction in RSV lung concentration. To scale up this molecule for RSV treatment in humans, the safety, tolerability, and pharmacokinetics were tested on healthy adults, demonstrating its safety and tolerance in human subjects [154]. In the human clinical trials of ALN-RSV01, healthy subjects were grouped and administered either a placebo or ALN-RSV01 nasal spray for RSV. There was 44% reduction in the RSV infection in the subjects who received ALN-RSV01 without any adverse effect. Thus, this study in real terms has established a unique "proof-of-concept" for an RNAi therapeutic agent in RSV treatment [155]. ALN-RSV01 proved to be safe and was effective against RSV even in a complex clinical situation like lung transplants, which was a remarkable achievement [156]. It is the product of Alnylam Pharmaceuticals, Inc. and has completed phase IIb clinical trials.

Another approach of RNAi treatment to combat RSV is to decelerate the adverse effects of RSV mediated Th2 type immune response because the aggravated host immune response is more harmful than RSV infection itself.

Particularly in neonates, RSV bronchiolitis increases IL-4α levels which results in increased Th2 response, so an antisense oligomer was synthesised for the local silencing of the IL-4α gene. Intranasal application of the antisense oligomers into a neonatal murine model reduced the Th2 type mediated pulmonary pathological signs of inflammation and lung dysfunction [203]. A combinatorial approach of the anti-sense oligomer against RSV and IL-4α would control RSV infection and the adverse effects of RSV mediated inflammation.

Phosphorodiamidate morpholino oligomers (PMOs) are the oligomers where the nucleobases are covalently attached with the morpholine ring replacing the deoxyribose sugar while the phosphodiester bond is replaced by the phosphorodiamidate linkage [157]. Morpholino chemical modification of RNA can be used as antisense with the advantages of specificity, in vivo stability, and targeted delivery. PMOs block the target complementary RNA and the target RNA fails to interact with the proteins and thus the RNA function is hindered. This is specifically true for mRNA and its translation. This phenomenon is different from antisense as the RNase H is not involved [204]. Hence PMOs can be designed containing the initiation codon against viruses to be anti-viral. Better intake into the cell and in vivo systems can be facilitated by conjugating cell penetrating peptide(s) (arginine-rich peptide (RXR) 4XB) to it. This approach was attempted against the RSV-L gene to inhibit RSV in cellular and murine models [205].

4.5. Ethnobotanicals. Various natural and synthetic chemical compounds have been screened for their application to treat RSV infections. Plants are rich sources of alkaloids, steroids, flavonoids, and other complex compounds that have medicinal value and medicinal plants contribute significantly to the traditional Indian and Chinese medicine. Ancient Chinese literature has descriptions of plant extract against respiratory diseases [206]. The exact active compound or action of the traditional formulations is not understood [207]. Now, modern assays have made it possible to get an insight into the mechanism and purification of active plant based compound. Extracts of *Lonicera deflexicalyx* (*Chin Jinyinhua*) were tested against RSV. The active compound from the extract was 3, 5-dicaffeoylquinic acid (CJ 4-16-4), which was isolated, and purified by a series of chromatographic processes. It is suggested that CJ 4-16-4 is a fusion inhibitor and the in vitro and in vivo studies suggest that it is a more effective RSV inhibitor than ribavirin [208]. Cytopathic effect (CPE) assay based screening showed that 27 of 44 herbs had moderate or potent anti-RSV activity [206]. The plant extracts from *Cinnamomum cassia*, *Cimicifuga foetida*, Sheng-Ma-Ge-Gen-Tang (SMGGT) (Shoma-kakkon-to), Xiao-Qing-Long-Tang (Sho-seiryu-to, so-cheong-ryong-tang), Ge-Gen-Tang, and Ginger (*Zingiber officinale*) show anti-RSV activity. These extracts probably inhibit RSV infection by blocking the F protein binding to the cell and some of the extracts even stimulate IFN-β production [158–163]. Some decoctions like modified Dingchuan (consists of *Salviae miltiorrhizae* radix, *Scutellariae radix*, *Farfarae flos* and *Ephedrae herba*), Liu-He-Tang (consist of 13 plant extract), and water extracts of Licorice (*Radix glycyrrhizae* and *Radix glycyrrhizae* Preparata) have shown effectiveness against RSV *in vitro*. Moreover, the modified Dingchuan decoction (MDD) exhibited anti-inflammatory and anti-viral effect in mice (SPF ICR mice) infected with RSV. MDD suppressed eotaxin, IL-4 and IFN-γ level in serum, and mRNA expression of TLR4 and NF-κB in lungs of RSV infected mice [164].

5. Challenges in the Diagnosis, Prevention, and Treatment of RSV

Technology has provided enough capabilities for the detection of RSV in various sample types at various stages of infection using an array of techniques, but the challenge is, availability of these facilities at the correct time for a reasonable cost. The most important perspective of RSV diagnosis is the strategic management of choice between the point of care testing and central laboratory testing [36]. Though the point of care testing gives rapid detection advantage, it suffers from lower sensitivity and thus is a problem to be dealt with. Rapid diagnostic tools like RT-PCR/ESI-MS, microarray based semiautomated respiratory virus nucleic acid test (VRNAT) and the fully automated respiratory virus nucleic acid test SP (RVNATSP) (Nanosphere, Northbrook, IL) have proved their efficiency, but their application in routine clinical practice is still a challenge.

There are numerous molecules that can be potential antiviral drugs, but the screening of a vast number of compounds is cumbersome; hence, high throughput filtering is an essential part of drug development. As an example, when 313,816 compounds from the Molecular Libraries Small Molecule Repository were screened against RSV in HEp-2 cell line, only 409 compounds showed 50% inhibition of the cytopathic effects [209]. The challenge after this sophisticated screening is the translation into drugs by clearing the phases of *in vivo* animal studies and human trials. A setback for the development of therapies against RSV is the lack of a good animal model as they do not truly manifest effects of RSV infection as in humans. RSV experiments in various animal models like BALB/c mice, cotton rats, macaques, African green monkeys, owl monkeys, cebus monkeys, bonnet monkeys, olive baboons, and chimpanzees are evident in the literature. Small animal models like BALB/c mice and cotton rats are commonly used due to ease of handling and low cost, whereas the primate studies are conducted with more stringent regulations and bear heavy expenses [210, 211]. There are many aspects that need to be addressed in the challenges for vaccine development programs and the technological interventions to deal with RSV [212]. These challenges include safety issues concerning the subjects involved in clinical trials, as evident by the failure of formalin inactivated RSV vaccine and motavizumab at the clinical trial levels which resulted in undesired immunogenic responses in the patients involved [34, 116].

Though the present data conclude that RSV is one of the leading causes of morbidity and mortality in children and elders, there is no significant correlation between increased disease severity, respiratory deaths, and detection of any of the respiratory viruses [6, 213]. The situation is more intricate as some authors report that coinfection with non-RSV respiratory viruses tends to increase RSV severity [214] and also there is a hypothesis of a synergistic association of RSV with other viral or bacterial infections [215]. Thus, this topic is a subject of debate and these contradictions need clarification and consensus for proper treatment options. A link between atopy, asthma, and RSV was suspected for a very long time, but now there is supporting evidence for this possible relationship [216–218] and is also ascribed partly to genetic factors of RSV and the host [70, 71, 219, 220]. The activity of RSV in the community may be affected by many factors, including climate, air pollution [221], race/ethnicity [222], and social behavior of the population. Under these complex factors associated with RSV, it is established that early detection of risk factors and medical intervention can reduce the incidence of RSV [223]. A balance of execution and/or abeyance of prophylactic measures is critical with respect to the above discussed determinants [224]. There is limited data to correlate RSV global transmission dynamics with climate and population [2], due to which it is difficult to develop strategies for RSV prevention and treatment.

6. Conclusions

Currently, there is no vaccine or effective treatment against RSV, but the rapid and sensitive RSV detection is possible. The detection techniques are ameliorated by incorporating one or more methods and with the advancement in material science and biophysical capabilities, it has reinforced the development and design of RSV detection systems. However, an effective detection technique can be transformed into effective diagnosis by integrating it into the community health monitoring program at a reasonable cost. Prevention of RSV infection at present is limited to only high risk individuals with a limited efficacy. New preventive measures research like DNA vaccines, subunit vaccines, and nanovaccines have reached animal trials. On the other hand, the RSV treatment approaches using antisense oligomers, fusion inhibitors, and benzimidazole drug have proceeded into clinical trials. The challenges associated with RSV management are categorically numerous. However, at the current pace of scientific research and development and with the implementation of scientific, commercial, and program recommendations to develop epidemiological strategies, it seems optimistic to have an effective diagnosis, prevention, and treatment solution for RSV in near future.

Acknowledgments

The authors acknowledge the National Science Foundation Grant NSF-CREST (HRD-1241701) and HBCU-UP (HRD-1135863). They thank Eva Dennis for the figures used in this paper.

References

[1] H. Nair, E. A. F. Simões, I. Rudan et al., "Global and regional burden of hospital admissions for severe acute lower respiratory infections in young children in 2010: a systematic analysis," *The Lancet*, vol. 381, pp. 1380–1390, 2013.

[2] K. Bloom-Feshbach, W. J. Alonso, V. Charu et al., "Latitudinal variations in seasonal activity of influenza and respiratory syncytial virus (RSV): a global comparative review," *PLoS ONE*, vol. 8, no. 2, Article ID e54445, 2013.

[3] H. Henrickson, "Cost-effective use of rapid diagnostic techniques in the treatment and prevention of viral respiratory infections," *Pediatric Annals*, vol. 34, no. 1, pp. 24–31, 2005.

[4] L. R. Krilov, "Respiratory syncytial virus disease: update on treatment and prevention," *Expert Review of Anti-Infective Therapy*, vol. 9, no. 1, pp. 27–32, 2011.

[5] A. Galindo-Fraga, A. A. Ortiz-Hernández, A. Ramírez-Venegas et al., "Clinical characteristics and outcomes of influenza and other influenza-like illnesses in Mexico city," *International Journal of Infectious Diseases*, vol. 17, no. 7, pp. 510–517, 2013.

[6] D. N. Tran, T. M. H. Pham, M. T. Ha et al., "Molecular epidemiology and disease severity of human respiratory syncytial virus in Vietnam," *PLoS ONE*, vol. 8, no. 1, Article ID e45436, 2013.

[7] P. L. Collins, "Respiratory syncytial virus—human (Paramyxoviridae)," in *Encyclopedia of Virology*, G. Allan and G. W. Robert, Eds., pp. 1479–1487, Elsevier, Oxford, UK, 2nd edition, 1999.

[8] D. Hacking and J. Hull, "Respiratory syncytial virus—viral biology and the host response," *Journal of Infection*, vol. 45, no. 1, pp. 18–24, 2002.

[9] P. L. Ogra, "Respiratory syncytial virus: the virus, the disease and the immune response," *Paediatric Respiratory Reviews*, vol. 5, pp. S119–S126, 2004.

[10] H. M. Costello, W. C. Ray, S. Chaiwatpongsakorn, and M. E. Peeples, "Targeting RSV with vaccines and small molecule drugs," *Infectious Disorders*, vol. 12, no. 2, pp. 110–128, 2012.

[11] X. Zhao, M. Singh, V. N. Malashkevich, and P. S. Kim, "Structural characterization of the human respiratory syncytial virus fusion protein core," *Proceedings of the National Academy of Sciences of the United States of America*, vol. 97, no. 26, pp. 14172–14177, 2000.

[12] P. L. Collins and B. S. Graham, "Viral and host factors in human respiratory syncytial virus pathogenesis," *Journal of Virology*, vol. 82, no. 5, pp. 2040–2055, 2008.

[13] D. L. Kasper, *Harrison's Principles of Internal Medicine*, McGraw-Hill, 2005.

[14] J. M. Langley, J. C. LeBlanc, E. E. L. Wang et al., "Nosocomial respiratory syncytial virus infection in Canadian pediatric hospitals: a pediatric investigators collaborative network on infections in Canada study," *Pediatrics*, vol. 100, no. 6, pp. 943–946, 1997.

[15] A. Simon, A. Müller, K. Khurana et al., "Nosocomial infection: a risk factor for a complicated course in children with respiratory syncytial virus infection—results from a prospective multicenter German surveillance study," *International Journal of Hygiene and Environmental Health*, vol. 211, no. 3–4, pp. 241–250, 2008.

[16] A. R. Falsey, P. A. Hennessey, M. A. Formica, C. Cox, and E. E. Walsh, "Respiratory syncytial virus infection in elderly and high-risk adults," *The New England Journal of Medicine*, vol. 352, no. 17, pp. 1749–1759, 2005.

[17] R. Singleton, N. Etchart, S. Hou, and L. Hyland, "Inability to evoke a long-lasting protective immune response to respiratory syncytial virus infection in mice correlates with ineffective nasal antibody responses," *Journal of Virology*, vol. 77, no. 21, pp. 11303–11311, 2003.

[18] R. D. Pockett, D. Campbell, S. Carroll, F. Rajoriya, and N. Adlard, "Rotavirus, respiratory syncytial virus and non-rotaviral gastroenteritis analysis of hospital readmissions in England and Wales," *Acta Paediatrica*, vol. 102, no. 4, pp. e158–e163, 2013.

[19] A. R. Falsey, "Respiratory syncytial virus: a global pathogen in an aging world," *Clinical Infectious Diseases*, vol. 57, no. 8, pp. 1078–1080, 2013.

[20] C. B. Hall, "Respiratory syncytial virus and parainfluenza virus," *The New England Journal of Medicine*, vol. 344, no. 25, pp. 1917–1928, 2001.

[21] C. Rebuffo-Scheer, M. Bose, J. He et al., "Whole genome sequencing and evolutionary analysis of human respiratory syncytial virus A and B from Milwaukee, WI 1998-2010," *PLoS ONE*, vol. 6, no. 10, Article ID e25468, 2011.

[22] C. S. Khor, I. C. Sam, P. S. Hooi, and Y. F. Chan, "Displacement of predominant respiratory syncytial virus genotypes in Malaysia between 1989 and 2011," *Infection, Genetics and Evolution*, vol. 14, pp. 357–360, 2013.

[23] S. Svraka, E. Rosario, E. Duizer, H. Van Der Avoort, M. Breitbart, and M. Koopmans, "Metagenomic sequencing for virus identification in a public-health setting," *Journal of General Virology*, vol. 91, no. 11, pp. 2846–2856, 2010.

[24] C. F. Shafik, E. W. Mohareb, and F. G. Youssef, "Comparison of direct fluorescence assay and real-time RT-PCR as diagnostics for respiratory syncytial virus in young children," *Journal of Tropical Medicine*, vol. 2011, Article ID 781919, 3 pages, 2011.

[25] M. L. Landry and D. Ferguson, "SimulFluor respiratory screen for rapid detection of multiple respiratory viruses in clinical specimens by immunofluorescence staining," *Journal of Clinical Microbiology*, vol. 38, no. 2, pp. 708–711, 2000.

[26] A. Agrawal, R. A. Tripp, L. J. Anderson, and S. Nie, "Real-time detection of virus particles and viral protein expression with two-color nanoparticle probes," *Journal of Virology*, vol. 79, no. 13, pp. 8625–8628, 2005.

[27] P. K. Chattopadhyay, D. A. Price, T. F. Harper et al., "Quantum dot semiconductor nanocrystals for immunophenotyping by polychromatic flow cytometry," *Nature Medicine*, vol. 12, no. 8, pp. 972–977, 2006.

[28] R. A. Tripp, R. Alvarez, B. Anderson, L. Jones, C. Weeks, and W. Chen, "Bioconjugated nanoparticle detection of respiratory syncytial virus infection," *International Journal of Nanomedicine*, vol. 2, no. 1, pp. 117–124, 2007.

[29] E. L. Bentzen, F. House, T. J. Utley, J. E. Crowe Jr., and D. W. Wright, "Progression of respiratory syncytial virus infection monitored by fluorescent quantum dot probes," *Nano Letters*, vol. 5, no. 4, pp. 591–595, 2005.

[30] D. Mattanovich and N. Borth, "Applications of cell sorting in biotechnology," *Microbial Cell Factories*, vol. 5, article 12, 2006.

[31] M. Chen, J. S. Chang, M. Nason et al., "A flow cytometry-based assay to assess RSV-specific neutralizing antibody is reproducible, efficient and accurate," *Journal of Immunological Methods*, vol. 362, no. 1-2, pp. 180–184, 2010.

[32] P. J. Santangelo, B. Nix, A. Tsourkas, and G. Bao, "Dual FRET molecular beacons for mRNA detection in living cells," *Nucleic Acids Research*, vol. 32, no. 6, article e57, 2004.

[33] A. Jayagopal, K. C. Halfpenny, J. W. Perez, and D. W. Wright, "Hairpin DNA-functionalized gold colloids for the imaging of

mRNA in live cells," *Journal of the American Chemical Society*, vol. 132, no. 28, pp. 9789–9796, 2010.

[34] Centre for Drug Evaluation and Research, Division of Anti-Viral Products Advisory Committee Briefing document, FDA, 2010.

[35] M. Rochelet, S. Solanas, C. Grossiord et al., "A thin layer-based amperometric enzyme immunoassay for the rapid and sensitive diagnosis of respiratory syncytial virus infections," *Talanta*, vol. 100, pp. 139–144, 2012.

[36] P. von Lode, "Point-of-care immunotesting: approaching the analytical performance of central laboratory methods," *Clinical Biochemistry*, vol. 38, no. 7, pp. 591–606, 2005.

[37] R. J. Harbeck, J. Teague, G. R. Crossen, D. M. Maul, and P. L. Childers, "Novel, rapid optical immunoassay technique for detection of group A streptococci from pharyngeal specimens: comparison with standard culture methods," *Journal of Clinical Microbiology*, vol. 31, no. 4, pp. 839–844, 1993.

[38] W. K. Aldous, K. Gerber, E. W. Taggart, J. Rupp, J. Wintch, and J. A. Daly, "A comparison of Thermo Electron RSV OIA to viral culture and direct fluorescent assay testing for respiratory syncytial virus," *Journal of Clinical Virology*, vol. 32, no. 3, pp. 224–228, 2005.

[39] R. Slinger, R. Milk, I. Gaboury, and F. Diaz-Mitoma, "Evaluation of the QuickLab RSV test, a new rapid lateral-flow immunoassay for detection of respiratory syncytial virus antigen," *Journal of Clinical Microbiology*, vol. 42, no. 8, pp. 3731–3733, 2004.

[40] D. Gregson, T. Lloyd, S. Buchan, and D. Church, "Comparison of the RSV Respi-Strip with direct fluorescent-antigen detection for diagnosis of respiratory syncytial virus infection in pediatric patients," *Journal of Clinical Microbiology*, vol. 43, no. 11, pp. 5782–5783, 2005.

[41] A. P. Borek, S. H. Clemens, V. K. Gaskins, D. Z. Aird, and A. Valsamakis, "Respiratory syncytial virus detection by remel Xpect, Binax Now RSV, direct immunofluorescent staining, and tissue culture," *Journal of Clinical Microbiology*, vol. 44, no. 3, pp. 1105–1107, 2006.

[42] R. Selvarangan, D. Abel, and M. Hamilton, "Comparison of BD Directigen EZ RSV and Binax NOW RSV tests for rapid detection of respiratory syncytial virus from nasopharyngeal aspirates in a pediatric population," *Diagnostic Microbiology and Infectious Disease*, vol. 62, no. 2, pp. 157–161, 2008.

[43] K. Shirato, H. Nishimura, M. Saijo et al., "Diagnosis of human respiratory syncytial virus infection using reverse transcription loop-mediated isothermal amplification," *Journal of Virological Methods*, vol. 139, no. 1, pp. 78–84, 2007.

[44] X. Wang, Q. Zhang, F. Zhang et al., "Visual detection of the human metapneumovirus using reverse transcription loop-mediated isothermal amplification with hydroxynaphthol blue dye," *Virology Journal*, vol. 9, article 138, 2012.

[45] J. Mahony, S. Chong, D. Bulir, A. Ruyter, K. Mwawasi, and D. Waltho, "Development of a sensitive loop-mediated isothermal amplification (LAMP) assay providing specimen-to-result diagnosis of RSV infections in thirty minutes," *Journal of Clinical Microbiology*, vol. 51, no. 8, pp. 2696–2701, 2013.

[46] T. Jartti, M. Söderlund-Venermo, K. Hedman, O. Ruuskanen, and M. J. Mäkelä, "New molecular virus detection methods and their clinical value in lower respiratory tract infections in children," *Paediatric Respiratory Reviews*, vol. 14, no. 1, pp. 38–45, 2013.

[47] A. R. Falsey, M. A. Formica, and E. E. Walsh, "Diagnosis of respiratory syncytial virus infection: comparison of reverse transcription-PCR to viral culture and serology in adults with respiratory illness," *Journal of Clinical Microbiology*, vol. 40, no. 3, pp. 817–820, 2002.

[48] A. Hu, M. Colella, J. S. Tam, R. Rappaport, and S.-M. Cheng, "Simultaneous detection, subgrouping, and quantitation of respiratory syncytial virus A and B by real-time PCR," *Journal of Clinical Microbiology*, vol. 41, no. 1, pp. 149–154, 2003.

[49] G. Dewhurst-Maridor, V. Simonet, J. E. Bornand, L. P. Nicod, and J. C. Pache, "Development of a quantitative TaqMan RT-PCR for respiratory syncytial virus," *Journal of Virological Methods*, vol. 120, no. 1, pp. 41–49, 2004.

[50] M. Gueudin, A. Vabret, J. Petitjean, S. Gouarin, J. Brouard, and F. Freymuth, "Quantitation of respiratory syncytial virus RNA in nasal aspirates of children by real-time RT-PCR assay," *Journal of Virological Methods*, vol. 109, no. 1, pp. 39–45, 2003.

[51] R. Mentel, U. Wegner, R. Bruns, and L. Gürtler, "Real-time PCR to improve the diagnosis of respiratory syncytial virus infection," *Journal of Medical Microbiology*, vol. 52, no. 10, pp. 893–896, 2003.

[52] L. J. R. van Elden, A. M. van Loon, A. van der Beek et al., "Applicability of a real-time quantitative PCR assay for diagnosis of respiratory syncytial virus infection in immunocompromised adults," *Journal of Clinical Microbiology*, vol. 41, no. 9, pp. 4378–4381, 2003.

[53] J. Kuypers, N. Wright, and R. Morrow, "Evaluation of quantitative and type-specific real-time RT-PCR assays for detection of respiratory syncytial virus in respiratory specimens from children," *Journal of Clinical Virology*, vol. 31, no. 2, pp. 123–129, 2004.

[54] X. S. Chi, F. Li, J. S. Tam, R. Rappaport, and S.-M. Cheng, "Semi-quantitative one-step RT-PCR for simultaneous identification of human influenza and respiratory syncytial viruses," *Journal of Virological Methods*, vol. 139, no. 1, pp. 90–92, 2007.

[55] M. L. Choudhary, S. P. Anand, M. Heydari et al., "Development of a multiplex one step RT-PCR that detects eighteen respiratory viruses in clinical specimens and comparison with real time RT-PCR," *Journal of Virological Methods*, vol. 189, no. 1, pp. 15–19, 2013.

[56] K. Alby, E. B. Popowitch, and M. B. Miller, "Comparative evaluation of the Nanosphere Verigene RV+ assay with the Simplexa Flu A/B & RSV Kit for the detection of influenza and respiratory syncytial viruses," *Journal of Clinical Microbiology*, vol. 51, no. 1, pp. 352–353, 2012.

[57] M. Xu, X. Qin, M. L. Astion et al., "Implementation of FilmArray respiratory viral panel in a core laboratory improves testing turnaround time and patient care," *American Journal of Clinical Pathology*, vol. 139, no. 1, pp. 118–123, 2013.

[58] C. M. Niemeyer, M. Adler, and R. Wacker, "Immuno-PCR: high sensitivity detection of proteins by nucleic acid amplification," *Trends in Biotechnology*, vol. 23, no. 4, pp. 208–216, 2005.

[59] J. W. Perez, E. A. Vargis, P. K. Russ, F. R. Haselton, and D. W. Wright, "Detection of respiratory syncytial virus using nanoparticle amplified immuno-polymerase chain reaction," *Analytical Biochemistry*, vol. 410, no. 1, pp. 141–148, 2011.

[60] M. J. Heller, "DNA microarray technology: devices, systems, and applications," *Annual Review of Biomedical Engineering*, vol. 4, pp. 129–153, 2002.

[61] A. Djikeng, R. Halpin, R. Kuzmickas et al., "Viral genome sequencing by random priming methods," *BMC Genomics*, vol. 9, article 5, 2008.

[62] D. Wang, L. Coscoy, M. Zylberberg et al., "Microarray-based detection and genotyping of viral pathogens," *Proceedings of the*

National Academy of Sciences of the United States of America, vol. 99, no. 24, pp. 15687–15692, 2002.

[63] C. Y. Chiu, A. Urisman, T. L. Greenhow et al., "Utility of DNA microarrays for detection of viruses in acute respiratory tract infections in children," *Journal of Pediatrics*, vol. 153, no. 1, pp. 76–83, 2008.

[64] A. L. Greninger, E. C. Chen, T. Sittler et al., "A metagenomic analysis of pandemic influenza a (2009 H1N1) infection in patients from North America," *PLoS ONE*, vol. 5, no. 10, Article ID e13381, 2010.

[65] M. B. Townsend, E. D. Dawson, M. Mehlmann et al., "Experimental evaluation of the FluChip diagnostic microarray for influenza virus surveillance," *Journal of Clinical Microbiology*, vol. 44, no. 8, pp. 2863–2871, 2006.

[66] E. C. Chen, S. A. Miller, J. L. Derisi, and C. Y. Chiu, "Using a pan-viral microarray assay (virochip) to screen clinical samples for viral pathogens," *Journal of Visualized Experiments*, vol. 27, no. 50, Article ID e2536, 2011.

[67] C. W. Wong, C. L. W. Heng, L. Wan Yee et al., "Optimization and clinical validation of a pathogen detection microarray," *Genome Biology*, vol. 8, no. 5, article R93, 2007.

[68] Y. C. T. Huang, Z. Li, X. Hyseni et al., "Identification of gene biomarkers for respiratory syncytial virus infection in a bronchial epithelial cell line," *Genomic Medicine*, vol. 2, no. 3-4, pp. 113–125, 2008.

[69] R. Janssen, J. Pennings, H. Hodemaekers et al., "Host transcription profiles upon primary respiratory syncytial virus infection," *Journal of Virology*, vol. 81, no. 11, pp. 5958–5967, 2007.

[70] L. B. Bacharier, R. Cohen, T. Schweiger et al., "Determinants of asthma after severe respiratory syncytial virus bronchiolitis," *Journal of Allergy and Clinical Immunology*, vol. 130, no. 1, pp. 91–100, 2012.

[71] N. Krishnamoorthy, A. Khare, T. B. Oriss et al., "Early infection with respiratory syncytial virus impairs regulatory T cell function and increases susceptibility to allergic asthma," *Nature Medicine*, vol. 18, no. 10, pp. 1525–1530, 2012.

[72] X. Yu, N. Schneiderhan-Marra, and T. O. Joos, "Protein microarrays for personalized medicine," *Clinical Chemistry*, vol. 56, no. 3, pp. 376–387, 2010.

[73] H. Zhu, S. Hu, G. Jona et al., "Severe acute respiratory syndrome diagnostics using a coronavirus protein microarray," *Proceedings of the National Academy of Sciences of the United States of America*, vol. 103, no. 11, pp. 4011–4016, 2006.

[74] L. Yang, S. Guo, Y. Li, S. Zhou, and S. Tao, "Protein microarrays for systems biology," *Acta Biochimica et Biophysica Sinica*, vol. 43, no. 3, pp. 161–171, 2011.

[75] R. Sampath, K. L. Russell, C. Massire et al., "Global surveillance of emerging Influenza virus genotypes by mass spectrometry," *PLoS ONE*, vol. 2, no. 5, article e489, 2007.

[76] K. F. Chen, L. Blyn, R. E. Rothman et al., "Reverse transcription polymerase chain reaction and electrospray ionization mass spectrometry for identifying acute viral upper respiratory tract infections," *Diagnostic Microbiology and Infectious Disease*, vol. 69, no. 2, pp. 179–186, 2011.

[77] M. C. Demirel, S. Kao, N. Malvadkar et al., "Bio-organism sensing via surface enhanced Raman spectroscopy on controlled metal/polymer nanostructured substrates," *Biointerphases*, vol. 4, no. 2, pp. 35–41, 2009.

[78] R. A. Tripp, R. A. Dluhy, and Y. Zhao, "Novel nanostructures for SERS biosensing," *Nano Today*, vol. 3, no. 3-4, pp. 31–37, 2008.

[79] S. D. Hudson and G. Chumanov, "Bioanalytical applications of SERS (surface-enhanced Raman spectroscopy)," *Analytical and Bioanalytical Chemistry*, vol. 394, no. 3, pp. 679–686, 2009.

[80] H. Liu, L. Zhang, X. Lang et al., "Single molecule detection from a large-scale SERS-active Au79Ag21 substrate," *Scientific Reports*, vol. 1, article 112, 2011.

[81] J. M. Benevides, S. A. Overman, and G. J. Thomas Jr., "Raman, polarized Raman and ultraviolet resonance Raman spectroscopy of nucleic acids and their complexes," *Journal of Raman Spectroscopy*, vol. 36, no. 4, pp. 279–299, 2005.

[82] S. L. Hennigan, J. D. Driskell, R. A. Dluhy et al., "Detection of *Mycoplasma pneumoniae* in simulated and true clinical throat swab specimens by nanorod array-surface-enhanced raman spectroscopy," *PLoS ONE*, vol. 5, no. 10, Article ID e13633, 2010.

[83] J. D. Driskell, Y. Zhu, C. D. Kirkwood, Y. Zhao, R. A. Dluhy, and R. A. Tripp, "Rapid and sensitive detection of rotavirus molecular signatures using surface enhanced raman spectroscopy," *PLoS ONE*, vol. 5, no. 4, Article ID e10222, 2010.

[84] S. Shanmukh, L. Jones, Y.-P. Zhao, J. D. Driskell, R. A. Tripp, and R. A. Dluhy, "Identification and classification of respiratory syncytial virus (RSV) strains by surface-enhanced Raman spectroscopy and multivariate statistical techniques," *Analytical and Bioanalytical Chemistry*, vol. 390, no. 6, pp. 1551–1555, 2008.

[85] J. B. McGivney, E. Bishop, K. Miller et al., "Evaluation of a synthetic peptide as a replacement for the recombinant fusion protein of respiratory syncytial virus in a potency ELISA," *Journal of Pharmaceutical and Biomedical Analysis*, vol. 54, no. 3, pp. 572–576, 2011.

[86] G. Siuzdak, "Probing viruses with mass spectrometry," *Journal of Mass Spectrometry*, vol. 33, no. 3, pp. 203–211, 1998.

[87] B. Bothner and G. Siuzdak, "Electrospray ionization of a whole virus: analyzing mass, structure, and viability," *ChemBioChem*, vol. 5, no. 3, pp. 258–260, 2004.

[88] R. Aebersold and M. Mann, "Mass spectrometry-based proteomics," *Nature*, vol. 422, no. 6928, pp. 198–207, 2003.

[89] Y. P. Ho and P. Muralidhar Reddy, "Identification of pathogens by mass spectrometry," *Clinical Chemistry*, vol. 56, no. 4, pp. 525–536, 2010.

[90] T. C. Chou, W. Hsu, C.-H. Wang, Y.-J. Chen, and J.-M. Fang, "Rapid and specific influenza virus detection by functionalized magnetic nanoparticles and mass spectrometry," *Journal of Nanobiotechnology*, vol. 9, article 52, 2011.

[91] Z. P. Yao, P. A. Demirev, and C. Fenselau, "Mass spectrometry-based proteolytic mapping for rapid virus identification," *Analytical Chemistry*, vol. 74, no. 11, pp. 2529–2534, 2002.

[92] A. M. Caliendo, "Multiplex PCR and emerging technologies for the detection of respiratory pathogens," *Clinical Infectious Diseases*, vol. 52, no. 4, pp. S326–S330, 2011.

[93] R. Sampath, N. Mulholland, L. B. Blyn et al., "Comprehensive biothreat cluster identification by PCR/electrospray-ionization mass spectrometry," *PLoS ONE*, vol. 7, no. 6, Article ID e36528, 2012.

[94] A. E. Casiano-Colón, B. B. Hulbert, T. K. Mayer, E. E. Walsh, and A. R. Falsey, "Lack of sensitivity of rapid antigen tests for the diagnosis of respiratory syncytial virus infection in adults," *Journal of Clinical Virology*, vol. 28, no. 2, pp. 169–174, 2003.

[95] A. K. Shetty, E. Treynor, D. W. Hill, K. M. Gutierrez, A. Warford, and E. J. Baron, "Comparison of conventional viral cultures with direct fluorescent antibody stains for diagnosis of community-acquired respiratory virus infections in hospitalized children," *The Pediatric Infectious Disease Journal*, vol. 22, no. 9, pp. 789–794, 2003.

[96] J. Aslanzadeh, X. Zheng, H. Li et al., "Prospective evaluation of rapid antigen tests for diagnosis of respiratory syncytial virus and human metapneumovirus infections," *Journal of Clinical Microbiology*, vol. 46, no. 5, pp. 1682–1685, 2008.

[97] S. A. Ali, J. E. Gern, T. V. Hartert et al., "Real-world comparison of two molecular methods for detection of respiratory viruses," *Virology Journal*, vol. 8, article 332, 2011.

[98] P. J. Jannetto, B. W. Buchan, K. A. Vaughan et al., "Real-time detection of influenza A, influenza B, and respiratory syncytial virus A and B in respiratory specimens by use of nanoparticle probes," *Journal of Clinical Microbiology*, vol. 48, no. 11, pp. 3997–4002, 2010.

[99] C. S. Thaxton, D. G. Georganopoulou, and C. A. Mirkin, "Gold nanoparticle probes for the detection of nucleic acid targets," *Clinica Chimica Acta*, vol. 363, no. 1-2, pp. 120–126, 2006.

[100] C. B. Hall and R. G. Douglas Jr., "Modes of transmission of respiratory syncytial virus," *Journal of Pediatrics*, vol. 99, no. 1, pp. 100–103, 1981.

[101] M. G. Ottolini, S. R. Curtis, A. Mathews, S. R. Ottolini, and G. A. Prince, "Palivizumab is highly effective in suppressing respiratory syncytial virus in an immunosuppressed animal model," *Bone Marrow Transplantation*, vol. 29, no. 2, pp. 117–120, 2002.

[102] K. Huang, L. Incognito, X. Cheng, N. D. Ulbrandt, and H. Wu, "Respiratory syncytial virus-neutralizing monoclonal antibodies motavizumab and palivizumab inhibit fusion," *Journal of Virology*, vol. 84, no. 16, pp. 8132–8140, 2010.

[103] C. Harkensee, M. Brodlie, N. D. Embleton, and M. Mckean, "Passive immunisation of preterm infants with palivizumab against RSV infection," *Journal of Infection*, vol. 52, no. 1, pp. 2–8, 2006.

[104] Committee on Infectious Diseases, "Modified recommendations for use of Palivizumab for prevention of respiratory syncytial virus infections," *Pediatrics*, vol. 124, no. 6, pp. 1694–1701, 2009.

[105] L. Zeitlin, O. Bohorov, N. Bohorova et al., "Prophylactic and therapeutic testing of Nicotiana-derived RSV-neutralizing human monoclonal antibodies in the cotton rat model," *MAbs*, vol. 5, no. 2, pp. 263–269, 2013.

[106] X. Li, S. Sambhara, C. X. Li et al., "Plasmid DNA encoding the respiratory syncytial virus G protein is a promising vaccine candidate," *Virology*, vol. 269, no. 1, pp. 54–65, 2000.

[107] X. Li, S. Sambhara, C. X. Li et al., "Protection against respiratory syncytial virus infection by DNA immunization," *The Journal of Experimental Medicine*, vol. 188, no. 4, pp. 681–688, 1998.

[108] N. Ternette, B. Tippler, K. Überla, and T. Grunwald, "Immunogenicity and efficacy of codon optimized DNA vaccines encoding the F-protein of respiratory syncytial virus," *Vaccine*, vol. 25, no. 41, pp. 7271–7279, 2007.

[109] L. Martinez-Sobrido, N. Gitiban, A. Fernandez-Sesma et al., "Protection against respiratory syncytial virus by a recombinant Newcastle disease virus vector," *Journal of Virology*, vol. 80, no. 3, pp. 1130–1139, 2006.

[110] H. Wu, V. A. Dennis, S. R. Pillai, and S. R. Singh, "RSV fusion (F) protein DNA vaccine provides partial protection against viral infection," *Virus Research*, vol. 145, no. 1, pp. 39–47, 2009.

[111] A. M. Talaat, R. Lyons, and S. A. Johnston, "A combination vaccine confers full protection against co-infections with influenza, herpes simplex and respiratory syncytial viruses," *Vaccine*, vol. 20, no. 3-4, pp. 538–544, 2001.

[112] S. M. Bueno, P. A. González, K. M. Cautivo et al., "Protective T cell immunity against respiratory syncytial virus is efficiently induced by recombinant BCG," *Proceedings of the National Academy of Sciences of the United States of America*, vol. 105, no. 52, pp. 20822–20827, 2008.

[113] C. Xie, J. S. He, M. Zhang et al., "Oral respiratory syncytial virus (RSV) DNA vaccine expressing RSV F protein delivered by attenuated *Salmonella typhimurium*," *Human Gene Therapy*, vol. 18, no. 8, pp. 746–752, 2007.

[114] Y. H. Fu, J. S. He, X. B. Wang et al., "A prime-boost vaccination strategy using attenuated *Salmonella typhimurium* and a replication-deficient recombinant adenovirus vector elicits protective immunity against human respiratory syncytial virus," *Biochemical and Biophysical Research Communications*, vol. 395, no. 1, pp. 87–92, 2010.

[115] M. B. Elliott, T. Chen, N. B. Terio et al., "Alphavirus replicon particles encoding the fusion or attachment glycoproteins of respiratory syncytial virus elicit protective immune responses in BALB/c mice and functional serum antibodies in rhesus macaques," *Vaccine*, vol. 25, no. 41, pp. 7132–7144, 2007.

[116] H. W. Kim, J. G. Canchola, C. D. Brandt et al., "Respiratory syncytial virus disease in infants despite prior administration of antigenic inactivated vaccine," *American Journal of Epidemiology*, vol. 89, no. 4, pp. 422–434, 1969.

[117] P. J. M. Openshaw and J. S. Tregoning, "Immune responses and disease enhancement during respiratory syncytial virus infection," *Clinical Microbiology Reviews*, vol. 18, no. 3, pp. 541–555, 2005.

[118] J. F. Valarcher and G. Taylor, "Bovine respiratory syncytial virus infection," *Veterinary Research*, vol. 38, no. 2, pp. 153–180, 2007.

[119] S. van Drunen Littel-van den Hurk, J. W. Mapletoft, N. Arsic, and J. Kovacs-Nolan, "Immunopathology of RSV infection: prospects for developing vaccines without this complication," *Reviews in Medical Virology*, vol. 17, no. 1, pp. 5–34, 2007.

[120] S. R. Singh, V. A. Dennis, C. L. Carter et al., "Immunogenicity and efficacy of recombinant RSV-F vaccine in a mouse model," *Vaccine*, vol. 25, no. 33, pp. 6211–6223, 2007.

[121] P. Subbarayan, H. Qin, S. Pillai et al., "Expression and characterization of a multivalent human respiratory syncytial virus protein," *Molekuliarnaia Biologiia*, vol. 44, no. 3, pp. 477–487, 2010.

[122] M. R. Murawski, L. W. McGinnes, R. W. Finberg et al., "Newcastle disease virus-like particles containing respiratory syncytial virus G protein induced protection in BALB/c mice, with no evidence of immunopathology," *Journal of Virology*, vol. 84, no. 2, pp. 1110–1123, 2010.

[123] T. Takimoto, J. L. Hurwitz, C. Coleclough et al., "Recombinant Sendai virus expressing the G glycoprotein of respiratory syncytial virus (RSV) elicits immune protection against RSV," *Journal of Virology*, vol. 78, no. 11, pp. 6043–6047, 2004.

[124] B. G. Jones, R. E. Sealy, R. Rudraraju et al., "Sendai virus-based RSV vaccine protects African green monkeys from RSV infection," *Vaccine*, vol. 30, no. 5, pp. 959–968, 2012.

[125] T. M. Ward, V. Traina-Dorge, K. A. Davis, and W. L. Gray, "Recombinant simian varicella viruses expressing respiratory syncytial virus antigens are immunogenic," *Journal of General Virology*, vol. 89, no. 3, pp. 741–750, 2008.

[126] M. Chen, K.-F. Hu, B. Rozell, C. Örvell, B. Morein, and P. Liljeström, "Vaccination with recombinant alphavirus or immune-stimulating complex antigen against respiratory syncytial virus," *Journal of Immunology*, vol. 169, no. 6, pp. 3208–3216, 2002.

[127] S. Boyoglu, K. Vig, S. Pillai et al., "Enhanced delivery and expression of a nanoencapsulated DNA vaccine vector for respiratory syncytial virus," *Nanomedicine*, vol. 5, no. 4, pp. 463–472, 2009.

[128] S. B. Barnum, P. Subbarayan, K. Vig et al., "Nano-Encapsulated DNA and/or protein boost immunizations increase efficiency of DNA vaccine protection against RSV," *Journal of Nanomedicine and Nanotechnology*, vol. 3, article 312, 2012.

[129] E. Eroglu, P. M. Tiwari, A. B. Waffo et al., "A nonviral pHEMA+chitosan nanosphere-mediated high-efficiency gene delivery system," *International Journal of Nanomedicine*, vol. 8, pp. 1403–1415, 2013.

[130] X. Roux, C. Dubuquoy, G. Durand et al., "Sub-nucleocapsid nanoparticles: a nasal vaccine against respiratory syncytial virus," *PLoS ONE*, vol. 3, no. 3, Article ID e1766, 2008.

[131] D. M. Lindell, S. B. Morris, M. P. White et al., "A novel inactivated intranasal respiratory syncytial virus vaccine promotes viral clearance without TH2 associated Vaccine-Enhanced disease," *PLoS ONE*, vol. 6, no. 7, Article ID e21823, 2011.

[132] G. Smith, R. Raghunandan, Y. Wu et al., "Respiratory syncytial virus fusion glycoprotein expressed in insect cells form protein nanoparticles that induce protective immunity in cotton rats," *PLoS ONE*, vol. 7, no. 11, Article ID e50852, 2012.

[133] G. M. Glenn, G. Smith, L. Fries et al., "Safety and immunogenicity of a Sf9 insect cell-derived respiratory syncytial virus fusion protein nanoparticle," *Vaccine*, vol. 31, no. 3, pp. 524–532, 2013.

[134] F. M. Ferolla, D. R. Hijano, P. L. Acosta et al., "Macronutrients during pregnancy and life-threatening respiratory syncytial virus infections in children," *American Journal of Respiratory and Critical Care Medicine*, vol. 187, no. 9, pp. 983–990, 2013.

[135] N. Zang, X. Xie, Y. Deng et al., "Resveratrol-mediated gamma interferon reduction prevents airway inflammation and airway hyperresponsiveness in respiratory syncytial virus-infected immunocompromised mice," *Journal of Virology*, vol. 85, no. 24, pp. 13061–13068, 2011.

[136] X. H. Xie, N. Zang, S. M. Li et al., "Reseveratrol inhibits respiratory syncytial virus-induced IL-6 production, decreases virla replication, and downregulates TRIF expression in airway epithelial cells," *Inflammation*, vol. 35, no. 4, pp. 1392–1401, 2012.

[137] M. E. Belderbos, M. L. Houben, B. Wilbrink et al., "Cord blood vitamin D deficiency is associated with respiratory syncytial virus bronchiolitis," *Pediatrics*, vol. 127, no. 6, pp. e1513–e1520, 2011.

[138] S. Hansdottir, M. M. Monick, N. Lovan, L. Powers, A. Gerke, and G. W. Hunninghake, "Vitamin D decreases respiratory syncytial virus induction of NF-κB-linked chemokines and cytokines in airway epithelium while maintaining the antiviral state," *Journal of Immunology*, vol. 184, no. 2, pp. 965–974, 2010.

[139] N. Al-Sonboli, N. Al-Aghbari, A. Al-Aryani et al., "Micronutrient concentrations in respiratory syncytial virus and human metapneumovirus in yemeni children," *Annals of Tropical Paediatrics*, vol. 29, no. 1, pp. 35–40, 2009.

[140] S. F. Dowell, Z. Papic, J. S. Bresee et al., "Treatment of respiratory syncytial virus infection with vitamin A: a randomized, placebo-controlled trial in Santiago, Chile," *The Pediatric Infectious Disease Journal*, vol. 15, no. 9, pp. 782–786, 1996.

[141] E. Chiba, Y. Tomosada, M. G. Vizoso-Pinto et al., "Immunobiotic Lactobacillus rhamnosus improves resistance of infant mice against respiratory syncytial virus infection," *International Immunopharmacology*, vol. 17, no. 2, pp. 373–382, 2013.

[142] J. Z. Wu, C. C. Lin, and Z. Hong, "Ribavirin, viramidine and adenosine-deaminase-catalysed drug activation: implication for nucleoside prodrug design," *Journal of Antimicrobial Chemotherapy*, vol. 52, no. 4, pp. 543–546, 2003.

[143] R. G. Gish, "Treating HCV with ribavirin analogues and ribavirin-like molecules," *Journal of Antimicrobial Chemotherapy*, vol. 57, no. 1, pp. 8–13, 2006.

[144] B. Langhans, H. D. Nischalke, S. Arndt et al., "Ribavirin exerts differential effects on functions of CD4+Th1, Th2, and regulatory T cell clones in hepatitis C," *PLoS ONE*, vol. 7, no. 7, Article ID e42094, 2012.

[145] W. Markland, T. J. Mcquaid, J. Jain, and A. D. Kwong, "Broad-spectrum antiviral activity of the IMP dehydrogenase inhibitor VX-497: a comparison with ribavirin and demonstration of antiviral additivity with alpha interferon," *Antimicrobial Agents and Chemotherapy*, vol. 44, no. 4, pp. 859–866, 2000.

[146] J. Chapman, E. Abbott, D. G. Alber et al., "RSV604, a novel inhibitor of respiratory syncytial virus replication," *Antimicrobial Agents and Chemotherapy*, vol. 51, no. 9, pp. 3346–3353, 2007.

[147] L. Ni, L. Zhao, Y. Qian et al., "Design and characterization of human respiratory syncytial virus entry inhibitors," *Antiviral Therapy*, vol. 10, no. 7, pp. 833–840, 2005.

[148] M. K. Pastey, T. L. Gower, P. W. Spearman, J. E. Crowe Jr., and B. S. Graham, "A RhoA-derived peptide inhibits syncytium formation induced by respiratory syncytial virus and parainfluenza virus type 3," *Nature Medicine*, vol. 6, no. 1, pp. 35–40, 2000.

[149] X. A. Wang, C. W. Cianci, K. L. Yu et al., "Respiratory syncytial virus fusion inhibitors. Part 5: optimization of benzimidazole substitution patterns towards derivatives with improved activity," *Bioorganic and Medicinal Chemistry Letters*, vol. 17, no. 16, pp. 4592–4598, 2007.

[150] A. A. Nikitenko, Y. E. Raifeld, and T. Z. Wang, "The discovery of RFI-641 as a potent and selective inhibitor of the respiratory syncytial virus," *Bioorganic and Medicinal Chemistry Letters*, vol. 11, no. 8, pp. 1041–1044, 2001.

[151] L. Sun, A. K. Singh, K. Vig, S. R. Pillai, and S. R. Singh, "Silver nanoparticles inhibit replication of respiratory syncytial virus," *Journal of Biomedical Nanotechnology*, vol. 4, no. 2, pp. 149–158, 2008.

[152] S. R. Singh, P. M. Tiwari, and V. A. Dennis, "Anti-respiratory syncytial virus peptide functionalized gold nanoparticles," US Patent Office, 2012.

[153] R. Alvarez, S. Elbashir, T. Borland et al., "RNA interference-mediated silencing of the respiratory syncytial virus nucleocapsid defines a potent antiviral strategy," *Antimicrobial Agents and Chemotherapy*, vol. 53, no. 9, pp. 3952–3962, 2009.

[154] J. DeVincenzo, J. E. Cehelsky, R. Alvarez et al., "Evaluation of the safety, tolerability and pharmacokinetics of ALN-RSV01, a novel RNAi antiviral therapeutic directed against respiratory syncytial virus (RSV)," *Antiviral Research*, vol. 77, no. 3, pp. 225–231, 2008.

[155] J. DeVincenzo, R. Lambkin-Williams, T. Wilkinson et al., "A randomized, double-blind, placebo-controlled study of an RNAi-based therapy directed against respiratory syncytial virus," *Proceedings of the National Academy of Sciences of the United States of America*, vol. 107, no. 19, pp. 8800–8805, 2010.

[156] M. R. Zamora, M. Budev, M. Rolfe et al., "RNA interference therapy in lung transplant patients infected with respiratory syncytial virus," *American Journal of Respiratory and Critical Care Medicine*, vol. 183, no. 4, pp. 531–538, 2011.

[157] J. Summerton and D. Weller, "Morpholino antisense oligomers: design, preparation, and properties," *Antisense and Nucleic Acid Drug Development*, vol. 7, no. 3, pp. 187–195, 1997.

[158] K. C. Wang, J. S. Chang, L. C. Chiang, and C.-C. Lin, "Sheng-Ma-Ge-Gen-Tang (Shoma-kakkon-to) inhibited cytopathic effect of human respiratory syncytial virus in cell lines of human respiratory tract," *Journal of Ethnopharmacology*, vol. 135, no. 2, pp. 538–544, 2011.

[159] K. C. Wang, J. S. Chang, L. C. Chiang, and C. C. Lin, "*Cimicifuga foetida* L. inhibited human respiratory syncytial virus in HEp-2 and A549 cell lines," *The American Journal of Chinese Medicine*, vol. 40, no. 1, pp. 151–162, 2012.

[160] J. S. Chang, C. F. Yeh, K. C. Wang, D. E. Shieh, M. H. Yen, and L. C. Chiang, "Xiao-Qing-Long-Tang (Sho-seiryu-to) inhibited cytopathic effect of human respiratory syncytial virus in cell lines of human respiratory tract," *Journal of Ethnopharmacology*, vol. 147, no. 2, pp. 481–487, 2013.

[161] J. S. Chang, K. C. Wang, C. F. Yeh, D. E. Shieh, and L. C. Chiang, "Fresh ginger (*Zingiber officinale*) has anti-viral activity against human respiratory syncytial virus in human respiratory tract cell lines," *Journal of Ethnopharmacology*, vol. 145, no. 1, pp. 146–151, 2013.

[162] J. S. Chang, K. C. Wang, D. E. Shieh, F. F. Hsu, and L. C. Chiang, "Ge-Gen-Tang has anti-viral activity against human respiratory syncytial virus in human respiratory tract cell lines," *Journal of Ethnopharmacology*, vol. 139, no. 1, pp. 305–310, 2012.

[163] C. F. Yeh, J. S. Chang, K. C. Wang, D. E. Shieh, and L. C. Chiang, "Water extract of *Cinnamomum cassia* Blume inhibited human respiratory syncytial virus by preventing viral attachment, internalization, and syncytium formation," *Journal of Ethnopharmacology*, vol. 147, no. 2, pp. 321–326, 2013.

[164] L. Li, C. H. Yu, H. Z. Ying, and J. M. Yu, "Antiviral effects of modified Dingchuan decoction against respiratory syncytial virus infection in vitro and in an immunosuppressive mouse model," *Journal of Ethnopharmacology*, vol. 147, no. 1, pp. 238–244, 2013.

[165] J. H. Connor, M. O. McKenzie, G. D. Parks, and D. S. Lyles, "Antiviral activity and RNA polymerase degradation following Hsp90 inhibition in a range of negative strand viruses," *Virology*, vol. 362, no. 1, pp. 109–119, 2007.

[166] R. Geller, R. Andino, and J. Frydman, "Hsp90 inhibitors exhibit resistance-free antiviral activity against respiratory syncytial virus," *PLoS ONE*, vol. 8, no. 2, Article ID e56762, 2013.

[167] C. Wild, T. Greenwell, and T. Matthews, "A synthetic peptide from HIV-1 gp41 is a potent inhibitor of virus-mediated cell-cell fusion," *AIDS Research and Human Retroviruses*, vol. 9, no. 11, pp. 1051–1053, 1993.

[168] E. Wang, X. Sun, Y. Qian, L. Zhao, P. Tien, and G. F. Gao, "Both heptad repeats of human respiratory syncytial virus fusion protein are potent inhibitors of viral fusion," *Biochemical and Biophysical Research Communications*, vol. 302, no. 3, pp. 469–475, 2003.

[169] K. N. Bossart, B. A. Mungall, G. Crameri, L. F. Wang, B. T. Eaton, and C. C. Broder, "Inhibition of Henipavirus fusion and infection by heptad-derived peptides of the Nipah virus fusion glycoprotein," *Virology Journal*, vol. 2, article 57, 2005.

[170] M. Porotto, L. Doctor, P. Carta et al., "Inhibition of Hendra virus fusion," *Journal of Virology*, vol. 80, no. 19, pp. 9837–9849, 2006.

[171] M. Porotto, P. Carta, Y. Deng et al., "Molecular determinants of antiviral potency of paramyxovirus entry inhibitors," *Journal of Virology*, vol. 81, no. 19, pp. 10567–10574, 2007.

[172] M. Porotto, B. Rockx, C. C. Yokoyama et al., "Inhibition of Nipah virus infection in vivo: targeting an early stage of paramyxovirus fusion activation during viral entry," *PLoS Pathogens*, vol. 6, no. 10, Article ID e1001168, 2010.

[173] C. Deffrasnes, M.-È. Hamelin, G. A. Prince, and G. Boivin, "Identification and evaluation of a highly effective fusion inhibitor for human metapneumovirus," *Antimicrobial Agents and Chemotherapy*, vol. 52, no. 1, pp. 279–287, 2008.

[174] D. M. Lambert, S. Barney, A. L. Lambert et al., "Peptides from conserved regions of paramyxovirus fusion (F) proteins are potent inhibitors of viral fusion," *Proceedings of the National Academy of Sciences of the United States of America*, vol. 93, no. 5, pp. 2186–2191, 1996.

[175] K. Andries, M. Moeremans, T. Gevers et al., "Substituted benzimidazoles with nanomolar activity against respiratory syncytial virus," *Antiviral Research*, vol. 60, no. 3, pp. 209–219, 2003.

[176] A. Lundin, T. Bergström, L. Bendrioua, N. Kann, B. Adamiak, and E. Trybala, "Two novel fusion inhibitors of human respiratory syncytial virus," *Antiviral Research*, vol. 88, no. 3, pp. 317–324, 2010.

[177] K. L. Yu, Y. Zhang, R. L. Civiello et al., "Fundamental structure-activity relationships associated with a new structural class of respiratory syncytial virus inhibitor," *Bioorganic and Medicinal Chemistry Letters*, vol. 13, no. 13, pp. 2141–2144, 2003.

[178] K. L. Yu, Y. Zhang, R. L. Civiello et al., "Respiratory syncytial virus inhibitors. Part 2: benzimidazol-2-one derivatives," *Bioorganic and Medicinal Chemistry Letters*, vol. 14, no. 5, pp. 1133–1137, 2004.

[179] K. L. Yu, X. A. Wang, R. L. Civiello et al., "Respiratory syncytial virus fusion inhibitors. Part 3: water-soluble benzimidazol-2-one derivatives with antiviral activity in vivo," *Bioorganic and Medicinal Chemistry Letters*, vol. 16, no. 5, pp. 1115–1122, 2006.

[180] K. D. Combrink, H. B. Gulgeze, J. W. Thuring et al., "Respiratory syncytial virus fusion inhibitors. Part 6: an examination of the effect of structural variation of the benzimidazol-2-one heterocycle moiety," *Bioorganic and Medicinal Chemistry Letters*, vol. 17, no. 17, pp. 4784–4790, 2007.

[181] N. Sin, B. L. Venables, K. D. Combrink et al., "Respiratory syncytial virus fusion inhibitors. Part 7: structure-activity relationships associated with a series of isatin oximes that demonstrate antiviral activity in vivo," *Bioorganic and Medicinal Chemistry Letters*, vol. 19, no. 16, pp. 4857–4862, 2009.

[182] A. Nikitenko, Y. Raifeld, B. Mitsner, and H. Newman, "Pyrimidine containing RSV fusion inhibitors," *Bioorganic and Medicinal Chemistry Letters*, vol. 15, no. 2, pp. 427–430, 2005.

[183] D. C. Pryde, T. D. Tran, I. Gardner et al., "Non-benzimidazole containing inhibitors of respiratory syncytial virus," *Bioorganic and Medicinal Chemistry Letters*, vol. 23, no. 3, pp. 827–833, 2013.

[184] S. Galdiero, A. Falanga, M. Vitiello, M. Cantisani, V. Marra, and M. Galdiero, "Silver nanoparticles as potential antiviral agents," *Molecules*, vol. 16, no. 10, pp. 8894–8918, 2011.

[185] P. M. Tiwari, K. Vig, V. A. Dennis, and S. R. Singh, "Functionalized gold nanoparticles and their biomedical applications," *Nanomaterials*, vol. 1, no. 1, pp. 31–63, 2011.

[186] S. Vardharajula, S. Z. Ali, P. M. Tiwari et al., "Functionalized carbon nanotubes: biomedical applications," *International Journal of Nanomedicine*, vol. 7, pp. 5361–5374, 2012.

[187] J. W. Stone, N. J. Thornburg, D. L. Blum, S. J. Kuhn, D. W. Wright, and J. E. Crowe Jr., "Gold nanorod vaccine for respiratory syncytial virus," *Nanotechnology*, vol. 24, no. 29, Article ID 295102, 2013.

[188] N. S. Lee, T. Dohjima, G. Bauer et al., "Expression of small interfering RNAs targeted against HIV-1 rev transcripts in human cells," *Nature Biotechnology*, vol. 20, no. 5, pp. 500–505, 2002.

[189] S. Barik, "Control of nonsegmented negative-strand RNA virus replication by siRNA," *Virus Research*, vol. 102, no. 1, pp. 27–35, 2004.

[190] Y. L. Zhang, T. Cheng, Y. J. Cai et al., "RNA interference inhibits hepatitis B virus of different genotypes in vitro and in vivo," *BMC Microbiology*, vol. 10, article 214, 2010.

[191] S. Jairath, P. Brown Vargas, H. A. Hamlin, A. K. Field, and R. E. Kilkuskie, "Inhibition of respiratory syncytial virus replication by antisense oligodeoxyribonucleotides," *Antiviral Research*, vol. 33, no. 3, pp. 201–213, 1997.

[192] V. Bitko, A. Musiyenko, O. Shulyayeva, and S. Barik, "Inhibition of respiratory viruses by nasally administered siRNA," *Nature Medicine*, vol. 11, no. 1, pp. 50–55, 2005.

[193] V. Bitko and S. Barik, "Phenotypic silencing of cytoplasmic genes using sequence-specific double-stranded short interfering RNA and its application in the reverse genetics of wild type negative-strand RNA viruses," *BMC Microbiology*, vol. 1, article 34, 2001.

[194] K. Vig, N. Lewis, E. G. Moore, S. Pillai, V. A. Dennis, and S. R. Singh, "Secondary RNA structure and its role in RNA interference to silence the respiratory syncytial virus fusion protein gene," *Molecular Biotechnology*, vol. 43, no. 3, pp. 200–211, 2009.

[195] M. Kumar, S. S. Mohapatra, A. K. Behera et al., "Intranasal gene transfer by chitosan-DNA nanospheres protects BALB/c mice against acute respiratory syncytial virus infection," *Human Gene Therapy*, vol. 13, no. 12, pp. 1415–1425, 2002.

[196] W. Zhang, H. Yang, X. Kong et al., "Inhibition of respiratory syncytial virus infection with intranasal siRNA nanoparticles targeting the viral NS1 gene," *Nature Medicine*, vol. 11, no. 1, pp. 56–62, 2005.

[197] B. Dong and R. H. Silverman, "2-5A-dependent RNase molecules dimerize during activation by 2-5A," *The Journal of Biological Chemistry*, vol. 270, no. 8, pp. 4133–4137, 1995.

[198] N. M. Cirino, G. Li, W. Xiao, P. F. Torrence, and R. H. Silverman, "Targeting RNA decay with $2',5'$ oligoadenylate-antisense in respiratory syncytial virus-infected cells," *Proceedings of the National Academy of Sciences of the United States of America*, vol. 94, no. 5, pp. 1937–1942, 1997.

[199] D. L. Barnard, R. W. Sidwell, W. Xiao, M. R. Player, S. A. Adah, and P. F. Torrence, "2-5A-DNA conjugate inhibition of respiratory syncytial virus replication: effects of oligonucleotide structure modifications and RNA target site selection," *Antiviral Research*, vol. 41, no. 3, pp. 119–134, 1999.

[200] M. R. Player, D. L. Barnard, and P. F. Torrence, "Potent inhibition of respiratory syncytial virus replication using a 2-5A-antisense chimera targeted to signals within the virus genomic RNA," *Proceedings of the National Academy of Sciences of the United States of America*, vol. 95, no. 15, pp. 8874–8879, 1998.

[201] Z. Xu, M. Kuang, J. R. Okicki, H. Cramer, and N. Chaudhary, "Potent inhibition of respiratory syncytial virus by combination treatment with 2-5A antisense and ribavirin," *Antiviral Research*, vol. 61, no. 3, pp. 195–206, 2004.

[202] D. W. Leaman, F. J. Longano, J. R. Okicki et al., "Targeted therapy of respiratory syncytial virus in African green monkeys by intra nasally administered 2-5A antisense," *Virology*, vol. 292, no. 1, pp. 70–77, 2002.

[203] M. J. Ripple, D. You, S. Honnegowda et al., "Immunomodulation with IL-4Rα antisense oligonucleotide prevents respiratory syncytial virus-mediated pulmonary disease," *Journal of Immunology*, vol. 185, no. 8, pp. 4804–4811, 2010.

[204] J. Summerton, "Morpholino antisense oligomers: the case for an RNase H-independent structural type," *Biochimica et Biophysica Acta*, vol. 1489, no. 1, pp. 141–158, 1999.

[205] S. H. Lai, D. A. Stein, A. Guerrero-Plata et al., "Inhibition of respiratory syncytial virus infections with morpholino oligomers in cell cultures and in mice," *Molecular Therapy*, vol. 16, no. 6, pp. 1120–1128, 2008.

[206] S. C. Ma, J. Du, P. P.-H. But et al., "Antiviral Chinese medicinal herbs against respiratory syncytial virus," *Journal of Ethnopharmacology*, vol. 79, no. 2, pp. 205–211, 2002.

[207] R. Vaidya, "Observational therapeutics: scope, challenges, and organization," *Journal of Ayurveda and Integrative Medicine*, vol. 2, no. 4, pp. 165–169, 2011.

[208] J. O. Ojwang, Y. H. Wang, P. R. Wyde et al., "A novel inhibitor of respiratory syncytial virus isolated from ethnobotanicals," *Antiviral Research*, vol. 68, no. 3, pp. 163–172, 2005.

[209] D. H. Chung, B. Moore, D. Matharu et al., "A cell based high-throughput screening approach for the discovery of new inhibitors of respiratory syncytial virus," *Virology Journal*, vol. 10, article 19, 2013.

[210] U. F. Power, "Respiratory syncytial virus (RSV) vaccines-two steps back for one leap forward," *Journal of Clinical Virology*, vol. 41, no. 1, pp. 38–44, 2008.

[211] J. F. Papin, R. F. Wolf, S. D. Kosanke et al., "Infant Baboons infected with respiratory syncytial virus develop clinical and pathologic changes that parallel those of human infants," *American Journal of Physiology—Lung Cellular and Molecular Physiology*, vol. 304, no. 8, pp. L530–L539, 2013.

[212] B. S. Graham, "Biological challenges and technological opportunities for respiratory syncytial virus vaccine development," *Immunological Reviews*, vol. 239, no. 1, pp. 149–166, 2011.

[213] L. J. Stockman, W. A. Brooks, P. K. Streatfield et al., "Challenges to evaluating respiratory syncytial virus mortality in Bangladesh, 2004-2008," *PLoS ONE*, vol. 8, no. 1, Article ID e53857, 2013.

[214] Y. Harada, F. Kinoshita, L. M. Yoshida et al., "Does respiratory virus co-infection increase the clinical severity of acute respiratory infection among children infected with respiratory syncytial virus?" *The Pediatric Infectious Disease Journal*, vol. 32, no. 5, pp. 441–445, 2013.

[215] A. A. T. M. Bosch, G. Biesbroek, K. Trzcinski, E. A. M. Sanders, and D. Bogaert, "Viral and bacterial interactions in the upper respiratory tract," *PLoS Pathogens*, vol. 9, no. 1, Article ID e1003057, 2013.

[216] P. G. Holt and P. D. Sly, "Interactions between RSV infection, asthma, and atopy: unraveling the complexities," *The Journal of Experimental Medicine*, vol. 196, no. 10, pp. 1271–1275, 2002.

[217] J. Han, K. Takeda, and E. W. Gelfand, "The role of RSV infection in asthma initiation and progression: findings in a mouse model," *Pulmonary Medicine*, vol. 2011, Article ID 748038, 8 pages, 2011.

[218] S. M. Szabo, A. R. Levy, K. L. Gooch et al., "Elevated risk of asthma after hospitalization for respiratory syncytial virus infection in infancy," *Paediatric Respiratory Reviews*, vol. 13, no. 2, pp. 70161–70166, 2013.

[219] L. I. Tapia, S. Ampuero, M. A. Palomino et al., "Respiratory syncytial virus infection and recurrent wheezing in Chilean infants:

a genetic background?" *Infection, Genetics and Evolution*, vol. 16, no. 0, pp. 54–61, 2013.

[220] S. F. Thomsen, S. van der Sluis, L. G. Stensballe et al., "Exploring the association between severe respiratory syncytial virus infection and asthma: a registry-based twin study," *American Journal of Respiratory and Critical Care Medicine*, vol. 179, no. 12, pp. 1091–1097, 2009.

[221] S. Vandini, L. Corvaglia, R. Alessandroni et al., "Respiratory syncytial virus infection in infants and correlation with meteorological factors and air pollutants," *Italian Journal of Pediatrics*, vol. 39, article 1, 2013.

[222] M. K. Iwane, S. S. Chaves, P. G. Szilagyi et al., "Disparities between black and white children in hospitalizations associated with acute respiratory illness and laboratory-confirmed influenza and respiratory syncytial virus in 3 US Counties-2002-2009," *American Journal of Epidemiology*, vol. 177, no. 7, pp. 656–665, 2013.

[223] A. A. El Kholy, N. A. Mostafa, S. A. El-Sherbini et al., "Morbidity and outcome of severe respiratory syncytial virus infection," *Pediatrics International*, vol. 14, no. 10, pp. 283–288, 2013.

[224] R. C. Welliver Sr., "Temperature, humidity, and ultraviolet B radiation predict community respiratory syncytial virus activity," *The Pediatric Infectious Disease Journal*, vol. 26, supplement 11, pp. S29–S35, 2007.

4

Measles Virus: Identification in the M Protein Primary Sequence of a Potential Molecular Marker for Subacute Sclerosing Panencephalitis

Hasan Kweder,[1,2,3,4,5] Michelle Ainouze,[1,2,3,4,5] Joanna Brunel,[1,2,3,4,5] Denis Gerlier,[1,2,3,4,5] Evelyne Manet,[1,2,3,4,5] and Robin Buckland[1,2,3,4,5]

[1]CIRI, International Center for Infectiology Research, Université de Lyon, 69007 Lyon, France
[2]Inserm, U1111, 69007 Lyon, France
[3]Ecole Normale Supérieure de Lyon, 69007 Lyon, France
[4]Centre International de Recherche en Infectiologie, Université Lyon 1, 69007 Lyon, France
[5]CNRS, UMR 5308, Lyon, France

Correspondence should be addressed to Hasan Kweder; hasanisla@hotmail.com and Evelyne Manet; evelyne.manet@ens-lyon.fr

Academic Editor: Robert C. Gallo

Subacute Sclerosing Panencephalitis (SSPE), a rare lethal disease of children and young adults due to persistence of measles virus (MeV) in the brain, is caused by wild type (wt) MeV. Why MeV vaccine strains never cause SSPE is completely unknown. Hypothesizing that this phenotypic difference could potentially be represented by a molecular marker, we compared glycoprotein and matrix (M) genes from SSPE cases with those from the Moraten vaccine strain, searching for differential structural motifs. We observed that all known SSPE viruses have residues P64, E89, and A209 (PEA) in their M proteins whereas the equivalent residues for vaccine strains are either S64, K89, and T209 (SKT) as in Moraten or PKT. Through the construction of MeV recombinants, we have obtained evidence that the wt MeV-M protein PEA motif, in particular A209, is linked to increased viral spread. Importantly, for the 10 wt genotypes (of 23) that have had their M proteins sequenced, 9 have the PEA motif, the exception being B3, which has PET. Interestingly, cases of SSPE caused by genotype B3 have yet to be reported. In conclusion, our results strongly suggest that the PEA motif is a molecular marker for wt MeV at risk to cause SSPE.

1. Introduction

Subacute Sclerosing Panencephalitis (SSPE) is caused by measles virus (MeV), a member of the genus *Morbillivirus* in the family *Paramyxoviridae*. The enveloped MeV virion contains a nonsegmented negative-strand RNA genome encoding six structural proteins: N, P, M, F, H, and L. The glycoproteins H (hemagglutinin) and F (fusion) project from the virion membrane as spikes. The H protein is responsible for attachment to the cellular receptors and the F protein for the consequent fusion of the virion membrane with the host cell's plasma membrane [1]. In the infected cell, the glycoproteins accumulate in the plasma membrane where they interact with cellular receptors on neighboring uninfected cells to cause cell-cell fusion (syncytia formation). The matrix protein M is believed to line the inner surface of the plasma membrane of the infected cell, interacting with the cytoplasmic tails of the glycoproteins [2, 3]. As far as cellular receptors for MeV are concerned, the wt strains use SLAM (CD150) whereas the vaccine and laboratory strains use both SLAM and CD46 [4]. Although CD46 is ubiquitously expressed in the human body, the expression of SLAM is restricted to cells of the immune system [4]. A long searched for third receptor, nectin-4, has been recently identified [5, 6] which allows MeV infection via epithelial cells. Intriguingly, of the three MeV receptors, only CD46 is expressed in the CNS [7].

SSPE is a rare fatal disease of children and young adults that is due to a persistent MeV infection of the brain. SSPE symptoms appear several years after an apparently banal wt MeV infection; SSPE cases caused by vaccine strains of MeV have never been reported. Although more than 80 years has passed since the disease was described by Dawson [8], no mechanisms responsible for the pathogenesis of SSPE have been identified (for a review see [9–12]). However, MeVs from SSPE neurological tissue are characterized by multiple mutations in the H, F, and M proteins of the viral envelope that impairs their function [13–16]. Measles inclusion body encephalitis (MIBE) also infects the human brain but is caused by both wt and vaccine strains of MV [17, 18]. Moreover, in contrast to SSPE, MIBE only occurs in immunocompromised individuals, such as AIDS patients, and presents within weeks of infection rather than years.

In a previous study, we showed that vaccine/laboratory strains of MeV are susceptible, like wt MeV strains, to accumulate mutations in their H, F, and M proteins under immune pressure [19]. However, it would appear that, unlike wt MeV, they lack a phenotypic marker that allows spread and persistence in the CNS.

The working hypothesis for the present study was thus that the capacity to spread and persist in the human brain is somehow intrinsic to wt MeV. Potentially this could be due to sequence differences with laboratory/vaccine MeV strains. We therefore decided to compare the primary sequences of SSPE MeV genomes with those from vaccine strains searching for a molecular marker common to all SSPE MeV strains. Initially, we compared primary sequences encoding the H protein but as our previous study [19] had indicated that mutations in both the F protein and M protein can affect the 3D conformation of the H we extended our search to the primary sequences encoding these latter proteins. Comparing SSPE MeV genomes with that from the Moraten vaccine strain of MeV we observed a triresidue motif PEA (P64, E89, and A209) that is always present in the M proteins from SSPE cases but which is, respectively, SKT (S64, K89, and T209) in the Moraten vaccine M protein. Moreover, these residue identities are the same for all vaccine/laboratory strains except that in some residue 64 is proline (hence PKT).

We hypothesize that to cause SSPE a wt MeV strain (or wt MeV genotype for that matter) should have the PEA motif in its M protein. By consulting published sequences available for the 23 wt MeV, we found that only 10 of these include the sequence for the M gene (obtained by direct RT-PCR amplification from brain tissues [20]). Interestingly, all have the PEA motif except genotype B3, which is PET. This intrigued us because B3 is the most prevalent genotype in Sub-Saharan Africa [21], where, despite the hyperendemicity of MeV, the reported prevalence of SSPE cases is unexpectedly low [22]. Even though it is possible that there has been an underreporting of SSPE cases in this region, B3 cases have occurred elsewhere in the world, including the USA. Importantly for our hypothesis, despite an extensive search in the literature, we were unable to find a single SSPE case involving this genotype.

Serendipitously, two B3 genotype strains are available in our laboratory, Lys-1 [23] and G954 [24], so that we were able to compare their capacities for cell-cell fusion and virus production, with a PEA motif-containing D4 genotype virus, Lys05/06. The results suggested that B3 genotype strains produce less virus than PEA motif-containing genotype strains, but to confirm that this was due to the nature of the M protein tri-residue motif, we turned to reverse genetics.

MeV recombinants were thus constructed in which elements of the wt PEA motif were introduced into the vaccine strain's SKT motif within the gene encoding the M protein. By comparing the phenotypes of these different recombinants with regard to their capacities for cell-cell fusion and virus assembly, we obtained results that strongly support the hypothesis that the M protein triresidue motif PEA is important for the spread of wt MV and hence SSPE pathogenesis.

2. Materials and Methods

2.1. Cells. Vero cells and vero/hSLAM (vero cells constitutively expressing human SLAM) were maintained in Dulbecco's modified Eagle's medium (DMEM) supplemented with 10% fetal bovine serum (FBS), 2 mM L-glutamine, 100 U/mL penicillin, 0.1 mg/mL streptomycin, and 10 mM HEPES. CHO/hSLAM (Chinese hamster ovary cells constitutively expressing human SLAM) were maintained in F12 medium containing 10% fetal bovine serum (FBS), 100 U/mL penicillin, 0.1 mg/mL streptomycin, and 1X MEN nonessential amino acids. The human "helper" cell line 293-3-46 stably expressing the N and P proteins and T7 RNA polymerase were maintained in DMEM medium with 10% FBS, 2 mM L-glutamine, 1.2 mg/mL G418, and 10 mM HEPES.

2.2. Viruses. Three wt MeVs were used in this study: Lys-1 and G954 (both B3 genotype), Lys05/06 (D4 genotype). Eight recombinants were built and rescued using a Moraten vaccine strain reverse genetics system (a kind gift from Roberto Cattaneo).

2.3. Production of Moraten M Gene Mutants. Mutations in residues 64, 86, 89, and 209 of the M gene of Moraten strain of MeV were introduced separately or in combination using QuickChange kit (Stratagene) according to the manufacturer's instructions. These mutations were introduced into the gene encoding MeV-M cloned into a shuttle plasmid p588 containing the N, P, M, and F genes of Moraten MeV. Then the M gene of this plasmid was cloned, using the In-Fusion HD cloning kit (Clontech), into another plasmid, p698, which contained the totality of the Moraten genome except for the deleted M gene. This plasmid was used in the production of recombinant viruses. All mutations were verified by DNA sequencing.

2.4. Production of Moraten Recombinants. 293-3-46 cells cultured overnight in 6-well tissue culture plates were transfected with 10 μg of p698 containing the mutated M gene, together with 40 ng of the plasmid pEMC-La, which encodes the MeV polymerase L protein, using the Promega transfection kit (Mammalian Transfection System, calcium phosphate). 16 h after transfection, the medium was replaced with antibody-free medium. 1 h later, 293-3-46 cells were subjected to thermal shock at 42°C for 3 h. After 48 h, cells were gently detached by squidging using the medium and added to 100 mm culture vessel containing vero/hSLAM cells. After 2 to 3 days, syncytia in overlaid vero/hSLAM cells were individually picked and transferred to vero/hSLAM cell monolayers in 75 cm^2 flasks. Finally, we obtained 6 recombinant viruses named according to the amino acids occupying the triresidue motif (aa 64, 89, and 209) of the M protein: SKT (S64, K89, and T209); PKT (P64, K89, and T209); SET (S64, E89, and T209); SKA (S64, K89, and A209); PET (P64, E89, and T209); and PEA (P64, E89, and A209). Two additional recombinants were built and rescued: S(R)KT (S64, R86, K89, and T209) and P(R)ET (P64, R86, E89, and T209). It should be noted that residue 86 is K in the recombinant viruses SKT, PKT, SET, SKA, PET, and PEA. The M gene of all recombinants was sequenced to confirm the mutagenesis.

2.5. Virus Amplification and Titration. A virus stock was made following a second passage of amplification: cells with 2 mL of medium were frozen at −80°C overnight 2-3 days after infection when the majority of cells showed fusion/syncytium formation. Then the medium was thawed and harvested and the virus stock titrated. Cells in 96-well tissue culture plates were inoculated with 1/10 serially diluted culture medium samples for 1 h at 37°C. Then, the inocula were removed and new medium was added to each well. After 4 days, the number of infected wells was counted and the 50% tissue culture infective dose (TCID$_{50}$) and the plaque-forming unit (PFU) were calculated.

2.6. SLAM- and CD46-Dependent Fusion Assay. Each virus was studied to determine its capacity to induce the fusion in the presence of either SLAM or CD46 as cellular receptor by using CHO/hSLAM cells or vero cells (CD46+), respectively. The cells in 6-well plates were infected at a m.o.i of 0.01. Cell-cell fusion in infected cells was quantified as described previously [25]. Briefly, 30–36 h after infection, images of ten microscope fields were taken randomly and the proportion of nuclei in syncytia relative to the total number of nuclei was determined by counting.

2.7. Cell-Free Virus and Cell-Associated Virus Titrations. Vero cells or CHO/hSLAM cells were infected at a m.o.i of 0.1 of recombinant virus. 48 h after infection, the culture media were harvested, centrifuged at 3000 rpm for 5 min at 4°C, and stored at −80°C until being used for further analysis to determine cells-free virus. In addition, infected cells were also frozen at −80°C overnight. They were then thawed and harvested and the supernatant used to determine the level of cell-associated virus. Thereafter, the titration of cell-free virus and cell-associated virus was made as described above.

2.8. Confocal Microscope Study for the Localization of MeV Proteins, H, F, and M. Vero SLAM cells grown on glass cover slides in 12-well culture plate were infected with recombinant viruses at 37°C for 1 h. Then the medium was changed with medium containing the anti-MeV fusion tripeptide FIP [26]. Cells were subjected to immunofluorescence 24 h after infection. Three antibodies were used, anti-H mAb BH129,

anti-F mAb Y503, and anti-M mAb8910 (Millipore). The antibodies were labelled using Zenon Mouse IgG Labeling Kits (Molecular Probes). Anti-H mAb BH129 was stained with Alexa Fluor 488, anti-F mAb Y503 with Alexa Fluor 555, and anti-M mAb8910 with Alexa Fluor 647. Cells were first washed with PBS 1x. Then the live cells were incubated only with labelled anti-H and anti-F for 1 h at 4°C. Next, cells were washed with PBS, fixed with 3% PFA, and permeabilized with 0.1% Triton X-100 for 10 minutes at RT. Subsequently, cells were washed with blocking solution (0.2% Tween 20, 2% BSA, and 5% glycerol in PBS) and incubated in blocking solution for 10 minutes. Cells were first incubated with labelled anti-M for 1 h at 4°C and then the slides were prepared for confocal microscope study. Laser argon, laser 561, and laser 633 were used for H, F, and M, respectively. The specimens were studied in two steps, H and M in one step and F in another step to avoid interference between the emission signals of H, F, and M.

3. Results

3.1. Identification of a Potential Molecular Maker for SSPE in the wt MeV-M Protein. This study's starting point was the observation that SSPE is caused only by wt MeV, never by MeV vaccines [27]. This suggests that wt MeV strains possess a phenotypic marker that vaccine strains lack. Hypothesizing that such a phenotypic marker could be represented by an associated molecular marker we decided to compare H, F, and M sequences from SSPE cases with those from vaccine strains searching for differential structural motifs. We were unable to identify any type of molecular marker that differentiated SSPE glycoproteins from their vaccine counterparts except for the differences at residues 481 and 546 in the H protein that have been shown to play a role in allowing vaccine strains to use CD46 as a receptor in addition to SLAM [28, 29]. However, comparison of five SSPE case M protein primary sequences with the Moraten and Rubeovax vaccine M proteins [30] revealed the presence of a triresidue motif at residues 64, 89, and 209 that appears to differentiate SSPE and vaccine M proteins (Figure 1). All five SSPE cases have the residues proline, glutamate, and alanine (PEA) at these positions whereas the vaccine strain M proteins have serine, lysine, and threonine (SKT). Extending our search to the totality of published SSPE sequences we were unable to find a single case where the M protein did not contain the PEA motif. However, in making a similar search of vaccine M proteins we found that they all have the SKT or PKT motif.

The present-day attenuated MeV vaccine strains were produced by passaging the original wt Edmonston strain and its derivatives on various nonhost animal cell lines [31]. Unfortunately, the original wt Edmonston strain is no longer available so that we can only speculate that the triresidue motif in its M protein was PEA. However, it can be concluded that this motif was at least PET as this is the nature of the motif in the minimally passaged "wt Edmonston." It is interesting that modification of the triresidue motif appears to coincide with attenuation. Although circumstantial, this

```
  1 MTEIYDFDKS AWDIKGSIAP IQPTTYSDGR LVPQVRVIDP GLGDRKDECF MYMFLLGVVE SSPE 1
  1 MTEIYDFDKS AWDIKGSIAP IQPTTYSDGR LVPQVRVIDP GLGDRKDECF MYMFLLGVVE SSPE 2
  1 MTEIHDFDKS AWDIKGSITP TQPTTYSDGR LVPQVRVIDP GLGDRKDECF MYMSLLGVVE SSPE 3
  1 MTEIYDFDKS AWDIKGSIAP TQPTTYSDGR LVPQVRVIDP GLGDRKDECS TYMFPLGVVE SSPE 4
  1 MTEIYDFDKS AWDIKGSIAP IQPTTYSDGR LVPQVRVIDP GLGDRKDECL MYMFLLGAVE SSPE 5
  1 MTEIYDFDKS AWDIKGSIAP IQPTTYSDGR LVPQVRVIDP GLGDRKDECF MYMFLLGVVE Moraten
  1 MTETYDFDKS AWDIKGSIAP IQPTTYSDGR LVPQVRVIDP GLGDRKDECF MYMFLLGVVE Rubeovax

 61 DSDPLGPPIG RAFGSLPLGV GRSTAKPEEL LKEATELDIV VRRTAELNEK LVFYNNTPLT SSPE 1
 61 DSDPLGPPIG RAFGSLPSGV GRSTAKPEEL LKEATELDIV ARRTAGLNEK PVFYNNTPPT SSPE 2
 61 DSDPPGPPIG RAFGSPPLGV GRSTAKPEEL LKEATELDIV ARRTAGLNEK LVFHNSTPST SSPE 3
 61 DSDPPGPPIG RALGSLPLGV GRSTAKPEEL LKEATEPDIV VRRTAGLNEK LVFYNNTPPT SSPE 4
 61 DSDPLGPPIG RAPGSLPLGA GRSTAKPEEL LKEATELDTA VRRTAGLNEK LVFYNNTPPT SSPE 5
 61 DSDSLGPPIG RAFGFLPLGV GRSTAKPEKL LKEATELDIV VRRTAGLNEK LVFYNNTPLT Moraten
 61 DSDSLGPPIG RAFGSLPLGV GRSTAKPEKL LKEATELDIV VRRTAGLNEK LVFYNNTPLT Rubeovax
            ↑                    ↑

121 LLTPWRKVLT TGSVFNANQV CNAVNLIPLD TPQRFRVVYM SITRLSDNGH YTVPRRMLEF SSPE 1
121 LLIPWRKVQT TGSVLNANQV CNAVNPLPLD TPQRFRVVYM SITRLSDNGY YTVPRRMLEF SSPE 2
121 LLTPWRKVPT TGSVFNANQA CNAVNLIPLD TPQRFRVVYM SITRPSDNGH YTVPRRMPEF SSPE 3
121 LLTPWRKVPT TGSVFNANQV CNAVNLIPLD TPQRLRAVYM SITRPSDNGH YTAPRRMLEF SSPE 4
121 LLTPWRKVPT TGSVFNANQV CNAVNLIPLD TPQRFRVVYM SITRLSDNGY YTVPRRMLEF SSPE 5
121 LLTPWRKVLT TGSVFNANQV CNAVNLIPLD TPQRFRVVYM SITRLSDNGY YTVPRRMLEF Moraten
121 LLTPWRKVLT TGSVFNANQV CNAVNLIPLD TPQRFRVVYM SITRLSDNGY YTVPRRMLEF Rubeovax

181 RSVNAVAFNL LVTLRIDKAI GPGKIIDNAE QLPEATFMVY IGNFRRKKSE VYSADYCKMK SSPE 1
181 RSVNAVAFNL LVTPRIDKAI GPGKIIDNAE QLPEAISMVH IGNLRRKKSE VHSADHCKMK SSPE 2
181 RSVNAVAFNL LVTLRIDKAI GPGKIIDNAE QLPEATSMVH IGNFRRKKSE VYSADYCKMK SSPE 3
181 RSVNAVAFNL LVTLRIDKAI GPGKIIDNAE QLPEATSMVH IGNFRRKKSE VHSADHCKMK SSPE 4
181 RSVNAVAFNL LVTLRIDKAI GPGKIIDNAE QLPEATFMVH IGNFRRKKSE VYSADYCKMK SSPE 5
181 RSVNAVAFNL LVTLRIDKAI GPGKIIDNTE QLPEATFMVH IGNFRRKKSE VYSADYCKMK Moraten
181 RSVNAVAFNL LVTLRIDKAI GPGKIIDNTE QLPEATFMVH IGNFRRKKSE VYSADYCKMK Rubeovax
                              ↑

241 IEKMGLVFAL GGIGGTSLHI RSTGKMSKTL HAQLGFKKTL CYPLMDINED LNRLLWRSRC SSPE 1
241 IEKMGLVSAL GGIGGTSPHI RSTGKMSKTL HAQLGFKKTL CYPLMDINED LNRLLWRSRC SSPE 2
241 IEKMGLVFAL GGIGGTSLHI RSTGKMSKTL HAQLGFKKTL CYPLMDINED LNRLLWRSRC SSPE 3
241 IEKMGPVPAP GGIGGTSPHT RSTGKMSKTL HAQLGFKKTL CYPLMDINED PNRLLWRSRC SSPE 4
241 IEKMGLVFAL GGIGGTSLHI RSTGKMSKTL HAQLGFKKTL CYPLMDINED LNRLLWRSRC SSPE 5
241 IEKMGLVFAL GGIGGTSLHI RSTGKMSKTL HAQLGFKKTL CYPLMDINED LNRLLWRSRC Moraten
241 IEKMGLVFAL GGIGGTSLHI RSTGKMSKTL HAQLGFKKTL CYPLMDINED LNRLLWRSRC Rubeovax

301 KIVRIQAVLQ PSVPQESRIY DDVIINDDQG LFKVL SSPE 1
301 KIVRIQAVLQ PSVPQEFRIY DDVIINDDQG FKVVL SSPE 2
301 KIVRIQAVLQ PSVPQEFCIY DDVIINDDQG LFKVL SSPE 3
301 KIARIQAVLQ PPVPQELRIY DDAITNDDQG LFKVL SSPE 4
301 KIVRIQAVLQ PSVPQEFRIY DDVITNDDQG LFKVL SSPE 5
301 KIVRIQAVLQ PSVPQEFRIY DDVIINDDQG LFKVL Moraten
301 KIVRIQAVLQ PSVPQEFRIY DDVIINDDQG LFKVL Rubeovax
```

FIGURE 1: Primary sequence comparison of M protein genes from five SSPE cases [20] and two MeV vaccine strains. Accession numbers SSPE1 (London) AF503528; SSPE2 (Nottingham) AF503530; SSPE3 (Cardiff) AF503531; SSPE4 (Belfast87) AF503526; SSPE5 (Belfast88) AF503524; Moraten (vaccine) AF266287; Rubeovax (vaccine) AF266289.

could suggest that replacement of the PEA or PET motif with SKT is involved, at least in part, in loss of virulence.

3.2. The B3 Genotype Lys-1 MeV Strain Has a Lowered Capacity for Virus Production.
Interestingly, while all SSPE cases appear to be caused by wt MeV with the PEA motif in their M proteins, not all wt MeVs are PEA. Of the 23 wt MeV genotypes only 10 have had their M genes sequenced (Table 1). All have the PEA motif except the B3 genotype, which has PET. That the B3 genotype has the motif PET was of great interest to us for two reasons: (i) B3 is

the prevalent genotype in Sub-Saharan Africa [21] and (ii) it has been observed [22] that, for unknown reasons, few cases of SSPE have been notified in this vast region where MeV is hyperendemic.

Hypothesizing that the PET motif could potentially reduce the capability of the B3 genotype to spread within the human body, we compared two B3 genotype viruses (Lys-1 and G954) with a PEA motif-containing D4 genotype virus (Lys05/06) and the vaccine strain Moraten (SKT), for their cell-cell fusion and virus production capacities. As far as cell-cell fusion was concerned, we found little difference

TABLE 1: Nature of the triresidue motif found in the gene encoding the matrix protein in the different wt MeV genotypes.

Genotype	M protein triresidue motif
A	SKT or PKT
B1	?
B2	?
B3	PET
C1	?
C2	PEA
D1	?
D2	?
D3	PEA
D4	PEA
D5	PEA
D6	PEA
D7	PEA
D8	PEA
D9	?
D10	?
E	?
F	?
G1	?
G2	?
G3	PEA
H1	PEA
H2	?
dl1	?

between the four viruses (Figure 2(a)). Although the fusion capacity of one of the PET motif-containing B3 strains (G954) was reduced by 17% in comparison with the PEA motif-containing D4 genotype virus Lys05/06 (Figure 2(a)), this was not statistically significant. For production of cell-associated virus, the three wt strains produced less than the vaccine Moraten strain (Figure 2(b)) and, comparing the B3 strains with the D4 strain, G954 had a 13% less production than Lys05/06 but again this reduction is not statistically important. However, for Lys-1 B3 the reduction in cell-associated virus compared to Lys05/06 was 32% ($P < 0.025$). Moreover, although there was only a slight (7%) reduction for G954, there was a significant reduction (82%; $P < 0.001$) in cell-free virus production for the Lys-1 B3 strain compared to the Lys05/06 D4 strain (Figure 2(c)).

Taken together, these results suggest that the Lys-1 B3 strain has a less productive phenotype than the Lys05/06 D4 strain but this is not the case for the G954 B3 strain. As both B3 strains contain the PET motif in their M proteins, this could thus suggest that the M protein PET motif has no influence on the phenotype of wt MeV strains in terms of virus production. However, upon sequencing the M gene of G954 we observed that residue 86, just three residues upstream of E89, was arginine (R) rather than lysine (K). This change both increases the positive charge of residue 86 [32] and introduces the possibility of cation-π interactions with aromatic residues [33] that could potentially play

a compensatory role if, as has been previously suggested [34], the K89E mutation abrogates an electrostatic interaction between the M protein and the cytoplasmic tails of the glycoproteins which favors virus assembly.

In effect, our results show that the Moraten strain, which has the M protein motif SKT, exhibits much higher virus production levels than the PET or PEA motif-containing wt strains (Figures 2(b) and 2(c)). But is this difference related to the PEA/PET motifs? Evidently, to relate virus assembly differences between vaccine and wt MeV strains to the identity of a triresidue motif in the M protein, when variation also exists in other MeV proteins, in particular the H is pure speculation. We therefore undertook the construction of MeV recombinants to investigate the potential role, if any, that the M protein motif PEA plays in wt MeV production.

3.3. Production of Recombinant Moraten Viruses Differing according to the Nature of the Triresidue Motif in Their M Proteins. Recombinant Moraten viruses were built and rescued in which the triresidue SKT motifs present in the M protein of this vaccine strain were systematically replaced by elements from the wt motif PEA (Figure 3). These recombinant viruses are named according to the amino acids occupying the triresidue motif (aa 64, 89, and 209) of the M protein: SKT, PKT, SET, SKA, PET, and PEA. Two additional recombinants were made to test the effect of the K86R mutation: S(R)KT and P(R)ET (Figure 3). The phenotypes of these different recombinants were investigated with regard to their cell-cell fusion and virus assembly capacities. We used the Moraten reverse genetics system to do this study as a vaccine strain affords the possibility to test both the fusion and viral assembly capacities of recombinants in terms of differential receptor usage.

3.4. Substitution of the SKT Motif of the Moraten M Protein with Elements of the wt PEA Motif Results in an Increase in Cell-Cell Fusion. The cell-cell fusion capacity of the different recombinants which have various permutations of the SKT and PEA motifs in their M proteins was compared using both CD46-expressing cells (vero) and SLAM-expressing cells (CHO-SLAM). Surprisingly, the results suggest that the identity of three particular residues in the M protein (at positions 64, 89, and 209) can have an effect on MeV cell-cell fusion levels. The results shown in Figure 4(a), comparing the levels of cell-cell fusion obtained for the various recombinants, suggest that whenever an element of the (vaccine) SKT motif is substituted by an element from the (wt) PEA motif, individually or in combination, there is an increase in cell-cell fusion even if we found that this is only statistically important for PET ($P < 0.005$) and PEA ($P < 0.001$). Interestingly, the highest values for both CD46-dependent cell-cell fusion and SLAM-dependent cell-cell fusion were obtained with the MeV recombinant containing the PEA motif in its M protein. This was particularly true for CD46-dependent fusion; the amount of cell-cell fusion generated by

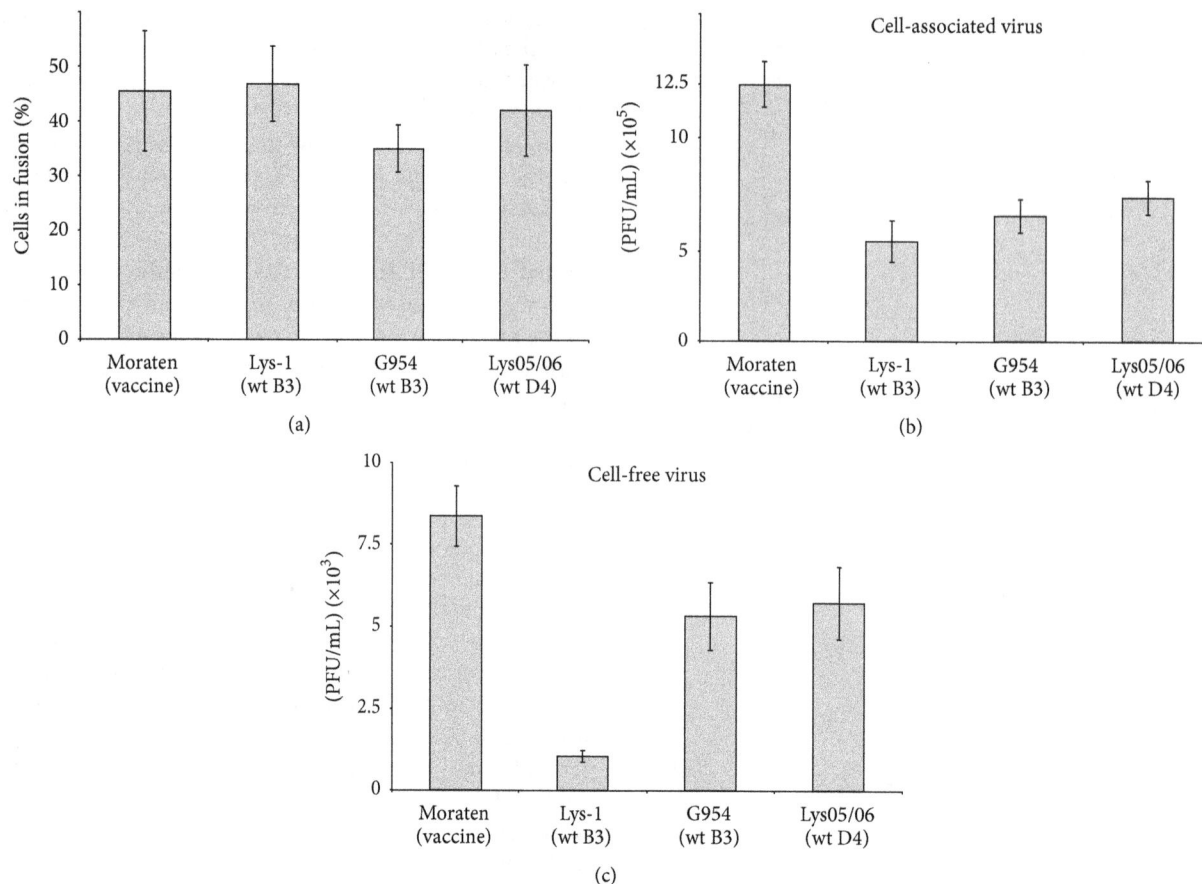

FIGURE 2: (a) Comparison of the cell-cell fusion capacity of Moraten (genotype A), Lys-1 (genotype B3), G954 (genotype B3), and Lys05/06 (genotype D4) strains. CHO-SLAM cells were infected with the different strains and the fusion levels analysed 30–36 h after infection. The histogram data represent the mean percentages ± standard deviations for three experiments. (b) Cell-associated virus production of the Moraten (genotype A), Lys-1 (genotype B3), G954 (genotype B3), and Lys05/06 (genotype D4) strains. CHO-SLAM cells infected with the different viruses were analysed for the production of cell-associated viral particles 48 h after infection. The histogram data represent the mean percentages ± standard deviations for three experiments. (c) Cell-free virus production of the Moraten (genotype A), Lys-1 (genotype B3), G954 (genotype B3), and Lys05/06 (genotype D4) strains. CHO-SLAM cells infected with the different viruses were analysed for the production of cell-free viral particles 48 h after infection. The histogram data represent the mean percentages ± standard deviations for three experiments.

the PEA mutant was more than twice that generated by the SKT mutant (Figure 4(a)).

As a looser interaction between the M protein and the H and F proteins has been proposed to increase cell-cell fusion rates [2, 3], a possible explanation for our results is that these PEA-based substitutions in the Moraten M protein have loosened its interaction with the glycoprotein cytoplasmic tails. The substitution S64P does not appear to cause any increase in cell-cell fusion (comparing SKT with PKT) but the K89E substitution when allied with the S64P substitution (PET and PEA) increases both CD46- and SLAM-dependent fusion substantially.

3.5. A PET Motif in the Moraten M Protein Results in a >40% Reduction in Both Cell-Associated and Cell-Free Virus Production irrespective of Receptor Usage. We next examined the capacity of each recombinant to produce cell-associated and cell-free virus. As a tight interaction between the M

protein and the glycoproteins has been reported to favor virus assembly [2, 3], the use of this assay should indicate the state of this interaction for each recombinant. The results obtained for CD46- and SLAM-dependent cell-associated virus production 2 days after infection (Figures 4(b) and 4(d)) suggest that PEA motif substitutions, with one exception, only have a slight negative effect on Moraten virus assembly. An exception however is the PET recombinant whose cell-associated virus production was reduced by more than 40% (Figures 4(b) and 4(d)). Very similar results were obtained for CD46- and SLAM-dependent cell-free virus production (Figures 4(c) and 4(e)).

The results obtained with the PET recombinant are in perfect accordance with our previous observation that Lys-1, a B3 genotype virus, gives very little cell-free virus (Figure 2(c)). This strongly suggests that the PET M protein motif in B3 genotype wt MeV viruses has a negative effect on virus production, possibly via modulation of assembly.

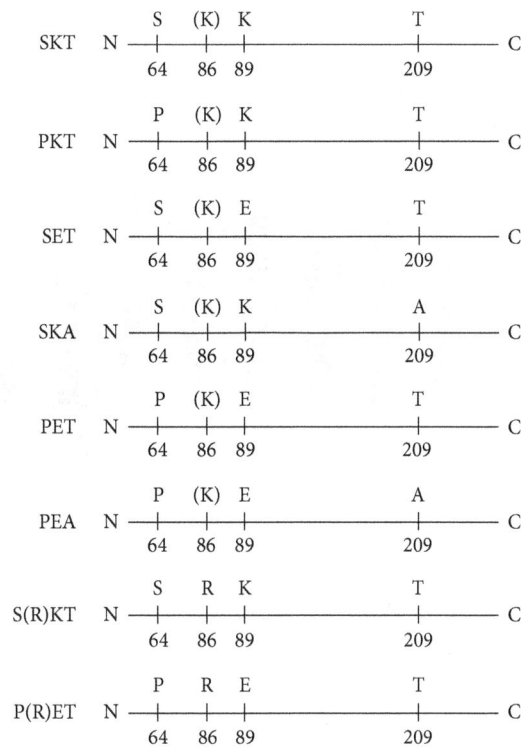

FIGURE 3: Nature of the M protein motif present in the different Moraten recombinants.

We also tested the P(R)ET recombinant, which differs from the PET recombinant only in having the K86R substitution, to determine whether this change is responsible for the higher level of virus production observed with the G954 B3 strain compared to the Lys-1 B3 strain.

3.6. Adding the Mutation K86R to the PET Recombinant Increases Virus Production but Lowers Fusion. We found that the P(R)ET recombinant indeed has compared with the PET mutant, a higher capacity ($P < 0.001$) for cell-free and cell-associated virus production with both SLAM-expressing cells and CD46-expressing cells (Figures 4(b), 4(c), 4(d), and 4(e)) and a lower fusion capacity (both SLAM-dependent and CD46-dependent) (Figure 4(a); $P < 0.025$). In fact, the P(R)ET recombinant has fusion and production properties identical to that of the parental S(K)KT Moraten virus. On the other hand, adding the K86R mutation to the SKT recombinant, S(R)KT, had little effect on virus production or cell-cell fusion (Figures 4(a), 4(b), 4(c), 4(d), and 4(e)). If the interaction between the Moraten M protein and the glycoprotein's cytoplasmic tails is indeed loosened by the K89E substitution, it is tempting to speculate that the K86R mutation restores the level of basic charge required for this interaction and that this tighter interaction is reflected in increased virus production and lower cell-cell fusion.

3.7. Confocal Microscopy Studies Show That MeV-M Proteins Associate with the H and F Glycoproteins irrespective of the Nature of the Triresidue Motif They Possess. To investigate whether substituting the Moraten SKT motif with elements of the wt PEA motif had an effect on the colocalization of the M protein with the H and F proteins, we made confocal microscopy studies. The colocalization of the Moraten M, H, and F proteins was not found to be affected by any of the PEA substitutions (Figure 5). This suggests that even if particular PEA substitutions have the effect to lessen or increase the interaction between the M protein and the cytoplasmic tails of the glycoproteins, these effects are probably subtle as they are not accompanied by topological displacement of the participating proteins.

4. Discussion

The reason why SSPE is caused exclusively by wt MeV and never by vaccine strains is not known. However, our results suggest that the capacity of wt MeV strains to cause SSPE results from their elevated capacity to spread and that this is due, at least in part, to a triresidue motif, PEA, in their M proteins. We thus propose the PEA motif as a molecular marker of wt MeV that risk causing SSPE.

Indeed, all SSPE cases reported in the literature have the PEA motif in their M proteins and we show that replacing the SKT motif in the Moraten vaccine M protein with the wt PEA motif increases fusion whilst maintaining virus production capacity. Moreover, changing this motif to PET via the single mutation A209T results in a significant reduction in virus production. Importantly, lowered virus production could hamper efficient viral spread in the CNS.

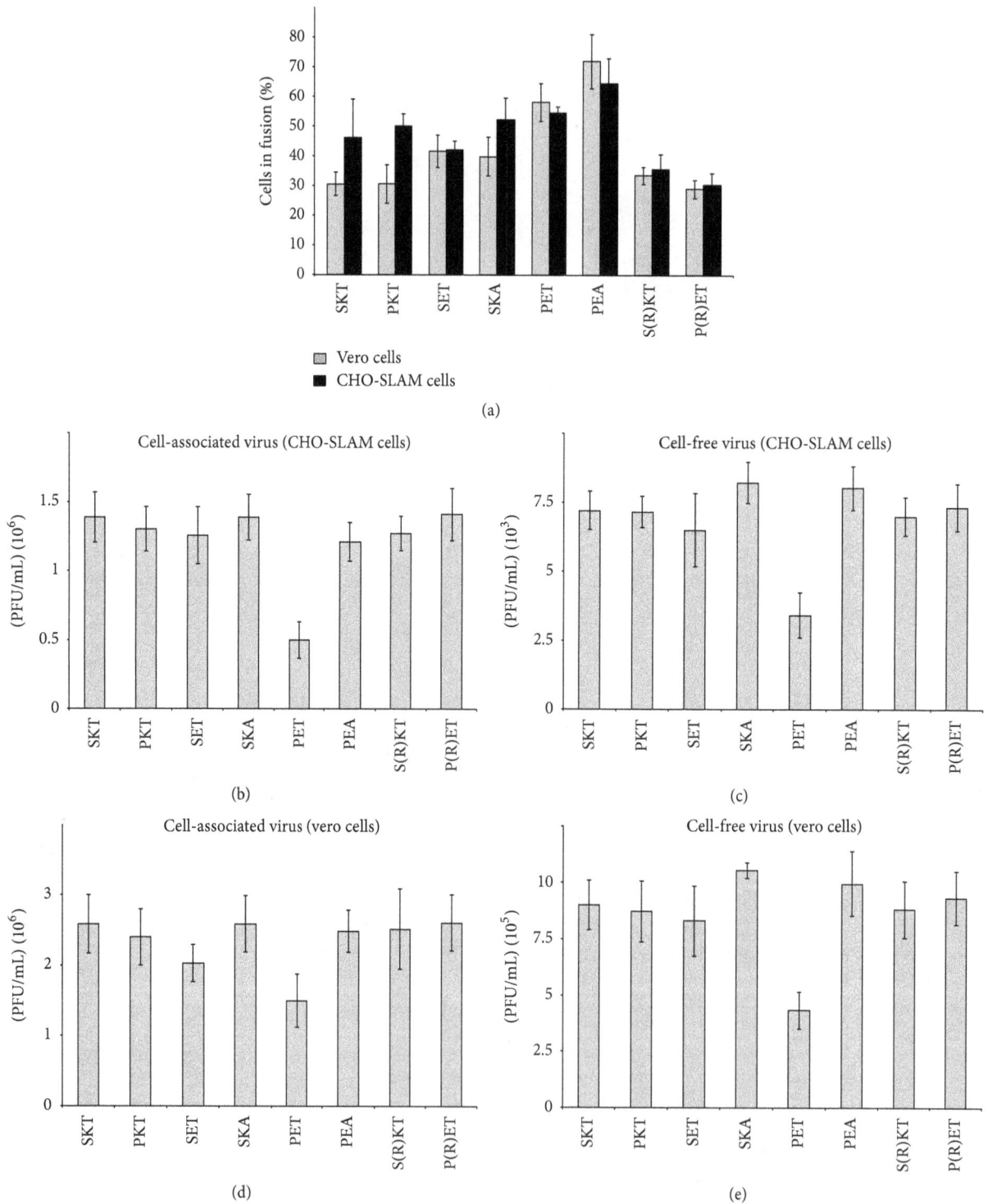

FIGURE 4: Assay of Moraten strain recombinants cell-cell fusion and virus assembly capacities. Vero cells and CHO-SLAM cells infected with Moraten recombinants were analysed 30 h and 48 h after infection for their cell-cell fusion and virus assembly capacities, respectively. Histogram data represent the mean percentages ± standard deviations for three experiments. (a) CD46- and SLAM-dependent fusion; ((b) and (c)) SLAM-dependent production of cell-associated and cell-free virus, respectively; ((d) and (e)) CD46-dependent production of cell-associated and cell-free virus, respectively.

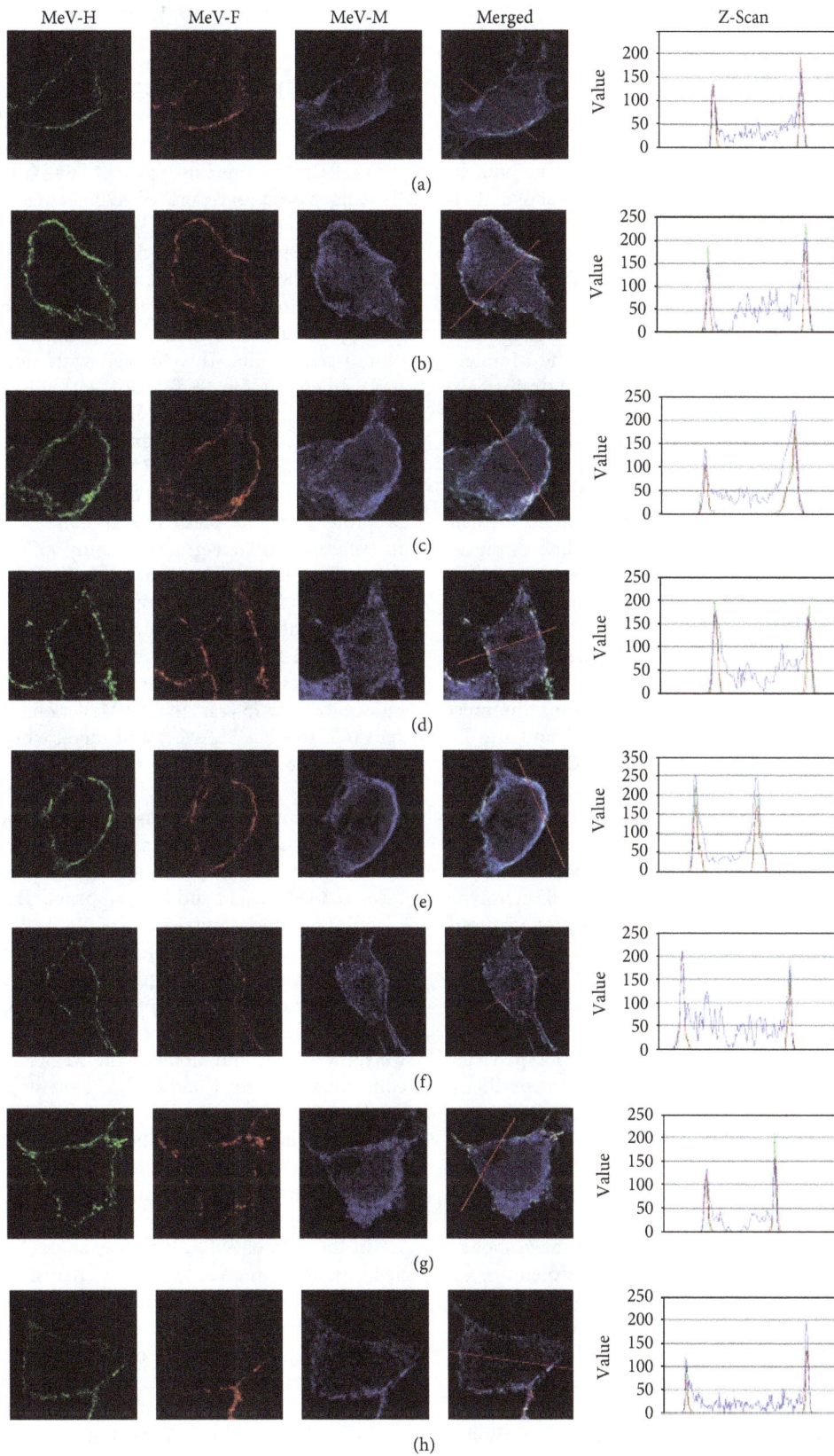

FIGURE 5: Confocal study of localization of MeV proteins, H, F, and M, of Moraten recombinant viruses. Vero-SLAM cells infected with these recombinant viruses were stained with anti-H labeled with Alexa Fluor 488, anti-F mAb Y503 with Alexa Fluor 555, and anti-M mAb8910 with Alexa Fluor 647. (a) SKT; (b) PKT; (c) SET; (d) SKA; (e) PET; (f) PEA; (g) S(R)KT; (h) P(R)ET.

Such attenuation could lessen the risk of B3 genotype viruses, which carry the PET motif rather than the PEA motif, to cause SSPE. Moreover, the general attenuated phenotype of vaccine strains would thus appear to preclude them from causing SSPE.

Importantly, our results suggest that the triresidue motif SKT (or PKT) in their M proteins could contribute to the attenuation of vaccine strains over and above their incapacity to act against the host innate immune response. But are vaccine strains with SKT more attenuated than those with PKT? Some MeV vaccine strains (Moraten, Schwarz, Rubeovax, and CAM70) have the SKT motif in their M proteins, whereas others (AIK-C, Zagreb, Leningrad16, Shanghai191, and Changchun47) have PKT. The Moraten, Schwarz, Rubeovax, AIK-C, and Zagreb vaccine strains have been adapted from the Edmonston-Enders strain, but with different passage histories [31]. As it has been shown that the AIK-C and Zagreb vaccine strains are more virulent than Moraten [35], it is tempting to speculate that this is due to the presence of a proline residue at position 64 in their M proteins rather than serine. It would be interesting to determine whether the introduction of the P64S mutation into the M proteins of these vaccine strains would lessen their virulence and thereby increase their safety.

Although Nigeria has had the highest rates for measles morbidity and mortality in the world [36], it was not until 1999 that field isolates of measles in this country, the most populous nation in Africa, were studied [37] and the B3 genotype was found to be prevalent. The B3 genotype is now the predominant genotype in Sub-Saharan Africa [21]. However, despite MeV's hyperendemicity, few SSPE cases have been reported in this vast region [22]. Moreover, despite an extended search of the literature and data banks, we have been unable to find a single SSPE case involving a B3 genotype wt virus [38]. Our results show that the M protein motif PET is associated with the reduced virus production of the B3 genotype virus Lys-1 possibly because of less efficient assembly. However, the PRET recombinant reveals that the additional mutation K86R is responsible for the normal levels of viral production exhibited by the other B3 genotype virus, G954. Although the G954 virus and its equivalent PRET recombinant exhibit lowered cell-cell fusion compared to the D4 genotype virus Lys05/06 and the PEA recombinant, we predict that if ever a B3 genotype virus case of SSPE appears, it will likely possess the K86R mutation in addition to PET. The chance of this occurring however is probably slight as G954 is the only B3 genotype virus (out of four) that has been shown, as yet, to possess the K86R mutation in the M protein.

Loosening the interaction between the M protein and the cytoplasmic tail of the H and F glycoproteins results in increased cell-cell fusion and less virus assembly; tightening this interaction leads to lower fusion and increased virus assembly [2, 3]. Moreover, in a study of the adaptation of wt MeV to vero cells, the introduction of an M gene coming from a vaccine strain (and thus the P64S, E89K, and A209T changes) into a wt MeV recombinant allowed growth in these CD46+/SLAM− cells, albeit with a low entry efficiency and no cell-cell fusion [39]. Observing that the essential difference between the vaccine M and the wt M was the identity at

residues 64, 89, and 209, these authors then made wt MeV recombinants with the P64S, E89K, and A209T changes present both individually and in combination. They found that only the P64S and E89K changes allowed the wt MeV to grow well in vero cells; A209T had no effect. This is supported by PCR studies that have shown that when wt MeV adapts to vero cells, mutations can appear at P64 [40] and E89 [41], but there have been no reports of A209 being mutated.

In the following study [39], the Yanagi group (who discounted A209T) present results suggesting that the mutations P64S and E89K substitutions allow a strong interaction of the M protein with the cytoplasmic tail of the H protein and thereby an enhancement of virus assembly at the expense of cell-cell fusion. This fit well with our results, in that replacement of the SKT motif with PEA motif elements has the effect of increasing fusion although we do not see large decreases in virus production, except of course for the PET mutant. Using confocal microscopy, we did not obtain evidence for any change in the colocalization of H, F, and M proteins at the plasma membrane for any M mutant. In contrast, the Yanagi group study [39] found that some delocalization occurs but their results appear to reflect the topology of the three proteins on intracellular membranes rather than at the plasma membrane.

It would appear that adaptation of wt MeV to vero cells and SSPE pathogenesis have much in common. It has been known for over 25 years that wt MeV strains can adapt to vero cells (CD46+ but SLAM−) and thereby become attenuated. The virus enters by as yet unknown mechanism and replicates efficiently but, at least at first, there is no sign of cell-cell fusion despite multiple "blind" passages. Such persisting viruses have been shown to have mutations in the M protein, at either residue P64 [40] or residue E89 [41]. However, in some cases cell-cell fusion suddenly appears. This is triggered by mutations in the H protein such as N481Y [42] that allows the virus to use CD46 as receptor [28]. Presumably, the N481Y mutation changes the 3D conformation of the H protein so as to allow this change in receptor usage. We speculate that the PEA motif has its effect by a similar mechanism. Previously we showed that mutations in the M protein can affect the conformation of the F and H glycoproteins and thus facilitate resistance to neutralizing anti-MV sera [19]. Presumably, the PEA motif in the M proteins of wt MV strains induces a particular 3D conformation in the glycoproteins that allows an elevated immune escape and hence an increased potential for persistence.

In the case of SSPE, the virus enters brain cells such as neurons, again by an unknown mechanism, where it persists, accumulating mutations, mainly in the M protein but also in the H and F glycoproteins. After a long persistence usually lasting years, the symptoms of SSPE suddenly appear. However, perhaps not in all cases, it has been found, via autopsy examinations, that MeV "commonly" persists in the brains of healthy individuals [43]. We hypothesize that the "trigger" for this event is mutations that break the interaction between the M protein and the glycoprotein tails leading to accelerated spread of the virus in the brain and a resultant destructive inflammatory response. Such mutations, truncation of the M protein and/or F protein cytoplasmic tail, are found in the majority, if not all, SSPE cases where the pertinent sequences are available [38]. A possible means of MeV entry not involving fusion is macropinocytosis. Evidence is accumulating that MeV can enter cells by this endocytic mechanism [44, 45] but intriguingly, both for adaptation to vero cells and entry into neurons, the only known MeV receptor available is CD46.

Considering that the triresidue MeV-M protein motif PEA seems to be a molecular marker for SSPE, it would appear to be a priority to sequence the M proteins from the B1, B2, C1, D1, D2, D9, D10, E, F, G1, G2, H2, and d11 genotypes of wt MeV, to determine the nature of their triresidue motif and to make epidemiological correlations with the frequency of SSPE cases according to the circulating genotype. However, the underlying molecular mechanisms that allow wt MeV viruses with the PEA M molecular marker to slowly invade the CNS and give rise to SSPE remain unknown.

5. Conclusions

This study seeks to discover why SSPE is always caused by wt MeV, never by vaccine MeV. Our results suggest that SSPE can only be caused by MeV strains that contain a particular triresidue structural motif (PEA) in the primary sequence of their M proteins. Only wt MeV-M proteins contain the PEA motif. Results obtained from the construction of MeV recombinants suggest that the absence of this motif, in all MeV vaccine strains but also the wt B3 genotype, lowers the capacity of MeV to spread. Hence, we propose the MeV-M protein PEA motif as a molecular marker for MeV strains that risk causing SSPE.

Acknowledgments

The authors thank Roberto Cattaneo and Patricia Devaux for the Moraten recombinant system, Yusuke Yanagi for CHO-SLAM cells, Branka Horvat for the Lys05/06 wt MeV strain from the Inserm U758 CRB MeV Bank, Olivier Duc and Christophe Chamot (Platim microscope facilities of SRF Biosciences Gerland-Lyon-Sud U58/UM53444), as well as Olivier Reynaud for advice on confocal microscopy, Fabian Wild for helpful discussions, and Inserm for funding. Denis Gerlier and Joanna Brunel are supported by the ANR Grant "Physico-Chimie du Vivant" (ANR-08-PCVI-0020-01). Evelyne Manet and Robin Buckland are CNRS scientists.

References

[1] T. F. Wild and R. Buckland, "Functional aspects of envelope-associated measles virus proteins," in *Measles Virus*, V. ter Meulen and M. Billeter, Eds., vol. 191 of *Current Topics in Microbiology and Immunology*, pp. 51–64, Springer, Berlin, Germany, 1995.

[2] T. Cathomen, H. Y. Naim, and R. Cattaneo, "Measles viruses with altered envelope protein cytoplasmic tails gain cell fusion competence," *Journal of Virology*, vol. 72, no. 2, pp. 1224–1234, 1998.

[3] T. Cathomen, B. Mrkic, D. Spehner et al., "A matrix-less measles virus is infectious and elicits extensive cell fusion: consequences for propagation in the brain," *The EMBO Journal*, vol. 17, no. 14, pp. 3899–3908, 1998.

[4] Y. Yanagi, M. Takeda, S. Ohno, and T. Hashiguchi, "Measles virus receptors," *Current Topics in Microbiology and Immunology*, vol. 329, pp. 13–30, 2009.

[5] M. D. Mühlebach, M. Mateo, P. L. Sinn et al., "Adherens junction protein nectin-4 is the epithelial receptor for measles virus," *Nature*, vol. 480, no. 7378, pp. 530–533, 2011.

[6] R. S. Noyce, D. G. Bondre, M. N. Ha et al., "Tumor cell marker pvrl4 (nectin 4) is an epithelial cell receptor for measles virus," *PLoS Pathogens*, vol. 7, no. 8, Article ID e1002240, 2011.

[7] M. Takeda, M. Tahara, N. Nagata, and F. Seki, "Wild-type measles virus is intrinsically dual-tropic," *Frontiers in Microbiology*, vol. 2, article 279, 7 pages, 2012.

[8] J. R. Dawson Jr., "cellular inclusions in cerebral lesions of epidemic encephalitis: second report," *Archives of Neurology & Psychiatry*, vol. 31, no. 4, pp. 685–700, 1934.

[9] J. Gutierrez, R. S. Issacson, and B. S. Koppel, "Subacute sclerosing panencephalitis: an update," *Developmental Medicine & Child Neurology*, vol. 52, no. 10, pp. 901–907, 2010.

[10] B. K. Rima, "Molecular biological basis of measles virus strain differences," in *Measles and Poliomyelitis*, E. Kurstak, Ed., pp. 151–160, Springer, Heidelberg, Germany, 1993.

[11] B. K. Rima, "The pathogenesis of subacute sclerosing panencephalitis," *Reviews in Medical Virology*, vol. 4, no. 2, pp. 81–90, 1994.

[12] B. K. Rima and W. P. Duprex, "Molecular mechanisms of measles virus persistence," *Virus Research*, vol. 111, no. 2, pp. 132–147, 2005.

[13] M. A. Billeter, R. Cattaneo, P. Spielhofer et al., "Generation and properties of measles virus mutations typically associated with subacute sclerosing panencephalitis," *Annals of the New York Academy of Sciences*, vol. 724, pp. 367–377, 1994.

[14] R. Cattaneo, A. Schmid, M. A. Billeter, R. D. Sheppard, and S. A. Udem, "Multiple viral mutations rather than host factors cause defective measles virus gene expression in a subacute sclerosing panencephalitis cell line," *Journal of Virology*, vol. 62, no. 4, pp. 1388–1397, 1988.

[15] R. Cattaneo, A. Schmid, D. Eschle, K. Baczko, V. ter Meulen, and M. A. Billeter, "Biased hypermutation and other genetic changes in defective measles viruses in human brain infections," *Cell*, vol. 55, no. 2, pp. 255–265, 1988.

[16] R. Cattaneo, A. Schmid, P. Spielhofer et al., "Mutated and hypermutated genes of persistent measles viruses which caused lethal human brain diseases," *Virology*, vol. 173, no. 2, pp. 415–425, 1989.

[17] A. Bitnun, P. Shannon, A. Durward et al., "Measles inclusion-body encephalitis caused by the vaccine strain of measles virus," *Clinical Infectious Diseases*, vol. 29, no. 4, pp. 855 861, 1999.

[18] D. R. Hardie, C. Albertyn, J. M. Heckmann, and H. E. Smuts, "Molecular characterisation of virus in the brains of patients with measles inclusion body encephalitis (MIBE)," *Virology Journal*, vol. 10, article 283, 2013.

[19] H. Kweder, M. Ainouze, S. L. Cosby et al., "Mutations in the H, F, or M proteins can facilitate resistance of measles virus to

neutralizing human anti-MV sera," *Advances in Virology*, vol. 2014, Article ID 205617, 18 pages, 2014.

[20] L. Jin, S. Beard, R. Hunjan, D. W. G. Brown, and E. Miller, "Characterization of measles virus strains causing SSPE: a study of 11 cases," *Journal of NeuroVirology*, vol. 8, no. 4, pp. 335–344, 2002.

[21] P. A. Rota, K. Brown, A. Mankertz et al., "Global distribution of measles genotypes and measles molecular epidemiology," *Journal of Infectious Diseases*, vol. 204, supplement 1, pp. S514–S523, 2011.

[22] D. E. Griffin, "Emergence and re-emergence of viral diseases of the central nervous system," *Progress in Neurobiology*, vol. 91, no. 2, pp. 95–101, 2010.

[23] J. Fayolle, B. Verrier, R. Buckland, and T. Fabian Wild, "Characterization of a natural mutation in an antigenic site on the fusion protein of measles virus that is involved in neutralization," *Journal of Virology*, vol. 73, no. 1, pp. 787–790, 1999.

[24] D. Waku-Kouomou and T. F. Wild, "Adaptation of wild-type measles virus to tissue culture," *Journal of Virology*, vol. 76, no. 3, pp. 1505–1509, 2002.

[25] V. Guillaume, H. Aslan, M. Ainouze et al., "Evidence of a potential receptor-binding site on the Nipah virus G protein (NiV-G): identification of globular head residues with a role in fusion promotion and their localization on an NiV-G structural model," *Journal of Virology*, vol. 80, no. 15, pp. 7546–7554, 2006.

[26] C. D. Richardson, A. Scheid, and P. W. Choppin, "Specific inhibition of paramyxovirus and myxovirus replication by oligopeptides with amino acid sequences similar to those at the N-termini of the F1 or HA2 viral polypeptides," *Virology*, vol. 105, no. 1, pp. 205–222, 1980.

[27] W. J. Bellini, J. S. Rota, L. E. Lowe et al., "Subacute sclerosing panencephalitis: more cases of this fatal disease are prevented by measles immunization than was previously recognized," *Journal of Infectious Diseases*, vol. 192, no. 10, pp. 1686–1693, 2005.

[28] V. Lecouturier, J. Fayolle, M. Caballero et al., "Identification of two amino acids in the hemagglutinin glycoprotein of measles virus (MV) that govern hemadsorption, hela cell fusion, CD46 downregulation: Phenotypic markers that differentiate vaccine and wild-type MV strains," *Journal of Virology*, vol. 70, no. 7, pp. 4200–4204, 1996.

[29] N. Massé, T. Barrett, C. P. Muller, T. F. Wild, and R. Buckland, "Identification of a second major site for CD46 binding in the hemagglutinin protein from a laboratory strain of measles virus (MV): potential consequences for wild-type MV infection," *Journal of Virology*, vol. 76, no. 24, pp. 13034–13038, 2002.

[30] C. L. Parks, R. A. Lerch, P. Walpita, H.-P. Wang, M. S. Sidhu, and S. A. Udem, "Comparison of predicted amino acid sequences of measles virus strains in the Edmonston vaccine lineage," *Journal of Virology*, vol. 75, no. 2, pp. 910–920, 2001.

[31] J. S. Rota, Z.-D. Wang, P. A. Rota, and W. J. Bellini, "Comparison of sequences of the H, F, and N coding genes of measles virus vaccine strains," *Virus Research*, vol. 31, no. 3, pp. 317–330, 1994.

[32] R. L. Beardsley and J. P. Reilly, "Optimization of guanidination procedures for MALDI mass mapping," *Analytical Chemistry*, vol. 74, no. 8, pp. 1884–1890, 2002.

[33] J. P. Gallivan and D. A. Dougherty, "Cation-pi interactions in structural biology," *Proceedings of the National Academy of Sciences of the United States of America*, vol. 96, no. 17, pp. 9459–9464, 1999.

[34] M. Tahara, M. Takeda, and Y. Yanagi, "Altered interaction of the matrix protein with the cytoplasmic tail of hemagglutinin modulates measles virus growth by affecting virus assembly and cell-cell fusion," *Journal of Virology*, vol. 81, no. 13, pp. 6827–6836, 2007.

[35] A. Valsamakis, H. Kaneshima, and D. E. Griffin, "Strains of measles vaccine differ in their ability to replicate in and damage human thymus," *The Journal of Infectious Diseases*, vol. 183, no. 3, pp. 498–502, 2001.

[36] WHO, "Expanded programme on immunization—measles control in the WHO African region," *Weekly Epidemiological Record*, vol. 71, pp. 201–203, 1996.

[37] F. Hanses, A. T. Truong, W. Ammerlaan et al., "Molecular epidemiology of Nigerian and Ghanaian measles virus isolates reveals a genotype circulating widely in western and central Africa," *Journal of General Virology*, vol. 80, no. 4, pp. 871–877, 1999.

[38] R. Buckland and H. Kweder, Unpublished observations.

[39] M. Tahara, M. Takeda, and Y. Yanagi, "Contributions of matrix and large protein genes of the measles virus Edmonston strain to growth in cultured cells as revealed by recombinant viruses," *Journal of Virology*, vol. 79, no. 24, pp. 15218–15225, 2005.

[40] K. Takeuchi, N. Miyajima, F. Kobune, and M. Tashiro, "Comparative nucleotide sequence analyses of the entire genomes of B95a cell-isolated and vero cell-isolated measles viruses from the same patient," *Virus Genes*, vol. 20, no. 3, pp. 253–257, 2000.

[41] J. Druelle, C. I. Sellin, D. Waku-Kouomou, B. Horvat, and F. T. Wild, "Wild type measles virus attenuation independent of type I IFN," *Virology Journal*, vol. 5, article 22, 2008.

[42] K. Shibahara, H. Hotta, Y. Katayama, and M. Homma, "Increased binding activity of measles virus to monkey red blood cells after long-term passage in vero cell cultures," *Journal of General Virology*, vol. 75, no. 12, pp. 3511–3516, 1994.

[43] Y. Katayama, K. Kohso, A. Nishimura, Y. Tatsuno, M. Homma, and H. Hotta, "Detection of measles virus mRNA from autopsied human tissues," *Journal of Clinical Microbiology*, vol. 36, no. 1, pp. 299–301, 1998.

[44] C. Frecha, C. Lévy, C. Costa et al., "Measles virus glycoprotein-pseudotyped lentiviral vector-mediated gene transfer into quiescent lymphocytes requires binding to both SLAM and CD46 entry receptors," *Journal of Virology*, vol. 85, no. 12, pp. 5975–5985, 2011.

[45] O. Pernet, C. Pohl, M. Ainouze, H. Kweder, and R. Buckland, "Nipah virus entry can occur by macropinocytosis," *Virology*, vol. 395, no. 2, pp. 298–311, 2009.

5

Appearance of L90I and N205S Mutations in Effector Domain of NS1 Gene of pdm (09) H1N1 Virus from India during 2009–2013

Sachin Kumar,[1,2] Shashi Khare,[1] Bano Saidullah,[3] Inderjeet Gandhoke,[1] Hanu Ram,[2] Supriya Singh,[1,2] L. S. Chauhan,[1,2] and Arvind Rai[1]

[1] Division of Microbiology, National Centre for Disease Control, 22 Sham Nath Marg, Delhi 110054, India
[2] Division of Biotechnology, National Centre for Disease Control, 22 Sham Nath Marg, Delhi 110054, India
[3] Discipline of Life Science, School of Science, Indira Gandhi National Open University, Delhi 110068, India

Correspondence should be addressed to Arvind Rai; arvindrai.nicd@gmail.com

Academic Editor: Subhash Verma

In the present study, full length sequencing of NS gene was done in 91 samples which were obtained from patients over the time period of five years from 2009 to 2013. The sequencing of NS gene was undertaken in order to determine the changes/mutations taking place in the NS gene of A H1N1 pdm (09) since its emergence in 2009. Analysis has shown that the majority of samples belong to New York (G1 type) strain with valine at position 123. Effector domain of NS1 protein displays the appearance of three mutations L90I, I123V, and N205S in almost all the samples from 2010 onwards. Phylogenetic analysis of available NS1 sequences from India has grouped all the sequences into four clusters with mean genetic distance ranging from 12% to 24% between the clusters. Variability in length of NS1 protein was seen in sequences from these clusters, 230-amino-acid-residue NS1 for all strains from year 2007 to 2008 and for 21 strains from year 2009 and 219-residue products for 37 strains from year 2009 and all strains from year 2010 to 2013. Mutations like K62R, K131Q, L147R, and A202P were observed for the first time in NS1 protein and their function remains to be determined.

1. Introduction

Influenza viruses are responsible for acute respiratory infection and are a source of seasonal epidemics and occasional pandemics. Influenza A viruses are classified into subtypes based on the different types of HA and NA combinations that occur. So far 18 hemagglutinin (HA) and 11 neuraminidase (NA) subtypes have been reported from various organisms ranging between aquatic, avian, and human species [1, 2]. Segment 8 of influenza A (H1N1) encodes two proteins NS1 (nonstructural) protein and NEP (nuclear export protein) by alternative splicing. The mRNAs of both proteins share 56 nucleotides at the $5'$ end, resulting in both proteins sharing 10 amino acids at N terminal.

NS1 protein is encoded by the collinear mRNA from segment 8 of the influenza virus genome and has a strain specific length ranging from 230 to 237 amino acid residues. It is expressed exclusively in the infected cells [3]. NS1 could be divided into two functional domains: (i) N-terminal RNA binding domain (residues 1–73) and (ii) C-terminal effector domain, interacting with several host factors (residues 74–230) [3–6].

NS1 is a multifunctional protein involved in various functions of regulating immune responses. It functions as an interferon (IFN) antagonist, which allows efficient virus replication in IFN-competent hosts. NS1 targets both IFN-α/β production and the activation of IFN-induced antiviral genes [6]. The RNA binding domain (RBD) of NS1 binds to both ssRNA and dsRNA, thereby sequestering them and preventing their recognition by RIG1 (retinoic acid inducible gene), resulting in inhibition of IFN α and β expression [7, 8]. NS1 protein is also involved in inhibiting $3'$ end processing of host mRNA by binding to CPSF 30 (cleavage and polyadenylation specificity factor 30) and PABPN1 (poly(A) binding protein nuclear 1) [9]. Sequestering of dsRNA by RBD of NS1 from $2'$–$5'$ oligoadenylate synthetase (OAS) is essential for inhibition of ribonuclease L (RNase L) pathway, which is involved in the degradation of viral RNA. NS1 binds directly to the regulatory subunit of protein kinase R (PKR) and therefore regulates the effectors of IFN response and controls apoptosis, cell growth, cell proliferation, cytokine production, and signaling [10]. NS1 interacts with eIF4GI and PABP1 (poly(A) binding protein 1) and enhances viral protein synthesis in comparison to host cell protein. In this way, NS1 inhibits the innate immune response of the host by suppressing the interferon release. It also inhibits adaptive immunity by restricting human dendritic cells maturation and induction of T-cell response [11].

NS2 (NEP) is involved in the export of viral RNP from the nucleus to the cytoplasm through nuclear export signal and via interaction with Crml protein. NEP can be divided into a protease-sensitive N-terminal domain (amino acids 1–53) and a protease-resistant C-terminal domain (amino acids 54–121) [12]. Of the two domains N-terminal domain has been reported to contain nuclear export signal between residues 12 and 21 which interact with the nuclear export protein Crml and facilitate the export of viral RNPs [13].

In the present study, full length sequencing of pdm H1N1 (09) virus for NS gene was performed in samples collected from years 2009 to 2013 in order to determine the mutations taking place in the NS gene of pdm H1N1 (09) virus since its emergence in year 2009. Genetic and phylogenetic analyses of previously studied sequences reported from India and other countries were done, based on available literature in order to determine their phylogeny and sites under selection pressure (contributing towards the evolution of virus) and to study the possible effect of mutations on virulence and pathogenicity of influenza virus.

2. Materials and Methods

Samples (Nasal and Throat Swabs in viral transport media (VTM)) from years 2009 to 2013 (details given in Table 1) from patients with symptoms of fever, cough, sore throat, nasal catarrh, or shortness of breath were collected from hospitals of Delhi and outbreak samples from other states obtained for H1N1 testing at the National Centre for Disease Control (NCDC), New Delhi, India. The study was approved by the institutional ethical committee and all the samples were processed in a high containment facility (a biosafety level-3 laboratory) at NCDC, New Delhi. Viral RNA was extracted using QIAmp viral RNA mini kit (Qiagen, Germany) according to manufacturer's protocol. Finally RNA was eluted in 50 μL of elution buffer and stored at $-80°C$ until use. The initial detection of influenza viruses was done by RT PCR protocol for detection of influenza A (H1N1) pdm (09) by WHO/CDC [14, 15].

For sequencing, viral genes were amplified as described earlier [16, 17]. Nucleotide (nt) sequencing was carried out on Applied Biosystems 3130xl Genetic Analyzer (Applied Biosystems, Foster City, CA, USA), using gene specific primers. Nucleotide and protein sequence BLAST (Basic Local Alignment Search Tool) search was performed using the National Centre for Biotechnology Information (NCBI), National

Institute of Health, Bethesda, MD, BLAST server at GenBank database [18]. Sequences for phylogenetic analyses were retrieved and multiple sequence alignments were performed on the Influenza Virus Resource (IVR) at NCBI and the Influenza Research Database (IRD) at http://www.fludb.org/ [19, 20]. Phylogenetic analysis was done by MEGA v6.0, using maximum likelihood method and 500-replicate bootstrapping. Mean genetic distance within the cluster and between the clusters was determined by MEGA v6.0 [21].

Metadata-driven comparative analysis of study samples against all protein sequences in the influenza virus database at the Influenza Research Database (IRD) at http://www.fludb.org/ for NS1 protein till 21 December 2013 was performed by Meta-CATS tool [22] on the Influenza Research Database (IRD) at http://www.fludb.org/ at P value threshold of 0.05 (P value threshold is used as the maximum probability level for the likelihood that the position is different among the groups simply by chance) in order to identify significantly different sites between group 1 (database sequences) and group 2 of the study samples.

Selection pressure analysis acting on the codons of NS (nonstructural) gene of H1N1 pdm virus was carried out using HyPhy open-source software package available under the datamonkey web server (http://www.datamonkey.org/) [23]. Analysis was performed using reference sequences [$n =$ 72 (NS)] including Indian H1N1 pdm virus. A separate analysis for NS1 and NEP genes was also carried out by including 44 Indian H1N1 pdm viruses. The ratio of nonsynonymous (dN) to synonymous (dS) substitutions per site (dN/dS or v) was estimated using five different approaches, including single likelihood ancestor counting (SLAC), fixed effects likelihood (FEL), random effects method (REL), mixed effects model of evolution (MEME), and fast unbiased Bayesian approximation (FUBAR). The best nucleotide substitutions model for different data sets as determined through the available tool in datamonkey server was adopted in the analysis.

3. Results

3.1. Mutations Seen in the NS1 Gene of Influenza A H1N1 pdm (09). Total 48 nucleotide substitutions (27 synonymous and 21 nonsynonymous) were observed in 91 samples from the years ranging from 2009 to 2013 when compared with FJ969528 (A/California/07/2009) as shown in Tables 2 and 3. I123V mutation was seen in 88 samples, along with other common amino acid changes like E55Q, L90I, and N205S (30–50% samples). Changes like D53N, K62R, S73T, T94A, E96K, R108K, I111T, V129I, V129A, K131E, T143N, I145V, L147R, T151P, E172 K, A202P, and N209D were rare and observed only in few samples as given in Tables 2 and 3.

Mutations I43N, D53N, T94A, R108K, and E172K were observed only in 1–4 samples from year 2009, while two samples had mutation N209D and three had E55Q mutation similar to year 2010. E55Q mutation was also observed in a varied number of samples (given in Tables 2 and 3) from all the years except 2012. Mutation V129A was observed in a single sample each year from 2009 and 2011. Except three

TABLE 1: Details of patients with strain name, collection date, and genes sequenced per strain with accession number.

S. number	Strain name	Collection date	Sex	Age (years)	Accession number
1	A/Karnataka/001/2009	6-Aug-2009	F	21	KJ023091
2	A/Kerala/002/2009	14-Aug-2009	M	36	KJ023092
3	A/Bihar/003/2009	20-Aug-2009	M	5	KJ023093
4	A/West Bengal/004/2009	6-Dec-2009	F	4.6	KJ023094
5	A/Uttar Pradesh/005/2009	4-Oct-2009	F	17	KJ023095
6	A/Goa/006/2009	6-Oct-2009	M	47	KJ023096
7	A/Punjab/007/2009	26-Nov-2009	M	6	KJ023097
8	A/Gujarat/008/2009	8-Aug-2009	M	0.9	KJ023098
9	A/Rajasthan/009/2009	26-Nov-2009	M	1.6	KJ023099
10	A/Chhattisgarh/010/2009	30-Sep-2009	M	30	KJ023100
11	A/Madhya Pradesh/011/2009	10-Oct-2009	F	55	KJ023101
12	A/Tamil Naidu/012/2009	6-Nov-2009	M	15	KJ023102
13	A/Jammu Kashmir/013/2009	4-Dec-2009	M	14	KJ023103
14	A/Uttarakhand/014/2009	8-Dec-2009	M	5	KJ023104
15	A/Haryana/015/2009	16-Dec-2009	M	60	KJ023105
16	A/Delhi/016/2009	14-Dec-2009	F	6.6	KJ023106
17	A/Delhi/017/2009	13-Dec-2009	M	25	KJ023107
18	A/Delhi/018/2009	29-Jul-2009	F	74	KJ023108
19	A/Rajasthan/019/2010	10-Jan-2010	M	19	KJ023109
20	A/Assam/020/2010	28-Jan-2010	F	33	KJ023110
21	A/Jammu Kashmir/021/2010	19-Jul-2010	M	14	KJ023111
22	A/Jammu Kashmir/022/2010	15-Jan-2010	F	60	KJ023112
23	A/Kerala/023/2010	19-Jun-2010	F	21	KJ023113
24	A/Madhya Pradesh/024/2010	28-Jan-2010	M	47	KJ023114
25	A/Chhattisgarh/025/2010	7-Feb-2010	M	22	KJ023115
26	A/Chhattisgarh/026/2010	25-Sep-2010	F	30	KJ023116
27	A/Goa/027/2010	26-Aug-2010	M	33	KJ023117
28	A/Haryana/028/2010	19-Oct-2010	M	0.6	KJ023118
29	A/Haryana/029/2010	15-Aug-2010	M	56	KJ023119
30	A/Punjab/030/2010	21-Aug-2010	F	35	KJ023120
31	A/Punjab/031/2010	19-Oct-2010	F	4.6	KJ023121
32	A/Uttar Pradesh/032/2010	16-Sep-2010	M	27	KJ023122
33	A/Uttarakhand/033/2010	24-Sep-2010	M	2.6	KJ023123
34	A/Goa/034/2010	7-Aug-2010	M	32	KJ023124
35	A/Goa/035/2010	9-Oct-2010	F	35	KJ023125
36	A/Delhi/036/2010	20-Aug-2010	M	50	KJ023126
37	A/Delhi/037/2010	25-Aug-2010	M	18	KJ023127
38	A/Delhi/038/2010	22-Aug-2010	F	25	KJ023128
39	A/Delhi/039/2010	17-Sep-2010	M	14	KJ023129
40	A/Punjab/040/2011	21-Feb-2011	M	40	KJ023130
41	A/Punjab/041/2011	9-Mar-2011	F	20	KJ023131
42	A/Punjab/042/2011	17-Mar-2011	F	55	KJ023132
43	A/Punjab/043/2011	28-Mar-2011	F	27	KJ023133
44	A/Jammu Kashmir/044/2011	20-Jan-2011	F	25	KJ023134
45	A/Jammu Kashmir/045/2011	31-Jan-2011	M	40	KJ023135
46	A/Goa/046/2011	23-Jun-2011	M	60	KJ023136
47	A/Goa/047/2011	7-Jun-2011	F	30	KJ023137

TABLE 1: Continued.

S. number	Strain name	Collection date	Sex	Age (years)	Accession number
48	A/Goa/048/2011	19-May-2011	M	18	KJ023138
49	A/Haryana/049/2011	24-Mar-2011	M	19	KJ023139
50	A/Delhi/050/2011	19-May-2011	F	35	KJ023140
51	A/Delhi/051/2011	24-Feb-2011	F	36	KJ023141
52	A/Delhi/052/2011	24-Feb-2011	M	12	KJ023142
53	A/Delhi/053/2011	23-Jun-2011	M	10	KJ023143
54	A/Delhi/054/2011	1-Apr-2011	F	25	KJ023144
55	A/Delhi/055/2011	3-Apr-2011	F	16	KJ023145
56	A/Delhi/056/2011	5-Mar-2011	M	36	KJ023146
57	A/Delhi/057/2012	26-Jul-2012	M	3.6	KJ023147
58	A/Delhi/058/2012	14-Nov-2012	M	3	KJ023148
59	A/Delhi/059/2012	9-Oct-2012	M	34	KJ023149
60	A/Delhi/060/2012	13-Sep-2012	M	28	KJ023150
61	A/Delhi/061/2012	21-May-2012	F	58	KJ023151
62	A/Delhi/062/2012	16-Aug-2012	M	48	KJ023152
63	A/Delhi/063/2012	25-Sep-2012	M	32	KJ023153
64	A/Delhi/064/2012	10-Mar-2012	M	22	KJ023154
65	A/Delhi/065/2012	9-Mar-2012	F	25	KJ023155
66	A/Delhi/066/2012	24-Sep-2012	F	3	KJ023156
67	A/Delhi/067/2012	25-Sep-2012	M	32	KJ023157
68	A/Delhi/068/2012	20-Nov-2012	M	65	KJ023158
69	A/Delhi/069/2012	19-Dec-2012	M	48	KJ023159
70	A/Uttarakhand/070/2012	4-Oct-2012	M	26	KJ023160
71	A/Goa/071/2012	15-May-2012	F	60	KJ023161
72	A/Goa/072/2012	9-Oct-2012	F	40	KJ023162
73	A/Goa/073/2012	9-Jan-2012	M	20	KJ023163
74	A/Chhattisgarh/074/2012	10-May-2012	M	62	KJ023164
75	A/Chhattisgarh/075/2012	10-Aug-2012	F	21	KJ023165
76	A/Haryana/076/2012	11-Jun-2012	M	55	KJ023166
77	A/Haryana/077/2012	20-Nov-2012	F	24	KJ023167
78	A/Haryana/078/2012	17-Feb-2012	F	42	KJ023168
79	A/Haryana/079/2013	2-Jan-2013	F	75	KJ023169
80	A/Haryana/080/2013	2-May-2013	M	52	KJ023170
81	A/Haryana/081/2013	22-Jan-2013	M	70	KJ023171
82	A/Uttarakhand/082/2013	18-Feb-2013	F	47	KJ023172
83	A/Uttarakhand/083/2013	22-Feb-2013	F	30	KJ023173
84	A/Jammu Kashmir/084/2013	24-Feb-2013	M	52	KJ023174
85	A/Jammu Kashmir/085/2013	20-Feb-2013	F	31	KJ023175
86	A/Delhi/086/2013	31-Jan-2013	F	60	KJ023176
87	A/Delhi/087/2013	17-Feb-2013	M	13	KJ023177
88	A/Delhi/088/2013	18-Feb-2013	F	2	KJ023178
89	A/Delhi/089/2013	22-Feb-2013	F	55	KJ023179
90	A/Uttar Pradesh/090/2013	26-Feb-2013	F	25	KJ023180
91	A/Uttar Pradesh/091/2013	28-Feb-2013	M	62	KJ023181

samples (two from 2009 and one from 2010) all other 88 samples had I123V mutation in NS1 protein.

Mutations E96K (2 samples) and I145L and T151P (1 sample each) were detected only in samples from 2010. E55Q mutation was the second most common amino acid change after I123V and was seen in 16 samples of year 2010, while three mutations K62K, S73T, and I145V were common among 2010 and 2011 samples (given in Tables 2 and 3).

Mutations V129I (6 samples) and L147R (1 sample) were observed only in samples from 2011. Other mutations observed in a few samples in 2011 were E55Q (6 samples) and S73T (3 samples), while mutations K62R and I145V

TABLE 2: Year-wise samples showing amino acid changes in different domains of NS1 gene.

Domain	Position	Prevalence of mutation in samples each year				
		2009 (18)	2010 (21)	2011 (17)	2012 (22)	2013 (13)
RNA binding domain	I 43 N	1	—	—	—	—
	D 53 N	2	—	—	—	—
	E 55 Q	3	16	6	—	1
	E 55 K	—	—	—	—	1
	K 62 R	—	1	1	—	—
	S 73 T	—	1	3	—	—
Effector domain	L 90 I	—	—	8	all	all
	T 94 A	2	—	—	—	—
	E 96 K	—	2	—	—	—
	R 108 K	4	—	—	—	—
	I 123 V	16	20	all	all	all
	V 129 A	1	—	1	—	—
	V 129 I	—	—	6	—	—
	K 131 E	—	—	—	—	4
	K 131 Q	—	—	—	2	—
	I 145 L	—	1	—	—	—
	I 145 V	—	5	1	—	—
	L 147 R	—	—	1	—	—
	T 151 P	—	1	—	—	—
	E 172 K	2	—	—	—	—
	A 202 P	—	—	—	—	1
	N 205 S	—	—	13	all	all
	N 209 D	2	1	—	—	—

were found in single samples. The most common amino acid changes among 2011 samples were L90I (8 samples) and N205S (13 samples) which were noted for the first time in the samples of 2011 and then persisted thereon.

Study samples of 2012 were almost like the 2011 samples but a single difference at position K131Q was noted in 2 samples. In samples of the year 2013, a single sample had E55Q mutation which was similar to the mutation observed in 2010 samples. Mutations E55K, E55Q, and A202P in single samples and K131E in 4 samples were only detected in samples of the year 2013.

3.2. Mutations in RNA Binding Domain of NS1 Protein (Residues 1–73). The RNA binding domain is involved in binding and sequestering dsRNA from its recognition by RIG1 and OAS and thereby inhibiting the IFN response against the virus. In samples from 2009 to 2013, synonymous mutations were seen at ten nucleotide positions (14, 18, 27, 31, 36, 38, 44, 53, 68, and 71) and nonsynonymous mutations were noticed at positions I43N, D53N, E55Q, E55Q, K62R, and S73T as given in Table 2. Among these E55Q was the most common change found in 3 samples from 2009, most samples of 2010 (16), 6 samples from 2011, and 1 sample from 2013. Other changes were rare and only seen in two or three samples.

3.3. Mutations in Effector Domain of NS1 Gene. In total 17 synonymous mutations were observed at the following

positions 83, 85, 88, 99, 105, 125, 132, 138, 142, 143, 144, 152, 153, 163, 186, 214, and 217 from the year 2009 to 2013, while 17 nonsynonymous mutations were recognized at positions L90I, T94A, E96K, R108K, I111T, I123V, V129I, V129A, K131E, T143N, I145V, L147R, T151P, E172K, A202P, N205S, and N209D (in Table 2). I123V was the most common mutation seen in about 97% of samples, followed by E55Q, L90I, and N205S mutations which were noted in about 30% to 50% of samples. The remaining mutations were seen in 1% to 6% samples.

Metadata-driven comparative analysis tool (meta-CATS) of NS1 protein sequence between all database sequences and study sample sequences was performed for identification of amino acid positions that significantly differ between two or more groups of virus sequences. A total 79 sites were identified by Meta-CATS as sites having a significant nonrandom distribution between the specified groups (database sequences and study sequences). 18 of 79 sites identified by Meta-CATS were similar to sites with amino acid changes in study samples and most of the changes seen in the samples were common to sequences in the database. However, mutations like E96K and V129A were rare and viewed only in a limited number of samples in the database, while changes like K62R and K131Q were unique and seen only in one or two study samples.

3.4. Mutations in NEP Gene. NEP is reported to be involved in nuclear export of viral ribonucleoprotein (RNP) complexes

Table 3: NS1 protein positions with variable amino acids in study samples (2009–2013) with reference to A/California/07/2009 reference strain for pandemic H1N1 2009. Reference strain sequence is highlighted with bold font and study samples were arranged according to year of collection and separated with horizontal line.

	53	55	62	73	90	94	96	108	111	123	129	131	143	145	147	151	172	202	205	209
A/California/07/2009 (reference)	D	E	K	S	L	T	E	R	I	I	V	K	T	I	L	T	E	A	N	N
A/Karnataka/001/2009	N	·	·	·	·	·	·	K	·	V	·	·	N	·	·	·	·	·	·	D
A/Kerala/002/2009	·	·	·	·	·	A	·	·	·	V	·	·	·	·	·	·	K	·	·	·
A/Bihar/003/2009	·	·	·	·	·	·	·	·	·	·	·	·	·	·	·	·	·	·	·	·
A/West Bengal/004/2009	·	·	·	·	·	·	·	K	·	V	·	·	·	·	·	·	·	·	·	·
A/Uttar Pradesh/005/2009	·	·	·	·	·	·	·	·	·	·	·	·	·	·	·	·	·	·	·	·
A/Goa/006/2009	·	·	·	·	·	·	·	·	·	V	·	·	·	·	·	·	·	·	·	·
A/Punjab/007/2009	·	·	·	·	·	A	·	·	·	V	·	·	·	·	·	·	K	·	·	·
A/Gujarat/008/2009	·	·	·	·	·	·	·	·	·	V	·	·	·	·	·	·	·	·	·	·
A/Rajasthan/009/2009	N	·	·	·	·	·	·	K	·	V	·	·	N	·	·	·	·	·	·	D
A/Chhattisgarh/010/2009	·	·	·	·	·	·	·	·	·	V	·	·	·	·	·	·	·	·	·	·
A/Madhya Pradesh/011/2009	·	·	·	·	·	·	·	·	·	V	·	·	·	·	·	·	·	·	·	·
A/Tamil Naidu/012/2009	·	·	·	·	·	·	·	·	·	V	·	·	·	·	·	·	·	·	·	·
A/Jammu Kashmir/013/2009	·	·	·	·	·	·	·	·	·	V	A	·	·	·	·	·	·	·	·	·
A/Uttarakhand/014/2009	·	Q	·	·	·	·	·	K	·	V	·	·	·	·	·	·	·	·	·	·
A/Haryana/015/2009	·	·	·	·	·	·	·	·	·	V	·	·	·	·	·	·	·	·	·	·
A/Delhi/016/2009	·	Q	·	·	·	·	·	·	·	V	·	·	·	·	·	·	·	·	·	·
A/Delhi/017/2009	·	Q	·	·	·	·	·	·	·	V	·	·	·	·	·	·	·	·	·	·
A/Delhi/018/2009	·	·	·	·	·	·	·	·	·	V	·	·	·	·	·	·	·	·	·	·
A/Rajasthan/019/2010	·	·	·	·	·	·	·	·	·	V	·	·	·	V	·	·	·	·	·	·
A/Assam/020/2010	·	·	·	·	·	·	·	·	·	V	·	·	·	V	·	·	·	·	·	·
A/Jammu Kashmir/021/2010	·	·	·	·	·	·	K	·	·	V	·	·	·	L	·	·	·	·	·	·
A/Jammu Kashmir/022/2010	·	Q	·	·	·	·	K	·	·	V	·	·	·	V	·	·	·	·	·	·
A/Kerala/023/2010	·	·	·	·	·	·	·	·	·	V	·	·	·	V	·	·	·	·	·	·
A/Madhya Pradesh/024/2010	·	·	·	·	·	·	·	·	·	V	·	·	·	V	·	·	·	·	·	·
A/Chhattisgarh/025/2010	·	Q	·	·	·	·	·	·	·	V	·	·	·	·	·	·	·	·	·	·
A/Chhattisgarh/026/2010	·	Q	·	·	·	·	·	·	·	V	·	·	·	·	·	·	·	·	·	·
A/Goa/027/2010	·	Q	·	·	·	·	·	·	·	V	·	·	·	·	·	·	·	·	·	·
A/Haryana/028/2010	·	Q	·	·	·	·	·	·	·	V	·	·	·	·	·	·	·	·	·	·
A/Haryana/029/2010	·	Q	·	·	·	·	·	·	·	V	·	·	·	·	·	·	·	·	·	·
A/Punjab/030/2010	·	Q	·	·	·	·	·	·	·	V	·	·	·	·	·	·	·	·	·	·
A/Punjab/031/2010	·	Q	·	·	·	·	·	·	·	V	·	·	·	·	·	·	·	·	·	·
A/Uttar Pradesh/032/2010	·	Q	·	·	·	·	·	·	·	V	·	·	·	·	·	·	·	·	·	·
A/Uttarakhand/033/2010	·	Q	·	·	·	·	·	·	·	V	·	·	·	·	·	·	·	·	·	·
A/Goa/034/2010	·	Q	·	·	·	·	·	·	·	V	·	·	·	·	·	·	·	·	·	·
A/Goa/035/2010	·	Q	·	·	·	·	·	·	·	V	·	·	·	·	·	·	·	·	·	·
A/Delhi/036/2010	·	Q	·	·	·	·	·	·	·	V	·	·	·	·	·	·	·	·	·	·
A/Delhi/037/2010	·	Q	·	·	·	·	·	·	·	V	·	·	·	·	·	·	·	·	·	·
A/Delhi/038/2010	·	Q	R	·	·	·	·	·	·	·	·	·	·	·	·	P	·	·	·	D
A/Delhi/039/2010	·	Q	·	T	·	·	·	·	·	V	·	·	·	·	·	·	·	·	·	·
A/Punjab/040/2011	·	·	·	·	·	·	·	·	·	V	I	·	·	·	R	·	·	·	S	·
A/Punjab/041/2011	·	·	·	·	·	·	·	·	·	V	I	·	·	·	·	·	·	·	S	·
A/Punjab/042/2011	·	Q	·	T	·	·	·	·	·	V	·	·	·	·	·	·	·	·	·	·
A/Punjab/043/2011	·	Q	·	·	·	·	·	·	·	V	A	·	·	·	·	·	·	·	·	·
A/Jammu Kashmir/044/2011	·	Q	·	T	·	·	·	·	·	V	·	·	·	·	·	·	·	·	·	·

TABLE 3: Continued.

	53	55	62	73	90	94	96	108	111	123	129	131	143	145	147	151	172	202	205	209
A/Jammu Kashmir/045/2011	·	·	·	·	·	·	·	·	·	V	I	·	·	·	·	·	·	·	S	·
A/Goa/046/2011	·	Q	·	T	·	·	·	·	·	V	·	·	·	·	·	·	·	·	·	·
A/Goa/047/2011	·	Q	·	·	·	·	·	·	·	V	I	·	·	·	·	·	·	·	S	·
A/Goa/048/2011	·	Q	R	·	·	·	·	·	·	V	L	·	·	·	·	·	·	·	S	·
A/Haryana/049/2011	·	·	·	·	I	·	·	·	·	V	·	·	·	V	·	·	·	·	S	·
A/Delhi/050/2011	·	·	·	·	I	·	·	·	·	V	·	·	·	·	·	·	·	·	S	·
A/Delhi/051/2011	·	·	·	·	I	·	·	·	T	V	·	·	·	·	·	·	·	·	S	·
A/Delhi/052/2011	·	·	·	·	I	·	·	·	T	V	·	·	·	·	·	·	·	·	S	·
A/Delhi/053/2011	·	·	·	·	I	·	·	·	T	V	·	·	·	·	·	·	·	·	S	·
A/Delhi/054/2011	·	·	·	·	I	·	·	·	T	V	L	·	·	·	·	·	·	·	S	·
A/Delhi/055/2011	·	·	·	·	I	·	·	·	·	V	·	·	·	·	·	·	·	·	S	·
A/Delhi/056/2011	·	·	·	·	I	·	·	·	·	V	·	·	·	·	·	·	·	·	S	·
A/Delhi/057/2012	·	·	·	·	I	·	·	·	·	V	·	·	·	·	·	·	·	·	S	·
A/Delhi/058/2012	·	·	·	·	I	·	·	·	·	V	·	·	·	·	·	·	·	·	S	·
A/Delhi/059/2012	·	·	·	·	I	·	·	·	·	V	·	·	·	·	·	·	·	·	S	·
A/Delhi/060/2012	·	·	·	·	I	·	·	·	·	V	·	·	·	·	·	·	·	·	S	·
A/Delhi/061/2012	·	·	·	·	I	·	·	·	·	V	·	·	·	·	·	·	·	·	S	·
A/Delhi/062/2012	·	·	·	·	I	·	·	·	·	V	·	·	·	·	·	·	·	·	S	·
A/Delhi/063/2012	·	·	·	·	I	·	·	·	·	V	·	·	·	·	·	·	·	·	S	·
A/Delhi/064/2012	·	·	·	·	I	·	·	·	·	V	·	·	·	·	·	·	·	·	S	·
A/Delhi/065/2012	·	·	·	·	I	·	·	·	·	V	·	·	·	·	·	·	·	·	S	·
A/Delhi/066/2012	·	·	·	·	I	·	·	·	·	V	·	·	·	·	·	·	·	·	S	·
A/Delhi/067/2012	·	·	·	·	I	·	·	·	·	V	·	·	·	·	·	·	·	·	S	·
A/Delhi/068/2012	·	·	·	·	I	·	·	·	·	V	·	·	·	·	·	·	·	·	S	·
A/Delhi/069/2012	·	·	·	·	I	·	·	·	·	V	·	·	·	·	·	·	·	·	S	·
A/Uttarakhand/070/2012	·	·	·	·	I	·	·	·	·	V	·	·	·	·	·	·	·	·	S	·
A/Goa/071/2012	·	·	·	·	I	·	·	·	·	V	·	·	·	·	·	·	·	·	S	·
A/Goa/072/2012	·	·	·	·	I	·	·	·	·	V	·	·	·	·	·	·	·	·	S	·
A/Goa/073/2012	·	·	·	·	I	·	·	·	·	V	·	·	·	·	·	·	·	·	S	·
A/Chhattisgarh/074/2012	·	·	·	·	I	·	·	·	·	V	·	·	·	·	·	·	·	·	S	·
A/Chhattisgarh/075/2012	·	·	·	·	I	·	·	·	·	V	·	·	·	·	·	·	·	·	S	·
A/Haryana/076/2012	·	·	·	·	I	·	·	·	·	V	·	·	·	·	·	·	·	·	S	·
A/Haryana/077/2012	·	·	·	·	I	·	·	·	·	V	·	Q	·	·	·	·	·	·	S	·
A/Haryana/078/2012	·	·	·	·	I	·	·	·	·	V	·	Q	·	·	·	·	·	·	S	·
A/Haryana/079/2013	·	Q	·	·	I	·	·	·	·	V	·	·	·	·	·	·	·	·	S	·
A/Haryana/080/2013	·	K	·	T	I	·	·	·	·	V	·	E	·	·	·	·	·	·	S	·
A/Haryana/081/2013	·	·	·	·	I	·	·	·	·	V	·	E	·	·	·	·	·	·	S	·
A/Uttarakhand/082/2013	·	·	·	·	I	·	·	·	·	V	·	E	·	·	·	·	·	·	S	·
A/Uttarakhand/083/2013	·	·	·	·	I	·	·	·	·	V	·	E	·	·	·	·	·	·	S	·
A/Jammu Kashmir/084/2013	·	·	·	·	I	·	·	·	·	V	·	·	·	·	·	·	·	P	S	·
A/Jammu Kashmir/085/2013	·	·	·	·	I	·	·	·	·	V	·	·	·	·	·	·	·	·	S	·
A/Delhi/086/2013	·	·	·	·	I	·	·	·	·	V	·	·	·	·	·	·	·	·	S	·
A/Delhi/087/2013	·	·	·	·	I	·	·	·	·	V	·	·	·	·	·	·	·	·	S	·
A/Delhi/088/2013	·	·	·	·	I	·	·	·	·	V	·	·	·	·	·	·	·	·	S	·
A/Delhi/089/2013	·	·	·	·	I	·	·	·	·	V	·	·	·	·	·	·	·	·	S	·
A/Uttar Pradesh/090/2013	·	·	·	·	I	·	·	·	·	V	·	·	·	·	·	·	·	·	S	·
A/Uttar Pradesh/091/2013	·	·	·	·	I	·	·	·	·	V	·	·	·	·	·	·	·	·	S	·

TABLE 4: Selection pressure analysis of NS1 protein and NEP of H1N1 pdm (09) virus using SLAC, FEL, REL, MEME, and FUBAR methods at (http://www.datamonkey.org).

Protein	Codon*	SLAC dN-dS	SLAC P value	FEL dN-dS	FEL P value	REL dN-dS	REL Bayes factor	MEME ω+	MEME P value	FUBAR dN-dS	FUBAR post. pr.
	55	17.614	0.131	73.794	0.056	3.898	3779.510	>100	0.069	3.151	0.989
	108	**5.330**	**0.459**	**19.133**	**0.208**	**3.149**	**11.971**	**>100**	**0.251**	**0.286**	**0.769**
	123	**5.399**	**0.447**	**24.424**	**0.142**	**3.558**	**22.722**	**>100**	**0.219**	**0.542**	**0.813**
NS1	129	13.333	0.142	62.522	0.043	3.978	19238.300	>100	0.074	2.993	0.992
	145	**8.103**	**0.630**	**37.071**	**0.318**	**3.776**	**514.262**	**>100**	**0.328**	**1.285**	**0.924**
	147	**7.491**	**0.232**	**34.083**	**0.084**	**3.828**	**45.133**	**>100**	**0.101**	**1.081**	**0.880**
	205	**8.870**	**0.438**	**33.193**	**0.162**	**3.928**	**2940.990**	**>100**	**0.171**	**1.042**	**0.922**
NEP	49	6.729	0.457	134.502	0.158	0.260	2.013	>100	0.183	1.846	0.926

Significance value (SLAC P value = 0.5, FEL P value = 0.25, REL Bayes factor = 50, MEME P value = 0.1, FUBAR posterior probability = 0.9).
*The sites found under positive selection by at least two methods are shown.
Sites present in NS1 host factor interaction domains are highlighted with bold font.

and is conserved in comparison to NS1. Seven synonymous and 4 nonsynonymous mutations were observed in NEP gene from year 2009 to 2013. Amino acid changes noticed in NEP were M14I, N29S, T48A, and S60N among which T48A is the most common change found in around 50% of samples. Other mutations seen were M14I (2 samples) from year 2009, N29S (2 samples) and S60N (7 samples) from year 2013.

Study samples of 2009 and 2010 were similar to A/California/07/2009. However, in two samples from year 2009 single amino acid change replacing methionine at position 14 with isoleucine (M14I) was noticed in nuclear export signal of NEP protein. T48A mutation first appeared in 2011 and persisted thereon. Samples from 2012 were similar to 2011 with no change. Samples from 2013 displayed two mutations: N29S in two samples and S60N in 50% of the samples.

3.5. Selection Pressure Analysis. Selection pressure analysis of NS gene of influenza A H1N1 pdm virus strain revealed 8 positively selected sites. Integrated analysis was performed for differential selection pressure acting on NS1 (219 codons) and NEP (121 codons) proteins (shown in Table 4). Out of seven NS1 sites, one was located in RBD and six in ED. Analysis of NEP protein gene revealed single position 49 to be under positive selection. A specific selection pressure analysis for Indian isolates (n = 44) for NS1 and (n = 21) for NEP gene revealed 3 sites in NS1 and 1 site in NEP gene under positive selection.

3.6. Analysis of Available NS1 Sequences from India. Full length NS1 gene sequences (120 sequences) available from India till 31 March 2014 were retrieved from the Influenza Research Database (IRD) at http://www.fludb.org/. Sequences were phylogenetically analyzed by maximum likelihood method which grouped all sequences into 4 clusters. These clusters were represented by a single representative strain of each cluster from hereon: A/KOL/507/2007 (KOL 507), A/KOL/596/2007 (KOL 596), A/KOL/989/2007 (KOL 989), and A/Pune/NIV 6196/2009 (NIV 6196) (shown in Figure 1). Among these,

KOL 507 and KOL 596 clusters had sequences from the years 2007 and 2009, while, in KOL 989 cluster, sequences from year 2007, 2008, and 2009 were seen. NIV 6196 cluster was noted to have sequences from year 2009 to 2013. KOL 596 like strains constituted the smallest group with 6 samples, while NIV 6196 like strains formed the largest group with 61 samples.

NS1 protein encoded by clusters KOL 507, KOL 596, and KOL 989 was of 230 amino acid residues in length, whereas NIV 6196 cluster encoding NS1 protein was of 219 amino acid residues. Due to difference in length of NS1 protein, 12 sites were only seen in clusters encoding 230 amino acid residues' NS1 protein. Terminal amino sequence of avian influenza A (H5N1) virus NS1 protein is reported to be associated with virulence and pathogenicity (30). NS1 protein encoded by clusters has different C-terminal amino acid sequence; KOL 507 and KOL 596 have RSEV, KOL 989 had RSKV, and NIV 6196 had PEQK.

Multiple sequence alignment of 120 sequences from 2007 to 2013 strain of all clusters (from India) showed differences in amino acid sequence at 100 sites between the clusters when compared with reference to KOL 507 cluster of which some sites were cluster specific (shown in Table 5), while others were common between clusters (Shown in Table 6).

KOL 507 and KOL 596 clusters have no year specific distribution of mutations or signature sequence within the cluster. KOL 989 cluster have one such pattern in sequences from year 2009, which has arginine at position 135 and glycine at position 139 in place of serine and aspartic acid. The NIV 6196 cluster has isoleucine at position 90 and serine at position 205 in place of leucine and arginine in the majority of the samples from 2011 to 2013.

Mean distance in NS1 protein sequence between clusters with reference to KOL 507 cluster was approximately 12% for KOL 596 cluster, 17% for KOL 989 cluster, and 25% for NIV 6196 cluster. All clusters have maximum sequence dissimilarity of 1% between the sequences within the cluster except NIV 6196 which has the maximum dissimilarity of 2% within the cluster. It has been observed that all study samples (2009–2013) belonged to NIV 6196 cluster and no circulation

TABLE 5: Cluster specific amino acid changes were seen in all the sequences of cluster and were absent from sequences of all other clusters.

Clusters	Sites	Cluster specific amino acid changes
KOL 507 (2007, 2009)	13	D26, N53, C59, L85, V95, *N143* (99%), T145, N171, T209, F214, T216, T217, T226
KOL 596 (2007, 2009)	7	P3, R59, A60, S103, I106, A171, I226
KOL 989 (2007–2009)	19	V23, R41, A56, H59, K67, V82, T84, I95, L98, N101, E112, *M129* (80%), I144, V145, I171, K196, R224, A226, K229
NIV 6196 (2009–2013)	28	M6, F22, *N25* (99%), G26, L59, W67, S74, T76, R78, I81, T86, S91, *R108* (99%), *I111* (90%), I112, L119, *V123* (85%), V129, N139, Y171, I198, *N205* (77%), C206, D207, S213, P215, E217, *220

Some cluster specific changes were seen in variable percentage of sequences in cluster and those absent from all other clusters are highlighted with italic font.

TABLE 6: Amino acid positions common between clusters.

Clusters	KOL 596	KOL 989	NIV 6196
KOL 507	18, 48, 67, 112, 125, 129, 197, 224, 229	21, 166, 178, 211	44, 101, 117
KOL 596	—	26, 44, 117, 217, 221	21, 84, 95, 145, 166, 178, 211
KOL 989	26, 44, 117, 217, 221	—	18, 48, 125, 197

Amino acid position seen in 100% to 65% of samples between a pair of clusters.

of strain similar to KOL 507, KOL 596, and KOL 989 like strains has been seen in the last four years.

Mutations in two functional domains of NS1 protein were observed between various clusters which affect their function. RNA binding domain of NS1 protein has mutations at positions 41, 44, and 67 which are involved in binding dsRNA. Mutation at positions 41 and 44 were noticed in KOL 596 and KOL 989 clusters, whereas change at position 67 was seen in clusters KOL 989 and NIV 6196. Effector domain of NS1 protein has mutation at 12 positions: 91, 95, 98, 101, 117, 119, 123, 125, 135, 144, 145, and 145 which may affect its interaction with host protein. Mutations at positions 95, 143, and 145 were seen in all the clusters. Some changes were cluster specific: 98, 135, and 144 in KOL 989 cluster, 91, 119, and 123 in NIV 6196 cluster while others were common between two clusters 101 and 117 in KOL 596 and KOL 989 clusters, 125 in KOL 989 and NIV 6196 clusters.

4. Discussion

NS1 protein is responsible for regulation of antiviral immune response in the host cells and a number of NS1 molecular markers are reported to be associated with increased virulence and pathogenicity like R38, F103, and M106 [7, 24]. NS1 protein is functionally divided into two domains: RNA binding domain (RBD) and effector domain (ED). RBD is mainly involved in sequestering of dsRNA from OAS and RIG1. In the present study, analysis of sequencing data from NS1 gene showed relatively conserved RBD in comparison to ED (shown in Table 2). Only five amino acid changes were seen in RBD, out of which E55Q was the most common change in comparison to other mutations which were rare and occurred in two or three samples only. None of the changes occurred in positions reported to be involved in RNA binding [7].

Effector domain is involved in interactions with the host factors, associated with cell signaling and immune response.

I123V mutation was seen in ED of almost all study samples (shown in Tables 2 and 3) and was categorized into New York (G1 type) strains [25]. Apart from this mutation, L90I and N205S mutations were found to occur over three years in a large number of samples. Other mutations which were detected in 20% or more samples were R108K (year 2009), I145V (year 2010), V129I (year 2011), and K131E (year 2013). The rest of the mutations were seen to occur only in one or two samples in all years.

Glutamate at position 96 is functionally important for binding of NS1 to CPSF30 and necessary for interaction with TRIM25, a ubiquitin ligase which mediates the ubiquitination of the RIG-1 (a viral RNA sensor) in order to facilitate IFN production. It has been reported that E96A mutants were ineffective in blocking TRIM25 mediated IFN response [8, 26]. E96K substitution was noted in 2 samples from 2010, while the rest of 89 samples have E96, which shows that the majority of viruses in circulation with E96 are competent enough to inhibit TRIM25 mediated immune response and replicate efficiently in host cell.

It has been reported that interaction of NS1 residues 123–127 with PKR results in inhibition of eIF2α phosphorylation and viral protein synthesis, indicating that NS1-PKR binding is necessary and sufficient to block PKR activation in influenza A virus-infected cells [27]. In the present study eighty-eight samples were seen to have I123V mutation in this region. I123V mutation may therefore affect the inactivation of PKR by NS1 protein.

It has also been reported that NS1 protein with R108, E125, and G189 is unable to block the host gene expression resulting in inefficient replication of virus. This inhibitory effect could be restored by replacing above residue with residues corresponding to the human H1N1 virus consensus sequence [28]. One of these mutations R108K was seen in 4 samples from year 2009.

It has been reported that the influenza A (H5N1) NS1 protein interacts with eukaryotic translation initiation factor

4GI (eIF4GI) via eIF4GI binding domain (residues 81–113) resulting in the preferential translation of the viral mRNA in comparison to host mRNA [29]. Therefore, mutation in this domain may result in impaired ability of virus to inhibit interferon production which may result in inefficient virus replication. L90I and T94A mutations may, therefore, affect interferon response and virus replication. Similarly, in ferrets it has been reported that human (H5N1) virus with arginine (N) at position 205 of NS1 protein enhances the type I IFN antagonistic property of the host cell leading to high virulence in ferrets [30]. In the present study samples, we have seen N205S mutation in all samples from 2011 onwards.

In this study the nuclear export signal of NEP displays M14I mutation in 2 samples from 2009, while the C-terminal domain of NEP was reported to interact with the nuclear localization signal of the viral matrix protein M1 [31] which has shown two mutations, T48A in almost all samples from 2011–2013, S60N in 50% of samples from 2013. These mutations may affect nuclear transport and release of virus from cell.

Selection pressure analysis of NS gene of influenza A H1N1 pdm virus strain revealed 8 positively selected sites (shown in Table 4). Positions 108, 123, 145, 147, and 205 were noted to be situated in NS1 protein host factor interaction domains. Analysis of NEP protein gene revealed single position 49 to be under positive selection. A specific selection pressure analysis for Indian isolates revealed 3 sites in NS1 and 1 site in NEP gene to be under positive selection. Positions 55, 129, and 145 in NS1 gene were found to be common between India specific isolates and reference strain isolates. This showed that positive selection on NS1 gene was stronger than that on NEP, of which a large number of sites were located in influenza host factor interaction domains, which are reported to be associated with virulence and pathogenicity of influenza virus [3, 26–29].

Phylogenetic analysis of study samples on the basis of NS1 gene of influenza A (H1N1) virus broadly grouped all sequences into two major branches (shown in Figure 2). One (group one) is with samples from year 2009 to 2011 and the other (group two) is with samples from 2011 to 2013. In comparison to reference strain samples for year 2009 were most similar with mean distance of 0.9% followed by mean distance of 1.1% for 2011 samples, 1.7% for 2011 samples, 1.4% for 2012, and 1.6% for 2013 samples. This showed that samples from 2011 were most dissimilar to reference strain, followed by 2013 samples. Phylogenetic analysis showed that samples from year 2009 were almost identical to reference strain (A/California/07/2009). The majority of 2010 samples showed homology with A/Singapore/GP4138/2010 and A/Pennsylvania/17/2010, while 6 strains showed homology with A/Singapore/GP2892/2010 strain. Samples from each year formed a discrete branch on the tree except for samples from year 2011, which were seen in both groups. 2011 samples in group one were seen in three separate branches: one branch with four samples showing homology to A/India/P121778/2012 strain and another with three samples having homology with A/England/118/2010 strain and single samples with homology to A/Singapore/GP4138/2010

strain. However, 2011 samples in group two formed two separate branches: one at base of group two containing two samples and the other with four samples, which showed homology to A/Boston/DOA2-099/2012 strain. Samples from year 2012 showed homology to A/India/Nsk12388/2012 strain, while 2013 samples had homology with A/Helsinki/405/2013 and A/New Jersey/NHRC403730/2013 strains. Mutations I123V and N205S in NS1 protein observed in the present study have also been observed in a large number of sequences from Europe, America, Africa, and Asia. While L90I (NS1 protein) was seen in limited number of sample from Europe, America and Africa. T48A (NEP protein) was seen only in few samples from Europe, Asia, Africa, and America. An earlier study on H1N1 pdm (09) sequences from India involving 13 samples has also reported the mutation reported in this study [32]. However, in comparison to that study, the present study has used a larger number of samples and found additional mutation in NS1 gene.

Phylogenetic analyses of 120 full length NS1 sequences from India during the time period 2007–2013 (retrieved from the Influenza Research Database (IRD) at http://www.fludb.org/) were found to be grouped into four clusters as shown earlier in Figure 1. NS1 protein encoded by KOL 507, KOL 596, and KOL 989 cluster was seen to be of 230 amino acid residues in length, whereas NIV 6196 like strains were seen to encode 219 residue long NS1 protein. Our investigation reveals that influenza A (H1N1) is evolving and acquiring mutations, which could be noted, by observing the mean distance in NS1 protein sequence between the clusters, approximately 12% for KOL 596 cluster, 17% for KOL 989 cluster, and 25% for NIV 6196 cluster. NS1 protein of none of the clusters was seen to have ESEV, EPEV, and KSEV as their terminal amino acid sequence, which are reported to be associated with increased virulence in influenza A (H5N1) virus. All study samples belonged to NIV 6196 cluster and had loss of 11 amino acids at c-terminal end of NS1 protein. Analysis of NS1 protein shows that the four clusters were derived from three major reassortment events, with KOL 989 cluster derived from seasonal H3N2 virus, KOL 507 and KOL 596 clusters from prepandemic seasonal H1N1, and NIV 6196 cluster from H1N1 pdm (09) lineage. This has resulted in large mean distance between cluster and loss of terminal amino acid residue. High values for mean distance and loss of residue between NIV 619 cluster and KOL 507 could be explained by introduction of pandemic strain in year 2009 in human population. Circulation of three different clusters in period of four years from 2007 to 2009, with high mean distance between them, shows that influenza A virus has evolved rapidly (by antigenic shift and drift) in the past and it could do so in the future, which highlights the need of continuing surveillance and monitoring of influenza virus infection across the nation and worldwide.

5. Conclusions

Sequence analysis shows that NS1 protein is mutating more rapidly than NEP and that within NS1 protein RBD is more conserved than ED. The prominent change seen in RBD was

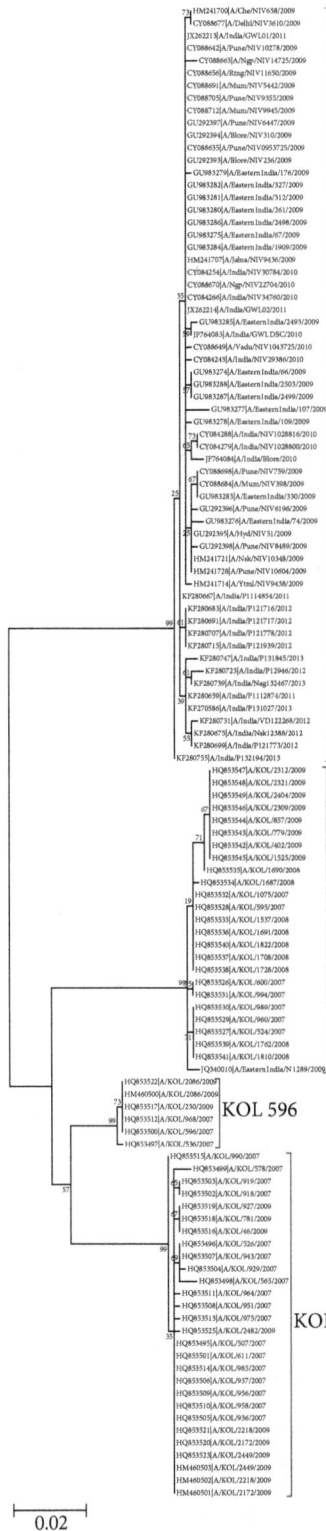

FIGURE 1: Phylogenetic tree of all sequences (120 sequences from India, 2007–2013) of NS1 gene of influenza A (H1N1) was constructed using maximum likelihood method in MEGA 6.0 software. Bootstrap values at 500 replications were shown at branches. Each node showing accession number with strain name. Phylogenetic tree is depicting 4 major clusters or groups represented by their cluster name given in front of each cluster.

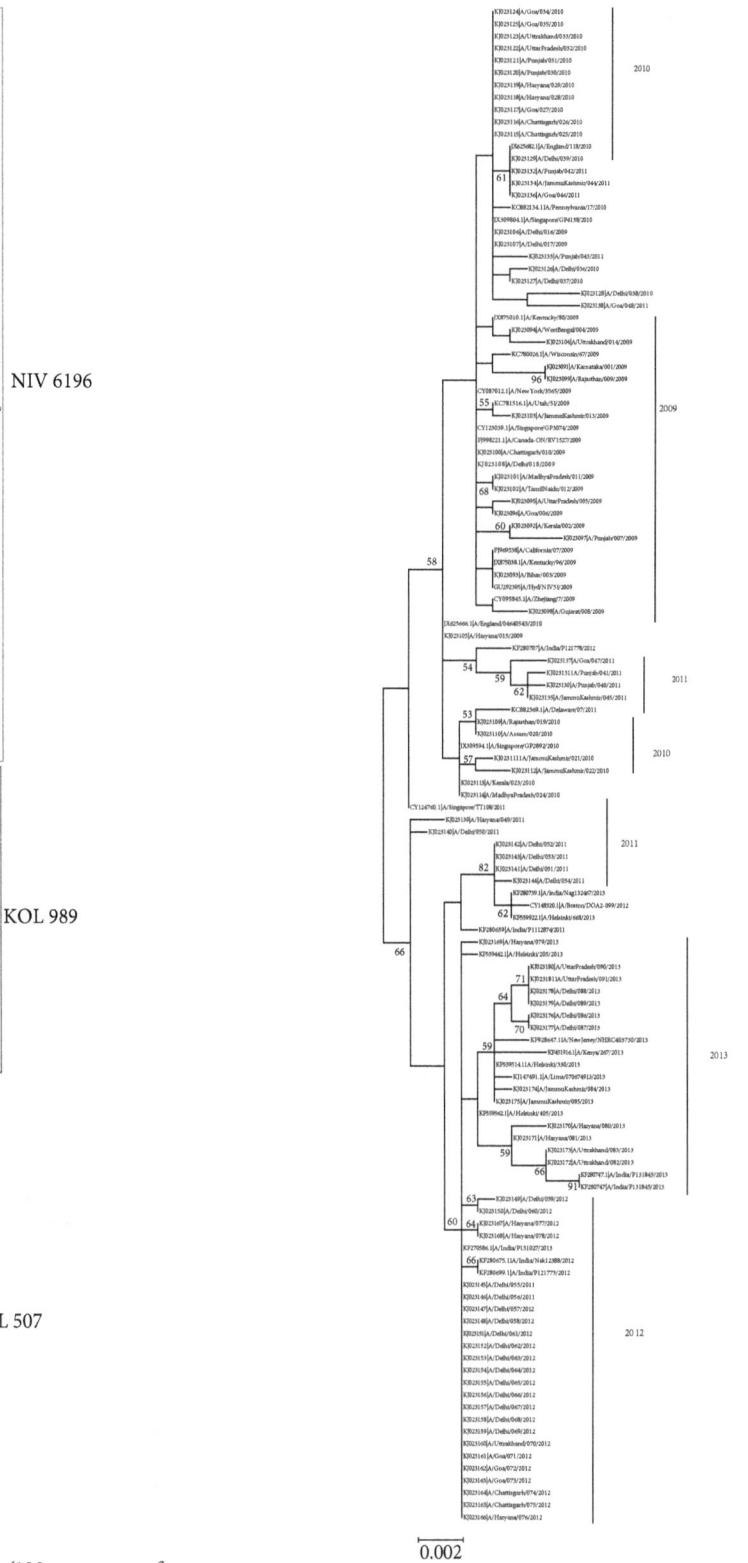

FIGURE 2: Phylogenetic tree of influenza A (H1N1) virus based on NS1 gene generated by the maximum likelihood method. Bootstrap support values (based on 500 replications) above 50% are shown at the branch node. Each branch is denoted by accession number and strain name.

E55Q in samples from 2009 to 2011. Within RBD, no change has been seen in sites reported to be involved in RNA binding, inhibiting IFN responses, and which are therefore believed to be efficient in sequestering dsRNA and inhibiting antiviral responses. Study of effector domain has displayed a number of changes in sites/domains reported to be associated with host factor interaction. Three major mutations identified in the ED were I123V which was seen in almost all samples from 2009 to 2013, while L90I and N205S mutations were found for the first time in samples from 2011 and then persisted onwards. It has been observed that the majority of the study samples were of New York (G1 type) with valine at position 123. Most of the mutations in the sequences observed in this study have also been reported from Asia, Europe, and America. Available NS1 sequences from India show that NS1 is evolving and acquiring mutations with the loss of terminal amino acid residues. On the basis of sequence similarity and available literature, we can say that the present circulating NS1 protein is an effective interferon antagonist. Mutations like K62R, K131Q, L147R, and A202P were seen for the first time in NS1 protein and their effect is yet to be determined.

Database

Nucleotide sequence data reported are available in the GenBank databases under the accession numbers "NCBI: KJ023091 to KJ023181."

Acknowledgments

Sachin Kumar acknowledges financial support of the Council for Scientific and Industrial Research (CSIR), Delhi, India, during the course of study and is thankful to the staff of Division of Biotechnology and Microbiology for their support.

References

[1] S. Tong, X. Zhu, Y. Li et al., "New world bats harbor diverse influenza A viruses," *PLoS Pathogens*, vol. 9, no. 10, Article ID e1003657, 2013.

[2] B. N. Fields, D. M. Knipe, and P. M. Howley, *Fields Virology*, Wolters Kluwer, Lippincott Williams & Wilkins, Philadelphia, Pa, USA, 5th edition, 2007.

[3] B. G. Hale, R. E. Randall, J. Ortin, and D. Jackson, "The multifunctional NS1 protein of influenza A viruses," *Journal of General Virology*, vol. 89, part 10, pp. 2359–2376, 2008.

[4] Z. A. Bornholdt and B. V. V. Prasad, "X-ray structure of influenza virus NS1 effector domain," *Nature Structural and Molecular Biology*, vol. 13, no. 6, pp. 559–560, 2006.

[5] Z. A. Bornholdt and B. V. V. Prasad, "X-ray structure of NS1 from a highly pathogenic H5N1 influenza virus," *Nature*, vol. 456, no. 7224, pp. 985–988, 2008.

[6] D. Lin, J. Lan, and Z. Zhang, "Structure and function of the NS1 protein of influenza A virus," *Acta Biochimica et Biophysica Sinica*, vol. 39, no. 3, pp. 155–162, 2007.

[7] A. Cheng, S. M. Wong, and Y. A. Yuan, "Structural basis for dsRNA recognition by NS1 protein of influenza A virus," *Cell Research*, vol. 19, no. 2, pp. 187–195, 2009.

[8] M. U. Gack, R. A. Albrecht, T. Urano et al., "Influenza A virus NS1 targets the ubiquitin ligase TRIM25 to evade recognition by the host viral RNA sensor RIG-I," *Cell Host and Microbe*, vol. 5, no. 5, pp. 439–449, 2009.

[9] K. Das, L.-C. Ma, R. Xiao et al., "Structural basis for suppression of a host antiviral response by influenza A virus," *Proceedings of the National Academy of Sciences of the United States of America*, vol. 105, no. 35, pp. 13093–13098, 2008.

[10] S. K. Dankar, S. Wang, J. Ping et al., "Influenza A virus NS1 gene mutations F103L and M106I increase replication and virulence," *Virology Journal*, vol. 8, article 13, 2011.

[11] A. Fernandez-Sesma, S. Marukian, B. J. Ebersole et al., "Influenza virus evades innate and adaptive immunity via the NS1 protein," *Journal of Virology*, vol. 80, no. 13, pp. 6295–6304, 2006.

[12] H. Akarsu, W. P. Burmeister, C. Petosa et al., "Crystal structure of the M1 protein-binding domain of the influenza A virus nuclear export protein (NEP/NS2)," *The EMBO Journal*, vol. 22, no. 18, pp. 4646–4655, 2003.

[13] K. Iwatsuki-Horimoto, T. Horimoto, Y. Fujii, and Y. Kawaoka, "Generation of influenza A virus NS2 (NEP) mutants with an altered nuclear export signal sequence," *Journal of Virology*, vol. 78, no. 18, pp. 10149–10155, 2004.

[14] WHO, "CDC protocol of realtime RTPCR for influenza A (H1N1)," WHO Guidance documents on pandemic (H1N1) October 2009, http://www.who.int/csr/resources/publications/swineflu/realtimeptpcr/en/.

[15] WHO, *WHO Information for Laboratory Diagnosis of Pandemic (H1N1) 2009 Virus in Humans—Revised*, WHO Guidance Documents on Pandemic (H1N1), 2009, http://www.who.int/csr/resources/publications/swineflu/diagnostic_recommendations/en/.

[16] E. Hoffmann, J. Stech, Y. Guan, R. G. Webster, and D. R. Perez, "Universal primer set for the full-length amplification of all influenza A viruses," *Archives of Virology*, vol. 146, no. 12, pp. 2275–2289, 2001.

[17] WHO, "WHO Guidance documents on pandemic (H1N1)," 2009, http://www.who.int/csr/resources/publications/swineflu/sequencing_primers/en/.

[18] S. F. Altschul, T. L. Madden, A. A. Schäffer et al., "Gapped BLAST and PSI-BLAST: a new generation of protein database search programs," *Nucleic Acids Research*, vol. 25, no. 17, pp. 3389–3402, 1997.

[19] R. B. Squires, J. Noronha, V. Hunt et al., "Influenza Research Database: an integrated bioinformatics resource for influenza research and surveillance," *Influenza and Other Respiratory Viruses*, vol. 6, no. 6, pp. 404–416, 2012.

[20] Y. Bao, P. Bolotov, D. Dernovoy et al., "The influenza virus resource at the National Center for Biotechnology Information," *Journal of Virology*, vol. 82, no. 2, pp. 596–601, 2008.

[21] K. Tamura, G. Stecher, D. Peterson et al., "MEGA6: molecular evolutionary genetics analysis version 6.0," *Molecular Biology and Evolution*, vol. 30, no. 12, pp. 2725–2729, 2013.

[22] B. E. Pickett, M. Liub, E. L. Sadat et al., "Metadata-driven comparative analysis tool for sequences (meta-CATS): an automated process for identifying significant sequence variations

that correlate with virus attributes," *Virology*, vol. 447, no. 1-2, pp. 45–51, 2013.

[23] S. L. Kosakovsky Pond and S. D. W. Frost, "Datamonkey: rapid detection of selective pressure on individual sites of codon alignments," *Bioinformatics*, vol. 21, no. 10, pp. 2531–2533, 2005.

[24] S. K. Dankar, E. Miranda, N. E. Forbes et al., "Influenza A/Hong Kong/156/1997(H5N1) virus NS1 gene mutations F103L and M106I both increase IFN antagonism, virulence and cytoplasmic localization but differ in binding to RIG-I and CPSF30," *Virology Journal*, vol. 10, article 243, 2013.

[25] C. Wang, Y. Zhang, B. Wu et al., "Evolutionary characterization of the pandemic H1N1/2009 influenza virus in humans based on non-structural genes," *PLoS ONE*, vol. 8, no. 2, Article ID e56201, 2013.

[26] B. G. Hale, P. S. Kerry, D. Jackson et al., "Structural insights into phosphoinositide 3-kinase activation by the influenza A virus NS1 protein," *Proceedings of the National Academy of Sciences of the United States of America*, vol. 107, no. 5, pp. 1954–1959, 2010.

[27] J. Y. Min, S. Li, G. C. Sen, and R. M. Krug, "A site on the influenza A virus NS1 protein mediates both inhibition of PKR activation and temporal regulation of viral RNA synthesis," *Virology*, vol. 363, no. 1, pp. 236–243, 2007.

[28] B. G. Hale, J. Steel, R. A. Medina et al., "Inefficient control of host gene expression by the 2009 pandemic H1N1 influenza A virus NS1 protein," *Journal of Virology*, vol. 84, no. 14, pp. 6909–6922, 2010.

[29] H. Zhou, J. Zhu, J. Tu et al., "Effect on virulence and pathogenicity of H5N1 influenza A virus through truncations of NS1 eIF4GI binding domain," *Journal of Infectious Diseases*, vol. 202, no. 9, pp. 1338–1346, 2010.

[30] H. Imai, K. Shinya, R. Takano et al., "The HA and NS genes of human H5N1 influenza a virus contribute to high virulence in ferrets," *PLoS Pathogens*, vol. 6, no. 9, Article ID e01106, 2010.

[31] D. Paterson and E. Fodor, "Emerging roles for the influenza A virus nuclear export protein (NEP)," *PLoS Pathogens*, vol. 8, no. 12, Article ID e1003019, 2012.

[32] M. Dakhave, A. Khirwale, K. Patil, A. Kadam, and V. Potdar, "Whole-genome sequence analysis of postpandemic Influenza A (H1N1) pdm09 virus isolates from India," *Genome Announcements*, vol. 1, article 5, 2013.

6

Investigation of Stilbenoids as Potential Therapeutic Agents for Rotavirus Gastroenteritis

Judith M. Ball,[1] **Fabricio Medina-Bolivar,**[2] **Katelyn Defrates,**[3] **Emily Hambleton,**[3] **Megan E. Hurlburt,**[3] **Lingling Fang,**[2] **Tianhong Yang,**[2] **Luis Nopo-Olazabal,**[2] **Richard L. Atwill,**[2] **Pooja Ghai,**[2] **and Rebecca D. Parr**[1,2,3]

[1]*Department of Pathobiology, Texas A&M University, College Station, TX 77843, USA*
[2]*Department of Biological Sciences & Arkansas Biosciences Institute, Arkansas State University, Jonesboro, AR 72401, USA*
[3]*Department of Biology, Stephen F. Austin State University, Nacogdoches, TX 75962, USA*

Correspondence should be addressed to Rebecca D. Parr; parrrl@sfasu.edu

Academic Editor: Finn S. Pedersen

Rotavirus (RV) infections cause severe diarrhea in infants and young children worldwide. Vaccines are available but cost prohibitive for many countries and only reduce severe symptoms. Vaccinated infants continue to shed infectious particles, and studies show decreased efficacy of the RV vaccines in tropical and subtropical countries where they are needed most. Continuing surveillance for new RV strains, assessment of vaccine efficacy, and development of cost effective antiviral drugs remain an important aspect of RV studies. This study was to determine the efficacy of antioxidant and anti-inflammatory stilbenoids to inhibit RV replication. Peanut (*A. hypogaea*) hairy root cultures were induced to produce stilbenoids, which were purified by high performance countercurrent chromatography (HPCCC) and analyzed by HPLC. HT29.f8 cells were infected with RV in the presence stilbenoids. Cell viability counts showed no cytotoxic effects on HT29.f8 cells. Viral infectivity titers were calculated and comparatively assessed to determine the effects of stilbenoid treatments. Two stilbenoids, trans-arachidin-1 and trans-arachidin-3, show a significant decrease in RV infectivity titers. Western blot analyses performed on the infected cell lysates complemented the infectivity titrations and indicated a significant decrease in viral replication. These studies show the therapeutic potential of the stilbenoids against RV replication.

1. Introduction

The mechanisms of RV-induced diarrhea are multifactorial and include both secretory and malabsorptive diarrhea components. Despite much effort, we do not have a complete understanding of RV pathophysiology [1]. Vaccine strategies against RV-associated diarrhea aim at stimulating the immune system using either attenuated live RV or RV proteins [2]. There are two licensed RV vaccines in the United States, RotaTeq, produced by Merck, and Rotarix, produced by GlaxoSmithKline. Both are effective in preventing severe diarrhea in vaccinated children [3, 4]. Recently, a new RV vaccine, Rotavac, was developed using a strain of the RV that was isolated, manufactured, and tested in India by Hyderabad-based Bharat Biotech International, Ltd. [5]. These vaccines are designed to protect against common RV strains and

therefore are dependent on the genetic stability of the viruses. Reassortment events are common and may lead to new virulent RV strains that may not be averted by the current vaccines [6]. Likewise, the zoonotic nature of RV infections supports the argument to continue to survey for emerging RV strains arising from interspecies transmission with potential of vaccine failures [7]. The licensed RV vaccines mentioned above are less efficacious in countries of sub-Saharan Africa and SE Asia where they are needed most [8–11]. Furthermore, high costs, limited availability, and poor logistics for the distribution of the vaccines are challenging problems for the developing world [3]. Consequently, the development of cost effective, easily distributed, novel, and host-oriented antiviral paradigms is needed that affect a wide range of RV strains and reduce the disease burden of RV infections. Taking advantage of the antioxidant and anti-inflammatory properties of a natural product to treat RV infections meets the principles of a novel therapeutic strategy that has an antiviral effect.

Stilbenoids are phenolic compounds derived from the phenylpropanoid/acetate pathway. Among these compounds, trans-resveratrol (t-Res) is the most extensively studied stilbenoid which demonstrates strong antioxidant and chemopreventive properties [12]. Additionally, studies examining resveratrol and its derivatives have demonstrated antiviral properties. Resveratrol strongly inhibits the replication of influenza virus in MDCK cells and improves survival and decreased pulmonary viral infectivity titers in influenza virus-infected mice. Furthermore, resveratrol exhibited no toxic effects in vitro or in vivo [13]. However, another study tested the effects of $20\,\mu M$ and $40\,\mu M$ concentrations of resveratrol incubated for 24 and 48 hours postinfection with polyomavirus in 3T3 and HL60 cells. The results showed cytotoxicity in a time- and dose-dependent manner and inhibition of polyomavirus DNA synthesis. There was no cytotoxic effect to either cell line with 0.02% dimethyl sulfoxide (DMSO) alone [14]. Another study identified resveratrol derivatives with potent anti-HSV-1 and HSV-2 activity. Several trimeric and tetrameric derivatives showed antiherpetic activity at single-digit micromolar concentrations [15].

Stilbenoids are produced by a group of plants which includes grapes, peanuts, and some berries [12, 16, 17]. trans-Piceatannol (t-PA) is a hydroxylated analog of resveratrol found in grapes and in minor quantities in peanuts. trans-Arachidin-1 (t-A1) is a prenylated (3-methyl-1-butenyl) analog of piceatannol, whereas trans-arachidin-3 (t-A3) is a prenylated (3-methyl-1-butenyl) analog of resveratrol (Figures 1(a)–1(d)). Both t-A1 and t-A3 are produced in peanuts upon fungal challenge. These stilbenoids can be extracted from some plants but are not suitable for many applications in the food/pharmaceutical sectors due to the overall low concentration of stilbenoids in the plant extracts. To deliver a highly defined and stilbenoid-enriched product, hairy root cultures of peanut (A. hypogaea) have been established in a bioproduction system that produces increased levels of stilbenoids, including t-A1 and t-A3, upon treatment with elicitors [18, 19]. t-PA and t-Res are commercially available, but t-A1 and t-A3 are still in an experimental stage resulting in an opportunity to explore new antiviral biological activity.

This study assessed the therapeutic potential of four stilbenoids, t-Res, t-PA, t-A1, and t-A3 (Figures 1(a)–1(d)), to inhibit RV infections in culture using a cloned human intestinal cell line, HT29.f8 [20]. The hypothesis for this work is that stilbenoids will modulate the viral load of RV generated during an infection. Two sets of experiments were performed in which different concentrations of the stilbenoids and two time points postinfection were evaluated. To determine the effect of the stilbenoids on the amount of virus produced during an infection, viral infectivity titers were determined using the supernatants for each of the different treatments ($10\,\mu M$ and $20\,\mu M$ stilbenoids) collected at 12 and 24 hours postinfection (hpi). The viral infectivity titers produced in cells treated with the stilbenoids using a focus forming unit (FFU) assay were compared to the virus infectivity titers generated from RV infections alone and RV infections with 0.02% DMSO. The results were reported as infectious virus

particles/mL. Western blot analyses using the cell lysates generated from these experiments demonstrated the presence of the nonstructural RV protein (NSP4 nonstructural protein 4), a multifunctional viral protein that is essential for virus replication and the production of infectious virus particles [21].

2. Materials and Methods

2.1. Cells, Virus, and Reagents. The objective of the study was to test the effect(s) of the four stilbenoids on RV replication in HT29.F8 cells with variable concentrations of the stilbenoids and different collection times. The dose was based on a previous study that assessed the effects of $20\,\mu M$ resveratrol and 0.02% DMSO on two cell lines infected with polyomaviruses [13, 14]. Another study utilized different concentrations of DMSO and higher concentrations of resveratrol (50, 100, and $200\,\mu M$) on influenza A-infected cell line. Fifty and $100\,\mu M$ concentrations of resveratrol were not cytotoxic to the cells [13, 14]. Based on these results, we choose to test $10\,\mu M$ and $20\,\mu M$ concentrations of each stilbenoid solubilized in DMEM with 0.02% DMSO. A total of five experimental sets were performed per stilbenoid. In the first experimental set, cells were infected with SA114F RV at a multiplicity of infection (MOI) of 2 as previously reported [22]. In the second experimental set, 0.02% DMSO was added to the RV infection to prove that 0.02% DMSO used to solubilize the stilbenoids had no effect on cell viability or production of RV. In the third and fourth experimental sets, $10\,\mu M$ or $20\,\mu M$ concentrations of the stilbenoids, respectively, were solubilized in 0.02% DMSO in DMEM, added to the RV inoculum, and used to infect the cells. The fifth set was uninfected HT29.f8 cells treated with the stilbenoids. The sixth set was uninfected HT29.f8 cells treated with 0.02% DMSO, and the seventh set was uninfected HT29.f8 cells alone (Figure 2). Each experimental set was tested in four wells of a 24-well tissue culture (TC) plate. The media from the four wells were pooled and centrifuged and the supernatants were stored at $-80°C$ and used to determine viral infectivity titers. The cells were collected in PBS, frozen, thawed 3 times, and centrifuged. The supernatants were collected as the cell lysates and stored at $-80°C$ until used in western blot assays. Viral infectivity titers were performed in triplicate using two assays, the focus forming units (FFU) and the plaque forming units (PFU) assays. Equal amounts of the cell lysates were used in western blot assays to resolve the viral proteins and probe for RV NSP4.

2.2. Bioproduction and Purification of the Stilbenoids. Hairy roots of peanut cv. Hull (line 3) were cultured in at least twenty 250 mL flasks, each containing 50 mL of MSV medium as previously described [18, 23]. At day nine of the hairy root culture, the spent medium from each flask was removed and replaced with elicitation medium (fresh MSV medium with 9 g/L methyl-β-cyclodextrin (Cavasol W7 M)) and incubated in the dark at $28°C$ for an additional 72 h to induce synthesis and secretion of stilbenoids into the culture medium as recently described [19]. After the elicitation period, the culture medium was removed from

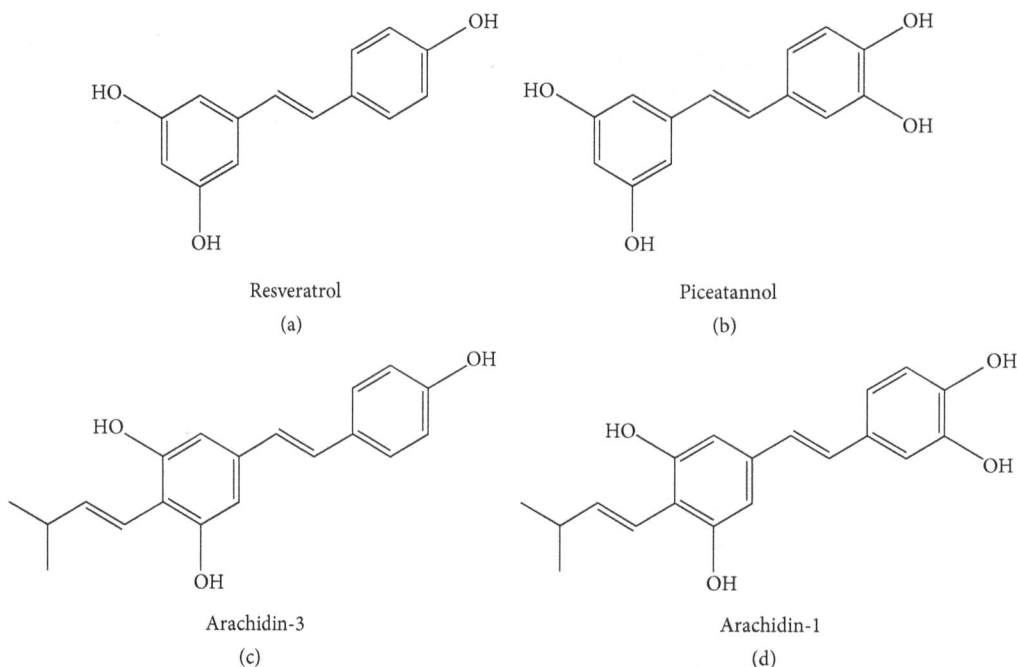

FIGURE 1: Chemical structures of the four stilbenoids tested. All compounds are shown in trans-isomers. (a) Resveratrol (t-Res). (b) Piceatannol (t-Pa). (c) Arachidin-3 (t-A3). (d) Arachidin-1 (t-A1).

each flask and combined. This pooled medium was mixed with an equal volume of ethyl acetate in a separatory funnel to extract the stilbenoids as described before [18]. The ethyl acetate phase was recovered and was dried in a Rotavapor (Buchi), and t-A1 and t-A3 were purified from the extract by HPCCC as follows. The dried ethyl acetate extract was resuspended in HPCCC solvent system (hexane : ethyl acetate : methanol : water (4 : 5 : 3 : 3)) and injected into a Spectrum (Dynamic Extractions) HPCCC system. The upper phase of the solvent system was used as the stationary phase and the chromatography was monitored at UV 340 nm. Fractions were collected every 30 s, dried in a SpeedVac, and analyzed by HPLC.

HPLC analyses were performed in a Dionex Summit system, equipped with a photodiode array (PDA) detector. The separation was performed on a SunFire C_{18}, 5 μm, 4.6 × 250 mm column (Waters) at 40°C at a flow rate of 1.0 mL/min. The mobile phase consisted of 2% formic acid in water (A) and methanol (B). The method started with 100% A for 1 min. Then a linear gradient was performed from 40% A and 60% B to 35% A and 65% B (1 to 20 min), followed by a linear gradient from 35% A and 65% B to 100% B (20 to 25 min). Then the column was washed with 100% A for 5 min (25 to 30 min). Elicited peanut seed-derived t-A1 and t-A3 were used as reference standards [23].

Purity of the fractions obtained after HPCCC was monitored by HPLC using UV absorbance at 280, 320, and 340 nm. Selected fractions also were checked for purity by mass spectrometry using an UltiMate 3000 ultrahigh performance liquid chromatography (UHPLC) system (Dionex, Thermo Scientific) coupled with a LTQ XL linear ion trap mass spectrometer (Thermo Scientific) as described in Marsh et

al. [24]. HPCCC fractions containing t-A1 and t-A3 with over 95% purity based on HPLC analysis (UV 340 nm) were combined, dried under a nitrogen stream, and used for viral assays. The dry mass of the purified stilbenoids was reconstituted in 0.02% DMSO with 1 μg/m trypsin (Worthington Biochemical, Lakewood, NJ) in DMEM medium. To compare the results between nonprenylated stilbenoids (t-Res and t-PA) and their prenylated analogs (t-A3 and t-A1, resp.) the synthetic/commercially available t-Res (Sigma-Aldrich) and t-PA (Alexis) were used in this study.

2.3. Cell Lines and Virus. MA104 cells were obtained from ATCC (Rockville, MD) and the HT29.F8 cells, a spontaneously polarizing cell line, were derived from the parent human adenocarcinoma (HT29) intestinal line [20]. The cell lines were confirmed to be free of mycoplasma contamination using the MycoFind mycoplasma PCR kit version 2.0 (Clongen Laboratories, LLC). RV SA11 clone 4F (P[1] and G[3] genotype) [25] was grown and titered in MA104 cells and stored at −80°C. Stilbenoid efficacy against RV was tested using HT29.f8 cells.

2.4. Viability Assay. The percentage of live/dead cells was calculated using the trypan blue dye exclusion assay as previously outlined [26]. Briefly, a cell suspension of ~10^6 cells/mL was diluted 1 : 1 with a 0.4% trypan blue solution and loaded onto a hemocytometer. The number of stained cells and total number of cells were counted, and the calculated percentage of unstained cells was reported as the percentage of viable cells. To determine if the 0.02% DMSO that was used to solubilize the hydrophobic stilbenoids adversely affected the

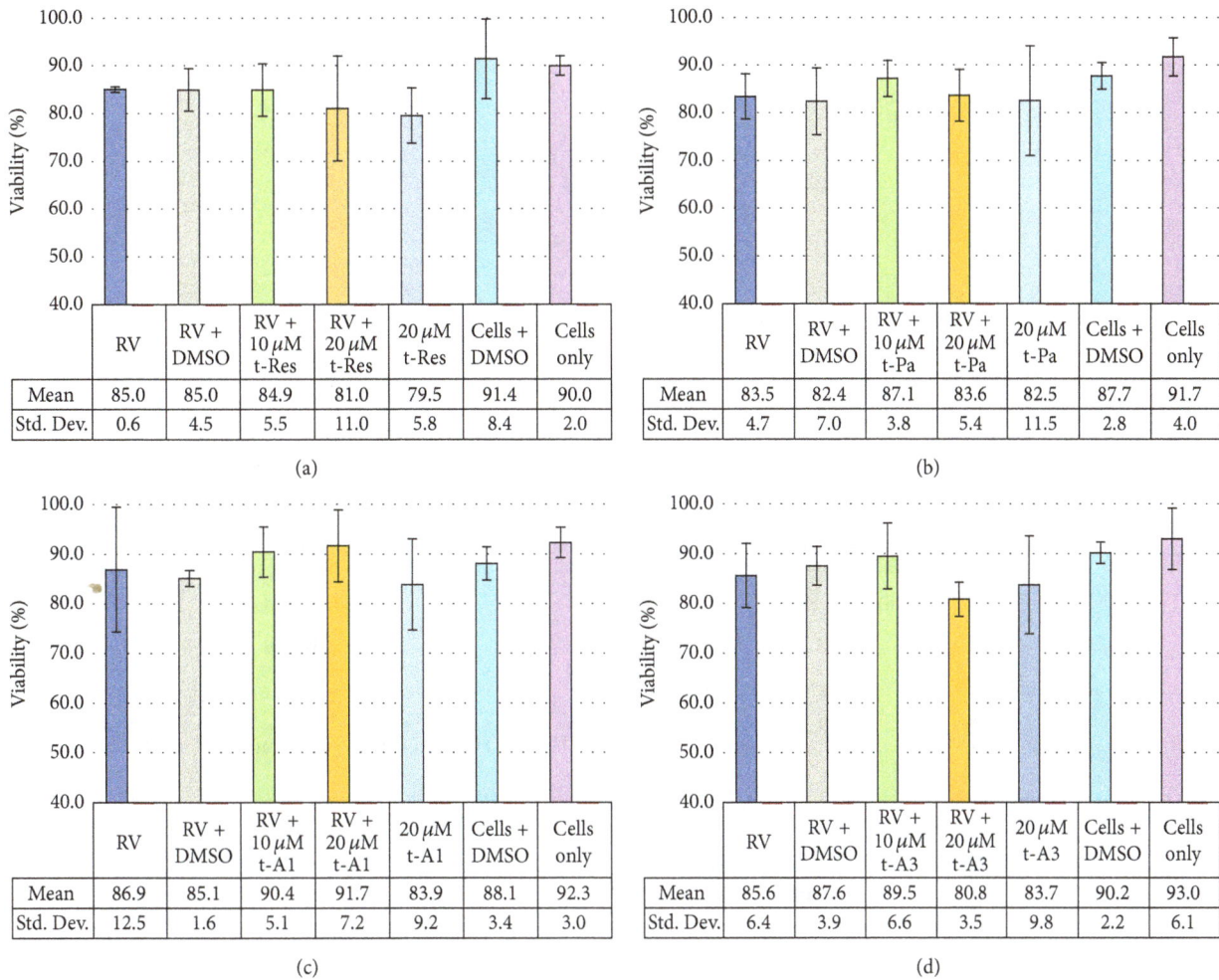

FIGURE 2: HT29.f8 Cell viability at 24 hpi with stilbenoids. (a) Resveratrol (t-Res). (b) Piceatannol (t-Pa). (c) Arachidin-1 (t-A1). (d) Arachidin-3 (t-A3).

life span of HT29.F8 cells, viability assays were performed with RV alone, RV with 0.02% DMSO, cells with 0.02% DMSO, cells alone, and RV with 0.02% DMSO with 10 μM and 20 μM stilbenoids using the trypan blue cell exclusion assay as described [26].

2.5. Virus Quantification. To test the biological activity of the stilbenoids on RV infections, both FFU and PFU assays were performed as previously described [27, 28]. MA104 cells were grown to 80% confluence in 24-well tissue culture plates (Corning Life Sciences), starved for fetal bovine sera 12 h prior to infection, and then infected with RV SA114F. Briefly, the SA114F RV stock was sonicated (5 min using a cup horn attachment and ice bath in a Misonix Sonicator 3000, Misonix, Inc., Farmingdale, NY) and incubated in serum-free DMEM with 1 μg/mL trypsin (Worthington Biochemical, Lakewood, NJ) for 30 min at 37°C. The activated viral inoculum was incubated with the cells for 1 h at 37°C in 5% CO$_2$ at an MOI of 2. The inoculum was replaced with serum-free DMEM supplemented with 1 μg/mL trypsin and incubated for 12 and 24 hpi. The supernatants were collected, clarified at 300 ×g for 5 min, and stored at −80°C. The cells were washed

in cold Dulbecco's PBS, 1X (Caisson Laboratories, Smithfield, UT), and released from the plates using a 0.25% trypsin-EDTA solution (1X) (Caisson Laboratories, Smithfield, UT). After the addition of DMEM with 5% FBS, the cells were resuspended in cold PBS and dilutions were prepared for live/dead cell counts (see Section 2.4). The balance of the cells was used to prepare cell lysates by subjecting them to repeated freeze-thaws three times, clarified at 300 ×g for 10 min. Media (supernatant) were collected, clarified at 300 ×g for 10 min, and stored at −80°C. Both the cell lysates and supernatants were stored at −80°C. Viral infectivity titers were done in triplicate by indirect immunofluorescent staining of MA104 monolayers infected with serial dilutions of the supernatants. The average number of fluorescent foci was calculated for three wells and used to determine the number of focus forming units/mL (FFU/mL) [29]. Since the RV viral infectivity titers are critical to our conclusion, two assays were used. The FFU data showed no difference in viral infectivity titers at 12 hpi using 10 μM concentrations (data not shown). Therefore, we chose to perform plaque forming unit (PFU) assays comparing the two controls, RV

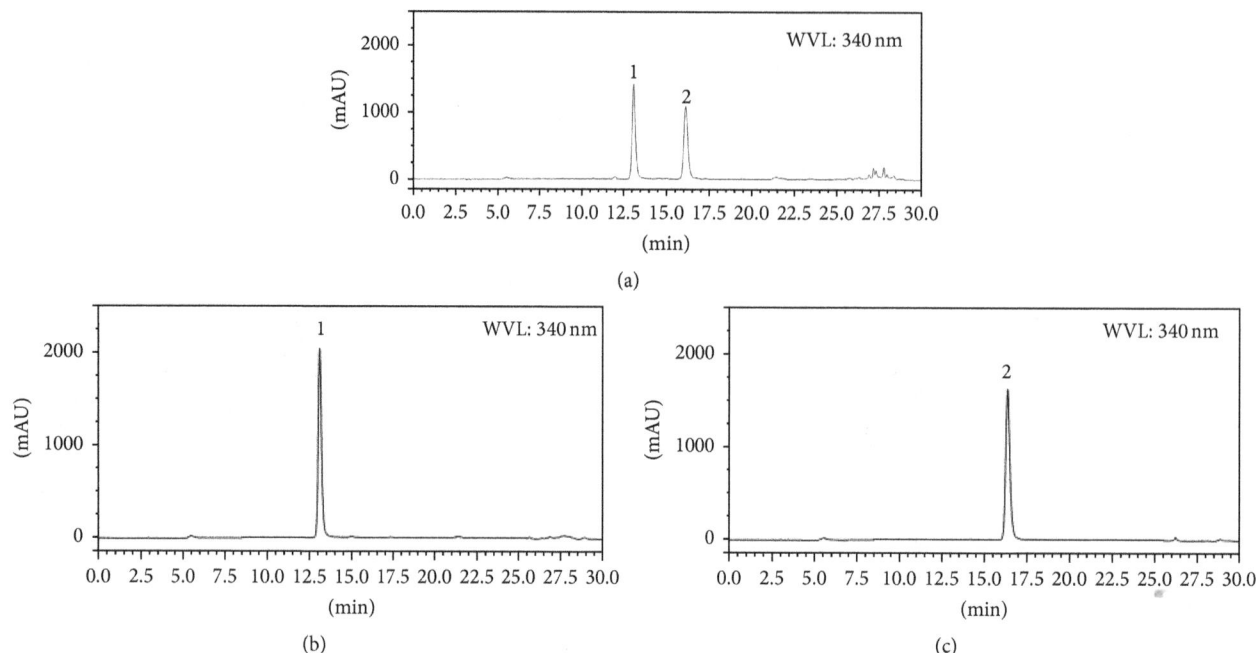

FIGURE 3: HPLC analysis of stilbenoids. The x-axis is time in minutes, and the y-axis is the absorbance at 340 nm. (a) HPLC chromatogram of ethyl acetate extract of the medium of hairy root culture of peanut treated with methyl-β-cyclodextrin for 72 h. Compounds: (1) arachidin-1 and (2) arachidin-3. (b) HPLC chromatogram of (1) arachidin-1 purified by HPCCC. (c) HPLC chromatogram of (2) arachidin-3 purified by HPCCC.

alone and RV with 0.02% DMSO to 20 μM of the stilbenoids at 24 hpi. Plaque forming assays were performed in triplicate as outlined above for the FFU assays, except after the 1-hour infection; the virus inoculum was replaced with 3 mL of a medium overlay (1:1 mixture of 1.2% agarose (Apex Low Melting Point Agarose, Genesee Scientific Inc.) and complete 2 × MEM containing 0.5 μg/mL trypsin) and incubated at 37°C in 5% CO_2 for 3 to 4 days or until plaques became visible. A neutral red overlay (1:1 mixture of 1.2% agarose with an equal volume of serum-free 2 × MEM containing 50 μg/mL neutral red) was prepared and 2 mL per well of stain overlay was added on top of the first agarose/medium overlay. The six-well plates were incubated at 37°C until plaques were visible (approximately 4 to 24 h). The individual plaques were counted, and the titers were calculated as follows: number of plaques × 1/dilution factor × 1/(mL of inoculum) = PFU/mL.

2.6. Statistical Analysis. Data are expressed as mean ± SD, and comparisons were statistically evaluated by analysis of variance (ANOVA) and Student's t-tests using Excel (significance level, $p \leq 0.05$).

2.7. Protein Quantification and Western Blot Assays. The micro bicinchoninic acid (BCA) protein assay was employed to quantify protein concentrations using bovine serum albumin as the standard per manufacturer's protocol (Thermo Scientific Pierce). One microgram of total protein from each sample was separated by 12.5% SDS-PAGE, electroblotted onto nitrocellulose membranes, and probed with NSP4 peptide-specific antibodies [30, 31] and reactive bands were

visualized by the addition of HRP-conjugated IgG and Super Signal West Pico chemiluminescent substrate (Pierce) followed by exposure to Kodak X-OMAT film [22, 32, 33].

3. Results and Discussion

3.1. Bioproduction of Stilbenoids in Hairy Root Cultures of Peanut. To produce the stilbenoids t-A1 and t-A3 we used our previously established hairy root line 3 from peanut cv. Hull. These hairy roots are capable of synthesizing and secreting t-Res, t-A1, and t-A3 into the culture medium upon treatment with the elicitor sodium acetate [18]. Depending on the period of elicitor treatment, the levels and types of stilbenoids found in the medium can be modified [18]. To study the effect of other elicitors on production of t-A1 and t-A3 we tested different elicitors, including methyl-β-cyclodextrin (CD). In preliminary experiments different doses of CD were added to the hairy root cultures for different periods between 0 and 96 h (data not shown). A 72 h treatment of 9 g/L CD was selected based on production of the highest levels of t-A1. As shown in Figures 3(a)–3(c), t-A1 and t-A3 were the major stilbenoids present in the culture medium. t-Res was present in very small amounts in these extracts. To purify t-A1 and t-A3, ethyl acetate extracts were made from the culture medium and subjected to HPCCC (high performance countercurrent chromatography). The solvent system was adapted from a previously used CPC (centrifugal partition chromatography) system which was effective in purifying t-A1 and t-A3 from hairy root culture medium extracts [16]. The only modifications were the replacement of heptane for hexane and ethanol for methanol. The separation was

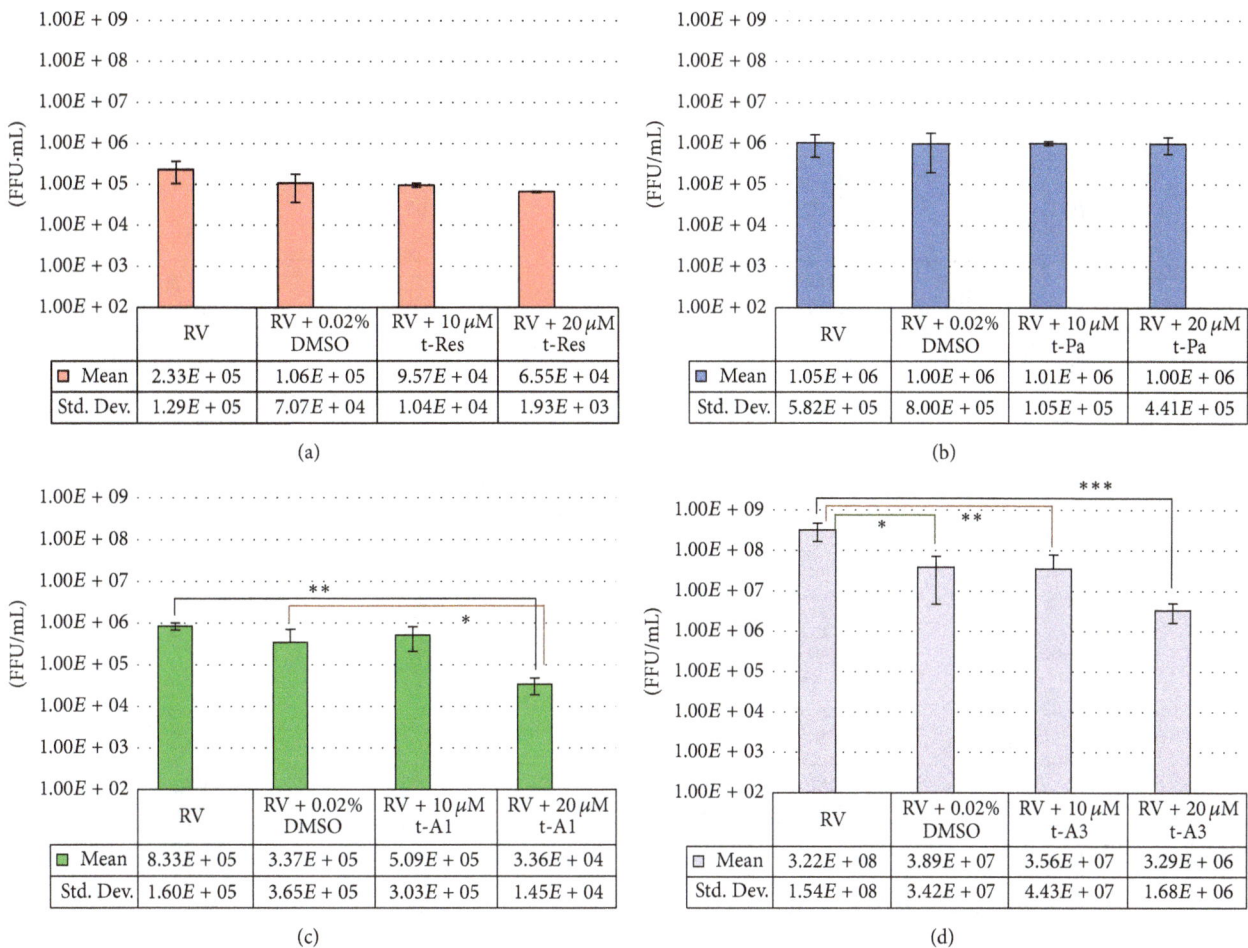

FIGURE 4: Quantification of progeny RV via focus forming units/mL (FFU/mL) at 24 hours postinfection. HT29.8 cells were infected with RV, with RV containing 0.02% DMSO, or 10 μM/20 μM of (a) resveratrol (t-Res). (b) Piceatannol (t-Pa). (c) Arachidin-1 (t-A1) $^*p$ = 0.02 and $^{**}p$ = 0.001 and (d) arachidin-3 (t-A3) $^*p$ = 0.04, $^{**}p$ = 0.04, and $^{***}p$ = 0.02.

effective and comparable to the one achieved before [16]. Thus high yields of highly purified fractions of t-A1 and t-A3 were achieved and were used in the antiviral assays.

3.2. Viability of HT29.F8 Cells in the Presence of 0.02% DMSO. The percentage of live/dead cells was calculated using the trypan blue exclusion dye assay (Figures 2(a)–2(d)). At 24 hpi, the cell viability between all groups tested (HT29.f8 cells with RV, HT29.f8 cells with RV and 0.02% DMSO, HT29.f8 cells with RV and 10 μM stilbenoids, HT29.f8 cells with RV and 20 μM stilbenoids, HT29.f8 cells with 20 μM stilbenoids only, HT29.f8 cells with 0.02% DMSO, and HT29.f8 cells only) was not statistically significantly different ($p < 0.05$). These data revealed that the addition of RV increases cell death, but not significantly in the time frame examined. Also, the addition of 20 μM concentrations of the stilbenoids decreased cell viability but not significantly, while the addition of 0.02% DMSO to the culture system did not adversely affect the viability of the HT29.f8 cells in culture or diminish viral replication (Figures 2(a)–2(d) and 6). These data demonstrate that HT29.f8 cells were not adversely affected by RV, 0.02% DMSO, or concentrations up to 20 μM of the four stilbenoids tested (t-Res, t-PA, t-A1, or t-A3).

3.3. The Effects of Stilbenoids on the Production of Infectious Rotavirus Particles. Viral infectivity titers were determined using FFU assays from the supernatants of RV-infected HT29.f8 cells treated with stilbenoids (10 μM and 20 μM t-Res, t-PA, t-A1, or t-A3). Supernatants collected at 12 hpi were equivalent to the RV-infected control cells (data not shown). Similarly, at 24 hpi, the 20 μM concentrations of the nonprenylated stilbenoids, t-Res and t-PA both, demonstrated no change in the virus titer when compared to the RV-infected control (Figures 4(a) and 4(b)). However at 24 hpi, the 10 μM concentrations of t-A1 generated a tenfold decrease in virus infectivity titer when compared to the RV-infected control supernatants ($p = 0.02$), and the 20 μM concentrations of t-A1 generated a twenty-fivefold decrease in virus infectivity titer when compared to the RV-infected control supernatants ($p = 0.001$) (Figure 4(c)), However, there was a statistical difference between RV and RV with DMSO with an eightfold decrease in virus infectivity titers with RV and DMSO ($p = 0.04$) (Figure 4(d)). The 10 μM concentrations of t-A3 generated a ninefold decrease in virus titer when compared to the RV-infected control supernatants ($p = 0.04$), and the 20 μM concentrations of t-A3 generated a

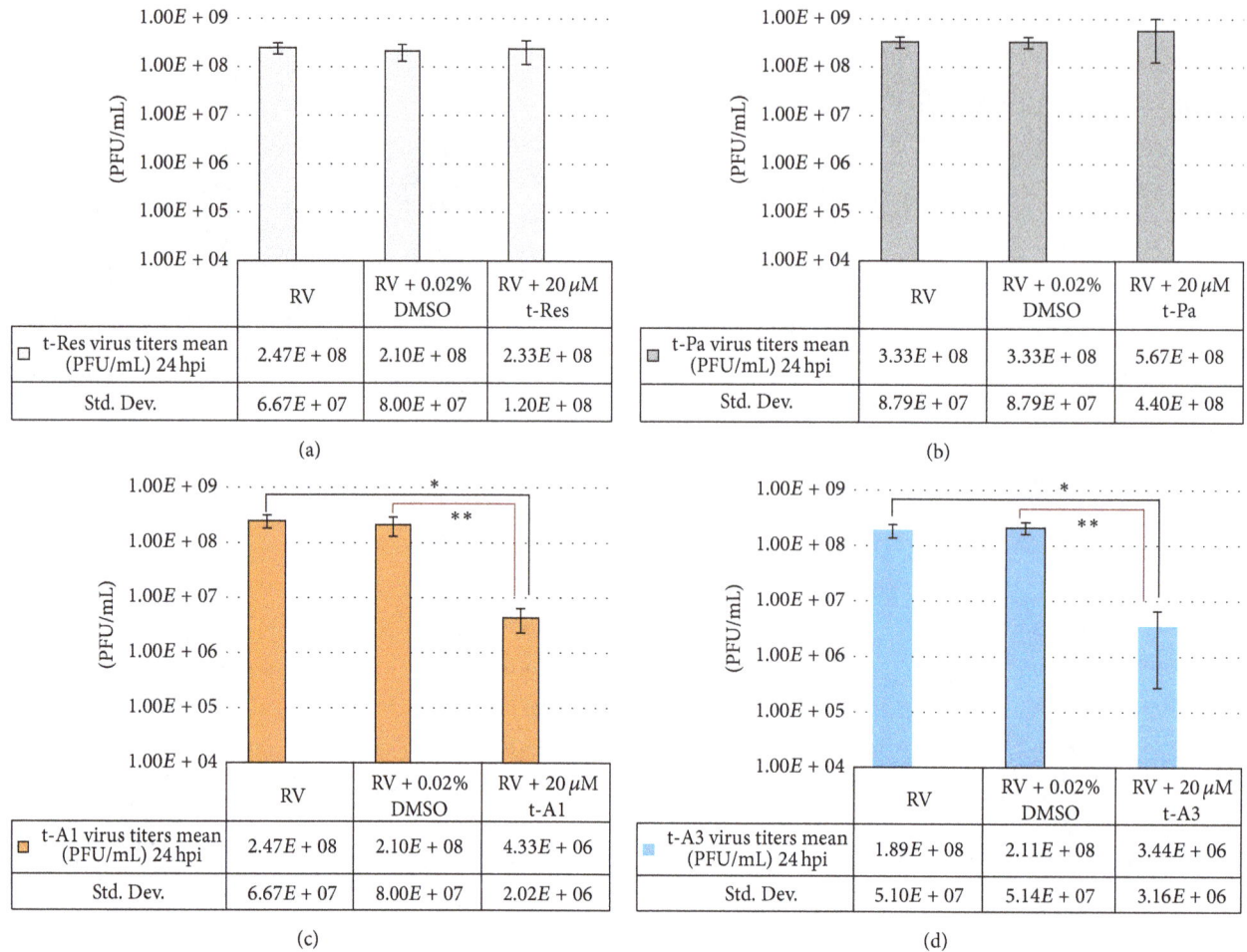

FIGURE 5: Quantification of progeny RV in plaque forming units/mL (PFU/mL) at 24 hpi. HT29.8 cells were infected with RV, with RV containing 0.02% DMSO, or 20 μM of (a) resveratrol (t-Res). (b) Piceatannol (t-Pa). (c) Arachidin-1 (t-A1). *Statistically significant $p = 0.02$. **Statistically significant $p = 0.04$. (d) Arachidin-3 (t-A3). *Statistically significant $p = 0.02$. **Statistically significant $p = 0.02$.

ninety-eightfold decrease in virus titer when compared to the RV-infected control supernatants ($p = 0.02$) (Figure 4(d)).

Since the data generated with the FFU assays were critical to test our hypothesis, plaque forming unit assays (PFU assays) were performed to corroborate the results obtained from the FFU assays. The PFU assays were performed using the same supernatants that were utilized for the FFU assays. Plaques were counted and the average of three experiments was calculated and graphed as PFU/mL (Figures 5(a)–5(d)).

The data produced using the PFU assays showed similar fold differences as shown with the FFU assays (Figures 5(a)–5(d) and 4(a)–4(d), resp.). Using the ANOVA and Student's t-test, the average and standard deviations were calculated and graphed (Figures 4(a)–4(d) and 5(a)–5(d)). The PFU experiments using t-Pa and t-Res showed no statistical differences between the controls, RV only and RV with DMSO, RV with 10 μM t-Pa/t-Res, or RV with 20 μM t-Pa/t-Res (Figures 5(a) and 5(b)). However, the experimental data from t-A1 PFU assays demonstrated a fifty-sevenfold difference from the control, RV only, and a forty-ninefold difference from the control, RV with DMSO that was statistically significant

($p = 0.02$ and 0.04, resp.) (Figure 5(c)). Likewise, the experimental data from 20 μM t-A3 PFU assays demonstrated a fifty-fivefold difference from the control, RV only, and a sixty-onefold difference from the control, RV with DMSO that was statistically significant ($p = 0.02$ and 0.02, resp.) (Figure 5(d)). Both assays show a significant decrease in RV infectivity titers in the presence of 20 μM t-A3.

3.4. Western Blot (WB) Analyses Imply Differences in RV Replication. To complement and visualize the differences demonstrated in the viral infectivity titers between RV alone and RV with DMSO, 20 μM t-A1, and 20 μM t-A3, western blot assays were performed as previously described [22, 32, 33]. Using equal amounts of protein of the corresponding cell lysates, the nonstructural viral protein 4, NSP4, was detected in all RV-infected cell lysates. The western blot data of the RV and RV with DMSO both demonstrated relatively equal amounts of multimeric forms and diglycosylated (fully glycosylated), monoglycosylated, and cleavage fragments of NSP4 and suggests that 0.02% DMSO does not affect the amount of NSP4 produced during a RV infection (Figure 6,

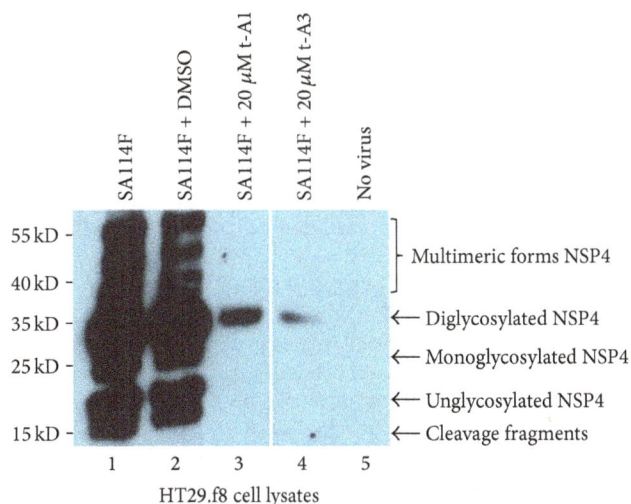

FIGURE 6: Western blot analysis of HT29.f8 cell lysates. Five micrograms of HT29.f8 cell lysates was separated on a 12.5% SDS-PAGE, electroblotted onto nitrocellulose membranes, probed with rabbit anti-NSP4$_{150-175}$ peptide-specific and goat anti-rabbit HRP-conjugated IgG, and visualized with Super Signal West Pico chemiluminescent substrate (Pierce) followed by exposure to Kodak X-OMAT film. (Lane 1) RV-infected HT29.f8 cells and (Lane 2) RV-infected HT29.f8 cells with 0.02% DMSO, respectively, show cleavage fragments and unglycosylated, mono-, diglycosylated, and multimeric forms of NSP4. (Lane 3) RV-infected HT29.f8 cells with 20 μM t-A1 and (Lane 4) RV-infected HT29.f8 cells with 20 μM t-A3 only show the diglycosylated form of NSP4. (Lane 5) HT29.f8 with no virus showed NSP4 banding pattern.

Lanes 1 and 2). The presence of multimeric forms of NSP4 has been previously studied [34–36]. The results for RV with t-A1 and t-A3 display a relatively small amount of the fully glycosylated form of NSP4 (Figure 6, Lanes 3 and 4). This indicates viral replication is negatively affected by 20 μM of both t-A1 and t-A3. Cell lysates without RV (Figure 6, Lane 5) reveal no bands and show the specificity of the anti-NSP4 antibodies.

4. Conclusions

Our data show a dose- and time-dependent decrease in viral progeny when RV and prenylated stilbenoids (t-A1 or t-A3) were incubated with the human intestinal cell line HT29.F8. The presence of the nonstructural viral protein NSP4 in the western blot assays confirms the RV infection and indicates the virus was replicating in the HT29.f8 cells. The prenylated stilbenoids, t-A1 and t-A3, significantly are more lipophilic than either of the nonprenylated t-Res or t-PA molecules. The prenylated side chain increases the lipophilicity of the molecules to which it is attached. Consequently, prenylation promotes association with and penetration through cell membranes. An increase in lipophilicity often correlates positively with increased biological activity within different groups of compounds of similar structure [37, 38]. Several delivery systems including emulsions and nanoparticles have been tested for the delivery of lipophilic, bioactive natural products [39]. Depending on the application, these delivery systems may be applicable to t-A1 and t-A3 and should be tested to advance their development as potential therapeutic agents.

Although the molecular mechanisms for the protective effect of t-A1 and t-A3 are not known, the inhibition of viral replication could be attributed to the antioxidative and anti-inflammatory properties of the constituent stilbenoids. In a previously published paper, t-A1 and t-A3 have been shown to modulate the cannabinoid receptors at micromolar levels [40]. The experimental data of this previous study show that t-A3 acts as a competitive cannabinoid receptor 1 (CB1R) antagonist, whereas t-A1 antagonizes CB1R agonists by both competitive and noncompetitive mechanisms [40]. It is interesting that the HT29 cell line, the parent cell line of HT29.f8, expresses cannabinoid receptors [41], and receptor expression should be investigated on the HT29.f8 cloned cells. These receptors are part of the endocannabinoid signaling system which is well known to regulate gastrointestinal functions, such as gastric emptying, secretion, and intestinal motility [42, 43]. Hence it is reasonable to propose a connection between the cannabinoid receptor functions and the mechanism of RV gastroenteritis. In a study on colorectal cancer, cannabinoid receptor (CB1 and CB2) agonists were shown to have an effect on apoptosis through a TNFα-mediated increase in ceramide production [44]. Another study on breast cancer shows the receptor agonists inhibit adenylyl cyclase activity, cAMP, and PKA activity resulting in the downregulation of gene transcription [45, 46]. A cAMP-dependent PKA mechanism also appears to be important in RV pathogenesis in a human intestinal cell line, Caco2 [47].

Recently, cannabinoid receptor antagonists were proposed as potential therapeutic agents against hepatitis C virus by modulating lipid homeostasis [48]. A study by Gaunt et al. [49] using RV-infected MA104 cells demonstrates a dose-dependent reduction in virus infectivity and viral RNA production with the addition of TOFA (5-(tetradecyloxy)-2-furoic acid), an inhibitor of the fatty acid synthase enzyme complex [49]. Further, the infectivity of RV in ACC1 knock-down cells was reduced by 8.5-fold (significant, $p = 0.01$) with siRNA directed against ACC1, the gene that encodes the enzyme catalyzing the rate-limiting step of the palmitoyl-CoA synthetic pathway. This strongly suggests that RV infectivity is mediated through fatty acid metabolism [49] or selected fatty acids.

Altogether, these data imply a possible antiviral mechanism for t-A1 and t-A3 through modulation of the cannabinoid receptors and subsequent alteration of fatty acid metabolism in the host cell. More studies are required to confirm and expand our knowledge of the RV reducing properties of t-A1 and t-A3. Thus, these compounds potentially could be used to design and develop more efficacious RV therapeutic agents.

Abbreviations

RV: Rotavirus
FFU: Focus forming units
PFU: Plaque forming units

WB: Western blots
NSP4: Nonstructural protein 4
hpi: Hours postinfection
DMSO: Dimethyl sulfoxide
FBS: Fetal bovine serum
DMEM: Dulbecco minimal essential
medium
t-A1: *trans*-Arachidin-1
t-A3: *trans*-Arachidin-3
t-PA: *trans*-Piceatannol
t-Rev: *trans*-Resveratrol
MOI: Multiplicity of infection
TC: Tissue culture
MSV: Modified Murashige and Skoog's
medium.

Acknowledgments

This work was supported by the Animal Formula Health Grant no. AH-9240 from the USDA Cooperative State Research, Education, and Extension Service. This work was supported by the Office of Research and Sponsored Programs at Stephen F. Austin State University (Research Pilot Study no. 107552-26112-150). This work was supported by the National Science Foundation-EPSCoR (Grant no. EPS-0701890; Center for Plant-Powered Production-P3), Arkansas ASSET Initiative, and the Arkansas Science and Technology Authority.

References

[1] U. Desselberger, "Rotaviruses," *Virus Research*, vol. 190, pp. 75–96, 2014.

[2] J. E. Tate and U. D. Parashar, "Rotavirus vaccines in routine use," *Clinical Infectious Diseases*, vol. 59, no. 9, pp. 1291–1301, 2014.

[3] E. Leshem, B. Lopman, R. Glass et al., "Distribution of rotavirus strains and strain-specific effectiveness of the rotavirus vaccine after its introduction: a systematic review and meta-analysis," *The Lancet Infectious Diseases*, vol. 14, no. 9, pp. 847–856, 2014.

[4] C. Yen, J. E. Tate, T. B. Hyde et al., "Rotavirus vaccines: current status and future considerations," *Human Vaccines & Immunotherapeutics*, vol. 10, no. 6, pp. 1436–1448, 2014.

[5] N. Bhandari, T. Rongsen-Chandola, A. Bavdekar et al., "Efficacy of a monovalent human-bovine (116E) rotavirus vaccine in Indian infants: a randomised, double-blind, placebo-controlled trial," *The Lancet*, vol. 383, no. 9935, pp. 2136–2143, 2014.

[6] G. A. Weinberg, E. N. Teel, S. Mijatovic-Rustempasic et al., "Detection of novel rotavirus strain by vaccine postlicensure surveillance," *Emerging Infectious Diseases*, vol. 19, no. 8, pp. 1321–1323, 2013.

[7] M. Martinez, M. E. Galeano, A. Akopov et al., "Whole-genome analyses reveals the animal origin of a rotavirus G4P[6] detected in a child with severe diarrhea," *Infection, Genetics and Evolution*, vol. 27, pp. 156–162, 2014.

[8] G. E. Armah, S. O. Sow, R. F. Breiman et al., "Efficacy of pentavalent rotavirus vaccine against severe rotavirus gastroenteritis in infants in developing countries in sub-Saharan Africa: a randomised, double-blind, placebo-controlled trial," *The Lancet*, vol. 376, no. 9741, pp. 606–614, 2010.

[9] S. A. Madhi, N. A. Cunliffe, D. Steele et al., "Effect of human rotavirus vaccine on severe diarrhea in African infants," *The New England Journal of Medicine*, vol. 362, no. 4, pp. 289–298, 2010.

[10] J. E. Tate, A. H. Burton, C. Boschi-Pinto, A. D. Steele, J. Duque, and U. D. Parashar, "2008 estimate of worldwide rotavirus-associated mortality in children younger than 5 years before the introduction of universal rotavirus vaccination programmes: a systematic review and meta-analysis," *The Lancet Infectious Diseases*, vol. 12, no. 2, pp. 136–141, 2012.

[11] K. Zaman, D. D. Anh, J. C. Victor et al., "Efficacy of pentavalent rotavirus vaccine against severe rotavirus gastroenteritis in infants in developing countries in Asia: a randomised, double-blind, placebo-controlled trial," *The Lancet*, vol. 376, no. 9741, pp. 615–623, 2010.

[12] V. S. Sobolev, S. I. Khan, N. Tabanca et al., "Biological activity of peanut (*Arachis hypogaea*) phytoalexins and selected natural and synthetic stilbenoids," *Journal of Agricultural and Food Chemistry*, vol. 59, no. 5, pp. 1673–1682, 2011.

[13] A. T. Palamara, L. Nencioni, K. Aquilano et al., "Inhibition of influenza A virus replication by resveratrol," *Journal of Infectious Diseases*, vol. 191, no. 10, pp. 1719–1729, 2005.

[14] V. Berardi, F. Ricci, M. Castelli, G. Galati, and G. Risuleo, "Resveratrol exhibits a strong cytotoxic activity in cultured cells and has an antiviral action against polyomavirus: potential clinical use," *Journal of Experimental and Clinical Cancer Research*, vol. 28, article 96, 2009.

[15] X. Chen, H. Qiao, T. Liu et al., "Inhibition of herpes simplex virus infection by oligomeric stilbenoids through ROS generation," *Antiviral Research*, vol. 95, no. 1, pp. 30–36, 2012.

[16] J. A. Abbott, F. Medina-Bolivar, E. M. Martin et al., "Purification of resveratrol, arachidin-1, and arachidin-3 from hairy root cultures of peanut (*Arachis hypogaea*) and determination of their antioxidant activity and cytotoxicity," *Biotechnology Progress*, vol. 26, no. 5, pp. 1344–1351, 2010.

[17] T. Michel, M. Halabalaki, and A.-L. Skaltsounis, "New concepts, experimental approaches, and dereplication strategies for the discovery of novel phytoestrogens from natural sources," *Planta Medica*, vol. 79, no. 7, pp. 514–532, 2013.

[18] J. Condori, G. Sivakumar, J. Hubstenberger, M. C. Dolan, V. S. Sobolev, and F. Medina-Bolivar, "Induced biosynthesis of resveratrol and the prenylated stilbenoids arachidin-1 and arachidin-3 in hairy root cultures of peanut: effects of culture medium and growth stage," *Plant Physiology and Biochemistry*, vol. 48, no. 5, pp. 310–318, 2010.

[19] T. Yang, L. Fang, C. Nopo-Olazabal et al., "Enhanced production of resveratrol, piceatannol, arachidin-1, and arachidin-3 in hairy root cultures of peanut co-treated with methyl jasmonate and cyclodextrin," *Journal of Agricultural and Food Chemistry*, vol. 63, no. 15, pp. 3942–3950, 2015.

[20] D. M. Mitchell and J. M. Ball, "Characterization of a spontaneously polarizing HT-29 cell line, HT-29/cl.f8," *In Vitro Cellular and Developmental Biology—Animal*, vol. 40, no. 10, pp. 297–302, 2004.

[21] J. M. Ball, D. M. Mitchell, T. F. Gibbons, and R. D. Parr, "Rotavirus NSP4: a multifunctional viral enterotoxin," *Viral Immunology*, vol. 18, no. 1, pp. 27–40, 2005.

[22] R. D. Parr, S. M. Storey, D. M. Mitchell et al., "The rotavirus enterotoxin NSP4 directly interacts with the caveolar structural protein caveolin-1," *Journal of Virology*, vol. 80, no. 6, pp. 2842–2854, 2006.

[23] F. Medina-Bolivar, M. Dolan, S. Bennett, J. Condori, and J. Hubstenberger, US Patent 7666677, 2010.

[24] Z. Marsh, T. Yang, L. Nopo-Olazabal et al., "Effect of light, methyl jasmonate and cyclodextrin on production of phenolic compounds in hairy root cultures of *Scutellaria lateriflora*," *Phytochemistry*, vol. 107, pp. 50–60, 2014.

[25] N. M. Mattion, J. Cohen, C. Aponte, and M. K. Estes, "Characterization of an oligomerization domain and RNA-binding properties on rotavirus nonstructural protein NS34," *Virology*, vol. 190, no. 1, pp. 68–83, 1992.

[26] R. I. Freshney, *Culture of Animal Cells: A Manual of Basic Technique*, Wiley-Liss, New York, NY, USA, 3rd edition, 1994.

[27] M. Arnold, J. T. Patton, and S. M. McDonald, "Culturing, storage, and quantification of rotaviruses," in *Current Protocols in Microbiology*, Unit 15C.3, John Wiley & Sons, 2009.

[28] K. A. Yakshe, Z. D. Franklin, and J. M. Ball, "Rotaviruses: extraction and isolation of RNA, reassortant strains, and NSP4 protein," in *Current Protocols in Microbiology*, John Wiley & Sons, 2015.

[29] N. Kitamoto, R. F. Ramig, D. O. Matson, and M. K. Estes, "Comparative growth of different rotavirus strains in differentiated cells (MA104, HepG2, and CaCo-2)," *Virology*, vol. 184, no. 2, pp. 729–737, 1991.

[30] H. Huang, F. Schroeder, M. K. Estes, T. McPherson, and J. M. Ball, "Interaction(s) of rotavirus non-structural protein 4 (NSP4) C-terminal peptides with model membranes," *Biochemical Journal*, vol. 380, no. 3, pp. 723–733, 2004.

[31] C. L. Swaggerty, H. Huang, W. S. Lim, F. Schroeder, and J. M. Ball, "Comparison of SIVmac239$_{(352–382)}$ and SIVsmmPBj41$_{(360–390)}$ enterotoxic synthetic peptides," *Virology*, vol. 320, no. 2, pp. 243–257, 2004.

[32] T. F. Gibbons, S. M. Storey, C. V. Williams et al., "Rotavirus NSP4: cell type-dependent transport kinetics to the exofacial plasma membrane and release from intact infected cells," *Virology Journal*, vol. 8, article 278, 2011.

[33] S. M. Storey, T. F. Gibbons, C. V. Williams, R. D. Parr, F. Schroeder, and J. M. Ball, "Full-length, glycosylated NSP4 is localized to plasma membrane caveolae by a novel raft isolation technique," *Journal of Virology*, vol. 81, no. 11, pp. 5472–5483, 2007.

[34] G. D. Bowman, I. M. Nodelman, O. Levy et al., "Crystal structure of the oligomerization domain of NSP4 from rotavirus reveals a core metal-binding site," *Journal of Molecular Biology*, vol. 304, no. 5, pp. 861–871, 2000.

[35] A. Didsbury, C. Wang, D. Verdon, M. A. Sewell, J. D. McIntosh, and J. A. Taylor, "Rotavirus NSP4 is secreted from infected cells as an oligomeric lipoprotein and binds to glycosaminoglycans on the surface of non-infected cells," *Virology Journal*, vol. 8, article 551, 2011.

[36] J. A. Taylor, J. A. O'Brien, and M. Yeager, "The cytoplasmic tail of NSP4, the endoplasmic reticulum-localized non-structural glycoprotein of rotavirus, contains distinct virus binding and coiled coil domains," *The EMBO Journal*, vol. 15, no. 17, pp. 4469–4476, 1996.

[37] T. Schultz, D. Nicholas, and T. Fisher, "Quantitative structure-activity relationships of stilbenes and related derivatives against wood-destroying fungi," in *Recent Research Developments in Agricultural & Food Chemistry*, vol. 1, pp. 289–299, 1997.

[38] K. Yazaki, K. Sasaki, and Y. Tsurumaru, "Prenylation of aromatic compounds, a key diversification of plant secondary metabolites," *Phytochemistry*, vol. 70, no. 15-16, pp. 1739–1745, 2009.

[39] F. Silva, A. Figueiras, E. Gallardo, C. Nerín, and F. C. Domingues, "Strategies to improve the solubility and stability of stilbene antioxidants: a comparative study between cyclodextrins and bile acids," *Food Chemistry*, vol. 145, pp. 115–125, 2014.

[40] L. K. Brents, F. Medina-Bolivar, K. A. Seely et al., "Natural prenylated resveratrol analogs arachidin-1 and -3 demonstrate improved glucuronidation profiles and have affinity for cannabinoid receptors," *Xenobiotica*, vol. 42, no. 2, pp. 139–156, 2012.

[41] K. Ihenetu, A. Molleman, M. E. Parsons, and C. J. Whelan, "Inhibition of interleukin-8 release in the human colonic epithelial cell line HT-29 by cannabinoids," *European Journal of Pharmacology*, vol. 458, no. 1-2, pp. 207–215, 2003.

[42] A. A. Coutts and A. A. Izzo, "The gastrointestinal pharmacology of cannabinoids: an update," *Current Opinion in Pharmacology*, vol. 4, no. 6, pp. 572–579, 2004.

[43] F. Massa and K. Monory, "Endocannabinoids and the gastrointestinal tract," *Journal of Endocrinological Investigation*, vol. 29, no. 3, pp. 47–57, 2006.

[44] F. Cianchi, L. Papucci, N. Schiavone et al., "Cannabinoid receptor activation induces apoptosis through tumor necrosis factor alpha-mediated ceramide de novo synthesis in colon cancer cells," *Clinical Cancer Research*, vol. 14, no. 23, pp. 7691–7700, 2008.

[45] M. Bifulco, A. M. Malfitano, S. Pisanti, and C. Laezza, "Endocannabinoids in endocrine and related tumours," *Endocrine-Related Cancer*, vol. 15, no. 2, pp. 391–408, 2008.

[46] C. Laezza, A. M. Malfitano, M. C. Proto et al., "Inhibition of 3-hydroxy-3-methylglutarylcoenzyme a reductase activity and of Ras farnesylation mediate antitumor effects of anandamide in human breast cancer cells," *Endocrine-Related Cancer*, vol. 17, no. 2, pp. 495–503, 2010.

[47] S. Martin-Latil, J. Cotte-Laffitte, I. Beau, A.-M. Quéro, M. Géniteau-Legendre, and A. L. Servin, "A cyclic AMP protein kinase A-dependent mechanism by which rotavirus impairs the expression and enzyme activity of brush border-associated sucrase-isomaltase in differentiated intestinal Caco-2 cells," *Cellular Microbiology*, vol. 6, no. 8, pp. 719–731, 2004.

[48] M. Shahidi, E. S. E. Tay, S. A. Read et al., "Endocannabinoid CB1 antagonists inhibit hepatitis C virus production, providing a novel class of antiviral host-targeting agents," *Journal of General Virology*, vol. 95, pp. 2468–2479, 2014.

[49] E. R. Gaunt, W. Cheung, J. E. Richards, A. Lever, and U. Desselberger, "Inhibition of rotavirus replication by downregulation of fatty acid synthesis," *Journal of General Virology*, vol. 94, no. 6, pp. 1310–1317, 2013.

7

Vitamin D-Regulated MicroRNAs: Are they Protective Factors against Dengue Virus Infection?

John F. Arboleda and Silvio Urcuqui-Inchima

Grupo Inmunovirología, Facultad de Medicina, Universidad de Antioquia (UdeA), Calle 70 No. 52-51, Medellín, Colombia

Correspondence should be addressed to Silvio Urcuqui-Inchima; silvio.urcuqui@udea.edu.co

Academic Editor: Subhash C. Verma

Over the last few years, an increasing body of evidence has highlighted the critical participation of vitamin D in the regulation of proinflammatory responses and protection against many infectious pathogens, including viruses. The activity of vitamin D is associated with microRNAs, which are fine tuners of immune activation pathways and provide novel mechanisms to avoid the damage that arises from excessive inflammatory responses. Severe symptoms of an ongoing dengue virus infection and disease are strongly related to highly altered production of proinflammatory mediators, suggesting impairment in homeostatic mechanisms that control the host's immune response. Here, we discuss the possible implications of emerging studies anticipating the biological effects of vitamin D and microRNAs during the inflammatory response, and we attempt to extrapolate these findings to dengue virus infection and to their potential use for disease management strategies.

1. Introduction

Activation of innate immune cells results in the release of proinflammatory mediators to initiate a protective local response against invading pathogens [1]. However, overactivated inflammatory activity could be detrimental since it can cause tissue damage and even death of the host. Therefore, negative feedback mechanisms are required to control the duration and intensity of the inflammatory response [1, 2]. Although little is known about the molecular mechanisms occurring during dengue virus (DENV) infection/disease, it has been suggested that the immune response initiated against the virus greatly contributes to pathogenesis. Indeed, several symptoms of the disease are tightly related to imbalanced immune responses, particularly to high production of proinflammatory cytokines [3, 4] suggesting an impairment of homeostatic mechanisms that control inflammation. Interestingly, vitamin D has been described as an important modulator of immune responses to several pathogens and as a key factor enhancing immunoregulatory mechanisms that avoid the damage that arises from excessive inflammatory responses [5, 6], as in dengue disease [7]. Mounting evidence obtained from human populations and experimental in vitro studies has suggested that this hormone can play a key role in the immune system's response to several viruses [8–14], thereby becoming a potential target of intervention to combat DENV infection and disease progression. Among several mechanisms, vitamin D activity has been associated with the expression of certain microRNAs (miRs) [15] that are one of the main regulatory switches operating at the translational level [16]. miRs constitute approximately 1% of the human genome and their sequences can be found within introns of other genes or can be encoded independently and transcribed in a similar fashion to mRNAs encoded by protein-coding genes [16]. A typical mature miR of 18–23 base pairs associates with the RNA-induced silencing complex (RISC) and moves towards the target mRNA [17]. Once there, the miR binds to the complementary sequence in the $3'$ untranslated region ($3'$UTR) of the mRNA, thereby inducing gene silencing through mRNA cleavage, translational repression, or deadenylation [16]. A single miR may directly regulate the expression of hundreds of mRNAs at once and several miRs can also target the same mRNA resulting in enhanced translation inhibition [18]. Targeting of specific genes involved in modulation of immune response pathways by miRs provides a finely tuned regulatory mechanism for the restoration of the host's resting inflammation state [19–21]. Since the association between vitamin D and miR activity may play a relevant role in ongoing DENV infections, here we provide an overview of DENV-induced inflammatory responses and the early evidence anticipating a possible participation of the vitamin D and miR interplay regulating antiviral and inflammatory responses during DENV infection/disease.

2. DENV and the Immune Response

DENV is an icosahedral-enveloped virus with a positive sense single-stranded RNA (ssRNA) genome that belongs to the family Flaviviridae, genus *Flavivirus*. There are four phylogenetically related but antigenically distinct viral serotypes (DENV 1–4) able to cause the full spectrum of the disease [22]. In addition, a sylvatic serotype (DENV-5), with no evidence regarding its ability to infect humans, has been recently reported [23]. DENV is transmitted by *Aedes* mosquitoes in tropical and subtropical areas where the disease has become a major public health threat and one of the most rapidly spreading vector-borne diseases in the world, with an increasing incidence of 30-fold in the past 50 years [24, 25]. An estimated 3.6 billion people live in high risk areas worldwide and it is estimated that over 390 million cases occur every year, of which 96 million suffer from dengue fever [26–28]. Although only a minor number of cases may progress to the severe forms of the disease, 21.000 deaths are reported annually [27]. Guidelines of the World Health Organization (WHO) recognize dengue as a clinical continuum from dengue fever (DF), a nonspecific febrile illness, to dengue with or without warning signs that can progress to dengue hemorrhagic fever (DHF) or dengue shock syndrome (DSS) [3]. These severe forms of the disease are characterized by a wide spectrum of symptoms, including the development of vascular permeability, plasma leakage, thrombocytopenia, focal or generalized hemorrhages, and tissue and/or organ damage that may lead to shock and death [29, 30]. Besides ecoepidemiology, host genetic variations, and virus virulence, the risk factor is increased mainly by secondary infections with different dengue serotypes, presumably through a mechanism known as antibody-dependent immune enhancement (ADE), whereby nonneutralizing antibodies from previous heterotypic infections enhance virus entry via receptors for immunoglobulins or Fc receptors (FcRs) [29, 31, 32].

Skin is the first barrier for the invading DENV and the site where innate immunity exerts the first line of defense [33]. Following the bite by an infected mosquito, local tissue resident dendritic cells (DCs) and macrophages are the main targets of the virus [34, 35]. The viral structural E protein binds to cellular receptors, such as DC-SIGN (Dendritic Cell-Specific Intercellular adhesion molecule-3-Grabbing Nonintegrin), CLEC5A (C-type lectin domain family 5, member A), and MR (mannose receptor), allowing internalization of the virus through receptor-mediated endocytosis [22, 36–38]. Once in the cytoplasm, DENV replication products, such as double-stranded RNA (dsRNA)

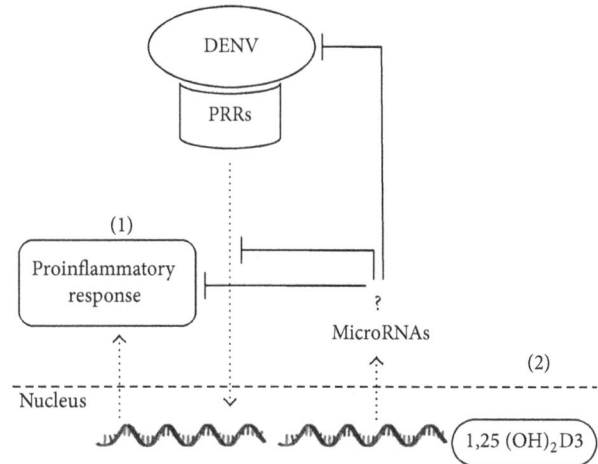

FIGURE 1: Potential link between vitamin D and miR controlling DENV-induced inflammatory response and antiviral activity. (1) DENV replication products and proteins are recognized by several PRRs whose signaling pathways promote the proinflammatory response. (2) Vitamin D activity induces transcription of microRNAs and other target genes that play a critical role in the control of inflammation-related signaling pathways and antiviral activity.

or genomic ssRNA, are sensed by several pattern recognition receptors (PRRs) (Figure 1), including TLR3, TLR7, TLR8, the cytosolic receptors RIG-I (Retinoic acid Inducible Gene-1), and MDA-5 (Melanoma Differentiation-Associated protein 5) [39–43]. Subsequently, this subset of PRRs triggers the activation of intracellular pathways, leading to the activation of transcription factors such as interferon regulatory factors 3 and 7 (IRF3 and IRF7) and the Nuclear Factor κB (NF-κB) and the later production of type I interferons and proinflammatory cytokines promoting an antiviral response [44, 45]. Additionally, the local activation of natural killer (NK) cells, neutrophils, and mast cells by the presence of the virus induces more proinflammatory mediators, complement activation, and the commitment of cellular and humoral immune responses to clear and control viral infection [46].

2.1. Inflammation and Cytokine Storm. Although the immune response is critical to combat and overcome invading pathogens, it is believed that the immune response greatly contributes to progression of dengue disease [31]. The pathogenesis and progression to the severe forms of dengue are still not completely understood; however, most cases are characterized by bleeding, hemorrhage, and plasma leakage that can progress to shock or organ failure [87, 88]. These physiological events are preceded by a hyperpermeability syndrome caused mainly by an imbalance between proinflammatory and anti-inflammatory cytokines produced in response to virus infection. The predominant proinflammatory mediators or "cytokine storm," secreted mainly by T cells, monocytes/macrophages, and endothelial cells (Table 1), promotes endothelial dysfunction by generating an endothelial "sieve" effect that leads to fluid and protein leakage. Increasing evidence suggests that endothelial

TABLE 1: Summary of the main cytokines associated with development of DHF/DSS and their biological function in relation to pathogenesis.

Cytokines	Biological function	Refs.
MCP-1	Monocyte chemoattractant protein-1 is critical to drive the extravasation of mononuclear cells into the inflamed, infected, and traumatized sites of infection. In addition, it promotes endothelial permeability increasing the vascular leakage as a result of dengue virus infection.	[47, 48]
IL-1	It induces tissue factor (TF) expression of endothelial cells (EC) and suppresses their cell surface anticoagulant activity. It may upregulate TNF-α production and activity. IL-1β mediates platelet-induced activation of ECs, which increases chemokine release and upregulates VCAM-1 enhancing adhesion of monocytes to the endothelium.	[43, 49]
IL-6	It has been described as a strong inducer of endothelial permeability resulting in vascular leakage. IL-6 potentiates the coagulation cascade and can downregulate production of TNF-α and its receptors. IL-6 may perform a synergistic role with some pyrogens such as IL-1 to induce fever.	[50, 51]
IL-8	Its systemic concentrations are increased by EC damage, which in turn induces endothelial permeability. Activation of the coagulation system results in increased expression of IL-6 and IL-8 by monocytes, while the APC anticoagulation pathway downregulates the production of IL-8 by ECs.	[49, 50, 52]
IL-10	It plays an immunosuppressive role that causes IFN resistance, followed by impaired immune clearance and a persistent infectious effect for acute viral infection. IL-10 also inhibits the expression of TF and inhibits fibrinolysis. IL-10 plasma levels have been associated with disease severity; however, its role in dengue pathogenesis has not been fully elucidated.	[53]
TNF-α	It is a potent activator of ECs; it enhances capillary permeability. TNF-α upregulates expression of TF in monocytes and ECs and downregulates expression of thrombomodulin on ECs. It also activates the fibrinolytic system and enhances expression of NO mediating activation-induced death of T cells, and it has therefore been implicated in peripheral T-cell deletion.	[49, 51, 54]
TGF-β	Early in infection, low levels of TGF-β may trigger secretion of IL-1 and TNF-β. However, later in infection, the cytokine inhibits the Th1 response and enhances production of Th2 cytokines such as IL-10. TGF-β increases expression of TF on ECs and upregulates expression and release of PAI-1 (plasminogen activator inhibitor-1).	[3]
VEGF	VEGF is a key driver of vascular permeability. It reduces EC occludins, claudins, and the VE-cadherin content, all of which are components of ECs junctions. Upon activation, VEGF stimulates expression of ICAM-1, VCAM-1, and E-selectin in ECs.	[3, 36]

integrity and vascular permeability are affected by proinflammatory cytokines through the induction of apoptosis and the modulation of tight junction molecules within endothelial cells [47, 52, 89, 90]. In addition, it has also been reported that these cytokines may often have synergistic effects and may induce expression of other cytokines, generating a positive feedback mechanism leading to further imbalanced levels of inflammatory mediators and higher permeability [4].

This oversustained inflammatory response may be due to an impairment of the regulatory mechanisms that control the duration and intensity of inflammation or cytokine production, especially through the regulation of PRR signaling activation [20]. Several studies have shown that alterations in proinflammatory cytokine production during DENV infection/disease can be attributed to variations in recognition and activation of TLR signaling, which contributes to progression of the disease (Figure 1) [91, 92]. It was recently reported that DENV NS1 proteins may be recognized by TLR2, TLR4, and TLR6 enhancing the production of proinflammatory cytokines and triggering the endothelial permeability that leads to vascular leakage [93, 94]. Interestingly, our group has recently shown a differential expression of TLRs in dendritic cells (DCs) of dengue patients depending on the severity of the disease [95]. Indeed, there was an increased expression of TLR3 and TLR9 in DCs of patients with DF in contrast to a poor stimulation of both receptors in DCs of patients with DHF. Conversely, a lower expression of TLR2 in DF patients compared to DHF patients was also observed. Additionally, IFN-α production was also altered via TLR9, suggesting that DENV may affect the type I IFN response through this signaling pathway [95]. Indeed, DENV has successfully evolved to overcome host immune responses, by efficiently subverting the IFN pathway and inhibiting different steps of the immune response through the expression of viral nonstructural proteins that antagonize several molecules of this activation pathway [96, 97]. Although DENV may evade immune recognition [42], cumulative data have shown that it is sensed by both TLR3 and TLR7/8 and activates signaling pathways upregulating IFN-α/β, TNF-α, human defensin 5 (HD5), and human β defensin 2 (HβD2) [39–41]. In addition, RIG-I and MDA-5 are also activated upon DENV infection and are essential for host defense against the virus [40]. Moreover, TLR3 controls DENV2 replication through NF-κB activation, suggesting that TLR3 agonists such as Poly (I : C) (Polyinosinic : Polycytidylic Acid) might work as immunomodulators of DENV infection [39]. Furthermore, besides DENV recognition and binding, C-type lectins such as the mannose receptor (MR) and CLEC5A may contribute to the inflammatory responses [98–100]. CLEC5A plays a critical role in the induction of NLRP3 inflammasome activation during DENV infection and enhances the release of IL-18 and IL-1β that are critical for activation of Th17 helper cells [99, 101].

While innate immune activation and proinflammatory cytokine production are being investigated during the course of DENV infections [53, 92, 102], vitamin D activity has gained special attention due to its importance in the modulation of the innate response. An increasing number of reports suggest that vitamin D activity is associated with the modulation of components implicated in antiviral immune responses and in the regulation of proinflammatory cytokine production through the modulation of miR expression [6, 13, 15, 103]. Although there is little information from observational studies and clinical trials demonstrating the role of vitamin D during dengue virus infection, here we postulate a potential role of vitamin D controlling progression of dengue disease and provide evidence of some vitamin D molecular mechanisms in support of our hypothesis.

3. Vitamin D: Antiviral and Anti-Inflammatory Activity

In addition to its well-known role in bone mineralization and calcium homeostasis, vitamin D is recognized as a pluripotent regulator of biological and immune functions [104]. A growing body of evidence suggests that it plays a major role during the immune system's response to microbial infection, thereby becoming a potential intervener to control viral infections and inflammation [13, 105, 106]. The term vitamin D refers collectively to the active form 1α-25-dihydroxyvitamin D$_3$ [1α-25(OH)$_2$D3] and the inactive form 25-hydroxyvitamin D$_3$ [25(OH)D$_3$] [107]. For their transport within the serum, vitamin D compounds bind to the vitamin D binding protein (DBP) and this complex is recognized by megalin and cubilin (members of low-density lipoprotein receptor family) that then internalize the complex by invagination [108]. Intracellular trafficking of vitamin D metabolites to specific destinations is performed by members of the HSP- (Heat Shock Proteins-) 70 family [104]. In addition, vitamin D metabolites are also lipophilic molecules that can easily penetrate cell membranes and translocate to the nucleus, where 1α-25(OH)$_2$D$_3$ binds to the vitamin D receptor (VDR), thereby inducing heterodimerization of VDR with an isoform of the retinoid X receptor (RXR) [109]. The VDR-RXR heterodimer binds to vitamin D response elements (VDRE) present in the promoter of hundreds of target genes, whose products play key roles in cellular metabolism, bone mineralization, cell growth, differentiation, and control of inflammation (Figure 1) [104, 110, 111]. Besides VDR, other related vitamin D metabolic components such as the hydrolase CYP27B1, the enzyme that catalyzes the synthesis of active 1α-25-dihydroxyvitamin D$_3$ from 25-hydroxyvitamin D$_3$, are present and induced in some cells of the immune system during immune responses [112]. Thus, an increasing number of studies have explored the relationship between vitamin D activity and the immune system, specifically, the mechanisms whereby vitamin D exerts its antimicrobial and immunoregulatory activity [14, 113, 114]. Here, we highlight those modulating antiviral and inflammatory responses.

Although controversial data have been reported, increasing clinical and observational studies have provided evidence supporting the protective features of vitamin D in viral infections, especially viral respiratory infections and HIV [13, 115, 116]. The activity of vitamin D in the innate immune system begins at the forefront of the body's defense against pathogens, the skin. Regardless of global serum

TABLE 2: Vitamin D-induced mechanisms/mediators associated with antiviral activity.

Mediator/mechanism	Virus	Refs.
Cathelicidin (LL-37)	VHS, influenza virus, HIV, retrovirus	[55–59]
HBD2	HIV	[60]
ROS	HCV	[61]
IFN response	HIV, HCV	[62–64]
Autophagy	HIV	[65, 66]
miR let-7	DENV	[67, 68]

TABLE 3: Vitamin D and miR targets associated with inflammatory response.

Target/mediator	Modulator	Refs.
TLR2/4	Vitamin D/miR155.miR146	[20, 69, 70]
TNF-α	Vitamin D/miR146	[70, 71]
IL-1β	Vitamin D/miR155	[19, 69]
IL-6	Vitamin D/let-7e	[72, 73]
MAPK	Vitamin D	[19]
NF-κB	Vitamin D/miR155, miR146	[20, 70, 74, 75]
IKK	Vitamin D	[76]
SOCS1	Vitamin D/miR155	[20]
TLR9	Vitamin D	[77]

vitamin D levels, sensing of microbial pathogens via PRRs induces upregulation of CYP27B1 and, as a consequence, local conversion of 1,25(OH)$_2$D$_3$ from 25(OH)D$_3$, enhancing VDR nuclear translocation and subsequent transcription of target genes to exert antimicrobial effects [113, 117–119]. This establishes a linkage between vitamin D status and the intracrine and paracrine modulation of cellular immune responses, in which VDR and CYP27B1 activity are of central importance [117, 118, 120]. Indeed, this link is also evidenced by studies in which pathogen susceptibility associated with vitamin D deficiency/insufficiency levels is reduced by correct supplementation [121, 122]. Furthermore, some vitamin D-induced antiviral mechanisms have been shown by preliminary reports (Table 2). Peptides such as cathelicidins are strongly upregulated by 1,25(OH)$_2$D$_3$ due to its VDR response elements. In humans, active cathelicidin is known as LL-37 and has a C-terminal cationic antimicrobial domain that can induce bacterial membrane disruption and inhibition of herpes simplex virus, influenza virus, and retroviral replication, among others [55–57]. In fact, very recent reports have suggested an association between vitamin D and the LL-37 antiviral activity to HIV and rhinovirus [58, 59]. Likewise, HBD-2 is also induced by 1,25(OH)$_2$D$_3$. Interestingly, a correlation between VDR and HBD-2 was found to be associated with natural resistance to HIV infection, suggesting the potential participation of vitamin D-induced resistance to the virus [60, 106]. Moreover, vitamin D can also induce reactive oxygen species (ROS) that associates with suppression of the replicative activity of some viruses, such as hepatitis C virus (HCV) [61]. Although the vitamin D-induced antiviral mechanisms are not fully elucidated and further studies are needed to fully understand their roles, many are possible due to the pleiotropic nature of vitamin D and the complex transcriptional modulation of hundreds of genes controlled by its activity.

Several studies have reported a link between VDR polymorphisms and severe outcomes of bronchiolitis and acute lower respiratory tract infections (RTIs) with respiratory syncytial virus (RSV) [105]. Indeed, vitamin D supplementation is associated with reduced RTI, vitamin D status, and serum concentrations in children [123]. Likewise, some vitamin D supplementation studies have reported a reduction in cold/influenza linked to seasonal sunlight exposure and skin pigmentation [124]. In HIV infection, associations have also been reported between vitamin D levels with progression of the disease, survival times of HIV patients, CD4$^+$ T cell counts, inflammatory responses, and potential impact of HAART (Highly Active Anti-Retroviral Therapy) treatments [125]. Finally, similar population and ecoepidemiological reports have associated the role of vitamin D in several viral infections, including DENV and other flaviviruses [10–13], not only highlighting inhibition of viral replication but also controlling the inflammatory response and progression of the disease.

In addition to viral control, vitamin D-induced immune mechanisms have important effects providing potential feedback modulation in pathways that regulate immune activation, avoiding excessive elaboration of the inflammatory responses and its potential risk for tissue homeostasis (Table 3) [5, 6, 126]. TLRs can both affect and be affected by VDR signaling and likewise some antimicrobial peptides associated with TLRs have demonstrated antiviral effects [6, 13, 127]. In this sense, and due to the interest in the modulatory effect of vitamin D on TLR expression and proinflammatory cytokine production, some authors have shown that vitamin D can induce hyporesponsiveness to PAMPs (Pathogen-Associated Molecular Patterns) by downregulating the expression of TLR2 and TLR4 on monocytes that in turn have been associated with impaired production of TNF-α, suggesting a critical role of vitamin D in regulating TLR driven inflammation [71]. Importantly, a link between the DENV NS1 protein and activation of the inflammatory response via TLR2 and TLR4 impacting the progression of the disease has very recently been described [93, 128]. DENV NS1 antigens may induce the activation of TLR2 and TLR4 inducing high secretion of proinflammatory mediators that enhance endothelial dysfunction and permeability [46, 94, 129, 130]. Interestingly, it was reported that 1,25(OH)$_2$D$_3$ significantly reduces the levels of TLR2/TLR4 expression and of proinflammatory cytokines (TNF-α, IL-6, IL-12p70, and IL-1β) produced by U937 cells after exposure to DENV [72]. The same approach used in primary human monocytes and macrophages led to similar results, consistent with data obtained in our laboratory [19]. It has been suggested that vitamin D may regulate proinflammatory cytokine levels by targeting TLR activation signaling molecules (Figure 1). Indeed, it has been reported that treatment of monocytes with 1,25(OH)$_2$D$_3$ regulates TLR expression via the NF-κB pathway and reduces signaling of the mitogen-activated protein kinases MAPKs/p38 and p42/44 [19]. One of the most critical steps in NF-κB regulation is IκBα proteasomal degradation mediated by IKK (I kappa B Kinase) that leads to the nuclear entry of the NF-κB heterodimer p65/p50 to trans-activate gene expression, resulting in a decrease of inflammatory genes. Accordingly, a novel molecular mechanism has recently been described in which 1,25(OH)$_2$D$_3$ binding to VDR attenuates NF-κB activation by directly interacting with the IKKβ protein to block its activity and, consequently, the NF-κB-dependent inflammatory response [76]. Besides TLR2 and TLR4, it has been shown that vitamin D can also downregulate the intracellular TLR9 expression and, subsequently, lead to less secretion of IL-6 in response to TLR9 stimulation [77]. Although intracellular downregulation of some PRRs such as TLR3, TLR7/8, and RIG-I/MDA5 may affect the potential antiviral response induced by type I IFN, various reports have shown that vitamin D treatment does not affect the type I IFN-induced antiviral response against various viruses [69, 131, 132]. In fact, it has been reported that porcine rotavirus (PRV) infection induces CYP27B1-dependent generation of 1,25(OH)$_2$D$_3$ which leads to an increased expression of TLR3 and RIG-I that consequently enhance the type I IFN-dependent antiviral response [76].

3.1. Vitamin D and miRs: Potential Implications for Inflammation Balance.

Although vitamin D may impact distinct pathways and molecules to modulate inflammatory responses, current evidence suggests TLRs and TLR signaling mediators as main targets by which vitamin D modulates inflammation (Table 3) [6, 113, 133, 134]. However, a novel regulatory vitamin D mechanism in which TLR signaling/activation and miR function are associated has been recently documented, suggesting a crucial role of vitamin D and miRs for the host immune system homeostasis [15, 135, 136]. The participation of miRs as general regulatory mechanisms of initiation, propagation, and resolution of immune responses has been widely reviewed elsewhere [21, 137, 138]. Therefore, we discuss here its potential relationship with vitamin D activity in the control of inflammatory responses, attempting to extrapolate these findings to DENV infection.

The ability of vitamin D to regulate miRs and their emerging relationship have been proposed by means of several experimental and clinical approaches; however, the implications of their impact on inflammatory responses have only been studied in in vitro models [15, 20, 135, 136, 139]. In patient trials with vitamin D supplementation, significant differences in miR expression profiles have been reported, suggesting that dietary vitamin D may also globally regulate miR levels [15]. Although several mechanisms may be involved in regulating such a global effect, some authors have found that chromatin states may be altered by VDR activity, determining accessibility for binding of the transcription and regulation of activation or inhibition of transcription [140, 141]. This in turn could be of relevance for canonical VDR-VDRE-mediated transcription regulation. In fact, VDR-induced regulation of miRs via VDRE has been demonstrated for some miRs such as miR-182 and let-7a whose pri-miRs (Primary miR) have multiple VDR/RXR

binding sites, suggesting that these miRs could potentially be regulated by vitamin D metabolites [67, 142]. Moreover, a negative feedback loop between some miRNAs and VDR signaling has been reported. This is the case of miR-125b whose overexpression can reduce VDR/RXR protein levels. Since miR-125b is commonly downregulated in cancer cells, it has been proposed that such a decrease in miR-125b may result in the upregulation of VDR and in increasing antitumor effects driven by vitamin D in cancer cell models [136].

Additionally, it has been reported that VDR signaling may attenuate TLR-mediated inflammation by enhancing a negative feedback inhibition mechanism (Figure 1). A recent report has shown that VDR inactivation leads to a hyper-inflammatory response in LPS-cultured mice macrophages through overproduction of miR-155 which in turns downregulates the suppressor of the cytokine signaling (SOCS) family of proteins that are key components of the negative feedback loop regulating the intensity, duration, and quality of cytokine signaling [2, 143, 144]. As feedback inhibitors of inflammation, SOCS proteins are upregulated by inflammatory cytokines, and, in turn, they block cytokine signaling by targeting the JAK/STAT (Janus Kinase/Signal Transducer and Activator of Transcription) pathway [2]. Evidence suggests that SOCS inhibits the proinflammatory pathways of cytokines such as TNF-α, IL-6, and IFN-γ and can inhibit the LPS-induced inflammatory response by directly blocking TLR4 signaling by targeting the IL-1R-associated kinases (IRAK) 1 and 4 [20, 144]. Consequently, deletion of miR-155 attenuates 1,25(OH)$_2$D$_3$ suppression of LPS-induced inflammation, confirming that vitamin D stimulates SOCS1 by downregulating miR-155 [20]. Taken together, these results highlight the importance of the VDR pathways controlling the inflammatory response by modulating miRNA-155-SOCS1 interactions. Finally, an additional reinforcing issue that may validate the link between vitamin D activity and miRs is the fact that 1,25(OH)$_2$D$_3$ deficiency has been related to reduced leukotriene synthetic capacity in macrophages [145, 146]. Recently, it was reported that leukotriene B4 (LTB$_4$) can upregulate macrophage MyD88 (Myeloid Differentiation primary response-88) expression by decreasing SOCS-1 stability that is associated with the expression of proinflammatory miRs, such as miR-155, miR-146b, and miR-125b, and TLR4 activation in macrophages [147]. miR-146 has been also shown as a modulator of inflammatory responses mediated by TLR4/NF-κB and TNF-α [70]. Importantly, this miR has been found downregulated in patients with autoimmune disorders in which low levels of vitamin D have also been reported [148, 149]. These results suggest that vitamin D can orchestrate miR diversity involved in TLR signaling, thereby regulating inflammatory responses and activation of immune responses.

4. Insights into Vitamin D and DENV Infection

Little is known about the link between DENV infection and vitamin D; however, since severe dengue is associated with imbalanced production of proinflammatory cytokines, it is very tempting to suggest that vitamin D could play an important role in modulating the inflammatory responses during ongoing DENV infections. Although only few studies can illustrate a link between vitamin D activity and DENV infection or disease, these reports have provided preliminary epidemiological evidence supporting this novel hypothesis. Initially, it was reported that heterozygosity in the VDR gene was correlated with progression of dengue. It was shown in a small Vietnamese population where dengue is endemic that the low frequency of a dimorphic (T/t) "t" allele in the VDR gene was associated with dengue disease severity, suggesting a protective role of VDR activity against dengue disease progression [12]. Variations in VDR have also been associated with susceptibility to osteoporosis in humans and with reduced risk of tuberculosis and persistent hepatitis B virus infections [150–152], highlighting the importance of VDR variations in signaling and immune protection. Accordingly, a study revealed the association of the "T" allele with DHF, by showing that the "T" allele codes for a longer length VDR that is the least active form of VDR. Since vitamin D is known to suppress TNF-α, it is possible that such inappropriate VDR signaling may contribute to higher levels of inflammation, enhancing the susceptibility to severity of the disease [10]. Although the modulatory effect of vitamin D during DENV infection and disease has not been widely tested in human populations, initial studies have associated the effect of oral 25(OH)D$_3$ supplementation with antiviral responses, resistance, and overcoming of the disease. Specifically, a study reported the case of five DF patients that ameliorated the signs and symptoms of the disease, improving the overall clinical conditions and reducing the risk of disease progression [11]. Interestingly, this may be linked to other clinical approaches where oral supplementation with vitamin D enhanced the antiviral response to HCV [63], another RNA virus belonging also to the family Flaviviridae.

The potential antiviral mechanism of vitamin D against DENV has yet not been fully explored; however, certain reports support the proposal that vitamin D could perform anti-DENV effects and immunoregulatory functions on innate immune responses [10–12]. In line with this, the effect of vitamin D treatment of human monocytic cell lines on DENV infection was recently reported [72]. The authors showed that cell exposure to 1,25(OH)$_2$D$_3$ resulted in a significant reduction of DENV-infected cells, a variable modulation of TLR2 and TLR4, and reduced levels of secreted proinflammatory cytokines such as TNF-α, IL-6, and IL-1β after infection [72]. The molecular mechanisms by which vitamin D can elicit an antiviral and anti-inflammatory role towards DENV have not been fully described, and although we observed that monocyte-derived macrophages differentiated in the presence of 1,25(OH)$_2$D$_3$ are less susceptible to DENV infection and express lower levels of mannose receptor restricting binding of DENV to target cells (manuscript in preparation), further studies are required to confirm that vitamin D treatment confers both anti-inflammatory and antiviral responses. Another interesting mechanism that could support the antiviral activity of vitamin D is the VDR-induced regulation of miRs via VDRE. This has been demonstrated for some miRs, such as let-7a (Table 2), whose

pri-miR has multiple VDR/RXR binding sites that could potentially be regulated by vitamin D [67, 142]. miR let-7a belongs to a highly conserved family of miRs that contains other miRs previously reported to inhibit DENV replicative activity, such as let-7c [68]. Besides the members of the let-7 family, other miRs have also been associated with suppression of DENV infection and the inflammatory responses against the virus, as discussed below.

4.1. MicroRNAs in DENV Infection. Viruses strictly depend on cellular mechanisms for their replication; therefore, there is an obligatory interaction between the virus and the host RNA silencing machinery. Although virus-derived small interfering RNAs may induce changes in cellular mRNA and miR expression profiles to induce replication, cellular miRs can also target viral sequences or induce antiviral protein expression to inhibit viral replication and translation [153]. Indeed, during DENV infection, several cellular miRs have been reported to have an effect on the replicative activity of the virus and the permissiveness of the host cells. Although some host miRs can also enhance DENV replication [81, 154], here we highlight the miRs affecting DENV replicative activity and modulating the immune response (Table 4).

The expression levels of different miRs regulated during DENV infection have been screened in the hepatic cell line Huh-7. This approach identified miR let-7c as a key regulator of the viral replicative cycle that affects viral replication and the oxidative stress immune response through the protein Heme Oxygenase-1 (HO-1) by activating its transcription factor BACH1 (Basic Leucine Zipper Transcription Factor-1) [68]. In addition, it was recently reported that, after DENV-2 infection of the C6/36 cell line, endogenous miR-252 is highly induced and associated with a decreased level of viral RNA copies. This antiviral effect was explained by the fact that miR-252 targets the DENV-2 E protein gene sequence, downregulating its expression and therefore acting as an antiviral regulator [78]. Although DENV can escape the immune system by decreasing the production of type I IFN due to DENV NS5 and NS4B activity [42, 97], DENV infection also induces the upregulation of the cellular miR-30e* that suppresses DENV replication by increasing IFN-β production. This antiviral effect of miR-30e* depends mainly on NF-κB activation by targeting the NF-κB inhibitor IκBα in DENV-permissive cells [79]. This antiviral effect induced by signaling of type I IFN is also promoted by miR-155 that has been reported to control virus-induced immune responses in models of infection with other members of the *Flavivirus* genus such as HCV [155–157]. In this latter model, the antiviral effect greatly depended on miR-155 targeting SOCS-1. This observation is in accordance with a study in which elevated expression of miR-150 in patients with DHF was correlated with suppression of SOCS-1 expression in monocytes [80] that in turn could be linked to the fact that vitamin D controls inflammatory responses through modulation of SOCS by downregulating miR-155 [20].

Although it has remained unclear whether endogenous miRs can interfere with viral replicative activity by targeting DENV sequences or viral mRNAs, some experimental approaches have shown the importance of miRs in restricting

TABLE 4: Summary of miRs regulating DENV-induced inflammatory response and viral replicative activity.

miRNA	Target	Cell line	Refs.
let-7e	$3'$-UTR of IL-6	Human peripheral blood mononuclear cells	[73]
let-7c	HO-1 protein and the transcription factor BACH1	Huh-7 human hepatic cell line	[68]
miR-252	DENV envelope E protein	*Aedes albopictus* C6/36 cell line	[78]
miR-30e*	IkBα in DENV-permissive cells and IFN-β production	Peripheral blood mononuclear cells and U937 and HeLa cell lines	[79]
miR-150	$3'$-UTR of SOCS-1	Peripheral blood mononuclear cells and monocytes	[80]
miR-122	$3'$-UTR of the DENV genome/mRNA	BHK-21, HepG2, and Huh-7 cell lines	[81]
miR-142	$3'$-UTR of the DENV genome/mRNA	Human dendritic cells and macrophages	[82]
miR-133a	$3'$-UTR of PTB; $3'$-UTR of the DENV genome/mRNA	Mouse C2C12 cells and Vero cells	[83, 84]
miR-548	$5'$-UTR SLA (Stem Loop A) DENV	U937 monocyte/macrophages	[85]
miR-223	Microtubule destabilizing protein stathmin 1 (STMN-1)	EA.hy926 endothelial cell line	[86]

viral replication through this mechanism [85, 158–160]. Some artificial miRs (amiRs) have been described as targeting the highly conserved regions of the DENV-2 genome and promoting efficient inhibition of virus replication [158]. Using DENV subgenomic replicons carrying the specific miR recognition element (MRE) for miR-122 in the $3'$-UTR of the DENV genome/mRNA, some authors have shown that the liver-specific miR-122 suppresses translation and replication of DENV by targeting this MRE sequence [81]. Likewise, the insertion of the MRE for the hematopoietic specific miR-142 into the DENV-2 genome restricts replication of the virus in DCs and macrophages, highlighting the importance of this hematopoietic miR in dissemination of the virus [82]. In addition, DENV replication is enhanced by the interaction of the viral genome $3'$-UTR and the host polypyrimidine tract binding (PTB) protein that translocates from the nucleus to the cytoplasm facilitating DENV replication [36, 161, 162]. However, the PTB mRNA $3'$-UTR contains MREs that can be targeted by miR-133a, providing a mechanism for the downregulation of the PTB protein expression levels [163]. Moreover, in our group, we found that miR-133a contains target sites in the $3'$-UTR sequence of the 4 DENV serotypes and that overexpression of miR-133a in Vero cells was associated with decreased DENV-2 replication activity [84]. All these data suggest a possible antiviral mechanism via miR-133a targeting the PTB protein mRNA and the DENV $3'$-UTR sequence. Furthermore, we also showed that miR-744 and miR-484 can downregulate DENV replication by targeting the $3'$UTR of the DENV RNA genome [Betancur et al., submitted]. In addition, the cellular miR-548g-3p has been identified as displaying antiviral activity by targeting the $5'$-UTR SLA (Stem Loop A) promoter of the four DENV serotypes, thus, repressing viral replication and expression of viral proteins, independently of interferon signaling [85]. Moreover, overexpression of miR-223 inhibited replication of DENV in an endothelial cell-like cell line. The authors showed that miR-223 inhibits DENV by negatively regulating the microtubule destabilizing protein stathmin 1 (STMN-1) that is crucial for reorganization of microtubules and later replication of the virus. In addition, this study identified that

the transcription factors C/EBP-α and EIF2 are regulators of miR-223 expression after DENV infection [86].

Although little is known regarding the variations in miR expression in DENV-infected individuals, a recent study showed the expression profile of the miRs in blood samples of DEN-infected patients. The authors report 12 miRs that were specifically altered upon acute dengue and 17 miRs that could potentially be associated with specific dengue-related complications [164]. In addition, another profiling study reported abundance changes in the expression of some miRs in DENV-infected peripheral blood monocytes. Importantly, let-7e was among the miRs with the most significant regulation which, besides anti-DENV activity, may be of crucial importance for the modulation of inflammatory responses. Specifically, let-7e shares matching sequences with the $3'$UTR mRNA of IL-6 and CCL3, as well as of other cytokines, highlighting a key role of miRs in immune response homeostasis during DENV infection (Figure 1) [67, 73, 86]. Likewise, miR-223 that also shares antiviral activity against DENV has been shown to have an important effect on the inflammatory response by regulating IL-β and IL-6 through IKKα and MKP-5 [86, 165, 166], stressing its potential contribution in DENV pathogenesis control. Since a link between vitamin D and miR expression has been established, but no reports discuss their combined implications for DENV antiviral and inflammatory response, we hypothesized here a vitamin D and miR interplay that could modulate DENV pathogenesis, opening new horizons in the therapeutic field of dengue disease.

5. Concluding Remarks and Future Perspectives

Severe dengue disease symptoms and DENV infection are characterized by overproduction of proinflammatory cytokines driven mainly by activation of several PRRs [29].

Here, we hypothesize that vitamin D may contribute to avoiding DENV infection and disease progression, especially through the modulation of miRs/TLRs that enhance the antiviral activity and regulate the inflammatory response. Although vitamin D's antiviral mechanism has not been fully elucidated, it may be linked to vitamin D's ability to control the permissiveness of DENV target cells and the virus-induced proinflammatory responses [72]. However, a better understanding of these mechanisms is required to provide interesting clues regarding DENV pathogenesis and dengue disease treatment. Certainly, epidemiological and experimental evidence describe an overall positive vitamin D-related immune effect in which increased levels of vitamin D and variants in the VDR receptor are associated with reduction of viral replication, decreased risk of infection, lower disease severity, and better outcome of the dengue symptoms [9–12, 72]. Additionally, the emerging relationships between vitamin D, the TLR signaling pathway, and its regulation by miRs are beginning to gain critical importance in infectious diseases. Indeed, as discussed above, several DENV infection studies have started to illustrate these vitamin D regulatory features that could be key mechanisms for the control of virus replication and homeostasis of the inflammatory response, thus making this hormone a special candidate for therapeutic strategies [127]. Although most of the studies have focused on the effects of vitamin D induced in dendritic cells and macrophages, others have also described the same immunoregulatory effects on other cell populations of the immune system such as CD8$^+$ T cells, NK cells, and B cells [167–169] suggesting their impact not only on DENV target cells but also at the level of cells associated with virus clearance. All the data discussed here suggest that vitamin D could constitute a strong potential strategy to modulate the "cytokine storm" that occurs during ongoing DENV infections and the progression to severe states of the disease. Although it is important to note that such a global effect on the inflammatory activity could weaken the host response to other opportunistic pathogens, it has been suggested that while vitamin D may reduce inflammatory markers during viral infections, it also exerts protective effects against coinfections with other opportunistic pathogens [14, 106]. Moreover, its clinical effectiveness has been tested by improving the overall physical condition of DENV patients and reducing the progression of the disease [11]. Although incoming supplementary trials are required to fully elucidate the therapeutic relevance of vitamin D, it is evident that this hormone may be an excellent alternative of a natural immune-regulatory agent capable of modulating the innate immune response against DENV, which will provide crucial information to understand and design strategies to treat and control progression of dengue disease. Although further experimental studies are required to boost the understanding of vitamin D in the regulation of inflammation and antiviral response against DENV infection, the information discussed above highlights the features of vitamin D in immune regulation as an exciting research field and as an efficient and low-cost therapeutic procedure against DENV and possibly other viral infections.

Acknowledgments

The authors thank Anne-Lise Haenni for reading the paper and for her constructive and valuable comments. This study was supported by COLCIENCIAS, Grant no. 111556933443, and Universidad de Antioquia, UdeA, CODI (Mediana Cuantía), Acta 624.

References

[1] B. A. Beutler, "TLRs and innate immunity," *Blood*, vol. 113, no. 7, pp. 1399–1407, 2009.

[2] A. Dalpke, K. Heeg, H. Bartz, and A. Baetz, "Regulation of innate immunity by suppressor of cytokine signaling (SOCS) proteins," *Immunobiology*, vol. 213, no. 3-4, pp. 225–235, 2008.

[3] B. E. E. Martina, "Dengue pathogenesis: a disease driven by the host response," *Science Progress*, vol. 97, part 3, pp. 197–214, 2014.

[4] T. Pang, M. J. Cardosa, and M. G. Guzman, "Of cascades and perfect storms: the immunopathogenesis of dengue haemorrhagic fever-dengue shock syndrome (DHF/DSS)," *Immunology and Cell Biology*, vol. 85, no. 1, pp. 43–45, 2007.

[5] E. van Etten, K. Stoffels, C. Gysemans, C. Mathieu, and L. Overbergh, "Regulation of vitamin D homeostasis: implications for the immune system," *Nutrition Reviews*, vol. 66, no. 2, pp. S125–S134, 2008.

[6] X. Guillot, L. Semerano, N. Saidenberg-Kermanac'h, G. Falgarone, and M.-C. Boissier, "Vitamin D and inflammation," *Joint Bone Spine*, vol. 77, no. 6, pp. 552–557, 2010.

[7] V. V. Costa, C. T. Fagundes, D. G. Souza, and D. M. M. Teixeira, "Inflammatory and innate immune responses in dengue infection: protection versus disease induction," *The American Journal of Pathology*, vol. 182, no. 6, pp. 1950–1961, 2013.

[8] Z. R. Brenner, A. B. Miller, L. C. Ayers, and A. Roberts, "The role of vitamin D in critical illness," *Critical Care Nursing Clinics of North America*, vol. 24, no. 4, pp. 527–540, 2012.

[9] S. Ahmed, J. L. Finkelstein, A. M. Stewart et al., "Micronutrients and dengue," *The American Journal of Tropical Medicine and Hygiene*, vol. 91, no. 5, pp. 1049–1056, 2014.

[10] K. Alagarasu, T. Honap, A. P. Mulay, R. V. Bachal, P. S. Shah, and D. Cecilia, "Association of vitamin D receptor gene polymorphisms with clinical outcomes of dengue virus infection," *Human Immunology*, vol. 73, no. 11, pp. 1194–1199, 2012.

[11] E. Sánchez-Valdéz, M. Delgado-Aradillas, J. A. Torres-Martínez, and J. M. Torres-Benítez, "Clinical response in patients with dengue fever to oral calcium plus vitamin D administration: study of 5 cases," *Proceedings of the Western Pharmacology Society*, vol. 52, pp. 14–17, 2009.

[12] H. Loke, D. Bethell, C. X. T. Phuong et al., "Susceptibility to dengue hemorrhagic fever in vietnam: evidence of an association with variation in the vitamin D receptor and FCγ receptor IIA genes," *The American Journal of Tropical Medicine and Hygiene*, vol. 67, no. 1, pp. 102–106, 2002.

[13] J. A. Beard, A. Bearden, and R. Striker, "Vitamin D and the antiviral state," *Journal of Clinical Virology*, vol. 50, no. 3, pp. 194–200, 2011.

[14] E. Borella, G. Nesher, E. Israeli, and Y. Shoenfeld, "Vitamin D: a new anti-infective agent?" *Annals of the New York Academy of Sciences*, vol. 1317, no. 1, pp. 76–83, 2014.

[15] A. A. Giangreco and L. Nonn, "The sum of many small changes: microRNAs are specifically and potentially globally altered by

vitamin D_3 metabolites," *The Journal of Steroid Biochemistry and Molecular Biology*, vol. 136, no. 1, pp. 86–93, 2013.

[16] L. He and G. J. Hannon, "MicroRNAs: small RNAs with a big role in gene regulation," *Nature Reviews. Genetics*, vol. 5, no. 7, pp. 522–531, 2004.

[17] J. Han, Y. Lee, K.-H. Yeom, Y.-K. Kim, H. Jin, and V. N. Kim, "The Drosha-DGCR8 complex in primary microRNA processing," *Genes & Development*, vol. 18, no. 24, pp. 3016–3027, 2004.

[18] L.-A. MacFarlane and P. R. Murphy, "MicroRNA: biogenesis, function and role in cancer," *Current Genomics*, vol. 11, no. 7, pp. 537–561, 2010.

[19] Y. Zhang, D. Y. M. Leung, B. N. Richers et al., "Vitamin D inhibits monocyte/macrophage proinflammatory cytokine production by targeting MAPK phosphatase-1," *The Journal of Immunology*, vol. 188, no. 5, pp. 2127–2135, 2012.

[20] Y. Chen, W. Liu, T. Sun et al., "1,25-dihydroxyvitamin D promotes negative feedback regulation of TLR signaling via targeting microRNA-155-SOCS1 in macrophages," *The Journal of Immunology*, vol. 190, no. 7, pp. 3687–3695, 2013.

[21] E. Sonkoly, M. Ståhle, and A. Pivarcsi, "MicroRNAs and immunity: novel players in the regulation of normal immune function and inflammation," *Seminars in Cancer Biology*, vol. 18, no. 2, pp. 131–140, 2008.

[22] S. R. S. Hadinegoro, "The revised WHO dengue case classification: does the system need to be modified?" *Paediatrics and International Child Health*, vol. 32, supplement 1, pp. 33–38, 2012.

[23] M. S. Mustafa, V. Rasotgi, S. Jain, and V. Gupta, "Discovery of fifth serotype of dengue virus (DENV-5): a new public health dilemma in dengue control," *Medical Journal Armed Forces India*, vol. 71, no. 1, pp. 67–70, 2015.

[24] B.-A. Coller, A. D. T. Barrett, S. J. Thomas, J. Whitehorn, and C. P. Simmons, "The pathogenesis of dengue," *Vaccine*, vol. 29, no. 42, pp. 7221–7228, 2011.

[25] A. Wilder-Smith, E.-E. Ooi, S. G. Vasudevan, and D. J. Gubler, "Update on dengue: epidemiology, virus evolution, antiviral drugs, and vaccine development," *Current Infectious Disease Reports*, vol. 12, no. 3, pp. 157–164, 2010.

[26] M. E. Wilson and L. H. Chen, "Dengue: update on epidemiology," *Current Infectious Disease Reports*, vol. 17, no. 1, p. 457, 2015.

[27] S. Bhatt, P. W. Gething, O. J. Brady et al., "The global distribution and burden of dengue," *Nature*, vol. 496, no. 7446, pp. 504–507, 2013.

[28] O. J. Brady, P. W. Gething, S. Bhatt et al., "Refining the global spatial limits of dengue virus transmission by evidence-based consensus," *PLoS Neglected Tropical Diseases*, vol. 6, no. 8, Article ID e1760, 2012.

[29] S. B. Halstead, "Controversies in dengue pathogenesis," *Paediatrics and International Child Health*, vol. 32, no. 1, pp. 5–9, 2012.

[30] S. Yacoub, J. Mongkolsapaya, and G. Screaton, "The pathogenesis of dengue," *Current Opinion in Infectious Diseases*, vol. 26, no. 3, pp. 284–289, 2013.

[31] B. E. E. Martina, P. Koraka, and A. D. M. E. Osterhaus, "Dengue virus pathogenesis: an integrated view," *Clinical Microbiology Reviews*, vol. 22, no. 4, pp. 564–581, 2009.

[32] J. Flipse, J. Wilschut, and J. M. Smit, "Molecular mechanisms involved in antibody-dependent enhancement of dengue virus infection in humans," *Traffic*, vol. 14, no. 1, pp. 25–35, 2013.

[33] T. Kawai and S. Akira, "The role of pattern-recognition receptors in innate immunity: update on Toll-like receptors," *Nature Immunology*, vol. 11, no. 5, pp. 373–384, 2010.

[34] S.-J. L. Wu, G. Grouard-Vogel, W. Sun et al., "Human skin Langerhans cells are targets of dengue virus infection," *Nature Medicine*, vol. 6, no. 7, pp. 816–820, 2000.

[35] Z. Kou, M. Quinn, H. Chen et al., "Monocytes, but not T or B cells, are the principal target cells for dengue virus (DV) infection among human peripheral blood mononuclear cells," *Journal of Medical Virology*, vol. 80, no. 1, pp. 134–146, 2008.

[36] E. G. Acosta, A. Kumar, and R. Bartenschlager, "Revisiting dengue virus-host cell interaction: new insights into molecular and cellular virology," *Advances in Virus Research*, vol. 88, pp. 1–109, 2014.

[37] S. Urcuqui-Inchima, C. Patiño, S. Torres, A.-L. Haenni, and F. J. Díaz, "Recent developments in understanding dengue virus replication," *Advances in Virus Research*, vol. 77, pp. 1–39, 2010.

[38] D. G. Nielsen, "The relationship of interacting immunological components in dengue pathogenesis," *Virology Journal*, vol. 6, article 211, 2009.

[39] Z. Liang, S. Wu, Y. Li et al., "Activation of toll-like receptor 3 impairs the dengue virus serotype 2 replication through induction of IFN-β in cultured hepatoma cells," *PLoS ONE*, vol. 6, no. 8, Article ID e23346, 2011.

[40] A. M. A. Nasirudeen, H. H. Wong, P. Thien, S. Xu, K.-P. Lam, and D. X. Liu, "RIG-I, MDA5 and TLR3 synergistically play an important role in restriction of dengue virus infection," *PLoS Neglected Tropical Diseases*, vol. 5, no. 1, article e926, 2011.

[41] S. Jensen and A. R. Thomsen, "Sensing of rna viruses: a review of innate immune receptors involved in recognizing RNA virus invasion," *Journal of Virology*, vol. 86, no. 6, pp. 2900–2910, 2012.

[42] J. Morrison, S. Aguirre, and A. Fernandez-Sesma, "Innate immunity evasion by dengue virus," *Viruses*, vol. 4, no. 3, pp. 397–413, 2012.

[43] A. Huerta-Zepeda, C. Cabello-Gutiérrez, J. Cime-Castillo et al., "Crosstalk between coagulation and inflammation during Dengue virus infection," *Thrombosis and Haemostasis*, vol. 99, no. 5, pp. 936–943, 2008.

[44] M. Yu and S. J. Levine, "Toll-like receptor 3, RIG-I-like receptors and the NLRP3 inflammasome: key modulators of innate immune responses to double-stranded RNA viruses," *Cytokine & Growth Factor Reviews*, vol. 22, no. 2, pp. 63–72, 2011.

[45] M. Yoneyama, M. Kikuchi, K. Matsumoto et al., "Shared and unique functions of the DExD/H-box helicases RIG-I, MDA5, and LGP2 in antiviral innate immunity," *Journal of Immunology*, vol. 175, no. 5, pp. 2851–2858, 2005.

[46] B. E. E. B. Martina, P. Koraka, and A. D. M. E. Osterhaus, "Dengue virus pathogenesis: an integrated view," *Clinical Microbiology Reviews*, vol. 22, no. 4, pp. 564–581, 2009.

[47] S. M. Stamatovic, R. F. Keep, S. L. Kunkel, and A. V. Andjelkovic, "Potential role of MCP-1 in endothelial cell tight junction 'opening': signaling via Rho and Rho kinase," *Journal of Cell Science*, vol. 116, no. 22, pp. 4615–4628, 2003.

[48] Y.-R. Lee, M.-T. Liu, H.-Y. Lei et al., "MCP1, a highly expressed chemokine in dengue haemorrhagic fever/dengue shock syndrome patients, may cause permeability change, possibly through reduced tight junctions of vascular endothelium cells," *The Journal of General Virology*, vol. 87, no. 12, pp. 3623–3630, 2006.

[49] J. F. Kelley, P. H. Kaufusi, and V. R. Nerurkar, "Dengue hemorrhagic fever-associated immunomediators induced via maturation of dengue virus nonstructural 4B protein in monocytes modulate endothelial cell adhesion molecules and human microvascular endothelial cells permeability," *Virology*, vol. 422, no. 2, pp. 326–337, 2012.

[50] Y. Huang, H. Lei, H. Liu, Y. Lin, C. Liu, and T. Yeh, "Dengue virus infects human endothelial cells and induces IL-6 and IL-8 production," *The American Journal of Tropical Medicine and Hygiene*, vol. 63, no. 1, pp. 71–75, 2000.

[51] H. Puerta-Guardo, A. Raya-Sandino, L. González-Mariscal et al., "The cytokine response of U937-derived macrophages infected through antibody-dependent enhancement of dengue virus disrupts cell apical-junction complexes and increases vascular permeability," *Journal of Virology*, vol. 87, no. 13, pp. 7486–7501, 2013.

[52] D. Talavera, A. M. Castillo, M. C. Dominguez, A. Escobar Gutierrez, and I. Meza, "IL8 release, tight junction and cytoskeleton dynamic reorganization conducive to permeability increase are induced by dengue virus infection of microvascular endothelial monolayers," *The Journal of General Virology*, vol. 85, no. 7, pp. 1801–1813, 2004.

[53] T.-T. Tsai, Y.-J. Chuang, Y.-S. Lin, S.-W. Wan, C.-L. Chen, and C.-F. Lin, "An emerging role for the anti-inflammatory cytokine interleukin-10 in dengue virus infection," *Journal of Biomedical Science*, vol. 20, no. 1, article 40, 2013.

[54] P. Liu, M. Woda, F. A. Ennis, and D. H. Libraty, "Dengue virus infection differentially regulates endothelial barrier function over time through type I interferon effects," *Journal of Infectious Diseases*, vol. 200, no. 2, pp. 191–201, 2009.

[55] C. Lee, O. Buznyk, L. Kuffova et al., "Cathelicidin LL-37 and HSV-1 corneal infection: peptide versus gene therapy," *Translational Vision Science & Technology*, vol. 3, no. 3, article 4, 2014.

[56] S. Tripathi, G. Wang, M. White, L. Qi, J. Taubenberger, and K. L. Hartshorn, "Antiviral activity of the human cathelicidin, LL-37, and derived peptides on seasonal and pandemic influenza A viruses," *PLoS ONE*, vol. 10, no. 4, Article ID e0124706, 2015.

[57] P. Bergman, L. Walter-Jallow, K. Broliden, B. Agerberth, and J. Söderlund, "The antimicrobial peptide LL-37 inhibits HIV-1 replication," *Current HIV Research*, vol. 5, no. 4, pp. 410–415, 2007.

[58] V. Tangpricha, S. E. Judd, T. R. Ziegler et al., "LL-37 concentrations and the relationship to vitamin D, immune status, and inflammation in HIV-infected children and young adults," *AIDS Research and Human Retroviruses*, vol. 30, no. 7, pp. 670–676, 2014.

[59] A. Schögler, R. J. Muster, E. Kieninger et al., "Vitamin D represses rhinovirus replication in cystic fibrosis cells by inducing LL-37," *The European Respiratory Journal*, vol. 47, no. 2, pp. 520–530, 2016.

[60] W. Aguilar-Jiménez, W. Zapata, A. Caruz, and M. T. Rugeles, "High transcript levels of vitamin D receptor are correlated with higher mRNA expression of human beta defensins and IL-10 in mucosa of HIV-1-exposed seronegative individuals," *PLoS ONE*, vol. 8, no. 12, Article ID e82717, 2013.

[61] M. Gal-Tanamy, L. Bachmetov, A. Ravid et al., "Vitamin D: an innate antiviral agent suppressing hepatitis C virus in human hepatocytes," *Hepatology*, vol. 54, no. 5, pp. 1570–1579, 2011.

[62] H. Farnik, J. Bojunga, A. Berger et al., "Low vitamin D serum concentration is associated with high levels of hepatitis B virus replication in chronically infected patients," *Hepatology*, vol. 58, no. 4, pp. 1270–1276, 2013.

[63] S. Yokoyama, S. Takahashi, Y. Kawakami et al., "Effect of vitamin D supplementation on pegylated interferon/ribavirin therapy for chronic hepatitis C genotype 1b: a randomized controlled trial," *Journal of Viral Hepatitis*, vol. 21, no. 5, pp. 348–356, 2014.

[64] B. Terrier, F. Carrat, G. Geri et al., "Low 25-OH vitamin D serum levels correlate with severe fibrosis in HIV-HCV co-infected patients with chronic hepatitis," *Journal of Hepatology*, vol. 55, no. 4, pp. 756–761, 2011.

[65] G. R. Campbell and S. A. Spector, "Autophagy induction by vitamin D inhibits both *Mycobacterium tuberculosis* and human immunodeficiency virus type 1," *Autophagy*, vol. 8, no. 10, pp. 1523–1525, 2012.

[66] G. R. Campbell and S. A. Spector, "Hormonally active vitamin D3 (1α,25-dihydroxycholecalciferol) triggers autophagy in human macrophages that inhibits HIV-1 infection," *The Journal of Biological Chemistry*, vol. 286, no. 21, pp. 18890–18902, 2011.

[67] H. Guan, C. Liu, Z. Chen et al., "1,25-dihydroxyvitamin D3 up-regulates expression of hsa-let-7a-2 through the interaction of VDR/VDRE in human lung cancer A549 cells," *Gene*, vol. 522, no. 2, pp. 142–146, 2013.

[68] M. Escalera-Cueto, I. Medina-Martínez, R. M. Del Angel, J. Berumen-Campos, A. L. Gutiérrez-Escolano, and M. Yocupicio-Monroy, "Let-7c overexpression inhibits dengue virus replication in human hepatoma Huh-7 cells," *Virus Research*, vol. 196, pp. 105–112, 2015.

[69] N. Fitch, A. B. Becker, and K. T. HayGlass, "Vitamin D [1,25(OH)$_2$D$_3$] differentially regulates human innate cytokine responses to bacterial versus viral pattern recognition receptor stimuli," *The Journal of Immunology*, vol. 196, no. 7, pp. 2965–2972, 2016.

[70] E.-A. Ye and J. J. Steinle, "miR-146a attenuates inflammatory pathways mediated by TLR4/NF-κB and TNFα to protect primary human retinal microvascular endothelial cells grown in high glucose," *Mediators of Inflammation*, vol. 2016, Article ID 3958453, 9 pages, 2016.

[71] K. Sadeghi, B. Wessner, U. Laggner et al., "Vitamin D3 down-regulates monocyte TLR expression and triggers hyporesponsiveness to pathogen-associated molecular patterns," *European Journal of Immunology*, vol. 36, no. 2, pp. 361–370, 2006.

[72] H. Puerta-Guardo, F. Medina, S. I. De la Cruz Hernández, V. H. Rosales, J. E. Ludert, and R. M. del Angel, "The 1α,25-dihydroxy-vitamin D3 reduces dengue virus infection in human myelomonocyte (U937) and hepatic (Huh-7) cell lines and cytokine production in the infected monocytes," *Antiviral Research*, vol. 94, no. 1, pp. 57–61, 2012.

[73] Y. Qi, Y. Li, L. Zhang, and J. Huang, "MicroRNA expression profiling and bioinformatic analysis of dengue virus-infected peripheral blood mononuclear cells," *Molecular Medicine Reports*, vol. 7, no. 3, pp. 791–798, 2013.

[74] V. Gonzalez-Pardo, N. D'Elia, A. Verstuyf, R. Boland, and A. Russo de Boland, "NFκB pathway is down-regulated by 1α,25(OH)$_2$-vitamin D$_3$ in endothelial cells transformed by Kaposi sarcoma-associated herpes virus G protein coupled receptor," *Steroids*, vol. 77, no. 11, pp. 1025–1032, 2012.

[75] D. Bhaumik, G. K. Scott, S. Schokrpur, C. K. Patil, J. Campisi, and C. C. Benz, "Expression of microRNA-146 suppresses NF-κB activity with reduction of metastatic potential in breast cancer cells," *Oncogene*, vol. 27, no. 42, pp. 5643–5647, 2008.

[76] Y. Chen, J. Zhang, X. Ge, J. Du, D. K. Deb, and Y. C. Li, "Vitamin D receptor inhibits nuclear factor κb activation by interacting with IκB kinase β protein," *The Journal of Biological Chemistry*, vol. 288, no. 27, pp. 19450–19458, 2013.

[77] L. J. Dickie, L. D. Church, L. R. Coulthard, R. J. Mathews, P. Emery, and M. F. McDermott, "Vitamin D_3 downregulates intracellular Toll-like receptor 9 expression and Toll-like receptor 9-induced IL-6 production in human monocytes," *Rheumatology*, vol. 49, no. 8, pp. 1466–1471, 2010.

[78] H. Yan, Y. Zhou, Y. Liu, Y. Deng, S. Puthiyakunnon, and X. Chen, "miR-252 of the Asian tiger mosquito *Aedes albopictus* regulates dengue virus replication by suppressing the expression of the dengue virus envelope protein," *Journal of Medical Virology*, vol. 86, no. 8, pp. 1428–1436, 2014.

[79] X. Zhu, Z. He, Y. Hu et al., "MicroRNA-30e* suppresses dengue virus replication by promoting NF-κB-dependent IFN production," *PLoS Neglected Tropical Diseases*, vol. 8, no. 8, p. e3088, 2014.

[80] R.-F. Chen, K. D. Yang, I.-K. Lee et al., "Augmented miR-150 expression associated with depressed SOCS1 expression involved in dengue haemorrhagic fever," *The Journal of Infection*, vol. 69, no. 4, pp. 366–374, 2014.

[81] T.-C. Lee, Y.-L. Lin, J.-T. Liao et al., "Utilizing liver-specific microRNA-122 to modulate replication of dengue virus replicon," *Biochemical and Biophysical Research Communications*, vol. 396, no. 3, pp. 596–601, 2010.

[82] A. M. Pham, R. A. Langlois, and B. R. tenOever, "Replication in cells of hematopoietic origin is necessary for dengue virus dissemination," *PLoS Pathogens*, vol. 8, no. 1, Article ID e1002465, 2012.

[83] P. L. Boutz, G. Chawla, P. Stoilov, and D. L. Black, "MicroRNAs regulate the expression of the alternative splicing factor nPTB during muscle development," *Genes & Development*, vol. 21, no. 1, pp. 71–84, 2007.

[84] J. A. Castillo, J. C. Castrillón, M. Diosa-Toro et al., "Complex interaction between dengue virus replication and expression of miRNA-133a," *BMC Infectious Diseases*, vol. 16, no. 1, article 29, 2015.

[85] W. Wen, Z. He, Q. Jing et al., "Cellular microRNA-miR-548g-3p modulates the replication of dengue virus," *The Journal of Infection*, vol. 70, no. 6, pp. 631–640, 2015.

[86] N. Wu, N. Gao, D. Fan, J. Wei, J. Zhang, and J. An, "miR-223 inhibits dengue virus replication by negatively regulating the microtubule-destabilizing protein STMN1 in EAhy926 cells," *Microbes and Infection*, vol. 16, no. 11, pp. 911–922, 2014.

[87] A. K. I. Falconar, "The dengue virus nonstructural-1 protein (NS1) generates antibodies to common epitopes on human blood clotting, integrin/adhesin proteins and binds to human endothelial cells: Potential implications in haemorrhagic fever pathogenesis," *Archives of Virology*, vol. 142, no. 5, pp. 897–916, 1997.

[88] A. L. Rothman, "Immunity to dengue virus: a tale of original antigenic sin and tropical cytokine storms," *Nature Reviews Immunology*, vol. 11, no. 8, pp. 532–543, 2011.

[89] T.-M. Yeh, S.-H. Liu, K.-C. Lin et al., "Dengue virus enhances thrombomodulin and ICAM-1 expression through the macrophage migration inhibitory factor induction of the MAPK and PI3K signaling pathways," *PLoS ONE*, vol. 8, no. 1, Article ID e55018, 2013.

[90] K. Ellencrona, A. Syed, and M. Johansson, "Flavivirus NS5 associates with host-cell proteins zonula occludens-1 (ZO-1) and regulating synaptic membrane exocytosis-2 (RIMS2) via an internal PDZ binding mechanism," *Biological Chemistry*, vol. 390, no. 4, pp. 319–323, 2009.

[91] M. D. de Kruif, T. E. Setiati, A. T. A. Mairuhu et al., "Differential gene expression changes in children with severe dengue virus

infections," *PLoS Neglected Tropical Diseases*, vol. 2, no. 4, article e215, 2008.

[92] S. Ubol, P. Masrinoul, J. Chaijaruwanich, S. Kalayanarooj, T. Charoensirisuthikul, and J. Kasisith, "Differences in global gene expression in peripheral blood mononuclear cells indicate a significant role of the innate responses in progression of dengue fever but not dengue hemorrhagic fever," *The Journal of Infectious Diseases*, vol. 197, no. 10, pp. 1459–1467, 2008.

[93] J. Chen, M. M.-L. Ng, and J. J. H. Chu, "Activation of TLR2 and TLR6 by dengue NS1 protein and its implications in the immunopathogenesis of dengue virus infection," *PLoS Pathogens*, vol. 11, no. 7, Article ID e1005053, 2015.

[94] P. R. Beatty, H. Puerta-Guardo, S. S. Killingbeck, D. R. Glasner, K. Hopkins, and E. Harris, "Dengue virus NS1 triggers endothelial permeability and vascular leak that is prevented by NS1 vaccination," *Science Translational Medicine*, vol. 7, no. 304, Article ID 304ra141, 2015.

[95] S. Torres, J. C. Hernández, D. Giraldo et al., "Differential expression of Toll-like receptors in dendritic cells of patients with dengue during early and late acute phases of the disease," *PLoS Neglected Tropical Diseases*, vol. 7, no. 2, Article ID e2060, 2013.

[96] S. Pagni and A. Fernandez-Sesma, "Evasion of the human innate immune system by dengue virus," *Immunologic Research*, vol. 54, no. 1–3, pp. 152–159, 2012.

[97] J. L. Muñoz-Jordán, M. Laurent-Rolle, J. Ashour et al., "Inhibition of alpha/beta interferon signaling by the NS4B protein of flaviviruses," *Journal of Virology*, vol. 79, no. 13, pp. 8004–8013, 2005.

[98] J. L. Miller, B. J. M. deWet, L. Martinez-Pomares et al., "The mannose receptor mediates dengue virus infection of macrophages," *PLoS Pathogens*, vol. 4, no. 2, article e17, 2008.

[99] M.-F. Wu, S.-T. Chen, A.-H. Yang et al., "CLEC5A is critical for dengue virus-induced inflammasome activation in human macrophages," *Blood*, vol. 121, no. 1, pp. 95–106, 2013.

[100] E. Schaeffer, V. Flacher, V. Papageorgiou et al., "Dermal CD14+ dendritic cell and macrophage infection by dengue virus is stimulated by interleukin-4," *Journal of Investigative Dermatology*, vol. 135, no. 7, pp. 1743–1751, 2015.

[101] M.-F. Wu, S.-T. Chen, and S.-L. Hsieh, "Distinct regulation of dengue virus-induced inflammasome activation in human macrophage subsets," *Journal of Biomedical Science*, vol. 20, no. 1, article 36, 2013.

[102] T. Dong, E. Moran, N. Vinh Chau et al., "High pro-inflammatory cytokine secretion and loss of high avidity cross-reactive cytotoxic T-cells during the course of secondary dengue virus infection," *PLoS ONE*, vol. 2, no. 12, Article ID e1192, 2007.

[103] F. Baeke, T. Takiishi, H. Korf, C. Gysemans, and C. Mathieu, "Vitamin D: modulator of the immune system," *Current Opinion in Pharmacology*, vol. 10, no. 4, pp. 482–496, 2010.

[104] M. Hewison, "Vitamin D and the intracrinology of innate immunity," *Molecular and Cellular Endocrinology*, vol. 321, no. 2, pp. 103–111, 2010.

[105] C. L. Greiller and A. R. Martineau, "Modulation of the immune response to respiratory viruses by vitamin D," *Nutrients*, vol. 7, no. 6, pp. 4240–4270, 2015.

[106] A. K. Coussens, A. R. Martineau, and R. J. Wilkinson, "Anti-inflammatory and antimicrobial actions of vitamin D in combating TB/HIV," *Scientifica*, vol. 2014, Article ID 903680, 13 pages, 2014.

[107] S. Christakos, D. V. Ajibade, P. Dhawan, A. J. Fechner, and L. J. Mady, "Vitamin D: metabolism," *Endocrinology and Metabolism Clinics of North America*, vol. 39, no. 2, pp. 243–253, 2010.

[108] S. Christakos, P. Dhawan, Q. Shen, X. Peng, B. Benn, and Y. Zhong, "New insights into the mechanisms involved in the pleiotropic actions of 1,25dihydroxyvitamin D3," *Annals of the New York Academy of Sciences*, vol. 1068, no. 1, pp. 194–203, 2006.

[109] A. S. Dusso and A. J. Brown, "Mechanism of vitamin D action and its regulation," *American Journal of Kidney Diseases*, vol. 32, no. 2, supplement 2, pp. S13–S24, 1998.

[110] J. S. Adams and M. Hewison, "Unexpected actions of vitamin D: new perspectives on the regulation of innate and adaptive immunity," *Nature Clinical Practice Endocrinology and Metabolism*, vol. 4, no. 2, pp. 80–90, 2008.

[111] R. F. Chun, J. S. Adams, and M. Hewison, "Back to the future: a new look at 'old' vitamin D," *The Journal of Endocrinology*, vol. 198, no. 2, pp. 261–269, 2008.

[112] M. Hewison, "Vitamin D and the immune system: new perspectives on an old theme," *Endocrinology and Metabolism Clinics of North America*, vol. 39, no. 2, pp. 365–379, 2010.

[113] M. Hewison, "Vitamin D and innate immunity," *Current Opinion in Investigational Drugs*, vol. 9, no. 5, pp. 485–490, 2008.

[114] M. Hewison, "Vitamin D and innate and adaptive immunity," *Vitamins and Hormones*, vol. 86, pp. 23–62, 2011.

[115] M. Etminani-Esfahani, H. Khalili, N. Soleimani et al., "Serum vitamin D concentration and potential risk factors for its deficiency in HIV positive individuals," *Current HIV Research*, vol. 10, no. 2, pp. 165–170, 2012.

[116] G. R. Campbell and S. A. Spector, "Vitamin D inhibits human immunodeficiency virus type 1 and *Mycobacterium tuberculosis* infection in macrophages through the induction of autophagy," *PLoS Pathogens*, vol. 8, no. 5, Article ID e1002689, 2012.

[117] P. T. Liu, S. Stenger, H. Li et al., "Toll-like receptor triggering of a vitamin D-mediated human antimicrobial response," *Science*, vol. 311, no. 5768, pp. 1770–1773, 2006.

[118] Y. Zhao, B. Yu, X. Mao et al., "Effect of 25-hydroxyvitamin D_3 on rotavirus replication and gene expressions of RIG-I signalling molecule in porcine rotavirus-infected IPEC-J2 cells," *Archives of Animal Nutrition*, vol. 69, no. 3, pp. 227–235, 2015.

[119] M. Reinholz and J. Schauber, "Vitamin D and innate immunity of the skin," *Deutsche Medizinische Wochenschrift*, vol. 137, no. 46, pp. 2385–2389, 2012.

[120] I. Szymczak and R. Pawliczak, "The active metabolite of vitamin D3 as a potential immunomodulator," *Scandinavian Journal of Immunology*, vol. 83, no. 2, pp. 83–91, 2016.

[121] H. A. Bischoff-Ferrari, B. Dawson-Hughes, A. Platz et al., "Effect of high-dosage cholecalciferol and extended physiotherapy on complications after hip fracture: a randomized controlled trial," *Archives of Internal Medicine*, vol. 170, no. 9, pp. 813–820, 2010.

[122] P. O. Lang, N. Samaras, D. Samaras, and R. Aspinall, "How important is vitamin D in preventing infections?" *Osteoporosis International*, vol. 24, no. 5, pp. 1537–1553, 2013.

[123] C. S. Maxwell, E. T. Carbone, and R. J. Wood, "Better newborn vitamin D status lowers RSV-associated bronchiolitis in infants," *Nutrition Reviews*, vol. 70, no. 9, pp. 548–552, 2012.

[124] J. R. Sabetta, P. DePetrillo, R. J. Cipriani, J. Smardin, L. A. Burns, and M. L. Landry, "Serum 25-hydroxyvitamin D and the incidence of acute viral respiratory tract infections in healthy adults," *PLoS ONE*, vol. 5, no. 6, Article ID e11088, 2010.

[125] L. Coelho, S. W. Cardoso, P. M. Luz et al., "Vitamin D_3 supplementation in HIV infection: effectiveness and associations with antiretroviral therapy," *Nutrition Journal*, vol. 14, article 81, 2015.

[126] C. Thota, T. Farmer, R. E. Garfield, R. Menon, and A. Al-Hendy, "Vitamin D elicits anti-inflammatory response, inhibits contractile-associated proteins, and modulates toll-like receptors in human myometrial cells," *Reproductive Sciences*, vol. 20, no. 4, pp. 463–475, 2013.

[127] M. Zasloff, "Fighting infections with vitamin D," *Nature Medicine*, vol. 12, no. 4, pp. 388–390, 2006.

[128] N. Modhiran, D. Watterson, D. A. Muller et al., "Dengue virus NS1 protein activates cells via Toll-like receptor 4 and disrupts endothelial cell monolayer integrity," *Science Translational Medicine*, vol. 7, no. 304, Article ID 304ra142, 2015.

[129] Y.-C. Chuang, H.-Y. Lei, H.-S. Liu, Y.-S. Lin, T.-F. Fu, and T.-M. Yeh, "Macrophage migration inhibitory factor induced by dengue virus infection increases vascular permeability," *Cytokine*, vol. 54, no. 2, pp. 222–231, 2011.

[130] N. A. Dalrymple and E. R. MacKow, "Roles for endothelial cells in dengue virus infection," *Advances in Virology*, vol. 2012, Article ID 840654, 8 pages, 2012.

[131] A. Nimer and A. Mouch, "Vitamin D improves viral response in hepatitis C genotype 2-3 naïve patients," *World Journal of Gastroenterology*, vol. 18, no. 8, pp. 800–805, 2012.

[132] G. R. Campbell and S. A. Spector, "Toll-like receptor 8 ligands activate a vitamin D mediated autophagic response that inhibits human immunodeficiency virus type 1," *PLoS Pathogens*, vol. 8, no. 11, Article ID e1003017, 2012.

[133] P. T. Liu, M. Schenk, V. P. Walker et al., "Convergence of IL-1β and VDR activation pathways in human TLR2/1-induced antimicrobial responses," *PLoS ONE*, vol. 4, no. 6, Article ID e5810, 2009.

[134] P. T. Liu, S. R. Krutzik, and R. L. Modlin, "Therapeutic implications of the TLR and VDR partnership," *Trends in Molecular Medicine*, vol. 13, no. 3, pp. 117–124, 2007.

[135] T. S. Lisse, R. F. Chun, S. Rieger, J. S. Adams, and M. Hewison, "Vitamin D activation of functionally distinct regulatory miRNAs in primary human osteoblasts," *Journal of Bone and Mineral Research*, vol. 28, no. 6, pp. 1478–1488, 2013.

[136] T. Mohri, M. Nakajima, S. Takagi, S. Komagata, and T. Yokoi, "MicroRNA regulates human vitamin D receptor," *International Journal of Cancer*, vol. 125, no. 6, pp. 1328–1333, 2009.

[137] E. Tsitsiou and M. A. Lindsay, "microRNAs and the immune response," *Current Opinion in Pharmacology*, vol. 9, no. 4, pp. 514–520, 2009.

[138] L. A. O'Neill, F. J. Sheedy, and C. E. McCoy, "MicroRNAs: the fine-tuners of Toll-like receptor signalling," *Nature Reviews Immunology*, vol. 11, no. 3, pp. 163–175, 2011.

[139] S. Essa, N. Denzer, U. Mahlknecht et al., "VDR microRNA expression and epigenetic silencing of vitamin D signaling in melanoma cells," *The Journal of Steroid Biochemistry and Molecular Biology*, vol. 121, no. 1-2, pp. 110–113, 2010.

[140] G. Disanto, G. K. Sandve, A. J. Berlanga-Taylor et al., "Vitamin d receptor binding, chromatin states and association with multiple sclerosis," *Human Molecular Genetics*, vol. 21, no. 16, Article ID dds189, pp. 3575–3586, 2012.

[141] F. Pereira, A. Barbáchano, P. K. Singh, M. J. Campbell, A. Muñoz, and M. J. Larriba, "Vitamin D has wide regulatory effects on histone demethylase genes," *Cell Cycle*, vol. 11, no. 6, pp. 1081–1089, 2012.

[142] E. L. Beckett, C. Martin, K. Duesing et al., "Vitamin D receptor genotype modulates the correlation between vitamin D and circulating levels of let-7a/b and vitamin D intake in an elderly cohort," *Journal of Nutrigenetics and Nutrigenomics*, vol. 7, no. 4–6, pp. 264–273, 2014.

[143] A. Yoshimura, T. Naka, and M. Kubo, "SOCS proteins, cytokine signalling and immune regulation," *Nature Reviews Immunology*, vol. 7, no. 6, pp. 454–465, 2007.

[144] I. Kinjyo, T. Hanada, K. Inagaki-Ohara et al., "SOCS1/JAB is a negative regulator of LPS-induced macrophage activation," *Immunity*, vol. 17, no. 5, pp. 583–591, 2002.

[145] M. Peters-Golden, C. Canetti, P. Mancuso, and M. J. Coffey, "Leukotrienes: underappreciated mediators of innate immune responses," *Journal of Immunology*, vol. 174, no. 2, pp. 589–594, 2005.

[146] M. J. Coffey, S. E. Wilcoxen, S. M. Phare, R. U. Simpson, M. R. Gyetko, and M. Peters-Golden, "Reduced 5-lipoxygenase metabolism of arachidonic acid in macrophages rrom 1,25-dihydroxyvitamin D_3-deficient rats," *Prostaglandins*, vol. 48, no. 5, pp. 313–329, 1994.

[147] Z. Wang, L. R. Filgueiras, S. Wang et al., "Leukotriene B4 enhances the generation of proinflammatory micrornas to promote MyD88-dependent macrophage activation," *The Journal of Immunology*, vol. 192, no. 5, pp. 2349–2356, 2014.

[148] M. A. Kriegel, J. E. Manson, and K. H. Costenbader, "Does vitamin D affect risk of developing autoimmune disease?: a systematic review," *Seminars in Arthritis and Rheumatism*, vol. 40, no. 6, pp. 512–531.e8, 2011.

[149] E. K. L. Chan, M. Satoh, and K. M. Pauley, "Contrast in aberrant microRNA expression in systemic lupus erythematosus and rheumatoid arthritis: Is microRNA-146 all we need?" *Arthritis and Rheumatism*, vol. 60, no. 4, pp. 912–915, 2009.

[150] S. W. Kim, J. M. Lee, J. H. Ha et al., "Association between vitamin D receptor polymorphisms and osteoporosis in patients with COPD," *International Journal of Chronic Obstructive Pulmonary Disease*, vol. 10, no. 1, pp. 1809–1817, 2015.

[151] J. Rashedi, M. Asgharzadeh, S. R. Moaddab et al., "Vitamin D receptor gene polymorphism and vitamin D plasma concentration: correlation with susceptibility to tuberculosis," *Advanced Pharmaceutical Bulletin*, vol. 4, supplement 2, pp. 607–611, 2014.

[152] Y.-W. Huang, Y.-T. Liao, W. Chen et al., "Vitamin D receptor gene polymorphisms and distinct clinical phenotypes of hepatitis B carriers in Taiwan," *Genes & Immunity*, vol. 11, no. 1, pp. 87–93, 2010.

[153] R. Lu, M. Maduro, F. Li et al., "Animal virus replication and RNAi-mediated antiviral silencing in *Caenorhabditis elegans*," *Nature*, vol. 436, no. 7053, pp. 1040–1043, 2005.

[154] S. Wu, L. He, Y. Li et al., "MiR-146a facilitates replication of dengue virus by dampening interferon induction by targeting TRAF6," *The Journal of Infection*, vol. 67, no. 4, pp. 329–341, 2013.

[155] S. Bala, M. Marcos, K. Kodys et al., "Up-regulation of MicroRNA-155 in macrophages contributes to increased tumor necrosis factor α (TNFα) Production via increased mRNA Half-life in alcoholic liver disease," *The Journal of Biological Chemistry*, vol. 286, no. 2, pp. 1436–1444, 2011.

[156] P. Wang, J. Hou, L. Lin et al., "Inducible microRNA-155 feedback promotes type I IFN signaling in antiviral innate immunity by targeting suppressor of cytokine signaling 1," *The Journal of Immunology*, vol. 185, no. 10, pp. 6226–6233, 2010.

[157] M. Jiang, R. Broering, M. Trippler et al., "MicroRNA-155 controls Toll-like receptor 3- and hepatitis C virus-induced immune responses in the liver," *Journal of Viral Hepatitis*, vol. 21, no. 2, pp. 99–110, 2014.

[158] P.-W. Xie, Y. Xie, X.-J. Zhang et al., "Inhibition of Dengue virus 2 replication by artificial micrornas targeting the conserved regions," *Nucleic Acid Therapeutics*, vol. 23, no. 4, pp. 244–252, 2013.

[159] J. L. Umbach and B. R. Cullen, "The role of RNAi and microRNAs in animal virus replication and antiviral immunity," *Genes & Development*, vol. 23, no. 10, pp. 1151–1164, 2009.

[160] B. L. Heiss, O. A. Maximova, and A. G. Pletnev, "Insertion of microRNA targets into the flavivirus genome alters its highly neurovirulent phenotype," *Journal of Virology*, vol. 85, no. 4, pp. 1464–1472, 2011.

[161] R. A. Agis-Juárez, I. Galván, F. Medina et al., "Polypyrimidine tract-binding protein is relocated to the cytoplasm and is required during dengue virus infection in Vero cells," *The Journal of General Virology*, vol. 90, part 12, pp. 2893–2901, 2009.

[162] L. Jiang, H. Yao, X. Duan, X. Lu, and Y. Liu, "Polypyrimidine tract-binding protein influences negative strand RNA synthesis of dengue virus," *Biochemical and Biophysical Research Communications*, vol. 385, no. 2, pp. 187–192, 2009.

[163] R. G. Fred, C. H. Bang-Berthelsen, T. Mandrup-Poulsen, L. G. Grunnet, and N. Welsh, "High glucose suppresses human islet insulin biosynthesis by inducing Mir-133a leading to decreased polypyrimidine tract binding protein-expression," *PLoS ONE*, vol. 5, no. 5, Article ID e10843, 2010.

[164] P. A. Tambyah, C. S. Ching, S. Sepramaniam, J. M. Ali, A. Armugam, and K. Jeyaseelan, "microRNA expression inblood of dengue patients," *Annals of Clinical Biochemistry*, 2015.

[165] F. Taïbi, V. Metzinger-Le Meuth, Z. A. Massy, and L. Metzinger, "MiR-223: an inflammatory oncomiR enters the cardiovascular field," *Biochimica et Biophysica Acta (BBA)—Molecular Basis of Disease*, vol. 1842, no. 7, pp. 1001–1009, 2014.

[166] S. Matsui and Y. Ogata, "Effects of miR-223 on expression of IL-1β and IL-6 in human gingival fibroblasts," *Journal of Oral Science*, vol. 58, no. 1, pp. 101–108, 2016.

[167] Y. Yuzefpolskiy, F. M. Baumann, L. A. Penny, G. P. Studzinski, V. Kalia, and S. Sarkar, "Vitamin D receptor signals regulate effector and memory CD8 T cell responses to infections in mice," *The Journal of Nutrition*, vol. 144, no. 12, pp. 2073–2082, 2014.

[168] A. Waddell, J. Zhao, and M. T. Cantorna, "NKT cells can help mediate the protective effects of 1,25-dihydroxyvitamin D_3 in experimental autoimmune encephalomyelitis in mice," *International Immunology*, vol. 27, no. 5, pp. 237–244, 2015.

[169] L. Rolf, A.-H. Muris, R. Hupperts, and J. Damoiseaux, "Vitamin D effects on B cell function in autoimmunity," *Annals of the New York Academy of Sciences*, vol. 1317, no. 1, pp. 84–91, 2014.

8

The Possible Role of TLR2 in Chronic Hepatitis B Patients with Precore Mutation

Malihe Moradzadeh,[1] **Sirous Tayebi,**[2] **Hossein Poustchi,**[2] **Kourosh Sayehmiri,**[3]
Parisa Shahnazari,[4] **Elnaz Naderi,**[1] **Ghodratollah Montazeri,**[2] **and Ashraf Mohamadkhani**[2]

[1] *Department of Modern Sciences and Technologies, School of Medicine, Mashhad University of Medical Sciences, Mashhad, Iran*
[2] *Liver and Pancreatobiliary Diseases Research Center, Digestive Diseases Research Institute, Tehran University of Medical Sciences, Tehran, Iran*
[3] *Psychosocial Injuries Research Centre, Ilam University of Medical Sciences, Ilam, Iran*
[4] *Monoclonal Antibody Research Centre, Avicenna Research Institute, ACECR, Tehran, Iran*

Correspondence should be addressed to Ashraf Mohamadkhani; mohamadkhani.ashraf@gmail.com

Academic Editor: Stefan Pöhlmann

Recognition mechanisms of innate immune response help to improve immunotherapeutic strategies in HBeAg-negative chronic hepatitis B (CHB). Toll-like receptor 2 (TLR2) is an important component of innate immunity. In this study, the frequency of precore mutations of the hepatitis B virus (HBV) and serum TLR2 were evaluated in CHB patients. Fifty-one patients with chronic hepatitis B, negative for HBeAg and detectable HBV DNA, were examined for the presence of mutations in pre-core region of HBV genome by direct sequencing. Serum TLR2 was measured by enzyme-linked immunosorbent assay. Interactions of truncated HBeAg and TLR2 proteins were evaluated with molecular docking software. The G1896A pre-core mutation were detected in 29 (57%) which was significantly associated with higher concentration of serum TLR2 in comparison with patients without this mutation (4.8 ± 2.9 versus 3.4 ± 2.2 ng/mL, $P = 0.03$). There was also a significant correlation between serum ALT and TLR-2 ($r = 0.46$; $P = 0.01$). Docking results illustrated residues within the N-terminus of truncated HBeAg and TLR2, which might facilitate the interaction of these proteins. These findings showed the dominance of G1896A pre-core mutation of HBV variants in this community which was correlated with serum TLR2. Moreover TLR2 is critical for induction of inflammatory cytokines and therefore ALT elevation.

1. Introduction

Hepatitis B virus (HBV) infection is an important cause of chronic hepatitis, cirrhosis, and hepatocellular carcinoma (HCC) [1]. The transmission of HBV from infected mothers to neonates causes persistent infection [2]. Chronic infection of HBV is a global health problem. However, the prevalence and genotype distribution of HBV are different among the geographical areas [3]. The majority of chronic hepatitis B patients lose HBe antigen (HBeAg) and develop anti-HBe antibody, which is generally associated with a decrease in serum HBV DNA levels and a gradual accumulation of precore or core promoter mutations [4]. HBeAg-negative chronic hepatitis B is the predominant type of CHB in Mediterranean inhabitants [3]. Two types of precore and core promoter HBV mutations that reduce HBeAg formation are more frequent in regions where patients are predominantly infected with HBV genotype D [4, 5]. Infection with wild-type strains of HBV often induces mild symptoms and responds well to interferon alpha therapy, but patients infected with precore mutant variants may show clinical evidence of elevated or fluctuating ALT and HBV DNA [6]. The reason that precore negative mutants become predominant in some patients during chronic hepatitis B infection is not clear. However, the host immune system has a functional role in the selection of precore mutant strains of HBV, and their appearance might reflect immunological control of infection [7, 8].

Infected hepatocytes are eliminated by vigorous CD4+ and CD8+ T-cell responses, and those who have insufficient cellular immune response will persist chronically infected [9].

The impact of innate immunity in liver damage also has been identified in several studies [10, 11]. Toll-like receptors (TLRs) describe a group of pattern recognition receptors (PRRs) playing critical roles in the host innate immune response [12]. These proteins are evolutionarily conserved from *Drosophila* to humans and important in controlling the activation of the adaptive immune response [13]. Various TLRs exhibit different patterns of expression [14]. Overactivation of TLRs plays a prominent role in the pathogenesis of a variety of acute and chronic inflammatory conditions [13]. A previous study reports that the HBeAg downregulates antiviral defenses of the host [15] and, in the absence of HBeAg, HBV replication is associated with upregulation of the TLR2 pathway, resulting in increased TNF-α production [16–19]. A wide range of microbial and viral components as well as several endogenous TLR ligands are recognized by TLR2 [14]. This receptor is expressed in peripheral blood leukocytes, mainly in monocytes, in lymph nodes, bone marrow, and spleen [20]. TLR2 is also released by normal monocytes and is present in serum and other biological fluids which mostly contain the TLR2 extracellular domain [20, 21].

The importance of diverse TLRs for the "*in vivo*" replication and pathogenesis of HBV have been evidenced in several reports [16, 17]. However, there are no reports of serum TLR2 in HBeAg negative chronic hepatitis B. In this study, the association of serum TLR2 with clinical findings in chronic hepatitis B patients especially in patients with G1896A stop codon mutation has been investigated.

2. Patients and Methods

2.1. Patients. A total of 51 chronic HBeAg negative patients with detectable HBV DNA and a range of normal to elevated ALT were evaluated during a period of 12 months. They perinatally acquired chronic infection as they had a clear history of familial HBV infection, without coinfection of human immunodeficiency virus (HIV), autoimmune hepatitis, and other hepatitis viruses. Blood samples were taken at the initial assessment before liver biopsy. Serum samples were stored at −70°C. No patient received anti-HBV therapy prior to liver biopsy. The protocol for the study was approved by the ethics committee of Shariati Hospital, Tehran, University of medical science.

2.2. Clinical Evaluation and HBV-DNA Quantification. The presence of HBsAg, HBeAg, anti-HBeAg, anti-HCV, anti-HDV, and anti-HIV were determined with commercial assay kit EIA, Dia.Pro diagnostic, Italy. Serum TLR2 was measured by toll-like Receptor2 ELISA Kit (Uscn Life Science, Wuhan, China) according to the manufacturer's specifications. Sensitivity of the ELISA for TLR2 was 0.312–20 ng/mL.

HBV DNA was extracted from 200 μL of serum using QIAamp DNA Blood Mini Kit (QIAGEN, USA), eluted in 50 μL of elution buffer, and then measured in the Light-Cycler (Roche) by RealARTTM HBV LC PCR (QIAGEN, Hilden, Germany) according to the manufacturer's instructions. This assay had a linear range of 10^2–10^9 copies/mL. Liver biopsies from all patients were assessed for the grade of histological

activity and stage of fibrosis using the modified histological activity index (HAI) scoring system [22].

2.3. Precore G1896A Mutation Detection and Direct Sequencing. The precore region was analyzed by hemi-nested PCR according to Gan et al. [23]. The sense primer PC5 5′-TCG CAT GGA GAC CAC CGT GA-3′ (nt. 204–223) and the antisense primer PC2 5′-GGC AAA AAC GAG AGT AAC TC-3′ (nt. 540–559) were performed as first round primers. An additional hemi-nested round of amplification was achieved using 2 μL of the first round product as template and the antisense primer 527 5′-GTA ACT CCA CAG WAG CTC C-3′ (nt. 528–546). The numbering organization for primer nucleotides was in accordance to the genome sequence HPBADR1CG [23]. The PCR products were purified using PCR purification kit from MO BIO Laboratories Inc. (Carlsbad, CA, USA) based on the manufacturer's instructions and eluted in 100 μL of elution buffer. The sequencing of PCR products was accomplished by the Big Dye Terminator Cycle sequencing Ready Reaction Kit Version 3.1 (Applied Biosystems, Foster City, CA, USA).

2.4. Statistical Analysis. Normality of data was assessed using One Sample Kolmogorov-Smirnov Test. Correlations between variables were analyzed using Pearson correlation coefficient (r). The independent samples t-test is used to compare the means ± standard deviation (SD) of data. The Mann-Whitney U test was utilized to test equality of TLR2 and ALT between patients with G1896A precore mutation and patients without mutation. A P value <0.05 was deemed statistically significant.

2.5. HBeAg and TLR2 Interaction Analysis. In order to identify the strongly associated functional of HBeAg, the protein-protein interaction solutions were mapped between HBeAg and TLR2. The sequence of truncated HBeAg amino acid, created as a result of a stop codon at position 28 of HBeAg and genomic mutation at base 1898 of HBV, was extracted from UniProt (P0C6H9). The tertiary structure of truncated HBeAg was built by Pepstr [24]. The Pepstr server predicts the tertiary structure of small peptides with sequence length varying of 7 to 25 (residues http://www.imtech.res.in/raghava/pepstr/). The X-ray crystal structure of the TLR2 (2Z80A) was retrieved from PDB (Protein Data Bank) [25]. The *PatchDock* web server and a refinement by *FireDock* evaluate the molecular docking of both proteins [26, 27].

3. Results

3.1. Demographic and Clinical Characteristics of the Patients. Demographic characteristics and frequency of the G1896A precore mutation along with the clinical and biochemical profiles of study subjects are summarized in Table 1. There were a total of 51 patients (mean age 37 ± 10 yr) including 16 females and 35 males. The quantification of HBV DNA was reported in log copies/mL with a mean value of 3.46±1.06 and 29 (57%) patients that showed the G1896A precore mutation. Total score of necroinflammatory grade and fibrosis stage

TABLE 1: Clinical and pathological data of 51 chronic hepatitis B patients with and without precore mutation G1896A.

Clinical factor[*]	All subjects ($n = 51$)	Patients with wild-type variant ($n = 22$)	Patients with G1896A mutation ($n = 29$)	P value
Age (years)	37 ± 10	36 ± 9	38 ± 11	0.4
log HBV DNA (copies/mL)	3.46 ± 1.06	3.54 ± 1.03	3.41 ± 1.10	0.6
ALT (IU/L)	57 ± 56	41 ± 27	68 ± 67	0.5
TLR2 (ng/mL)	4.2 ± 2.7	3.4 ± 2.2	4.8 ± 2.9	0.03[**]
Histological activity index (HAI) score	4.8 ± 2.3	4.2 ± 1.7	5.3 ± 2.6	0.076

[*] Mean ± SD, [**] P value computed using the Mann-Whitney test.

were measured based on the modified HAI system and serum ALT were 4.8 ± 2.3 and 57 ± 56 IU/l, respectively. The mean concentration of serum TLR2 was 4.2 ± 2.7 ng/mL.

3.2. Clinical Significance of G1896A Precore Mutation and Serum TLR2.

The concentration of serum TLR2 was higher in G1896A precore mutants than wild-type infected patients (4.8 ± 2.9 versus 3.4 ± 2.2 ng/mL, $P = 0.032$) (Table 1). There was no significant relationship between G1896A mutation variants with neither age nor sex. Patients infected with the wild-type HBV revealed similar mean of viral load compared to G1896A precore mutant patients (3.54 ± 1.02 versus 3.41 ± 1.1 log copies/mL). More likely, patients who harbored the precore mutant strains had higher levels of serum ALT and HIA score compared to those patients infected with wild-type strain of HBV, however it was not statistically significant (68 ± 67 versus 41 ± 27 IU/l and 5.3 ± 2.6 versus 4.2 ± 1.7) (Table 1). Serum TLR2 and ALT had significant correlation ($r = 0.46$; $P = 0.01$) which was more pronounced in patients with G1896A mutation ($r = 0.48$; $P = 0.008$) (Figure 1). Furthermore, there was statistically significant association between the log HBV DNA with serum ALT and with HIA score ($r = 0.36$; $P = 0.09$ and $r = 0.3$; $P = 0.03$). Serum ALT and HBV DNA did not show significant relationship to age, sex, and total grade of histological activity and stage of fibrosis.

3.3. TLR2 and Truncated HBeAg Interaction.

In this experiment, the protein-protein docking of HBeAg and TLR2 were studied by PatchDock bioinformatic docking tool. Then, high-throughput refinement of docking was selected by FireDock (Figure 2). Protein-protein docking and interaction simulations disclosed hydrogen and ionic bonds. The amino acid residues Cys14, Pro15, Thr16, Val17 and Gln18 from HBVD-truncated HBeAg bonded to Asp58, Leu59, Ser60, Asn61, Asn62 and Arg63 of TLR2, respectively.

4. Discussion

The dynamic state of chronic HBV infection is a result of interactions between the virus itself and the host immune response [3]. Accordingly, different phases in the natural course of HBV infection are observed. Patients in immune tolerant phase usually have high viral load and normal levels of serum ALT. However, during the immune clearance phase, patients have moderate levels of HBV replication and elevated

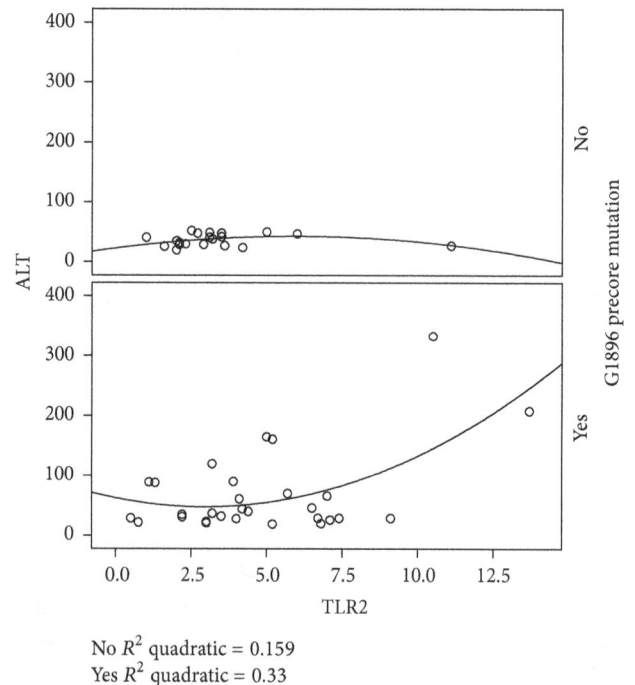

No R^2 quadratic = 0.159
Yes R^2 quadratic = 0.33

FIGURE 1: Estimating regression quadratic equation indicates that ALT sharply rises with increasing TLR2.

level of ALT [3, 28]. The HBeAg has been proposed as a viral approach to induce immunotolerance. The precore and basic core promoter (BCP) genetic variations of HBV lead to HBeAg loss and anti-HBe seroconversion [4, 29]. Principally, the stop codon mutation at base 1896 creates the truncated precore peptide that might represents an adaptation to immune pressure [30]. The function of innate immune response as the first line of defense against virus infection as well as its cooperation with adaptive immunity may induce the development of precore mutant variants of HBV [11, 31, 32]. Therefore, the clinical significance of these mutations appears to be linked to the function of host immune system.

The result of this study revealed the presence of HBV variants with precore mutations in 57% of anti-HBe-positive patients that explained the lack of HBeAg synthesis. Patients with precore mutants had higher concentrations of TLR2 compared to patients infected with wild-type HBV. Furthermore, increasing TLR2 is associated with serum ALT concentration in G1896A precore mutant patients. The increase of

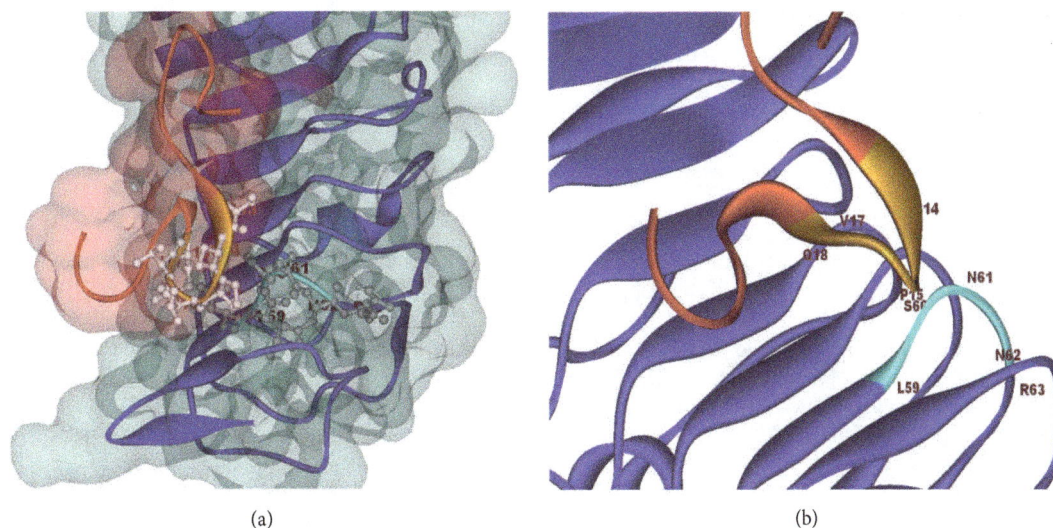

FIGURE 2: The docking result of HBeAg and TLR2. (a) The interaction of HBeAg with the accessible area of TLR2. (b) Docking of the a helices in the major groove of TLR2 with HBeAg as the ribbon form. The amino acids residues Cys14, Pro15, Thr16, Val17, and Gln18 of precore protein (HBeAg) colored in yellow bonding to Leu59, Ser60, Asn61, Asn62 and Arg63 of TLR2 that appeared in light blue.

serum ALT could reflect the hepatitis activity and the host immune response against HBV that induces apoptosis and necrosis [3]. These findings are consistent with a previous study reporting the association of HBV replication and activation of TLR2 in precore mutant patients [17]. The members of TLRs family play an essential role in the innate immune recognition [12, 16], are upregulated in response to microbial components [33, 34], and are critical for the development of effective immunity [14]. A downregulation of TLR2 receptor in peripheral blood monocytes of chronic patients infected with HBeAg positive variants has been shown previously [17]. Visvanathan et al. reported that wild-type (HBeAg-positive) and the precore stop codon (HBeAg-negative) HBV variants had different effects on TNF-α production. Moreover, they showed that the expression of TLR2 on hepatocytes, Kupffer cells, and peripheral monocytes were significantly increased in HBeAg-negative chronic hepatitis B [17]. Lian et al. also found that the expression of TLR2 was significantly upregulated in patients with liver cirrhosis and chronic hepatitis B patients [9]. Interestingly, the therapeutic TLR strategy by Isogawa et al. revealed that TLRs ligands except for TLR2 are able to induce antiviral cytokines (Interferon α/β) at the site of HBV replication [16]. We therefore speculate that the increased serum TLR2 levels in our patients may correspond to higher expression of cellular TLR2 and consequently elevated TLR2 signaling leading to expression of proinflammatory cytokines that explain the chronicity of HBV infection that usually accompanies an increase in ALT.

In conclusion, our data show that the concentration of serum TLR2 was significantly higher in HBeAg negative patients with G1896A mutation which in turn was associated with higher serum ALT in this group. These results indicate the interaction of precore mutant strain of HBV and TLR2. Additional studies with longitudinal follow-up of subjects are required to determine the precise impact of TLR2 in HBeAg negative chronic hepatitis B.

Authors' Contribution

All authors contributed both to the research and the discussion, and they have read and approved the final manuscript.

Acknowledgments

Authors of this article take this chance to appreciate all the patients which attended in this research program. This Study was supported by grants from the Digestive Disease Research Center, Tehran University of Medical Sciences.

References

[1] G. Montazeri, "Current treatment of chronic hepatitis B," *Archives of Iranian Medicine*, vol. 9, no. 1, pp. 1–10, 2006.

[2] P. Gerner, A. Hörning, S. Kathemann, K. Willuweit, and S. Wirth, "Growth abnormalities in children with chronic hepatitis B or C," *Advances in Virology*, vol. 2012, Article ID 670316, 5 pages, 2012.

[3] Y.-F. Liaw, M. R. Brunetto, and S. Hadziyannis, "The natural history of chronic HBV infection and geographical differences," *Antiviral Therapy*, vol. 15, supplement 3, pp. 25–33, 2010.

[4] H. Poustchi, A. Mohamadkhani, S. Bowden et al., "Clinical significance of precore and core promoter mutations in genotype D hepatitis B-related chronic liver disease," *Journal of Viral Hepatitis*, vol. 15, no. 10, pp. 753–760, 2008.

[5] S. M. Alavian, F. Fallahian, and K. B. Lankarani, "The changing epidemiology of viral hepatitis B in Iran," *Journal of Gastrointestinal and Liver Diseases*, vol. 16, no. 4, pp. 403–406, 2007.

[6] M. M. Mir-Nasseri, A. Mohammadkhani, H. Tavakkoli, E. Ansari, and H. Poustchi, "Incarceration is a major risk factor for

blood-borne infection among intravenous drug users: incarceration and blood borne infection among intravenous drug users," *Hepatitis Monthly*, vol. 11, no. 1, pp. 19–22, 2011.

[7] A. Mohamadkhani, M. Sotoudeh, S. Bowden et al., "Downregulation of HLA class II molecules by G1896A pre-core mutation in chronic hepatitis B virus infection," *Viral Immunology*, vol. 22, no. 5, pp. 295–300, 2009.

[8] A. Mohamadkhani, A. Pourdadash, S. Tayebi et al., "The potential role of APOBEC3G in limiting replication of hepatitis B virus," *Arab Journal of Gastroenterology*, vol. 13, no. 4, pp. 170–173, 2012.

[9] J.-Q. Lian, X.-Q. Wang, Y. Zhang, C.-X. Huang, and X.-F. Bai, "Correlation of circulating TLR2/4 expression with CD3$^+$/4$^+$/8$^+$ T cells and treg cells in HBV-related liver cirrhosis," *Viral Immunology*, vol. 22, no. 5, pp. 301–308, 2009.

[10] P. Carotenuto, A. Artsen, A. D. Osterhaus, and O. Pontesilli, "Reciprocal changes of naïve and effector/memory CD8$^+$ T lymphocytes in chronic hepatitis B virus infection," *Viral Immunology*, vol. 24, no. 1, pp. 27–33, 2011.

[11] A. Mohamadkhani, F. Bastani, M. Sotoudeh et al., "Influence of B cells in liver fibrosis associated with hepatitis B virus harboring basal core promoter mutations," *Journal of Medical Virology*, vol. 84, no. 12, pp. 1889–1896, 2012.

[12] M. Schnare, G. M. Barton, A. C. Holt, K. Takeda, S. Akira, and R. Medzhitov, "Toll-like receptors control activation of adaptive immune responses," *Nature Immunology*, vol. 2, no. 10, pp. 947–950, 2001.

[13] N. J. Gay and M. Gangloff, "Structure and function of toll receptors and their ligands," *Annual Review of Biochemistry*, vol. 76, pp. 141–165, 2007.

[14] E. E. Hamilton-Williams, A. Lang, D. Benke, G. M. Davey, K.-H. Wiesmüller, and C. Kurts, "Cutting edge: TLR ligands are not sufficient to break cross-tolerance to self-antigens," *Journal of Immunology*, vol. 174, no. 3, pp. 1159–1163, 2005.

[15] D. R. Milich, M. K. Chen, J. L. Hughes, and J. E. Jones, "The secreted hepatitis B precore antigen can modulate the immune response to the nucleocapsid: a mechanism for persistence," *Journal of Immunology*, vol. 160, no. 4, pp. 2013–2021, 1998.

[16] M. Isogawa, M. D. Robek, Y. Furuichi, and F. V. Chisari, "Toll-like receptor signaling inhibits hepatitis B virus replication in vivo," *Journal of Virology*, vol. 79, no. 11, pp. 7269–7272, 2005.

[17] K. Visvanathan, N. A. Skinner, A. J. V. Thompson et al., "Regulation of Toll-like receptor-2 expression in chronic hepatitis B by the precore protein," *Hepatology*, vol. 45, no. 1, pp. 102–110, 2007.

[18] M. K. Arababadi, A. A. Pourfathollah, A. Jafarzadeh, and G. Hassanshahi, "Serum levels of IL-10 and IL-17A in occult HBV-infected South-East Iranian patients," *Hepatitis Monthly*, vol. 10, no. 1, pp. 31–35, 2010.

[19] S. Tayebi and A. Mohamadkhani, "The TNF-α -308 promoter gene polymorphism and chronic HBV infection," *Hepatitis Research and Treatment*, vol. 2012, Article ID 493219, 6 pages, 2012.

[20] E. LeBouder, J. E. Rey-Nores, N. K. Rushmere et al., "Soluble forms of Toll-like receptor (TLR)2 capable of modulating TLR2 signaling are present in human plasma and breast milk," *Journal of Immunology*, vol. 171, no. 12, pp. 6680–6689, 2003.

[21] A.-C. Raby, E. Le Bouder, C. Colmont et al., "Soluble TLR2 reduces inflammation without compromising bacterial clearance by disrupting TLR2 triggering," *Journal of Immunology*, vol. 183, no. 1, pp. 506–517, 2009.

[22] K. Ishaka, A. Baptistab, L. Bianchic et al., "Histological grading and staging of chronic hepatitis," *Journal of Hepatology*, vol. 22, no. 6, pp. 696–699, 1995.

[23] R. B. Gan, M. J. Chu, L. P. Shen, S. W. Qian, and Z. P. Li, "The complete nucleotide sequence of the cloned DNA of hepatitis B virus subtype adr in pADR-1," *Scientia Sinica B*, vol. 30, no. 5, pp. 507–521, 1987.

[24] H. Kaur, A. Garg, and G. P. S. Raghava, "PEPstr: a de novo method for tertiary structure prediction of small bioactive peptides," *Protein and Peptide Letters*, vol. 14, no. 7, pp. 626–631, 2007.

[25] M. S. Jin, S. E. Kim, J. Y. Heo et al., "Crystal structure of the TLR1-TLR2 heterodimer induced by binding of a Tri-Acylated lipopeptide," *Cell*, vol. 130, no. 6, pp. 1071–1082, 2007.

[26] N. Andrusier, R. Nussinov, and H. J. Wolfson, "FireDock: fast interaction refinement in molecular docking," *Proteins*, vol. 69, no. 1, pp. 139–159, 2007.

[27] D. Schneidman-Duhovny, Y. Inbar, R. Nussinov, and H. J. Wolfson, "PatchDock and SymmDock: servers for rigid and symmetric docking," *Nucleic Acids Research*, vol. 33, supplement 2, pp. W363–W367, 2005.

[28] C.-M. Chu, C.-T. Yeh, C.-S. Lee, I.-S. Sheen, and Y.-F. Liaw, "Precore stop mutant in HBeAg-positive patients with chronic hepatitis B: clinical characteristics and correlation with the course of HBeAg-to-anti-HBe seroconversion," *Journal of Clinical Microbiology*, vol. 40, no. 1, pp. 16–21, 2002.

[29] S. A. Taghavi, M. Tabibi, A. Eshraghian, H. Keyvani, and H. Eshraghian, "Prevalence and clinical significance of hepatitis B basal core promoter and precore gene mutations in southern iranian patients," *Hepatitis Monthly*, vol. 10, no. 4, pp. 294–297, 2010.

[30] H. P. Dienes, G. Gerken, B. Goergen, K. Heermann, W. Gerlich, and K. H. M. zum Buschenfelde, "Analysis of the precore DNA sequence and detection of precore antigen in liver specimens from patients with anti-hepatitis B e-positive chronic hepatitis," *Hepatology*, vol. 21, no. 1, pp. 1–7, 1995.

[31] C.-M. Chu and Y.-F. Liaw, "Predictive factors for reactivation of hepatitis B following hepatitis B e antigen seroconversion in chronic hepatitis B," *Gastroenterology*, vol. 133, no. 5, pp. 1458–1465, 2007.

[32] C. P. Desmond, S. Gaudieri, I. R. James et al., "Viral adaptation to host immune responses occurs in chronic hepatitis B virus (HBV)infection, and adaptation is greatest in HBV e antigen-negative disease," *Journal of Virology*, vol. 86, no. 2, pp. 1181–1192, 2012.

[33] R. Romieu-Mourez, M. François, M.-N. Boivin, M. Bouchentouf, D. E. Spaner, and J. Galipeau, "Cytokine modulation of TLR expression and activation in mesenchymal stromal cells leads to a proinflammatory phenotype," *Journal of Immunology*, vol. 182, no. 12, pp. 7963–7973, 2009.

[34] T. Matsumura, T. Degawa, T. Takii et al., "TRAF6-NF-κB pathway is essential for interleukin-1-induced TLR2 expression and its functional response to TLR2 ligand in murine hepatocytes," *Immunology*, vol. 109, no. 1, pp. 127–136, 2003.

9

Direct Detection and Identification of Enteroviruses from Faeces of Healthy Nigerian Children using a Cell-Culture Independent RT-Seminested PCR Assay

Temitope Oluwasegun Cephas Faleye,[1,2] **Moses Olubusuyi Adewumi,**[1]
Bamidele Atinuke Coker,[3] **Felix Yasha Nudamajo,**[3] **and Johnson Adekunle Adeniji**[1,4]

[1]*Department of Virology, College of Medicine, University of Ibadan, Ibadan, Oyo State, Nigeria*
[2]*Department of Microbiology, Faculty of Science, Ekiti State University, Ado Ekiti, Ekiti, Nigeria*
[3]*Department of Microbiology, Faculty of Science, University of Ibadan, Ibadan, Oyo State, Nigeria*
[4]*WHO National Polio Laboratory, University of Ibadan, Ibadan, Oyo State, Nigeria*

Correspondence should be addressed to Johnson Adekunle Adeniji; adek1808@yahoo.com

Academic Editor: George N. Pavlakis

Recently, a cell-culture independent protocol for detection of enteroviruses from clinical specimen was recommended by the WHO for surveillance alongside the previously established protocols. Here, we investigated whether this new protocol will show the same enterovirus diversity landscape as the established cell-culture dependent protocols. Faecal samples were collected from sixty apparently healthy children in Ibadan, Nigeria. Samples were resuspended in phosphate buffered saline, RNA was extracted, and the VP1 gene was amplified using WHO recommended RT-snPCR protocol. Amplicons were sequenced and sequences subjected to phylogenetic analysis. Fifteen (25%) of the 60 samples yielded the expected band size. Of the 15 amplicons sequenced, 12 were exploitable. The remaining 3 had electropherograms with multiple peaks and were unexploitable. Eleven of the 12 exploitable sequences were identified as Coxsackievirus A1 (CVA1), CVA3, CVA4, CVA8, CVA20, echovirus 32 (E32), enterovirus 71 (EV71), EVB80, and EVC99. Subsequently, the last exploitable sequence was identified as enterobacteriophage baseplate gene by nucleotide BLAST. The results of this study document the first description of molecular sequence data on CVA1, CVA8, and E32 strains present in Nigeria. The result further showed that species A enteroviruses were more commonly detected in the region when cell-culture bias is bypassed.

1. Introduction

Enterovirus infections have been associated with an array of clinical manifestations that range from aseptic meningitis through type 1 diabetes to acute flaccid paralysis (AFP) among others [1]. However, these clinically manifest infections represent <10% of the actual burden of enterovirus infections and have been estimated to amount to about 10–15 million cases annually in the United States alone [2]. The remaining over 90% of such infections are asymptomatic [3].

Enteroviruses are nonenveloped viruses with a diameter of 20–30 nM. Within the virion is a positive sense, single stranded RNA genome that is approximately 7,500 nt long.

The genome has one open reading frame (ORF), the polyprotein product of which is autocatalytically cleaved into structural (VP1–VP4) and nonstructural (2A–3D) proteins. The ORF is flanked on both ends by untranslated regions (UTRs) and a poly-A tail at the $3'$-end.

Enteroviruses belong to the genus *Enterovirus* in the family Picornaviridae, order Picornavirales. Classification of enteroviruses used to be based on virion particle structure, tissue culture growth properties, and pathogenesis in humans and animals [4]. However, classification is now based on virus genomics [4] and most especially phylogeny of the VP1 protein [5–16]. Based on the recent classification (http://www.picornaviridae.com/), there are 12 species in the genus, four

(*Enterovirus* species A–species D [EVA-EVD]) of which were previously known as "human enteroviruses." At the time of writing, EVA contained 25 serotypes made up of 11 CVAs, 10 numbered enteroviruses, and four (4) enteroviruses isolated from nonhuman primates. EVB contained 63 serotypes consisting of one (1) CVA, six (6) CVBs, 28 echoviruses, 27 numbered enteroviruses, and one (1) enterovirus isolated from a nonhuman primate. EVC contained 23 serotypes consisting of nine (9) CVAs, three (3) poliovirus serotypes, and eleven (11) numbered enteroviruses. EVD contained five (5) serotypes consisting only of numbered enteroviruses (http://www.picornaviridae.com/).

Besides the fact that EVB has the highest number of serotypes, it is also the most commonly detected [15–21]. It has however been suggested that this phenomenon (called the EVB bias) might be an artefact of the strategy used for enterovirus isolation and might not be truly representative of the enterovirus diversity landscape [21, 22].

Almost all previous studies documenting enterovirus diversity in Nigeria [15, 16, 23, 24] clearly showed the preponderance of EVB. However, all such studies have been cell-culture based and mainly used the RD cell line which has been suggested to be the EVB bias [15–21], for enterovirus isolation. The only study that did differently [22] used MCF 7 and LLC-MK2 cell lines for enterovirus isolation and documented an increase in the detection rate of enterovirus species C (EVC) members.

Recently, Nix et al.'s [25] cell-culture independent protocol for direct detection of enteroviruses from clinical specimen was recommended [4] for enterovirus surveillance alongside the previously established protocols [26, 27]. In this study, we investigated whether this strategy will show the same enterovirus diversity landscape as the established cell-culture dependent protocols [26, 27] and document a preponderance of EVAs in Southwestern Nigeria.

2. Methodology

2.1. Sample Collection and Storage. Faecal samples were collected from sixty (male = 37, female = 23) apparently healthy children aged 1 to 10 years attending public primary schools in Ibadan, Nigeria. Samples were collected from the pupils after approval and consent were secured from the school administration and the guardian or parents of the children, respectively. Stool samples were collected from each of the children into appropriately labelled sterile collection bottles. Samples were then transported to the laboratory in the Department of Virology, College of Medicine, University College Hospital, Ibadan, Nigeria, in a cooler filled with ice packs to maintain a temperature of about 4°C. On arrival at the laboratory, the stool specimens were stored at −20°C until analysis.

2.2. Sample Processing. About one gram of each stool specimen was diluted in 3 mL phosphate buffered saline (PBS), 1 mL chloroform, and one gram of glass beads. The mixture was then vortexed for 20 minutes and thereafter centrifuged at 3000 rpm for 20 minutes. Subsequently, 2 mL of the supernatant was aliquoted in 1 mL volumes into cryovials. One vial was stored at −20°C while the other was analysed further.

2.3. RNA Extraction and cDNA Synthesis. JenaBioscience RNA extraction kit (Jena Bioscience, Jena, Germany) was used for viral RNA extraction according to the manufacturer's instructions. Script cDNA synthesis kit (Jena Bioscience, Jena, Germany) was used for cDNA synthesis according to the manufacturer's instructions. However, instead of random hexamers, primers AN32, AN33, AN34, and AN35 [25] were used for cDNA synthesis.

2.4. Enterovirus VP1 Gene Seminested PCR (snPCR) Assay. Primers were made in 25 μM concentrations and PCR was done in 30 μL reactions. The first-round PCR contained 2 μL of each of primers 224 and 222 (Nix et al., 2006), 6 μL of Red Load Taq, 10 μL of cDNA, and 10 μL of RNase-free water. Thermal cycling was done in a Veriti thermal cycler (Applied Biosystems, California, USA). Thermal cycling conditions were 94°C for 3 minutes followed by 45 cycles at 94°C for 30 seconds, 42°C for 30 seconds, and 60°C for 60 seconds with ramp of 40% from 42°C to 60°C. This was then followed by 72°C for 7 minutes and held at 4°C till being terminated. The second-round PCR was carried out with the first-round PCR product as template, with similar thermal cycling conditions except for the extension time that was reduced to 30 seconds, and the primers were substituted with AN89 and AN88 [25], respectively. Subsequently, PCR products were resolved on 2% agarose gel stained with ethidium bromide and viewed using a UV transilluminator.

2.5. Nucleotide Sequencing. All amplicons were shipped to Macrogen Inc., Seoul, South Korea, for purification and sequencing of only the bands of the expected size. Primers AN88 and AN89 were used for sequencing. Afterwards, the enterovirus genotyping tool [28] was used for enterovirus species and genotype determination.

2.6. Phylogenetic Analysis. To align the sequences described in this study with reference sequences downloaded from the GenBank, the ClustalW program in the MEGA 5 software [29] was used with default settings. Afterwards, neighbour-joining trees were constructed with the Kimura-2 parameter model [30] and 1,000 bootstrap replicates using the same MEGA 5 software.

2.7. Nucleotide Sequence Accession Numbers. All the sequences reported in this study have been deposited in GenBank under accession numbers KT717062–KT717072.

3. Results

3.1. RT-snPCR Result. A total of 15 (25%) of the sixty (60) stool samples screened yielded the expected band size for the enterovirus VP1 gene detection RT-snPCR screen (Table 1). Of the 37 and 23 samples collected from the male and female participants, respectively, 11 (29.73%) and four (17.39%) yielded the expected band size (Table 1).

3.2. Virus Identification. Of the 15 amplicons subjected to sequencing, only 12 were exploitable. The remaining 3 were

TABLE 1: Samples positive for the enterovirus VP1 nested RT-PCR screen and the identity of enteroviruses detected in these samples.

S. number	Sample ID	Gender	Age (years)	VP1 RT-PCR	Serotype	Species
1	5	F	3	Positive	Unexploitable	
2	10	M	5	Positive	Unexploitable	
3	11	M	6	Positive	CVA1	Species C
4	15	M	2	Positive	CVA8	Species A
5	16	M	5	Positive	EVB80	Species B
6	20	M	2	Positive	CVA8	Species A
7	36	M	4.5	Positive	CVA20	Species C
8	41	M	1.5	Positive	EV71	Species A
9	43	F	1.5	Positive	EV71	Species A
10	44	M	1.5	Positive	Unexploitable	
11	45	M	1.5	Positive	CVA4	Species A
12	46	F	3.5	Positive	CVA3	Species A
13	48	F	4.5	Positive	Phage baseplate	
14	50	M	10	Positive	E32	Species B
15	59	M	10	Positive	EVC99	Species C

unexploitable due to the presence of multiple peaks in their electropherograms. Eleven (11) of the 12 exploitable sequences were successfully typed by the enterovirus genotyping tool (EGT) as Coxsackievirus A1 (CVA1) (1 strain), CVA3 (1 strain), CVA4 (1 strain), CVA8 (2 strains), CVA20 (1 strain), echovirus 32 (E32) (1 strain), Enterovirus A71 (EVA71) (2 strains), EVB80 (1 strain), and EVC99 (1 strain) (Table 1). Subsequently, the last exploitable sequence was subjected to a BLAST search and found to be most similar to an enterobacteriophage baseplate gene (Table 1). Based on the eleven (11) typed strains, enterovirus species A, B, C, and D accounted for 54.55%, 18.18%, 27.27%, and 0% of the detected strains.

3.3. Phylogenetic Analysis. With respect to CVA1, the sequences obtained from GenBank and the one described in this study clustered into five different groups with strong bootstrap support (Figure 1(a)). The CVA1 sequence of Nigerian origin described in this study clustered with sequences from Eurasia (Figure 1(a)). In the CVA3 phylogram, there are three distinct clusters with strong bootstrap support (Figure 1(b)). Within cluster 2, the Nigerian CVA3 detected in this study clustered with another CVA3 previously detected in Nigeria in 2003 [15] (Figure 1(b)). Just like for CVA3, the Nigerian CVA4 detected in this study clustered with another CVA4 previously detected in Nigeria in 2003 [15] (Figure 1(c)).

The two CVA8 sequences described in this study clustered with one another, with strong bootstrap support. These CVA8 sequences did not appear to be too closely related to any of the CVA8 sequences in the phylogram (Figure 2(a)). The CVA20 sequence described in this study, on the other hand, did not cluster with that previously detected in the region in 2012 (Figure 2(b)). Rather, it clustered with other CVA20 sequences recently described in Central African Republic [19] and Cameroon [21] (Figure 2(b)).

The E32 sequence described in this study did not cluster with other E32 sequences recently described in Central African Republic [19] and Cameroon [21] (Figure 3(a)). Rather,

it clustered with E32 sequences recently described in India [31]. On the other hand, both EV71 sequences described in this study clustered together with strong bootstrap support in genotype E (Figure 3(b)). Though this genotype consisted only of sequences from sub-Saharan Africa [15, 19, 21], the EV71 sequences described in this study clustered with the EV71 from Cameroon [21] while the EV71 previously described in Nigeria in 2004 clustered with that from Central African Republic [19] (Figure 3(b)).

The single EVB80 sequence described in this study clustered, with strong bootstrap support, with others we recently found in 2014 in Nigerian children diagnosed with AFP (unpublished data). Contrary to the situation with EVB80, the EVC99 sequence described in this study was very different from the one we recently found in 2014 in Nigerian children diagnosed with AFP (unpublished data). Though they were both found in Nigeria in 2014, they appear to be most closely related to EVC99 sequences from Cameroon [21], but with different genotypes.

4. Discussion

4.1. Enterovirus Detection Rate. Considering that only eleven of the samples could be unequivocally shown to contain enteroviruses, the results of this study show enterovirus detection rate of 18.3% (11/60) in apparently healthy school aged children in Ibadan, Southwestern Nigeria. This is higher than the 5.5% and 10% described in previous studies from apparently healthy school aged children in Southwestern [15] and Northeastern [24] Nigeria, respectively. This might be a reflection of the impact of using different detection protocols. While direct detection of enterovirus genome from the clinical sample was used in this study, the other studies [15, 24] used a cell-culture based algorithm, particularly a combination of RD and L20b cell lines, as previously recommended by the WHO [27]. This might therefore suggest that the cell-culture independent protocol of Nix et al. [25] for direct detection of enteroviruses from clinical specimen might be

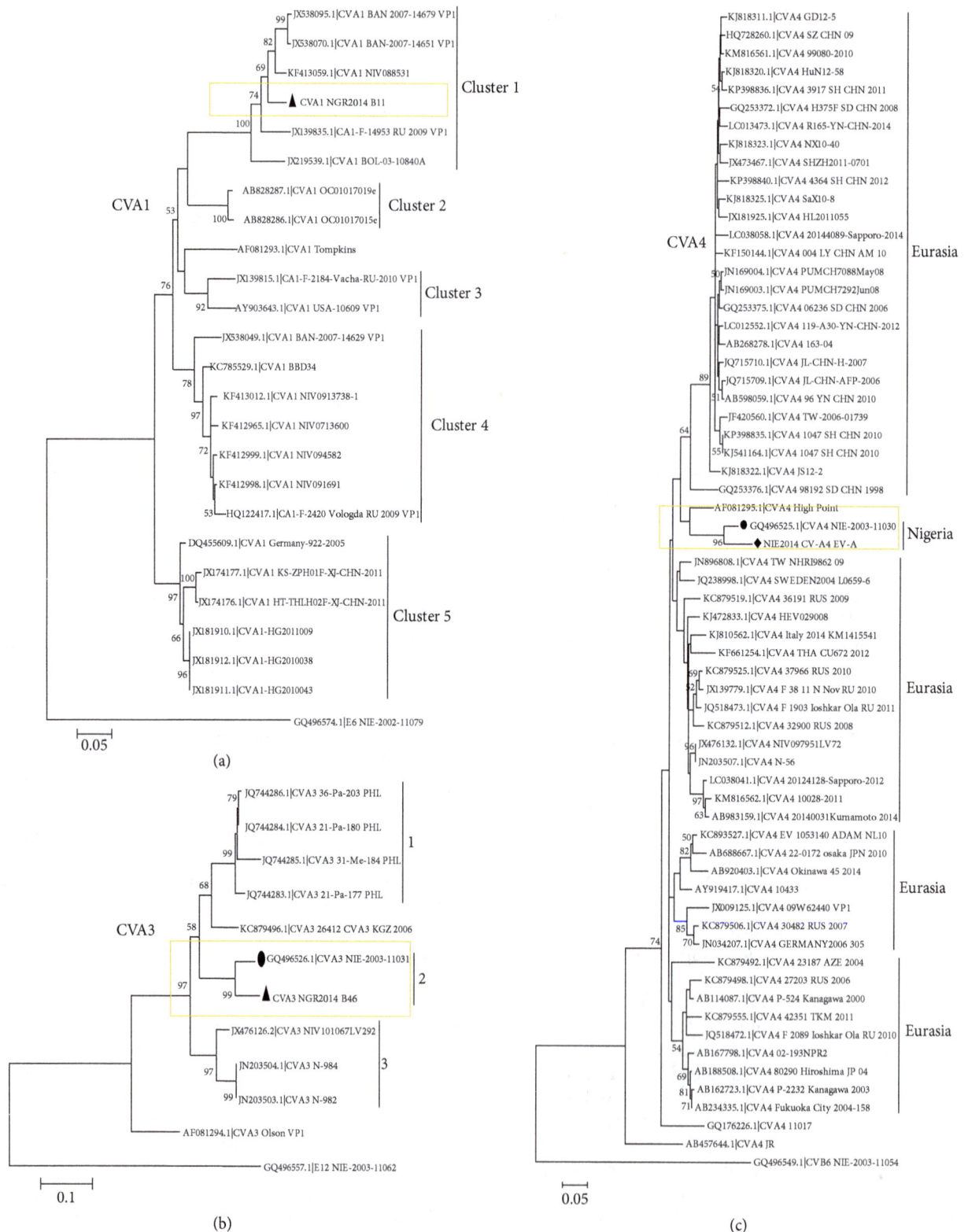

FIGURE 1: Phylogenetic relationship of recovered CVA1 (a), CVA3 (b), and CVA4 (c) strains. The phylogram is based on alignment of the partial VP1 sequences. The newly sequenced strains and previous strains from the region are highlighted with black triangles or diamonds and circles, respectively. The GenBank accession number of the strains is indicated in the phylogram. Bootstrap values are indicated if >50%.

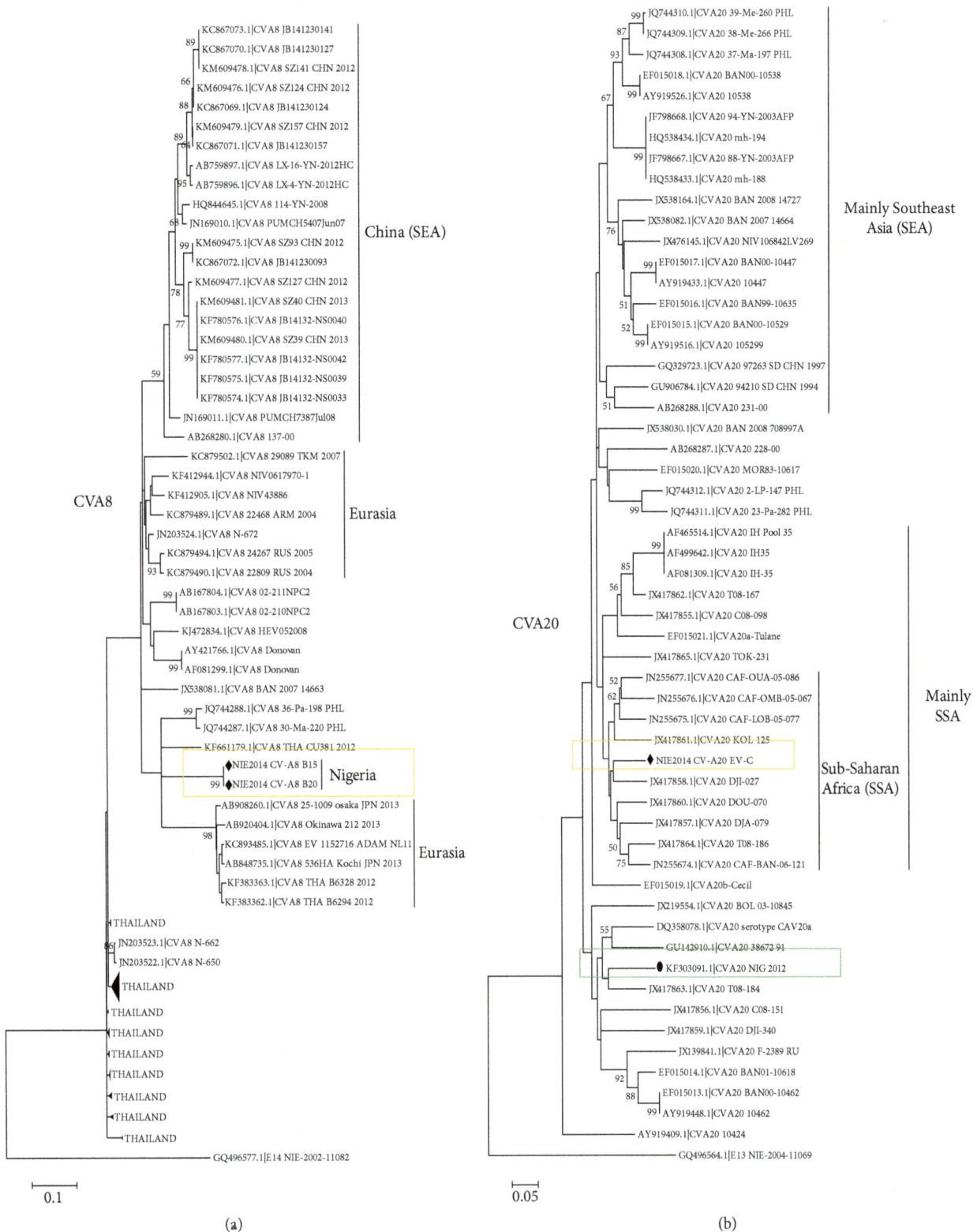

FIGURE 2: Phylogenetic relationship of recovered CVA8 (a) and CVA20 (b) strains. The phylogram is based on alignment of the partial VP1 sequences. The newly sequenced strains and previous strains from the region are highlighted with black triangles or diamonds and circles, respectively. The GenBank accession number of the strains is indicated in the phylogram. Bootstrap values are indicated if >50%.

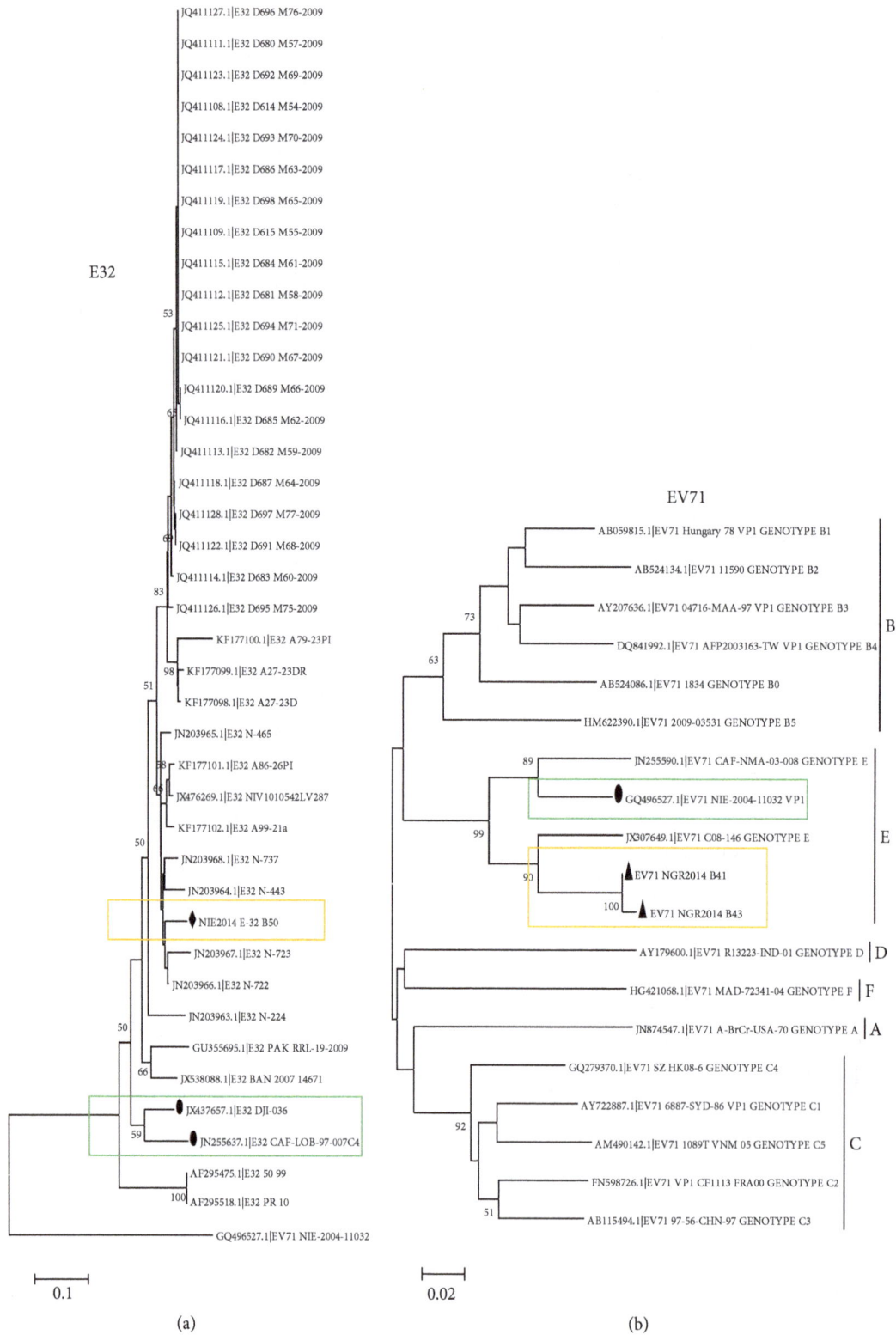

FIGURE 3: Phylogenetic relationship of recovered E32 (a) and EV71 (b) strains. The phylogram is based on alignment of the partial VP1 sequences. The newly sequenced strains and previous strains from the region are highlighted with black triangles or diamonds and circles, respectively. The GenBank accession number of the strains is indicated in the phylogram. Bootstrap values are indicated if >50%.

more sensitive than the WHO cell-culture based protocol [27]. However, such conclusion cannot be reached unequivocally, because the samples analysed in this study were not simultaneously screened using both the RD and L20b cell line based and the cell-culture independent protocols.

4.2. Enterovirus Species Diversity Landscape. The results of this study showed that species A enteroviruses were more commonly detected (54.55%) than members of the other enterovirus species (Table 1). This contradicts the findings of previous studies from the region [15, 16, 23, 24] which gave the impression that species B enteroviruses were the most commonly circulating. The true meaning and significance of this contradiction is difficult to determine considering that the same samples were not subjected to cell culture using the protocols previously documented in the region. However, this finding might be better descriptive of the enterovirus diversity landscape in the region because it bypasses cell-culture bias.

It can however be argued that the picture of the enterovirus diversity landscape painted by this cell-culture independent assay may just be a reflection of the primer specificities. Consequently, the tilt in the landscape towards species A members might not be a true reflection of the diversity landscape. However, considering that, as opposed to species A, C, and D which all individually have less than 30 serotypes documented, species B has over 60 serotypes documented (http://www.picornaviridae.com/), more species B members would have been considered during the primer design process [7, 9, 25, 32]. As a result, the primers should be biased towards species B rather than other species. Hence, the preponderance of species A enterovirus members in this population is unlikely to be as a result of primer bias.

4.3. Enterovirus Serotypes Detected. The results of this study showed the presence of nine (9) different serotypes of non-polio enteroviruses (CVA1, CVA3, CVA4, CVA8, CVA20, E32, EVA71, EVB80, and EVC99) in apparently healthy, school aged children in Ibadan, Southwestern Nigeria, in 2014 (Table 1). This study documents the first description of molecular sequence data on CVA1, CVA8, and E32 strains present in Nigeria. Though this is also the first publication of molecular sequence data of EVB80 and EVC99 from Nigeria, we had previously detected EVB80 and EVC99 in children with AFP in 2014 (unpublished data). Furthermore, we had previously described CVA20 in environmental samples in 2012 [33] and as for CVA3, CVA4, and EV71, molecular sequence data of strains circulating over ten (10) years ago were previously described [15].

4.4. Enterovirus Regional Confinement Hypothesis. The discovery of EV71 genotype F in Nigeria in 2004 [15, 23] and the subsequent detection of more members of the genotype in Central African Republic [19] and Cameroon [21] led to the hypothesis that certain enteroviruses strains circulating in sub-Saharan Africa might be confined to the region (the regional confinement hypothesis [RCH]). The recent discovery of EV71 genotype F in Madagascar [34] further

supports the RCH. It was postulated that paucity of data on enterovirus genotypes circulating in the region may be responsible for the delayed discovery of these EV71 genotypes [19, 21]. However, subsequent to the discovery of EV71 genotype F, it has recently been shown that both genotypes might have diverged from their independent, most recent common ancestors in the 1990s [34].

The EV71 strains detected in this study belonged to genotype E (Figure 3(b)), further confirming the RCH. Furthermore, CVA3 (Figure 1(b)), CVA4 (Figure 1(c)), CVA20 (Figure 2(b)), and EVC99 (Figure 4(b)) also showed evidence in support of the RCH. However, though regionally confined, the actual EV71 clade recovered in Nigeria in 2004 appears to have been replaced by a new clade (Figure 3(b)). In similar light, though regionally confined, the EVC99 strain detected in this study is different from that we recently detected in a child diagnosed with AFP (unpublished data) (Figure 4(b)). This therefore suggests the simultaneous circulation of two distinct clades of EVC99 in the country. On the other hand, genotype replacement has been observed for CVA20 (Figure 2(b)). However, the same cannot be said for CVA3 and CVA4 due to paucity of molecular sequence data from the region on these genotypes. Though the isolates of CVA3 and CVA4 detected in this study are similar to those detected over 10 years ago from the same region (Figures 1(b) and 1(c)), characterizing more isolates from the intervening years will help better understand the evolutionary dynamics of these serotypes.

The E32 isolate described in this study appeared to be more closely related to isolates from southeast Asia than those from sub-Saharan Africa (Figure 3(a)). On the one hand, this calls to question the RCH. However, on the other hand, it brings to the fore another salient underdiscussed issue concerning enterovirus identification that gives newcomers to the field some headache. Sequences of the VP1 gene are usually used for enterovirus identification. However, while the most appropriate strategy would be to amplify the entire VP1 gene, most protocols amplify either the $5'$- or the $3'$-end. For the newbie, it can be quite confusing to find out whether the partial VP1 gene is from the $5'$- or the $3'$-end of the gene. However, the enterovirus genotyping tool [28] helps to resolve this by giving a graphic view of the physical location on any VP1 gene (complete or partial) submitted as query sequence, thereby helping to determine whether the sequence in question is the complete gene or $5'$- or $3'$-end of the gene (Tables 2(a) and 2(b)).

Nix et al.'s [25] protocol is an upgrade of Oberste et al.'s [7, 9] protocol and amplifies the $5'$-end of the VP1 gene. Consequently, sequences generated using this protocol can only be compared to those generated using similar protocols that amplify the $5'$-end or complete VP1 gene. Sequences generated from protocols that amplify the $3'$-end of the VP1 gene like those of Oberste et al. [6], Casas et al. [10], and Caro et al. [12] or any iteration of these are of no value for phylogenetic analysis of VP1 gene sequences generated using Nix et al.'s [25] protocol. This is because it will be impossible to align partial VP1 sequences generated using Nix et al.'s [25] protocol with those from protocols that amplify the $3'$-end of the gene.

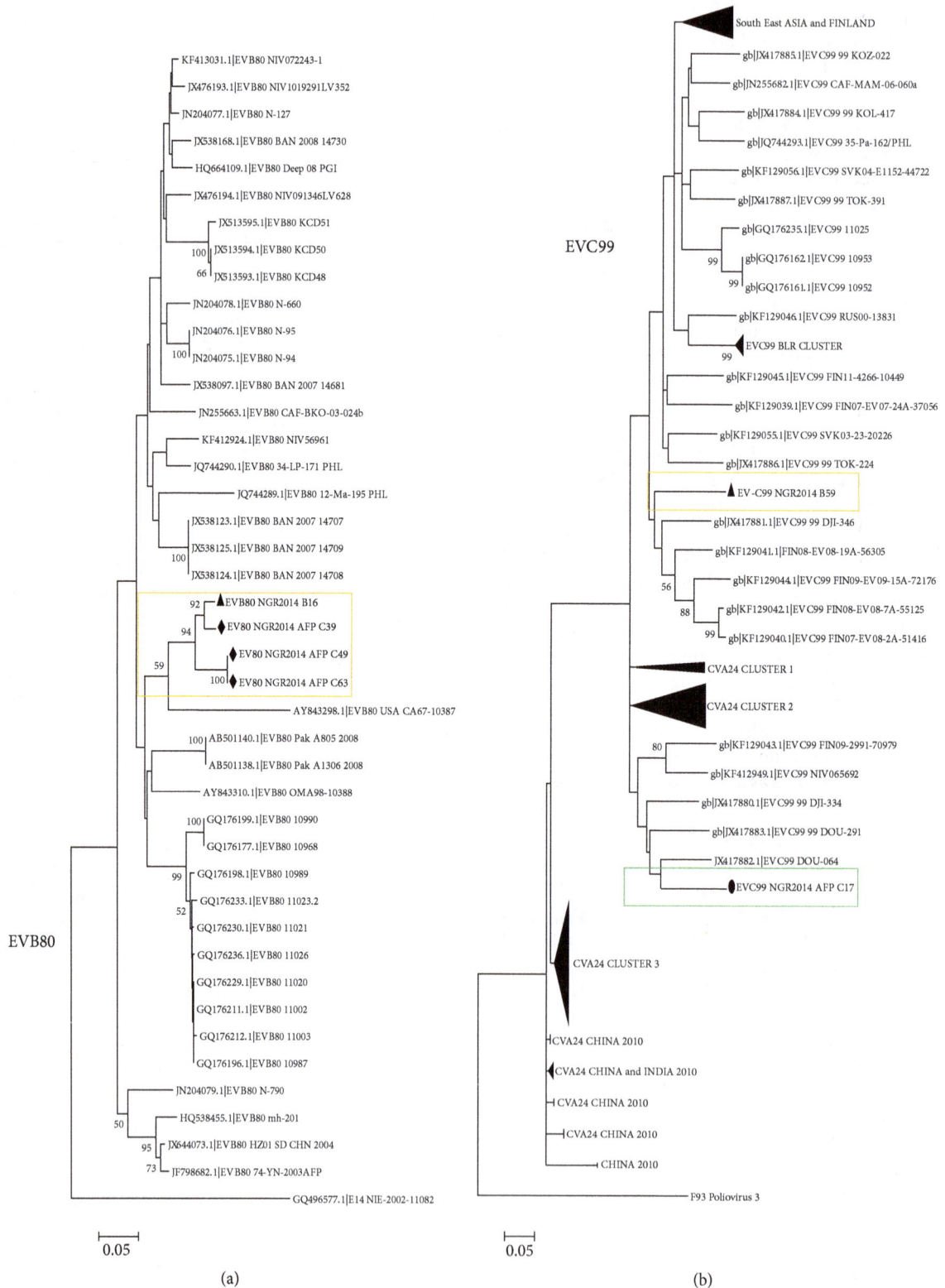

FIGURE 4: Phylogenetic relationship of recovered EVB80 (a) and EVC99 (b) strains. The phylogram is based on alignment of the partial VP1 sequences. The newly sequenced strains and previous strains from the region are highlighted with black triangles and circles, respectively. The GenBank accession number of the strains is indicated in the phylogram. Bootstrap values are indicated if >50%.

TABLE 2: Enterovirus genotyping tool (EGT) identification of isolates and graphic view of the region of the enterovirus genome represented by the query nucleotide sequence(s). This graphic view was generated by the enterovirus genotyping tool [28]. (a) shows CVA20 sequences while (b) shows E32. Column 1 shows the sequence ID (or accession number); column 2 shows the number of nucleotide sequences; column 3 shows the species to which the isolate belongs; column 4 shows the serotype of the isolates; column 5 shows a hyperlink (report) to more information on the alignment generated by the software (note: the hyperlink is generated afresh for every sequence data imputed into the enterovirus genotyping tool and is a temporary link that ceases to exist not long after a guest leaves the EGT); and column 6 shows a graphic view of the region of the enterovirus genome represented by the query sequence.

(a) CVA20

Name	Length	Genus/species	Serotype, subgenogroup	Report	Genome
DQ358078.1|CVA20_serotype_CAV2	7444	Enterovirus C	CVA20	Report	
KF303091.1|CVA20_NIG_2012	660	Enterovirus C	CVA20	Report	
NIE2014_CV-A20_EV-C	354	Enterovirus C	CVA20	Report	
JX426682.1|CVA20_T08-213	305	Enterovirus C	CVA20	Report	
JX426681.1|CVA20_T08-166	303	Enterovirus C	CVA20	Report	
JX426680.1|CVA20_T08-112	303	Enterovirus C	CVA20	Report	
JX426679.1|CVA20_MAR-252	303	Enterovirus C	CVA20	Report	
JX426678.1|CVA20_MAR-250	305	Enterovirus C	CVA20	Report	
JX426677.1|CVA20_MAR-249	305	Enterovirus C	CVA20	Report	

(b) E32

Name	Length	Genus/species	Serotype, subgenogroup	Report	Genome
JQ411108.1|E32_D614_M54-2009	259	Enterovirus B	E32	Report	
AF295475.1|E32_50_99	568	Enterovirus B	E32	Report	
AF295518.1|E32_PR_10	568	Enterovirus B	E32	Report	
HQ662326.1|E32_Mum-829	337	Enterovirus B	E32	Report	
HQ662323.1|E32_Mum-837	337	Enterovirus B	E32	Report	
JN203970.1|E32_N-990B	372	Enterovirus B	E32	Report	

(b) Continued.

Name	Length	Genus/species	Serotype, subgenogroup	Report	Genome
JN203969.1|E32_N-900	375	Enterovirus B	E32	Report	
JN203968.1|E32_N-737	876	Enterovirus B	E32	Report	

There are a significant number of sequences of this sort in GenBank and some of the sub-Saharan enterovirus sequences fall into this category alongside those from other world regions (Tables 2(a) and 2(b)). This dichotomy undermines the capacity to better investigate the RCH with the data at hand and necessitates the need to have a second look at the adoption and use of Nix et al.'s [25] protocol and other protocols that do not amplify the entire VP1 gene for studies focused on investigating the RCH. This dichotomy accounts for why the regional confinement of strains of E32, as well as CVA1, CVA8, and EVB80, detected in this study could not be exhaustively determined despite the availability of sequence data from sub-Saharan Africa strains in the nucleotide databases. Therefore, other cell-culture independent protocols for direct detection of enteroviruses from clinical samples like the ECRA recently described by Arita et al. [35], which have the capacity to amplify the complete VP1 gene, should be further investigated and developed to tools that are affordable and field deployable, especially in resource limited settings.

4.5. What Happens with Coinfections? Subsequent to the completion of this study, it was observed that, in cases of enterovirus coinfection, Nix et al.'s [25] protocol tends to amplify the most prevalent genome. For example, when further screened with enterovirus species specific primers, it was discovered that sample number 45 (Table 1) also had EVB88 in it (unpublished data). This was totally missed by Nix et al.'s [25] panenterovirus RT-snPCR screen. On the other hand, Nix et al.'s [25] panenterovirus RT-snPCR screen is not completely infallible. Failure on the part of the assay to amplify the gene of interest should not be considered with absolute certainty that the sample is negative for the virus of interest. For example, in another incident, Nix et al.'s [25] panenterovirus RT-snPCR screen failed to amplify the VP1 gene from an enterovirus isolate recovered on RD cell line in our laboratory. However, when species B and C specific RT-snPCR assays were used, echovirus 6 (E6) and poliovirus 1 (PV1) were detected, respectively. Hence, as valuable as this assay is, it also has its weaknesses. Consequently, strategies still have to be developed to improve its sensitivity as well as integrate it into already established enterovirus isolation protocols [26, 27]. In addition, Nix et al.'s [25] protocol consistently amplified an enterobacteriophage tail gene (Table 1) yielding a band that is similar in size to that expected for enteroviruses. Hence, the presence of a band in the expected range should be interpreted with caution pending the sequencing of the amplicon.

4.6. Conclusions. The results of this study showed the presence of CVA1, CVA3, CVA4, CVA8, CVA20, E32, EVA71, EVB80, and EVC99 in Ibadan, Southwestern Nigeria, in 2014. It thereby documents the first description of molecular sequence data on CVA1, CVA8, and E32 strains present in Nigeria. It further showed that species A enteroviruses were more commonly detected in the region when cell-culture bias is bypassed. The results of this study confirm that enteroviruses can be detected directly from faecal suspension using Nix et al.'s [25] protocol as proposed in the enterovirus surveillance guidelines [4]. Furthermore, the amplicons produced from Nix et al.'s [25] panenterovirus VP1 RT-snPCR assay are sufficient for sequencing and identification of the enteroviruses present in such samples. It further shows that Nix et al.'s [25] protocol tends to amplify the most prevalent genome when mixtures are present and failure on the part of the assay to amplify the gene of interest should not be considered with absolute certainty that the sample is negative for the virus of interest.

Authors' Contributions

(1) Study design was done by Temitope Oluwasegun Cephas Faleye, Moses Olubusuyi Adewumi, and Johnson Adekunle Adeniji. (2) Sample collection was carried out by Bamidele Atinuke Coker and Felix Yasha Nudamajo. (3) Acquisition of reagents and laboratory and data analysis were the responsibility of all authors. (4) Temitope Oluwasegun Cephas Faleye wrote the first draft of the paper. (5) All Authors revised the paper. (6) And all authors read and approved the final draft.

Acknowledgments

The authors would like to thank the study participants, their school administrators, and the guardians and/or parents for their cooperation throughout the period of the study.

References

[1] C. Tapparel, F. Siegrist, T. J. Petty, and L. Kaiser, "Picornavirus and enterovirus diversity with associated human diseases," *Infection, Genetics and Evolution*, vol. 14, no. 1, pp. 282–293, 2013.

[2] R. A. Strikas, L. J. Anderson, and R. A. Parker, "Temporal and geographic patterns of isolates of nonpolio enterovirus in the United States, 1970–1983," *Journal of Infectious Diseases*, vol. 153, no. 2, pp. 346–351, 1986.

[3] N. Nathanson and O. M. Kew, "From emergence to eradication: the epidemiology of poliomyelitis deconstructed," *American Journal of Epidemiology*, vol. 172, no. 11, pp. 1213–1229, 2010.

[4] World Health Organisation, *Enterovirus Surveillance Guidelines: Guidelines for Enterovirus Surveillance in Support of the Polio Eradication Initiative*, World Health Organisation, Geneva, Switzerland, 2015.

[5] D. R. Kilpatrick, B. Nottay, C.-F. Yang et al., "Serotype-specific identification of polioviruses by PCR using primers containing mixed-base or deoxyinosine residues at positions of codon degeneracy," *Journal of Clinical Microbiology*, vol. 36, no. 2, pp. 352–357, 1998.

[6] M. S. Oberste, K. Maher, D. R. Kilpatrick, and M. A. Pallansch, "Molecular evolution of the human enteroviruses: correlation of serotype with VP1 sequence and application to picornavirus classification," *Journal of Virology*, vol. 73, no. 3, pp. 1941–1948, 1999.

[7] M. S. Oberste, K. Maher, M. R. Flemister, G. Marchetti, D. R. Kilpatrick, and M. A. Pallansch, "Comparison of classic and molecular approaches for the identification of untypeable enteroviruses," *Journal of Clinical Microbiology*, vol. 38, no. 3, pp. 1170–1174, 2000.

[8] M. S. Oberste, D. Schnurr, K. Maher, S. al-Busaidy, and M. A. Pallansch, "Molecular identification of new picornaviruses and characterization of a proposed enterovirus 73 serotype," *Journal of General Virology*, vol. 82, no. 2, pp. 409–416, 2001.

[9] M. S. Oberste, W. A. Nix, K. Maher, and M. A. Pallansch, "Improved molecular identification of enteroviruses by RT-PCR and amplicon sequencing," *Journal of Clinical Virology*, vol. 26, no. 3, pp. 375–377, 2003.

[10] I. Casas, G. F. Palacios, G. Trallero, D. Cisterna, M. C. Freire, and A. Tenorio, "Molecular characterization of human enteroviruses in clinical samples: comparison between VP2, VP1, and RNA polymerase regions using RT nested PCR assays and direct sequencing of products," *Journal of Medical Virology*, vol. 65, no. 1, pp. 138–148, 2001.

[11] H. Norder, L. Bjerregaard, and L. O. Magnius, "Homotypic echoviruses share aminoterminal VP1 sequence homology applicable for typing," *Journal of Medical Virology*, vol. 63, no. 1, pp. 35–44, 2001.

[12] V. Caro, S. Guillot, F. Delpeyroux, and R. Crainic, "Molecular strategy for 'serotyping' of human enteroviruses," *Journal of General Virology*, vol. 82, no. 1, pp. 79–91, 2001.

[13] I. Thoelen, P. Lemey, I. Van der Donck, K. Beuselinck, A. M. Lindberg, and M. Van Ranst, "Molecular typing and epidemiology of enteroviruses identified from an outbreak of aseptic meningitis in Belgium during the summer of 2000," *Journal of Medical Virology*, vol. 70, no. 3, pp. 420–429, 2003.

[14] S. Blomqvist, A. Paananen, C. Savolainen-Kopra, T. Hovi, and M. Roivainen, "Eight years of experience with molecular identification of human enteroviruses," *Journal of Clinical Microbiology*, vol. 46, no. 7, pp. 2410–2413, 2008.

[15] O. G. Oyero, F. D. Adu, and J. A. Ayukekbong, "Molecular characterization of diverse species enterovirus-B types from children with acute flaccid paralysis and asymptomatic children in Nigeria," *Virus Research*, vol. 189, pp. 189–193, 2014.

[16] J. A. Adeniji and T. O. C. Faleye, "Isolation and identification of enteroviruses from sewage and sewage contaminated water in Lagos, Nigeria," *Food and Environmental Virology*, vol. 6, no. 2, pp. 75–86, 2014.

[17] E. J. Bell and B. P. Cosgrove, "Routine enterovirus diagnosis in a human rhabdomyosarcoma cell line," *Bulletin of the World Health Organization*, vol. 58, no. 3, pp. 423–428, 1980.

[18] V. V. Hamparian, A. C. Ottolenghi, and J. H. Hughes, "Enteroviruses in sludge: multiyear experience with four wastewater treatment plants," *Applied and Environmental Microbiology*, vol. 50, no. 2, pp. 280–286, 1985.

[19] M. Bessaud, S. Pillet, W. Ibrahim et al., "Molecular characterization of human enteroviruses in the Central African Republic: uncovering wide diversity and identification of a new human enterovirus A71 genogroup," *Journal of Clinical Microbiology*, vol. 50, no. 5, pp. 1650–1658, 2012.

[20] D. C. Rao, M. Ananda Babu, A. Raghavendra, D. Dhananjaya, S. Kumar, and P. P. Maiya, "Non-polio enteroviruses and their association with acute diarrhea in children in India," *Infection, Genetics and Evolution*, vol. 17, pp. 153–161, 2013.

[21] S. A. Sadeuh-Mba, M. Bessaud, D. Massenet et al., "High frequency and diversity of species C enteroviruses in Cameroon and neighboring countries," *Journal of Clinical Microbiology*, vol. 51, no. 3, pp. 759–770, 2013.

[22] J. A. Adeniji and T. O. Faleye, "Impact of cell lines included in enterovirus isolation protocol on perception of nonpolio enterovirus species C diversity," *Journal of Virological Methods*, vol. 207, pp. 238–247, 2014.

[23] O. G. Oyero and F. D. Adu, "Non-polio enteroviruses serotypes circulating in Nigeria," *African Journal of Medicine and Medical Sciences*, vol. 39, supplement, pp. 201–208, 2010.

[24] M. M. Baba, B. S. Oderinde, P. Z. Patrick, and M. M. Jarmai, "Sabin and wild polioviruses from apparently healthy primary school children in northeastern Nigeria," *Journal of Medical Virology*, vol. 84, no. 2, pp. 358–364, 2012.

[25] W. A. Nix, M. S. Oberste, and M. A. Pallansch, "Sensitive, seminested PCR amplification of VP1 sequences for direct identification of all enterovirus serotypes from original clinical specimens," *Journal of Clinical Microbiology*, vol. 44, no. 8, pp. 2698–2704, 2006.

[26] World Health Organisation, *Guidelines for Environmental Surveillance of Poliovirus Circulation*, World Health Organisation, Geneva, Switzerland, 2003.

[27] World Health Organzation, *Polio Laboratory Manual*, World Health Organzation, Geneva, Switzerland, 4th edition, 2004.

[28] A. Kroneman, H. Vennema, K. Deforche et al., "An automated genotyping tool for enteroviruses and noroviruses," *Journal of Clinical Virology*, vol. 51, no. 2, pp. 121–125, 2011.

[29] K. Tamura, D. Peterson, N. Peterson, G. Stecher, M. Nei, and S. Kumar, "MEGA5: molecular evolutionary genetics analysis using maximum likelihood, evolutionary distance, and maximum parsimony methods," *Molecular Biology and Evolution*, vol. 28, no. 10, pp. 2731–2739, 2011.

[30] M. Kimura, "A simple method for estimating evolutionary rates of base substitutions through comparative studies of nucleotide sequences," *Journal of Molecular Evolution*, vol. 16, no. 2, pp. 111–120, 1980.

[31] C. D. Rao, P. Yergolkar, and K. S. Subbanna, "Antigenic diversity of enteroviruses associated with nonpolio acute flaccid paralysis, India, 2007–2009," *Emerging Infectious Diseases*, vol. 18, no. 11, pp. 1833–1840, 2012.

[32] M. S. Oberste, K. Maher, A. J. Williams et al., "Species-specific RT-PCR amplification of human enterovirus: a tool for rapid species identification of uncharacterized enteroviruses," *Journal of General Virology*, vol. 87, no. 1, pp. 119–128, 2006.

[33] J. A. Adeniji and T. O. C. Faleye, "Enterovirus C strains circulating in Nigeria and their contribution to the emergence of recombinant circulating vaccine-derived polioviruses," *Archives of Virology*, vol. 160, no. 3, pp. 675–683, 2015.

[34] M. Bessaud, R. Razafindratsimandresy, A. Nougairède et al., "Molecular comparison and evolutionary analyses of VP1 nucleotide sequences of new African human enterovirus 71 isolates reveal a wide genetic diversity," *PLoS ONE*, vol. 9, no. 3, Article ID e90624, 2014.

[35] M. Arita, D. R. Kilpatrick, T. Nakamura et al., "Development of an efficient entire-capsid-coding-region amplification method for direct detection of poliovirus from stool extracts," *Journal of Clinical Microbiology*, vol. 53, no. 1, pp. 73–78, 2015.

10

Expression of Factor X in BHK-21 Cells Promotes Low Pathogenic Influenza Viruses Replication

Shahla Shahsavandi, Mohammad Majid Ebrahimi, Shahin Masoudi, and Hasan Izadi

Razi Vaccine & Serum Research Institute, P.O. Box 31975-148, Karaj 31976 19751, Iran

Correspondence should be addressed to Shahla Shahsavandi; s.shahsavandi@rvsri.ac.ir

Academic Editor: Subhash C. Verma

A cDNA clone for factor 10 (FX) isolated from chicken embryo inserted into the mammalian cell expression vector pCDNA3.1 was transfected into the baby hamster kidney (BHK-21) cell line. The generated BHK-21 cells with inducible expression of FX were used to investigate the efficacy of the serine transmembrane protease to proteolytic activation of influenza virus hemagglutinin (HA) with monobasic cleavage site. Data showed that the BHK-21/FX stably expressed FX after ten serial passages. The cells could proteolytically cleave the HA of low pathogenic avian influenza virus at multiplicity of infection 0.01. Growth kinetics of the virus on BHK-21/FX, BHK-21, and MDCK cells were evaluated by titrations of virus particles in each culture supernatant. Efficient multicycle viral replication was markedly detected in the cell at subsequent passages. Virus titration demonstrated that BHK-21/FX cell supported high-titer growth of the virus in which the viral titer is comparable to the virus grown in BHK-21 or MDCK cells with TPCK-trypsin. The results indicate potential application for the BHK-21/FX in influenza virus replication procedure and related studies.

1. Introduction

Influenza is one of the most economically important viral respiratory diseases of human as well as avian and animal species worldwide. The causative agent is belonging to Orthomyxoviridae, negative-sense, single-stranded RNA, which encodes at least eleven proteins. Type A viruses are subtyped on the basis of the two main surface glycoproteins hemagglutinin (HA) and neuraminidase (NA) and further classified as low pathogenic (LP) or highly pathogenic (HP) on the basis of specific molecular and pathogenesis criteria [1, 2]. Avian influenza viruses replicate mostly in the intestine while human influenza strains replicate in the upper respiratory tract [3, 4]. The infection cycle is initiated by the specific binding of viral HA to a terminal sialic acid-capped glycosylated molecule present on the surface of the host cells. Upon attachment to the cell, receptor-mediated endocytosis occurs either by α2–6Gal or by α2-3Gal linkage [5, 6]. Cellular tropism and the infectivity of influenza viruses are primarily determined by the distribution of these receptors in the cell surface. Also the presence of specific host cellular protease

(s) for posttranslational cleavage of HA0 precursor protein is essential for viral infectivity, pathogenicity, and tissue tropism [7, 8]. Influenza A viruses target a wide spectrum of tissues so replication of the viruses has been examined in a variety of cells [9–13]. The results demonstrated that susceptibility to the viruses varied significantly between the cells; and in particular the tracheal epithelial, MDCK, and A549 have been described as suitable permissive cells for the replication of both human and avian influenza viruses. Replication of influenza viruses can be attributed to expression of both 2-6 and 2-3 linked sialic acid receptors on the surfaces of the cells. Beside the viral tropism determined by virus-receptor interactions, local density of receptors, lipid raft microdomains, and host cell proteases activating the viral surface glycoproteins play major roles in influenza infectivity [14, 15]. The viruses have the ability to exploit a host virus-activating protease system to support own replication. Cellular host proteases such as transmembrane serine proteases (TMPRSS), an analogous protease from chicken allantoic fluid to the blood clotting factor 10 (FX), and plasmin were involved in the postentry stages of influenza A virus infection

[8, 16]. In particular, there are growing interests in the role of hemostasis during influenza virus infection lately. The fact that factor X might play a role in viral replication suggests that indeed hemostasis and coagulation might be deleterious for the host [17]. Avian influenza viruses reach high titers when grown within chicken-origin cells; however, the efficient replication and infectivity of LP viruses are achieved in the presence of supplemental trypsin [4]. The enzyme enhances the internalization of influenza virus into cells by cleavage of HA but did not improve the ability of the host cells to internalize the virus [18]. It is well documented that the LP viruses cannot be cleaved by ubiquitous intracellular proteases while they replicate efficiently in eggs because of the presence of a protease in allantoic fluid that can cleave HA [19]. Previous studies reveal that FX, a vitamin K-dependent serine protease in the prothrombin family, was induced upon virus infection. The viral activating protein cleaves the fusion proteins of Sendai virus, Newcastle disease virus, and influenza virus at a specific single arginine-containing site and plays a key role in the viral spreading in the allantoic sac [16, 17]. In this study, we cloned the FX mRNA in BHK-21 cell and evaluated the impact of the established BHK-21/FX cell on susceptibility and virus replication kinetics of a LP influenza virus strain to provide insights into the development of future influenza virus diagnostic approaches.

2. Materials and Methods

2.1. Cells and Virus Infection. MDCK and BHK-21 cells were grown in Dulbecco's modified Eagle's medium (DMEM) (Invitrogen) supplemented with 10% fetal bovine serum (FBS; Sigma Aldrich), 100 U/mL penicillin, and 100 mg/mL streptomycin, at 37°C with 5% CO_2. With the use of a dose-response test, the optimal L-(tosylamide-2-phenyl)ethyl chloromethyl ketone (TPCK) trypsin concentration was determined for each culture. The BHK-21 cell line had greater susceptibility to trypsin toxicity and received 0.1 mg/mL, while MDCK cell line was the most resistant to toxicity caused by trypsin and received 0.45 mg/mL. Monolayers of the cells at a concentration of 1×10^6 cells/mL were infected with a local isolate influenza H9N2 virus A/chicken/Iran SS8 (2011) at a multiplicity of infection (MOI) of 0.01 in the presence of supplemental trypsin. Following adsorption for 1 h at 37°C, the inoculum was removed and washed before DMEM was replaced. The cultures were incubated up to 72 hours post infection (hpi) and controlled by inverted light microscopy for cytopathic effect (CPE). For each cell, four different sets of tissue culture flasks were infected. Mock virus infected cells served as controls.

2.2. Expression Vector pcDNA3.1-FX and Transfection of BHK-21 Cells. The cDNA clone of FX in the pcDNA3.1 expression vector (Invitrogen) was constructed. The open reading frame of FX located in peptidase S1 domain (241–473 nt) isolated from chorioallantoic membrane of embryonated eggs was amplified with primers 5-<u>GGATCC</u>GATGAGTGT-CGTCCTGGTGA-3 and 5-<u>AAGCTT</u>AGCCACGCCACT-ACTACTTT-3, containing restriction sites for *Bam*HI and

*Hin*dIII (underlines), respectively. BHK-21 cells were transfected with pcDNA3.1-FX plasmid using Lipofectamine 2000 (Life Technologies) according to the manufacturer's instructions. The transfected cells were selected and cultivated in the presence of 10% FBS and 800 g/mL Geneticin (G418; Invitrogen). After two weeks, cells surviving the selection were pooled, passaged three times in 1.5 mg of G418/mL, and frozen in aliquots. The G418-resistant colonies were isolated and subjected to RT-PCR to verify the expression of FX at 232 bp lengths using the extracted RNA (High Pure RNA Extraction Kit, Roche, Germany) and the one-step RT-PCR (iNtRON Biotechnology, Korea).

2.3. BHK-21/FX Cell Screening during Multiple Cell Passages. The growth property and plating condition for the BHK-21/FX cells were assessed prior to virus infection. The cells were seeded into two 48-well cell culture plates and incubated overnight. When the monolayers were between 90% and 95% confluent, they were inoculated with the virus at MOI 0.01. One plate received trypsin-supplemented media and one plate received plain media. The cells were incubated for 4 days at 37°C with 5% CO_2 and checked microscopically for the presence of CPE. Mock cells were included in each experiment as controls. On days 1 and 4 post infection (pi), eight wells in each plate were immunostained using influenza NP specific antibody to assess the expression of the viral internal protein. The cells were examined for virus replication kinetics quantification at different times pi. Afterwards, infectious viral particles were quantified and expressed in $TCID_{50}$. Finally, the remaining harvested supernatant was used as inoculum for ten subsequent passages. The infectivity of influenza virus in each passage was determined using immunoassay procedure. Briefly, cells were washed once with phosphate buffered saline (PBS) and fixed with 4% paraformaldehyde in PBS for 15 min at room temperature. The fixed cells were washed and immunostained with mouse anti-influenza NP monoclonal antibody followed by FITC-labeled goat anti-mouse IgG (Dako, Glostrup, Denmark). Immunostained cells were examined under Nikon Eclipse E600 fluorescence microscope.

To evaluate the impact of producing influenza virus in BHK-21/FX cell, we have assessed the sensitivity assay by infecting an amount of 10^5 cells with the H9N2 virus for ten subsequent viral passages. At each passage, virus titer was estimated and the HA and NA nucleotide sequences were determined. In this case, the extracted virus RNA was amplified using the one-step RT-PCR for HA and NA segments at full-lengths [12]. The PCR products of viral genes at different passages were cloned and sequenced subsequently in both directions. Sequences were analyzed using the CLUSTAL W alignment method of the BioEdit sequence alignment editor version 7.0.9 software.

2.4. Statistical Analysis. The data are expressed as mean ± SD. Statistical correlation of data was checked for significance by ANOVA and Student's *t*-test. Differences with $P < 0.05$ were considered significant.

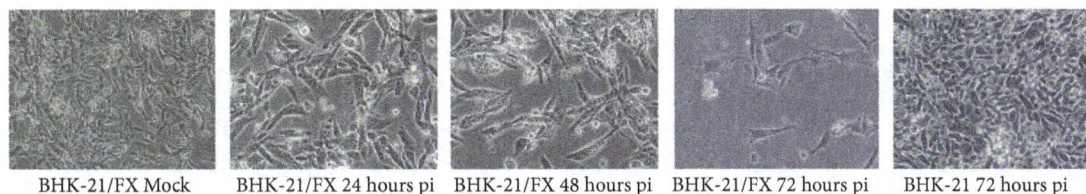

| BHK-21/FX Mock | BHK-21/FX 24 hours pi | BHK-21/FX 48 hours pi | BHK-21/FX 72 hours pi | BHK-21 72 hours pi |

FIGURE 1: Cytopathogenicity of BHK-21/FX cells to influenza virus infections (MOI 0.01) at interval hours post infection (200x magnification). The BHK-21 cells infected with influenza virus did not manifest cytopathic effects.

3. Results

3.1. Cell Lines Validation. The ability of H9N2 influenza virus to infect the MDCK and BHK-21 cell lines was assessed in the absence and presence of supplemental trypsin. The H9N2 virus replicated in both cells and moderate CPEs were evident 48 hpi only in the presence of trypsin. The virus failed to produce CPE in the absence of trypsin because the HA remains uncleaved and virus replication did not occur.

3.2. BHK-21/FX Cell Establishment and Screening. BHK-21/FX was established by transfection of plasmid encoding FX. Two weeks continuously under antibiotic selection, the surviving cultivated BHK-21/FX cell was assessed systematically for the sensitivity to primary influenza virus infection and permittivity for virus replication and spread. Specific band amplified from total RNA of BHK-21/FX cell in RT-PCR has confirmed the presence of FX gene in the cell. The morphology of BHK-21/FX cells was not different from BHK-21 cells. Following infection of BHK-21/FX cells with influenza virus, visible CPE was observed by 24 hpi with giant cell formation and massive detaching from the culture flask compared to BHK-21 cell (Figure 1). The BHK-21 cells infected by the virus developed a very light CPE at 72 hpi. Both cells showed an increase in virus titer from $10^{3.0}$ TCID$_{50}$ on the first passage which remained constant during serial passages for BHK-21 cell ($10^{5.5}$ TCID$_{50}$), while the H9N2 virus infected-BHK-21/FX cell exhibited the highest viral titer. The amount of infectious virus yield at the first passage in BHK-21/FX cells was increased ~2500 times ($10^{7.5}$ TCID$_{50}$) in fifth passage which was maintained up to seventh passage. This indicates clearly that the cell permits the production of infectious influenza virus particles (Figure 2). Production and spread of the virus were also monitored by detecting NP expression in immunofluorescence assay up to 72 hpi (Figure 3). From 8 hpi, the numbers of NP-expressed cells increased with time elapses. The virus infected-BHK-21 cells have exhibited the same panel in the presence of supplemental trypsin, while virus replication was not observed in the absence of trypsin. It may be due to protease activation of viral HA cleavage site which supports virus entry and replication.

Nucleotide sequences of the HA and NA genes of the virus passaged seven times in BHK-21/FX cells were analyzed and compared with the parental virus genome sequences. No amino acid change was observed at the cleavage site, in the receptor binding pocket, and within the N-glycosylation sites of HA protein. All of them showed conservation of residues

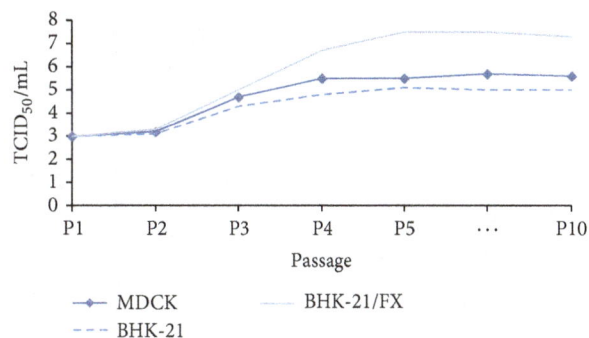

FIGURE 2: Replication of influenza virus in BHK-21/FX cells. The titer of virus in BHK-21/FX cell supernatants was assayed by TCID$_{50}$ in ten subsequent passages compared to the virus infected MDCK and BHK-21 cells supplemented with trypsin.

H^{183}, L^{190}, L^{226}, Q^{227}, and G^{228} in the receptor binding pocket and RSSR motif in cleavage site. The comparative sequence analysis of NA indicated that the amino acid sequences at the active sites of NA protein at positions ^{366} IRKDSRAG ^{373}, ^{399} DSDNRSGY ^{406}, and ^{431} PQE ^{433} were conserved. Three simultaneously nucleotide substitutions were found at the fifth passage which did not lead to amino acid changes. In the consequent passages of the virus on BHK-21/FX cell, mutation in the nucleotide sequences of viral genes that resulted in change in amino acid codon was not detected.

4. Discussion

Since 1997, several cases of human infections with different subtypes of avian influenza viruses have been identified and raised the pandemic potential of avian influenza virus in human population. Thus, early detection is very important to initiate efficient control programs. Virus isolation in embryonated eggs or in cell culture followed by subtyping is considered as the standard protocol for detecting avian influenza virus. Since previous decade some isolates have low growth properties in MDCK cells line [20]. Avian influenza viruses generally grow efficiently on embryonated eggs or primary chick embryo-originated cells due to the specific receptor distribution and host cell proteases. It has been demonstrated that blood-derived proteases promote influenza A virus replication outside the respiratory tract.

Here, we developed BHK-21 cell line that expresses FX, which could affect influenza virus HA activation in absence of supplemental trypsin. The main focus of our project was the

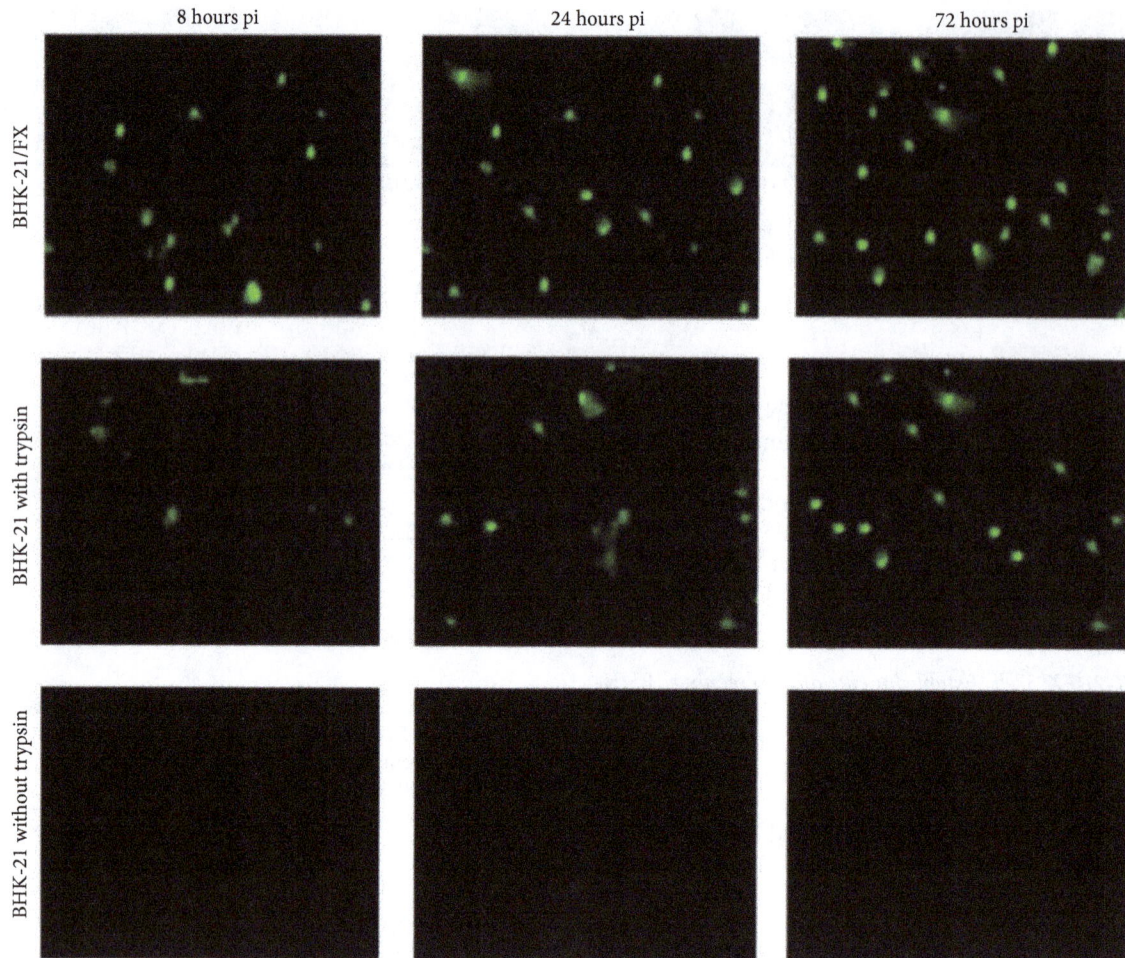

FIGURE 3: Immunofluorescent detection of influenza virus NP in infected-BHK-21/FX cells (×100). Fluorescence emission was detected in the infected cells at 24 and 72 hours post infection compared to the virus infected-BHK-21 cells supplemented with trypsin and without trypsin.

characterization of LP avian virus propagation in the genetically manipulated BHK-21 cell and potential use of BHK-21/FX cells for the isolation and replication of the viruses. RT-PCR confirmed the presence of the target gene expression in ten serially passaged cells. Growth kinetics of H9N2 virus on BHK-21/FX, BHK-21, and MDCK cells were evaluated by titrations of virus particles in each culture supernatant. The growth kinetic of the virus in the manipulated cell was comparable to those with MDCK or BHK-21 plus TPCK-trypsin ($P < 0.01$). Monitoring of virus titers and release of progeny viruses during multiple cycle infections is of great importance for process characterization and optimization. Based on the length of the interval between the virus inoculation and increasing viral infective titer, the duration of reproductive cycle of influenza virus was estimated as 5-6 hours. Other studies have shown that a single virus progeny production cycle of influenza virus requires 8–10 hours in MDCK cell, the same time in A549 cell, compared to 20 hours in Chang's conjunctival cell [10, 12, 21]. We detected progeny virus release in BHK-21/FX at 8, 18, and 24 hours pi, respectively. The overexpression of 2,6 sialic acid receptors in MDCK

cells transfected by type II transmembrane serine proteases members showed sufficient titer recommended for cell-based human influenza vaccine production [7]. Enhancing the virus receptor interaction leads to increasing the membrane fusion, the virion production, and binding rates. Viral entry and initiation infection procedure were also confirmed by detection of viral NP expression in BHK-21/FX cells in absence of exogenous trypsin. The internal viral protein potentially plays an important role in the early stages of the virus life cycle including encapsidating the segmented viral genome into ribonucleoproteins, vRNA synthesis, and interactions with the viral polymerase. The NP protein is an indicator for switching from transcription to replication. Thus, according to the immunofluorescence assay of the protein, the BHK-21/FX cell could cleave the HA protein of H9 subtype in absence of exogenous trypsin. The cleavage is a cell-associated process that leads to virus replication and production of high titer comparable to those in either MDCK or BHK-21 cells plus trypsin. Full-length amplification and sequencing patterns of the HA and NA genes at intervals passages demonstrated that the H9N2 virus at ten passages remained similar to the parent virus. These data indicate that BHK-21/FX is a permissive cell which is capable of optimizing avian influenza virus replication to obtain high titers.

Characterization of a modified cell line during growth and infection will provide the basis for cell culture-based influenza vaccine development, virus isolation, and diagnosis. In this study, we use FX, a Ca^{2+}-dependent serine protease to explore host factor involved in the influenza virus replication. FX is synthesized in liver, circulated in body as plasma protein, and activated into factor Xa via intrinsic and extrinsic pathways. The active site of factor Xa is divided into four subpockets as S1–S4. The S1 subpocket determines the major component of selectivity and binding [16]. Activation and binding of the protease at the surface of infected cells are an alternative mechanism for the proteolytic cleavage of HA which is perquisite for viral replication and pathogenicity [22]. Recent progress in understanding virus replication at the molecular level has revealed that many enveloped viruses, including influenza A viruses, incorporate cellular surface proteins into viral particles during virus-cell fusion and bud from the plasma membrane of their host cells following replication [23]. Among the host-encoded proteins in influenza virus particles, the annexin family is well represented. Annexins are a family of Ca^{2+}/lipid-binding proteins and act as membrane-membrane or membrane-cytoskeleton linkers. The proteins have been implicated in Ca^{2+}-regulated exocytotic events and certain aspects of endocytosis. Tissue specific expression of annexins suggests highly specialized function which correlates well with the ability of the proteins to interact with cellular membranes [24]. The role of annexin members as cellular interactant viral proteins in replication of influenza virus has been studied. As a virus-host interaction, expression of cellular protein annexin 2 (Anx2) is increased in response to influenza infection due to dysregulation of Ca^{2+} homeostasis. The upregulation of cell-surface Anx2 allows the recruitment of plasminogen which is activated by virion. Subsequently, HA cleaves through Anx2 which is incorporated into virus particles [25]. Anx6 expression significantly increased virus production, while its overexpression could reduce the titer of virus progeny, suggesting a negative regulatory role for Anx6 during influenza A virus infection [26]. Anx13 stimulates apical transport of influenza virus HA in MDCK cells [25]. It is assumed that Ann2 and Anx13 promote the Ca^{2+}-dependent association of lipid raft [24]. The rafts as platforms for intracellular sorting and signal transduction events [27] play a decisive role at several steps during virus replication including intracellular transport of viral proteins, assembly and budding of progeny virus at the plasma membrane, environmental stability of the virus particles, and fusion of viral and host cell endosomal membrane upon virus entry [28–30]. The lipid raft is used as a platform to concentrate HA binding receptors; thus, the virus exploits the signalling capacity of raft domains by mediating efficient fusion. Viral NA is concentrated in rafts for its normal incorporation into virions and budding [23, 28]. The raft association is also mediated by the binding of Anx5 to the plasma membrane. Anx5 is incorporated into influenza envelope during

the budding process and a substantial proportion of the protein is present in lipid rafts, the site of virus budding [30].

The interactions between viral and cellular factors determine host susceptibility to influenza infection so the absence of remarkable virus replication in some cells could obviously be due to the absence of specific proteases. On the bases of host-pathogen interactions the BHK-21/FX was developed using a cellular protease factor to support high-titer LP influenza virus replication. The impact of the cell on viral titers and sensitivity at subsequent viral passages at defined cell density and virus MOI were analyzed. Distribution of the specific influenza virus receptor, specificity to other subtypes, sensitivity to higher MOIs, cell stability, and the annexin and lipid raft cellular proteins interactions need to be investigated in future studies to establish influenza virus diagnosis or vaccine manufacturing platforms based on the cellular serine protease and engineered BHK-21 cells.

Disclosure

Shahla Shahsavandi as corresponding author of the paper would like to assure that neither the submitted materials nor portions have been published previously or are under consideration for publication elsewhere.

Authors' Contribution

Shahla Shahsavandi and other authors participated meaningfully in the study and have approved the final paper.

Acknowledgments

The authors would like to thank all members of our laboratory for helpful assistances. This study is funded by Grant 2-18-18-91128 from Razi Vaccine and Serum Research Institute.

References

[1] P. F. Wright, G. Neumann, Y. Kawaoka et al., "Orthomyxoviruses," in *Fields Virology*, B. N. Fields, D. M. Knipe, and P. M. Howley, Eds., Lippincott-Raven, Philadelphia, Pa, USA, 6th edition, 2013.

[2] D. E. Swayne and D. A. Helvorson, "Influenza," in *Diseases of Poultry*, Y. M. Saif, A. M. Fadly, J. R. Glisson, L. R. McDougald, L. K. Nolan, and D. E. Swayne, Eds., Iowa State University Press, Ames, Iowa, USA, 12th edition, 2008.

[3] C.-I. Shen, C.-H. Wang, S.-C. Shen, H.-C. Lee, J.-W. Liao, and H.-L. Su, "The infection of chicken tracheal epithelial cells with a H6N1 avian influenza virus," *PLoS ONE*, vol. 6, no. 5, Article ID e18894, 2011.

[4] S. Shahsavandi, M. M. Ebrahimi, A. Mohammadi, and N. Zarrin Lebas, "Impact of chicken-origin cells on adaptation of a low pathogenic influenza virus," *Cytotechnology*, vol. 65, no. 3, pp. 419–424, 2013.

[5] S. Bertram, I. Glowacka, I. Steffen, A. Kühl, and S. Pöhlmann, "Novel insights into proteolytic cleavage of influenza virus hemagglutinin," *Reviews in Medical Virology*, vol. 20, no. 5, pp. 298–310, 2010.

[6] M. N. Matrosovich, T. Y. Matrosovich, T. Gray, N. A. Roberts, and H.-D. Klenk, "Human and avian influenza viruses target different cell types in cultures of human airway epithelium," *Proceedings of the National Academy of Sciences of the United States of America*, vol. 101, no. 13, pp. 4620–4624, 2004.

[7] E. Böttcher, T. Matrosovich, M. Beyerle, H.-D. Klenk, W. Garten, and M. Matrosovich, "Proteolytic activation of influenza viruses by serine proteases TMPRSS2 and HAT from human airway epithelium," *Journal of Virology*, vol. 80, no. 19, pp. 9896–9898, 2006.

[8] T. H. Bugge, T. M. Antalis, and Q. Wu, "Type II transmembrane serine proteases," *The Journal of Biological Chemistry*, vol. 284, no. 35, pp. 23177–23181, 2009.

[9] K. A. Moresco, D. E. Stallknecht, and D. E. Swayne, "Evaluation and attempted optimization of avian embryos and cell culture methods for efficient isolation and propagation of low pathogenicity avian influenza viruses," *Avian Diseases*, vol. 54, no. 1, pp. 622–626, 2010.

[10] S. Shahsavandi, M. M. Ebrahimi, K. Sadeghi, S. Z. Mosavi, and A. Mohammadi, "Dose- and time-dependent apoptosis induced by avian H9N2 influenza virus in human cells," *BioMed Research International*, vol. 2013, Article ID 524165, 7 pages, 2013.

[11] L. Ollier, A. Caramella, V. Giordanengo, and J.-C. Lefebvre, "High permissivity of human HepG2 hepatoma cells for influenza viruses," *Journal of Clinical Microbiology*, vol. 42, no. 12, pp. 5861–5865, 2004.

[12] N. Z. Lebas, S. Shahsavandi, A. Mohammadi, M. M. Ebrahimi, and M. Bakhshesh, "Replication efficiency of influenza A virus H9N2: a comparative analysis between different origin cell types," *Jundishapur Journal of Microbiology*, vol. 6, no. 9, Article ID e8584, 2013.

[13] H. Ozaki, E. A. Govorkova, C. Li, X. Xiong, R. G. Webster, and R. J. Webby, "Generation of high-yielding influenza A viruses in African green monkey kidney (Vero) cells by reverse genetics," *Journal of Virology*, vol. 78, no. 4, pp. 1851–1857, 2004.

[14] H. Zhang, "Tissue and host tropism of influenza viruses: importance of quantitative analysis," *Science in China, Series C: Life Sciences*, vol. 52, no. 12, pp. 1101–1110, 2009.

[15] T. Suzuki and Y. Suzuki, "Virus infection and lipid rafts," *Biological and Pharmaceutical Bulletin*, vol. 29, no. 8, pp. 1538–1541, 2006.

[16] B. Gotoh, T. Ogasawara, T. Toyoda, N. M. Inocencio, M. Hamaguchi, and Y. Nagai, "An endoprotease homologous to the blood clotting factor X as a determinant of viral tropism in chick embryo," *The EMBO Journal*, vol. 9, no. 12, pp. 4189–4195, 1990.

[17] F. Berri, G. F. Rimmelzwaan, M. Hanss et al., "Plasminogen controls inflammation and pathogenesis of influenza virus infections via fibrinolysis," *PLoS Pathogens*, vol. 9, no. 3, Article ID e1003229, 2013.

[18] F. R. Zuhairi, Maharani, and M. I. Tan, "The role of trypsin in the internalization process of influenza H1N1 virus into vero and MDCK cells," *ITB Journal of Science*, vol. 44, no. 4, pp. 297–307, 2012.

[19] C. Chaipan, D. Kobasa, S. Bertram et al., "Proteolytic activation of the 1918 influenza virus hemagglutinin," *Journal of Virology*, vol. 83, no. 7, pp. 3200–3211, 2009.

[20] N. Asaoka, Y. Tanaka, T. Sakai, Y. Fujii, R. Ohuchi, and M. Ohuchi, "Low growth ability of recent influenza clinical isolates in MDCK cells is due to their low receptor binding affinities," *Microbes and Infection*, vol. 8, no. 2, pp. 511–519, 2006.

[21] A. Abdoli, H. Soleimanjahi, M. T. Kheiri, A. Jamali, and A. Jamaati, "Determining influenza virus shedding at different time points in Madin-Darby canine kidney cell line," *Cell Journal*, vol. 15, no. 2, pp. 130–135, 2013.

[22] K. Nagata, A. Kawaguchi, and T. Naito, "Host factors for replication and transcription of the influenza virus genome," *Reviews in Medical Virology*, vol. 18, no. 4, pp. 247–260, 2008.

[23] N. Chazal and D. Gerlier, "Virus entry, assembly, budding, and membrane rafts," *Microbiology and Molecular Biology Reviews*, vol. 67, no. 2, pp. 226–237, 2003.

[24] U. Rescher and V. Gerke, "Annexins—unique membrane binding proteins with diverse functions," *Journal of Cell Science*, vol. 117, no. 13, pp. 2631–2639, 2004.

[25] F. LeBouder, B. Lina, G. F. Rimmelzwaan, and B. Riteau, "Plasminogen promotes influenza A virus replication through an annexin 2-dependent pathway in the absence of neuraminidase," *Journal of General Virology*, vol. 91, no. 11, pp. 2753–2761, 2010.

[26] S. Lecat, P. Verkade, C. Thiele, K. Fiedler, K. Simons, and F. Lafont, "Different properties of two isoforms of annexin XIII in MDCK cells," *Journal of Cell Science*, vol. 113, no. 14, pp. 2607–2618, 2000.

[27] P. Scheiffele, A. Rietveld, T. Wilk, and K. Simons, "Influenza viruses select ordered lipid domains during budding from the plasma membrane," *The Journal of Biological Chemistry*, vol. 274, no. 4, pp. 2038–2044, 1999.

[28] M. Takeda, G. P. Leser, C. J. Russell, and R. A. Lamb, "Influenza virus hemagglutinin concentrates in lipid raft microdomains for efficient viral fusion," *Proceedings of the National Academy of Sciences of the United States of America*, vol. 100, no. 25, pp. 14610–14617, 2003.

[29] G. P. Leser and R. A. Lamb, "Influenza virus assembly and budding in raft-derived microdomains: a quantitative analysis of the surface distribution of HA, NA and M2 proteins," *Virology*, vol. 342, no. 2, pp. 215–227, 2005.

[30] T. Takahashi and T. Suzuki, "Role of membrane rafts in viral infection," *The Open Dermatology Journal*, vol. 3, pp. 178–194, 2009.

11

Mutations in the H, F, or M Proteins can Facilitate Resistance of Measles Virus to Neutralizing Human Anti-MV Sera

Hasan Kweder,[1,2,3,4] Michelle Ainouze,[1,2,3,4] Sara Louise Cosby,[5]
Claude P. Muller,[6] Camille Lévy,[1,2,3,4] Els Verhoeyen,[1,2,3,4] François-Loïc Cosset,[1,2,3,4]
Evelyne Manet,[1,2,3,4] and Robin Buckland[1,2,3,4]

[1] INSERM-U1111, 69007 Lyon, France
[2] ENS-Lyon, 69007 Lyon, France
[3] University of Lyon, UCB-Lyon1, 69007 Lyon, France
[4] LabEx Ecofect, University of Lyon, 69007 Lyon, France
[5] School of Medicine, Dentistry and Biomedical Sciences, Queen's University, BT7 1NN Belfast, UK
[6] Institute of Immunology, Public Research Center for Health/LNS, 20A rue Auguste Lumière, L-1950 Luxemburg,
  Grand-Duchy of Luxembourg, Luxembourg

Correspondence should be addressed to Evelyne Manet; evelyne.manet@ens-lyon.fr

Academic Editor: Finn S. Pedersen

Although there is currently no evidence of emerging strains of measles virus (MV) that can resist neutralization by the anti-MV antibodies present in vaccinees, certain mutations in circulating wt MV strains appear to reduce the efficacy of these antibodies. Moreover, it has been hypothesized that resistance to neutralization by such antibodies could allow MV to persist. In this study, we use a novel *in vitro* system to determine the molecular basis of MV's resistance to neutralization. We find that both wild-type and laboratory strain MV variants that escape neutralization by anti-MV polyclonal sera possess multiple mutations in their H, F, and M proteins. Cytometric analysis of cells expressing viral escape mutants possessing minimal mutations and their plasmid-expressed H, F, and M proteins indicates that immune resistance is due to particular mutations that can occur in any of these three proteins that affect at distance, rather than directly, the native conformation of the MV-H globular head and hence its epitopes. A high percentage of the escape mutants contain mutations found in cases of Subacute Sclerosing Panencephalitis (SSPE) and our results could potentially shed light on the pathogenesis of this rare fatal disease.

1. Introduction

Measles virus is (MV), a member of the genus *Morbillivirus* in the family Paramyxoviridae. The MV virion is enveloped and contains a nonsegmented negative-strand RNA genome encoding six structural proteins: N, P, M, F, H, and L. The genome is encapsidated by the N (nucleoprotein) which is associated with the P and L proteins (viral polymerase) to form the helical ribonucleoprotein complex (RNP). The glycoproteins H and F are embedded as spikes in the virion membrane. The H protein (hemagglutinin) is responsible for attachment to the cellular receptors of MV and the F protein is responsible for the consequent fusion of the virion

membrane with the host cell's plasma membrane whereby the RNP is delivered into the cytoplasm, and the matrix protein M lines the inner surface of the virion membrane [1]. In the infected cell, the glycoproteins accumulate in the plasma membrane. This allows the H protein to interact with cellular receptors on neighboring uninfected cells and cause cell-cell fusion (syncytia formation) through activation of the F protein. Moreover, in the case of the infected cell, evidence exists to suggest that the M protein interacts with the cytoplasmic tails of the glycoproteins H and F at the plasma membrane [2, 3]. As far as cellular receptors for MV are concerned, the wt strains have been shown to use Signaling

Lymphocyte Activation Molecule (SLAM; CD150) whereas the vaccine and laboratory strains use both SLAM and CD46 [4]. Expression of SLAM is restricted to cells of the human immune system whereas CD46 is expressed ubiquitously. Recently, a third receptor, the epithelial adherens junction protein nectin-4, has been identified [5, 6].

MV is a serologically monotypic virus and in theory, vaccination should provide life-long protection. However, the proportion of the population possessing only vaccine-induced immunity has increased over time with reduced exposure to wild-type MV infection and there is now evidence of resistance of recent measles virus wild-type isolates to antibody-mediated neutralization in vaccinees. This includes individuals with not only primary but also secondary vaccine failure [7, 8] and is a concern for global MV elimination. It is evident that a better understanding of the molecular basis of MV's escape from neutralizing antibody is required.

In a previous study, we used mutagenesis to allow lentiviral vectors pseudotyped with MV glycoproteins to escape neutralizing antibodies. The use of such vectors in in vivo use has been hampered by their susceptibility to anti-MV polyclonal antibodies that are present in the sera of most humans due to extensive vaccination. As the MV-H glycoprotein appears to be the principal target for these anti-MV sera [9], we introduced mutations into the major epitopes of the MV-H globular head to try to overcome this problem [10]. Although neutralization was partially reduced by the introduction of such mutations, we were able to increase protection by adding the mutation D416N that had the effect of providing an extra glycosylation site. This mutation, which is present in modern MeV strains that appear to better resist neutralization [11], presumably restricts access of anti-MV antibodies to the major MV-H epitopes. The possibility that this change has arisen in vivo in response to immune pressure suggested to us that generating viral escape mutants in vitro could be a potential means to identify other mutations affecting neutralization. Moreover, sequencing of membrane proteins from escape mutants and subsequent construction of mutant viral proteins potentially could allow the mutation responsible for immune escape to be identified.

We thus attempted to mimic immune selection in vivo by putting MV under selective pressure in vitro using polyclonal antibody sera. Hitherto, monoclonal antibodies (mAbs) have been used for making escape mutants [12] but polyclonal sera have also been shown to contain highly prevalent amounts of conformation-dependent antibodies [13]. We used sera from both healthy MV-vaccinated donors and a Subacute Sclerosing Panencephalitis (SSPE) patient. SSPE is a rare fatal disease of children and young adults caused by wild-type MV persisting in the human brain 8–10 years after an apparently banal acute infection (see [14] for a recent review). The SSPE serum was used because such sera have been shown to contain elevated levels of anti-MV antibodies [15]. Mutants isolated using this system were sequenced to identify the mutations they contained. As the globular head of the H protein contains the binding sites for the cellular receptors of MV, we were particularly interested in identifying mutations in this protein. However, we speculated that mutations

allowing immune escape do not necessarily have to reside in the MV-H globular head nor indeed within this protein. Our reasoning for this was based on the knowledge that both of the MV glycoproteins are required for fusion [16] and that they are associated physically in the endoplasmic reticulum before being transported to the plasma membrane of MV-infected cells [17]. Relatively minor conformational changes in the H protein, induced post-attachment to the cellular receptor, are believed to be transmitted to the F protein that then undergoes the major conformational changes that lead to the fusion process [18]. Moreover, it has been shown that the matrix protein (M) is associated with the cytoplasmic tails of both glycoproteins [2, 3]. Thus, the H, F, and M proteins of MV can be considered to form a trimeric protein complex in the infected cell's plasma membrane. Moreover, there is evidence that interaction of the M protein with the glycoprotein cytoplasmic tails can have an "inside-out" effect on their extracellular domains and hence their function [2, 3]. For example, recombinant MV that does not contain the M protein produces a higher rate of cell-cell fusion than the complete virus [3]. Such a requirement for the maintenance of a physical interaction between these three proteins could potentially allow mutations other than in the MV-H globular head to have an effect on MV-H epitopes.

Our results show that MV variants that escape in vitro immune neutralization possess multiple mutations not only in their H proteins, but also in their F and M proteins. In order to identify which mutations were allowing immune escape to occur, we focused on viral mutants with a minimal number of mutations. Cytometric analysis of cells expressing these minimal mutants and their individual H, F, and M proteins indicated that escape from neutralization by anti-MV sera can be due to mutations occurring in any of these three proteins: such mutations affect at distance, rather than directly, the native conformation of the MV-H globular head and hence its epitopes. Our results thus suggest that immune selection can occur in nature and reinforces calls for the constant monitoring of emerging wild-type MV strains. Interestingly, many of the mutations generated in our in vitro system are found in SSPE cases, which could suggest that immune selection potentially plays a role in SSPE pathogenesis. Although SSPE viruses—unlike our escape mutants—are nonfusogenic and hence noninfectious, it is probable that their lack of fusogenicity is due to the multiple mutations that they accumulate over several years within the human brain. To have had the capacity to cause the original infection, that later led to SSPE, the virus must have been fusogenic.

2. Materials and Methods

2.1. Cells. Vero cells and Vero/hSLAM (Vero cells constitutively expressing human SLAM) were maintained in Dulbecco's modified Eagle's medium (DMEM) supplemented with 10% fetal bovine serum (FBS), 2 mM L-glutamine, 100 U/mL penicillin, 0.1 mg/mL streptomycin, and 10 mM HEPES. Chinese hamster ovary cells constitutively expressing human SLAM (CHO/hSLAM) were maintained in F12 medium containing 10% fetal bovine serum (FBS), 100 U/mL penicillin, 0.1 mg/mL streptomycin, and 1x MEN nonessential amino acids.

2.2. Antibodies.

Antibodies used for flow cytometry studies included an SSPE serum (from a 1978 Belfast case), anti-MV-H mAbs 55 (SLAM-binding site) [19], BH129 (anti-NE epitote) [20], BH216 (anti-Noose epitope) [21], and anti-MV-F mAb Y503 (a gift from D. Gerlier, Inserm U758). Secondary antibodies were rabbit anti-human IgG and goat anti-mouse IgG (both from Millipore). For the western blot an anti-MV-H mAb BH195 [20] and an anti-MV-F mAb (a gift from Christian Buchholz, Paul Ehrlich Institute, Langen, Germany) were used together with secondary antibodies, polyclonal rabbit anti-mouse immunoglobulins/HRP, and polyclonal goat anti-rabbit immunoglobulins/HRP, respectively (Dako). Antibodies used for the confocal microscope studies were anti-MV-H mAb BH129, anti-MV-F mAb Y503, and anti-MV-M mAb 8910 (Millipore).

2.3. Viruses and Production of Escape Mutants.

Both a wild-type MV strain (G954) and a laboratory strain (Halle) were used to obtain the escape mutants. We used six different sera from healthy persons immunized against MV and one serum from a SSPE patient. Sera were heated at 56°C for 45 minutes to inactivate complement. In general, cells were infected by MV in the presence of low concentrations of serum. Medium was changed every 2 days, and then the virus was plaque purified in the presence of an anti-MV serum. The serum concentration was then increased gradually over 1-2 months of passaging. Subsequently, after 8–10 passages, two selection steps were made to isolate escape mutant clones. Vero/hSLAM cells were used to obtain SLAM-dependent escape mutants of the wild-type strain; Vero cells were used to obtain CD46-dependant escape mutants of the Halle strain and CHO/hSLAM to obtain SLAM-dependant escape mutants of the Halle strain. Escape mutants were amplified, and the viral titers were calculated.

2.4. Viral Amplification and Titration.

A virus stock was made following amplification: cells in 2 mL of medium were frozen at −80°C overnight 2-3 days after infection when most cells showed fusion/syncytium formation. The medium was then thawed and harvested and the virus stock titrated. Cells in 96-well tissue culture plates were inoculated with 1/10 serially diluted culture medium samples for 1 h at 37°C. The inocula were then removed and new medium added to each well. After 4 days, the number of infected wells was counted and the number of plaque-forming units (PFU) calculated.

2.5. RT-PCR: Subcloning of Hemagglutinin (H), Fusion (F), and Matrix (M) Genes and Sequencing.

Vero/hSLAM cells in 6-well tissue culture plates were infected with Halle and G954 escape mutants at a multiplicity of infection (m.o.i) of 0.01 for 1 h at 37°C using a virus stock diluted in 1 mL of nonsupplemented DMEM. Infection media were then removed and fresh medium was added. 36 h after infection an extraction of total RNA was made using the RNeasy kit (QIAGEN) and then RT-PCR and subcloning of H, F, and M genes into the phCMV plasmid were performed. All of the final constructs were fully sequenced.

2.6. Fusion Quantification.

To study fusion, Vero cells or CHO/hSLAM (CD46-expressing or SLAM-expressing, resp.) according to the mutant tested were infected in 6-well plates by the different escape mutants + (nonmutated) Halle at a m.o.i of 0.01. Fusion was quantified 30–36 h after-infection as described previously [22].

2.7. Transient Neutralization Assay.

The standard Plaque Reduction Neutralization Test ($PRNT_{50}$) was used: the concentration of serum that reduces the plaque number by at least 50% gives a measure of its neutralizing capacity—the $PRNT_{50}$ value. For each of the mutants + (nonmutated) Halle, serial two-fold dilutions of sera were mixed with the same volume of virus and incubated for 90 minutes at 37°C. The mixtures were then placed on Vero cells (at 80% confluence) at 37°C, 5% CO_2, for 60 minutes. The supernatant was then discarded and 30% carboxymethyl cellulose (CMC) was added to reduce the spread of viral particles. The number of syncytia/plaques was counted after 4 days of incubation: to visualize the plaques, the CMC was washed 1x with PBS and then 0.1% crystal violet solution was added.

2.8. Introduction of Mutations into the MeV-H, MeV-F, and MeV-M Genes Expressed From Plasmid PhCMV.

Some escape mutants in this study have more than one mutation in their H, F, or M genes. To determine the effect of these mutations separately, mutations were introduced separately in the different genes using the QuickChange mutagenesis kit (Stratagene) according to the manufacturer's instructions. These mutations were introduced in the genes encoding the H, F, and M proteins cloned into the plasmid phCMV. The mutated plasmids were amplified and purified. All mutations were verified by DNA sequencing.

2.9. Transfection.

Cells were transfected using jetPRIME (short protocol; Polyplus). For the flow cytometry, cells in 6-well plates were transfected using 2 μg of phCMV-H or 2 μg of phCMV-F for transfection of single plasmids. For double and triple transfections, 1 μg of phCMV-H, 1 μg of phCMV-F and 0.5 μg of phCMV-M were used. After 4 h incubation at 37°C, the medium was changed for medium containing the fusion inhibitor peptide (FIP) [23] to prevent cell-cell fusion. For transfections with a single plasmid, there was no requirement to change the medium or to use FIP. For the western blot assay, cells in 100 mm culture vessel were transfected using 10 μg of either phCMV-H or phCMV-F.

2.10. Viral Infection.

For the flow cytometry, cells in 6-well plates were infected at a m.o.i of 0.2 for 1 h at 37°C using a virus stock diluted in 1 mL of nonsupplemented DMEM. Infection media were then removed and fresh medium containing FIP was added. For western blot assay, 75 cm^2 flasks were used.

2.11. Flow Cytometry. 24 h after infection or transfection, cells were prepared for flow cytometry analysis. Generally, four sera were used as primary antibodies with the following concentrations: 1/2000 SSPE serum, 1/10 mAb 55 (anti-H), 1/100 mAb BH129 (anti-H), and 1/6000 mAb 503 (anti-F). Alternatively, 1/100 mAb BH216 (anti-H) was used.

2.12. Western Blot Assay. 24 h after infection or transfection, surface proteins were biotinylated then extracted, using the Pierce Cell Surface Protein Isolation Kit. Total protein was also extracted from the same samples. The proteins were separated using SDS-PAGE and then transferred to an Immobilon-P membrane (Millipore). To reveal the H and F proteins, 1/5000 mAb BH195 and 1/2000 mAb CB were used as primary antibodies, respectively.

2.13. Confocal Microscope Study for the Localization of MV Proteins, H, F, and M. Vero-SLAM cells, grown on glass cover slides in 12-well culture plates, were cotransfected with 0.5 μg phCMV-H, 0.5 μg phCMV-F, and 0.5 μg phCMV-M using jetPRIME (short protocol; Polyplus). Cells were subjected to immunofluorescence 24 h post-transfection. Three antibodies were used: anti-H mAb BH129, anti-F mAb Y503, and anti-M mAb8910 (Millipore). Antibodies were labelled using Zenon Mouse IgG Labeling Kits (Molecular Probes). Anti-H mAb BH129 was stained with Alexa Fluor 488, anti-F mAb Y503 with Alexa Fluor 555, and anti-M mAb8910 with Alexa Fluor 647. Cells were first washed with PBS 1x. Then live cells were incubated only with stained anti-H and anti-F for 1 h at 4°C. Next, cells were washed with PBS, fixed with 3% PFA, and permeabilized with 0.1% Triton X-100 for 10 minutes at RT. Subsequently, cells were washed with blocking solution (0.2% Tween 20, 2% BSA, 5% Glycerol in PBS) and incubated in blocking solution for 10 minutes. Cells were incubated with labelled anti-M for 1 h at 4°C and then the slides were prepared for confocal microscope study. Laser argon, laser 561, and laser 633 were used for H, F, and M respectively. The specimens were studied in two steps, H and M in one step and F in another step to avoid interference between the emission signals of H, F, and M.

3. Results

3.1. Wild-Type MV Escape Mutants Selected In Vitro with Anti-MV and SSPE Patient Polyclonal Sera Contain Multiple Mutations Distributed between Their H, F, and M Proteins. The wild-type (wt) MV escape mutants obtained with our *in vitro* selection system are listed in Table 1. These were obtained by putting the G954 strain of MV under selective pressure *in vitro* using polyclonal antibody sera from both healthy MV-vaccinated donors and an SSPE patient. SLAM-dependant escape mutants were selected using Vero-SLAM cells. Following selection, the H, F, and M genes of each escape mutant were sequenced in order to identify mutations. The escape mutants are listed in Table 1 according to the serum (from six MV vaccinees or an SSPE patient) used for their selection. Table 1 shows that the majority of wt MV escape mutants obtained with the *in vitro* selection

system contain multiple mutations that are spread within the different domains of the two glycoproteins and the matrix protein. Surprisingly, it was observed that 12 of the 14 escape mutants (86%) contain at least one "SSPE mutation" (in bold in Table 1)—that is, a mutation that has also been found in an SSPE case. 42% of the H proteins, 57% of the M proteins, and 36% of the F proteins from the escape mutants contain at least one "SSPE mutation."

Assuming that the viral variants escape neutralization by anti-MV antibodies principally via alterations in glycoprotein epitopes, we intended to make flow cytometry studies in order to identify those that manifested a reduction in relative fluorescence intensity in interaction with any of our anti-MV sera or antibodies. However, the presence of SSPE mutations in our wt MV escape mutants alerted us to the potential danger of their manipulation. We therefore repeated this experiment using the vaccine/laboratory MV strain Halle which is nonvirulent in man.

3.2. Vaccine Strain MV Escape Mutants Selected In Vitro with Anti-MV and SSPE Patient Polyclonal Sera Also Contain Multiple Mutations Distributed between Their H, F, and M Proteins. The Halle strain MV escape mutants obtained with our *in vitro* selection system are listed in Table 2. As with the wt G954 MV strain these were obtained by putting the Halle strain of MV under selective pressure *in vitro* using polyclonal antibody sera from both healthy MV-vaccinated donors and an SSPE patient. Table 2 shows that the majority of escape mutants obtained with the *in vitro* selection system contain multiple mutations that are again spread within the different domains of the two glycoproteins and the matrix protein. As was observed in the wt MV strain, a high percentage contained mutations that are found in SSPE cases: 31 of the 42 escape mutants (74%) contain at least one "SSPE mutation" (in bold in Table 2); 43% of the H proteins, 36% of the M proteins, and 21% of the F proteins from the escape mutants contain at least one "SSPE mutation."

To investigate whether the escape mutants possessed a reduced capacity to interact with anti-MV sera or antibodies, we next made flow cytometry studies in the presence of the antifusion peptide FIP [23]. As Table 2 indicates, 74% of the escape mutants demonstrated a reduction in relative fluorescence intensity or "negative shift" (cytometric data not shown). However, the problem of identifying the particular mutations responsible remained. To simplify this task, six escape mutants that caused a shift and contained a minimal number of mutations were selected (Table 2) for further characterization.

3.3. Study of the Capacities of the Selected Escape Mutants to Provoke Cell-Cell Fusion and to Resist Neutralization by Anti-MV Sera. In order to evaluate the fitness of the six mutants selected, their capacity to induce the fusion relative to (nonmutated) Halle was evaluated as well as their capacity to resist neutralization by the anti-MV sera.

All of the mutants have the capacity to provoke cell-cell fusion because this allowed their isolation. However, we wanted to investigate how their capacities to induce fusion

TABLE 1: Mutations present in the H, F, and M genes of MV-G954 escape mutants obtained by *in vitro* immune selection.

Escape mutant	MV-H mutations	MV-F mutations	MV-M mutations
G-954 without serum	No change	No change	No change
G-954 + serum 1 clone 1	I32V, F111L, I219M	G26S, V107I, H141R, V178A, S285G, Q359P, **E481G**	K92E, **Y114C**, S133G, N168D, **V192M, F217L**
G-954 + serum 1 clone 2	K185E, M222T, V322A, D442N, V525A	N380D	**V37A**, N168D, P314L
G-954 + serum 2 clone 1	D135G, **Y252H**, Q334R, C381R	I232V, N392S, **E481G**	S17P, I319T
G-954 + serum 2 clone 2	No change	P227S, R301L, **V550A**	S17P
G-954 + serum 2 clone 3	S335P	M15I	S17P
G-954 + serum 3 clone 1	**R62W**, V91A, S340P, G356S, **E379G**, M602V	P103S, I298V, G376R	S17P, **F50S, L122H, F276L**
G-954 + serum 3 clone 2	W336R, G591S	No change	S17P
G-954 + serum 3 clone 3	**N200S**, L470S	Q182R, R439W	R86K, N136S, E89K,**V183A**, R225G
G-954 + serum 4	T370A, L454P, T595A	**N186Y**, V322A, **D461G**	S17P, **M51V, M53V**, L90F, R297G
G-954 + serum 5 clone 1	M333V, E422G	L68P, S197P, I418T	**Premature termination codon** at amino acid W12
G-954 + serum 5 clone 2	N26S, D404G, E611G. **There are three additional amino acids:** QGC	I36V	S17P, **F54L**, E96G
G-954 + serum 6	H495Y	**F14L**, M49T, G467E	S17P, N208D, **premature termination codon** at amino acid K238
G-954 + SSPE serum clone 1	**I50V, H61R**, D128N, **Y232H**, D283N	M15K, A96T, V283A	S17P
G-954 + SSPE serum clone 2	**V485A**	No change	S17P

Escape mutants were selected using Vero-SLAM cells and their H, F, and M genes were then sequenced. The numbers 1–6 refer to the different sera from healthy MV-vaccinated donors used for selection; SSPE refers to the SSPE serum. Escape mutant mutations also found in SSPE cases are in bold. NB: for the G-954 control, two selection steps to isolate G-954 clones were made without sera. Three G-954 clones were isolated and sequencing showed no mutations in their H, F, or M genes. Strikingly, 12 of the 14 escape mutants (86%) contain at least one "SSPE mutation"; that is, a mutation that has also been found in an SSPE case. 42% of the H proteins, 57% of the M proteins, and 36% of the F proteins from the escape mutants contain at least one "SSPE mutation."

compared with (nonmutated) Halle virus. We previously showed that both glycoproteins are required in order for fusion to occur [16]; moreover, it has been shown that the fusion provoked by a particular MV is indirectly related to the strength of the interaction between the H and F glycoproteins [24]. The results (summarized in Table 3) show that fusion induced by five out of the six mutants was lower than that induced by Halle, suggesting that there is a tighter association between the two glycoproteins in these mutants. One mutant (Halle-SLAM 2.3) was shown to induce an elevated level of fusion compared to Halle, suggesting that in this case the two glycoproteins are more loosely associated.

We also employed a transient neutralization test to investigate whether the mutants were capable of resisting neutralization at increased concentrations of the anti-MV sera. The results, summarized in Table 4, show that this was indeed the case for all six mutants. It should be noted, however, that although the mutants we have isolated are, following convention, called "escape mutants," the adjective "escape" refers to their elevated resistance to anti-MV sera rather than an absolute capacity to escape neutralization.

We next investigated which of the mutations in each mutant was responsible for conferring the capacity to better resist neutralization.

3.4. Evidence of Immune Escape Resulting from a Mutation in the H Globular Head That Is Not in a Major Epitope

3.4.1. Escape Mutant Halle-SLAM 1.2. DNA sequencing showed that this SLAM-dependant escape mutant has two mutations in the H protein (D332G, T380I), two in the M protein (Y114H, I319T), but none in the F protein. It should be noted that one of the H protein mutations (T380I) localizes to the Noose epitope [21] on the globular head. The cytometric analysis of Vero-SLAM cells infected by this escape mutant revealed important reductions in the relative fluorescence intensity for the interaction with the polyclonal SSPE serum and the anti-SLAM binding site mAb 55 (56% and 57%, resp.; $P < 0.001$), and more modest reductions for the anti-H NE epitope mAb BH129 and the anti-F mAb Y503 (44% and 42%, resp.; $P < 0.005$) (Figures 1(a) and 1(b)).

TABLE 2: Mutations present in the H, F, and M genes of MV-Halle escape mutants obtained by *in vitro* immune selection.

Escape mutant	MV-H mutations	MV-F mutations	MV-M mutations	Shift
Normal Halle	No change	No change	No change	control
Halle-SLAM 1.1	**I473M**	M4V, S255G, N408S	No change	+
→ Halle-SLAM 1.2	D332G, T380I	No change	**Y114H**, I319T	+
Halle-SLAM 2.1	I25V, **R62Q**, V259G, **V604A**	D258G, D412G, **L538G**	No change	+
Halle-SLAM 2.2	D90G, **M163T**, **F382S**	**I13M**, I309K, C423R	R36G	+
→ Halle-SLAM 2.3	A158V	No change	M239I	+
Halle-SLAM 3	**K13E**	M4V, F116L, S229P, N380A	**V80A**, T95A	−
Halle-SLAM 4.1	L585P	I216V, K491R, **A515T**	**F73L**, Q310H	+
Halle-SLAM 4.2	L246S, Q311R	No change	Q310H	+
Halle-SLAM 4.3	M45L, **I50F**, **K147E**	No change	No change	−
Halle-SLAM 4.4	**F382S**	No change	Q310H	−
→ Halle-SLAM 5.1	No change	I329N	**T172I**	+
Halle-SLAM 5.2	No change	No change	**V173I**	−
Halle-SLAM 5.3	K364T, W472R	Q133R	K86R, **T172I**	+
Halle-SLAM 6	No change	T543K	No change	−
Halle-SLAM SSPE	H86R, R261G, **R348G**, D507G	S34P, I286V, Y440H, I511V	G245A	+
Halle-CD46 1.1	T93S, A182T, S244P	No change	No change	+
→ Halle-CD46 1.2	No change	M4I, V283A	No change	+
→ Halle-CD46 1.3	**T177I**, R533G	No change	I69T, I196V	+
Halle-CD46 1.4	H17R, **K295E**	No change	D290G	−
Halle-CD46 2.1	T469A	**Y401H**, R439G	**L291I** Premature codon stop at R299	+
Halle-CD46 2.2	No change	L263S, K399R	R181S, L295P	+
Halle-CD46 2.3	K477R	N95S, **R168G**, **N187D**, Y417H	No change	+
Halle-CD46 2.4	Y410C	A124V, R154S, T214S	**L178P**, **F317L**	+
Halle-CD46 3.1	R22G, F180L	S2N, **G40R**, I90F, Q142R, Y254H, F332Y	**I200M**	−
Halle-CD46 3.2	N26I	F332Y	No change	−
→ Halle-CD46 3.3	L136P	No change	K243E, **S312P**	+
Halle-CD46 4.1	R556K	N525S	No change	−
Halle-CD46 4.2	S119A, I407V, S429L, V539A	L68P, I167T, I329V, Q359R	**L280P**, **L291P**, **L294S**, **V303A**	+
Halle-CD46 4.3	No change	No change	N136S, S257N	−
Halle-CD46 5.1	V317D, K375E	V428A, T545I	No change	+
Halle-CD46 5.2	**Y66C**, S73R	Q434L, E458G, **K539E**	**Y5N**	+
Halle-CD46 5.3	No change	No change	D39G, **I260T**	−
Halle-CD46 5.4	**M288K**, S409T	V86A, R102G, I470T, **L553P**	No change	+
Halle-CD46 (6)	D135N, **S169P**	R51S, P319L, N331S	No change	+
Halle-CD46 SSPE 1	S48N, **I50T**, I55F, N262Y, L276F, S285G, **I346V**, **H448R**, K460R, E471A, A496T	R115G, L457W	R30G Premature codon **stop** at R102	+
Halle-CD46 SSPE 2	N77S, **M163T**, G196D	S2T, A129V, M354T	L107P, **Y232H**, **V303I**	+
Halle-CD46 SSPE 3	Q391R	V74I, G256R, V318A, L457W, L507F	Q34R, E93G	+
Halle-CD46 SSPE 4	**Q4L**, L205Q, S532P, D574G	L457W	Q34R	+

TABLE 2: Continued.

Escape mutant	MV-H mutations	MV-F mutations	MV-M mutations	Shift
Halle-CD46 SSPE 5	L95P, I118T, I427T I435T	**L21H**, V181A, S437G	Q34R	+
Halle-CD46 SSPE 1.1	E161G, **M163T**	I164V, A240V	P118S, V231A, **I302V**	+
Halle-CD46 SSPE 1.2	**K13R**, V80A, **M163T**	I164V, A240G, R535G	No change	+
Halle-CD46 SSPE 1.3	**M163T**, **R195S**, S244P, **F382L**, F571L	V39A, R102G, I164V, A240V	No change	+

CD46-dependant escape mutants were selected using Vero cells and SLAM-dependant escape mutants with CHO-SLAM cells and their H, F, and M genes were then sequenced. The numbers 1–6 refer to the different sera from healthy MV-vaccinated donors used for selection; SSPE refers to the SSPE. Escape mutant mutations also found in SSPE cases are in bold. The "Shift" column refers to flow cytometry studies made on the mutants. The Mean Fluorescence Intensity (MFI) values of the mutants were compared with that of Halle (non-mutated control). A plus sign in the "shift" column indicates those mutants having the capacity to induce a statistically important reduction (shift) in the MFI in interaction with any of the sera or specific anti-MV antibodies. Arrows indicate the six escape mutants with a minimal number of mutations that were selected for dissection studies. NB: for the Halle control, two selection steps to isolate Halle clones were made without sera. Three Halle clones were isolated and sequencing showed no mutations in their H, F, or M genes.

TABLE 3: Fusion capacities of the six escape mutants selected for further study relative to Halle.

Virus	Mutations	% fusion
Halle (control)		100
Halle-SLAM 1.2	H.D332G; H.T380I; M.Y114H; M.I319T	84
Halle-SLAM 2.3	H.A158V; M.M239I	180
Halle-SLAM 5.1	F.I329N; M.T172I	69
Halle-CD46 1.2	F.M4I; F.V283A	75
Halle-CD46 1.3	H.T177I; H.R533G; M.I69T; M.I196V	66
Halle-CD46 3.3	H.L136P; M.K243E; M.S312P	95

To study fusion, cells (either CD46-expressing or SLAM-expressing according to the mutant tested) were infected by the different escape mutants and the percentage fusion quantified relative to (non-mutated) Halle (set at 100%).

TABLE 4: Transient neutralization study of the six escape mutants selected for further study.

Virus	Serum 1	Serum 2	Serum 3	Serum 5	SSPE serum
Halle (+ve control)	1/80	1/80	1/400	1/40	1/2400
Halle-SLAM 1.2	1/20				1/600
Halle-SLAM 2.3		1/20			1/1200
Halle-SLAM 5.1				1/20	1/600
Halle-CD46 1.2	1/10				1/600
Halle-CD46 1.3	1/20				1/600
Halle-CD46 3.3			1/50		1/600

The standard Plaque Reduction Neutralization Test ($PRNT_{50}$) was used to give a measure of the neutralizing capacities of anti-MV sera regarding the different mutants. The neutralizing capacity of four sera (sera 1, 2, 3, and 5) obtained from persons immunized against measles virus and one serum from a SSPE patient was studied for the different mutants relative to (non-mutated) Halle. The $PRNT_{50}$ values obtained for each virus are indicated.

To investigate the contribution of the mutated H protein alone we transfected Vero-SLAM cells with the expression plasmid phCMV, expressing the mutated H protein (phCMV.H.D332G/T380I), and the cytometric analysis again showed significant decreases in the SSPE serum and mAb 55 (56% and 61%, resp.; $P < 0.001$) and a more modest reduction (44%; $P < 0.005$) in mAb BH129 (Figures 1(c) and 1(d)). The two mutations were then tested separately and surprisingly we found that D332G (Figures 1(e) and 1(f)) rather than T380I plays the essential role for the observed negative shifts (data not shown). As the mutation T380I localizes to the Noose epitope, we compared the Halle-SLAM 1.2 escape mutant virus and the mutated H proteins expressed from phCMV.H.D332G/T380I, phCMV.H.D332G, and phCMV.H.T380I for their capacity to be stained by the anti-Noose antibody, mAb BH216 [21]. All these constructions induced reductions in the relative fluorescence intensity except phCMV.H.T380I (Figures 1(g) and 1(h)).

Interestingly, although no mutations are present in the F protein of this mutant, a shift was observed with anti-F mAb Y503 (Figures 1(a) and 1(b)). We thus cotransfected phCMV.H.D332G/T380I, phCMV.H.D332G, or phCMV.H.T380I with the phCMV.F protein to investigate a possible influence on the latter. The subsequent cytometric analysis showed that the D332G mutation (compare (i,j) with (k,l) in Figure 1), not T380I (data not shown), is responsible for the shift with mAb Y503. It should be noted that although there are slight differences between the negative shifts obtained with phCMV.H.D332G/T380I and phCMV.H.D332G, they are not significant statistically. To determine whether the mutated M protein plays a role in the shifts observed, we cotransfected cells with phCMV.H + phCMV.F + phCMV.M.Y114H/I319T but no such shifts were observed with any of the antibodies (data not shown). Moreover, co-transfection of cells with phCMV.H.D332G/T380I + phCMV.F + either phCMV.M or phCMV.M.Y114H/I319T indicated that the two mutations in the M protein did not play a role in the observed shifts (data not shown).

To eliminate the possibility that the reductions in the relative fluorescence intensity of the glycoproteins are due to lowered cellular transport we systematically used a biotinylation/western blot assay to investigate their cell surface expression for each escape mutant studied. The results for escape mutant Halle-SLAM 1.2, suggest that both the nonmutated F protein and the mutated H protein of this escape mutant are present at the cell surface in amounts similar to those

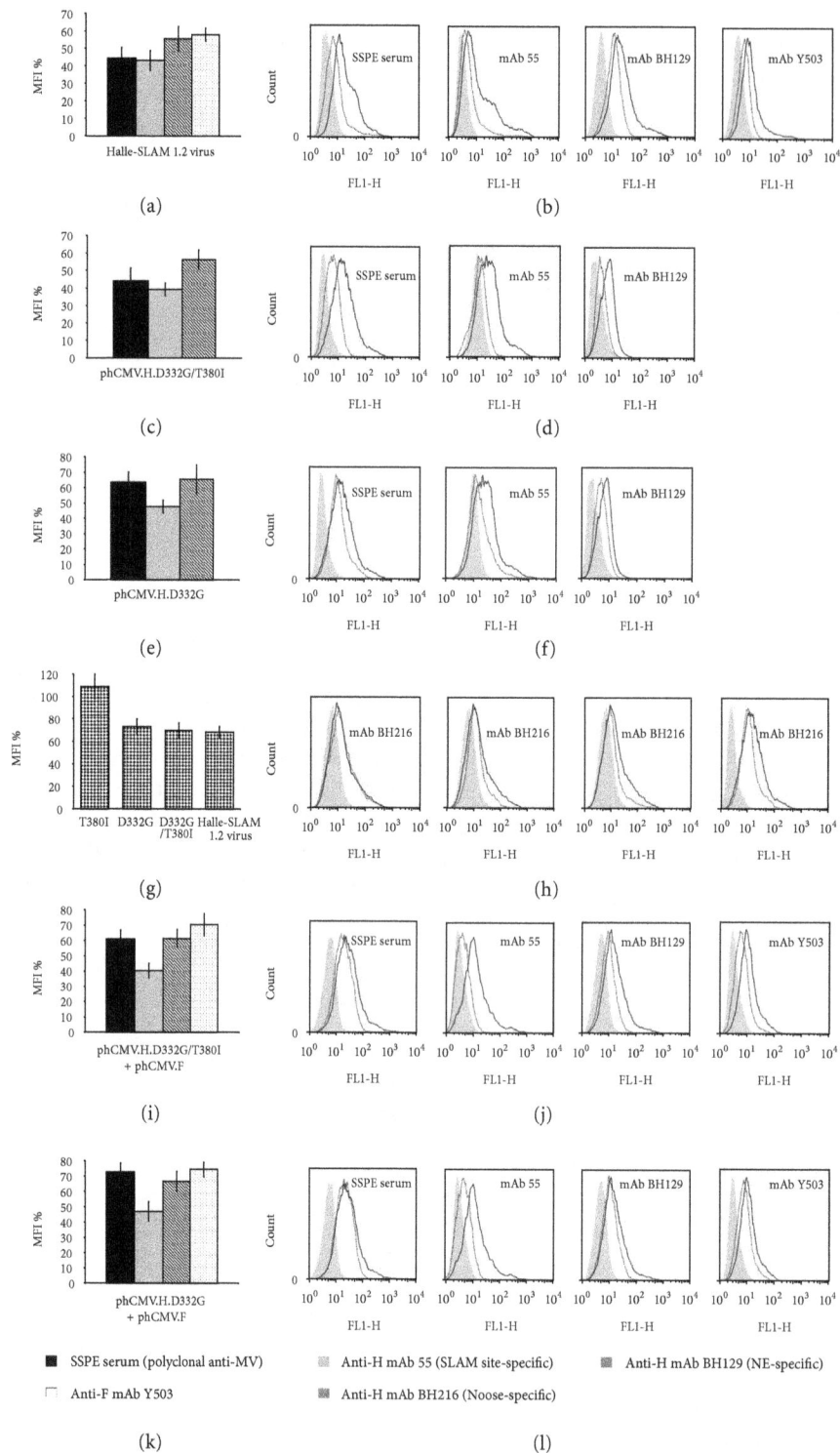

FIGURE 1: Flow cytometry analysis of escape mutant Halle-SLAM 1.2 (H: D332G, T380I; F: no change; M: Y114H, I319T). Vero-SLAM cells infected with the escape mutant virus or transfected with phCMV plasmids expressing the mutated H protein, either alone or with a phCMV plasmid expressing the nonmutated F protein, in the presence of antifusion tripeptide FIP were analysed 1 day after infection (or after transfection) using the indicated antibodies. Each overlay histogram plot represents one representative experiment; the MFI histogram data represent the mean percentages ± standard deviations for three such experiments. These values were obtained by using nonmutated Halle virus or expressed nonmutated Halle proteins as controls. (a) and (b) escape mutant Halle-SLAM 1.2-infected cells; ((c) and (d)) and ((e) and (f)), cells transfected with phCMV.H.D332G/T380I and phCMV.H.D332G, respectively; (g) and (h) cells transfected with different phCMV.H mutants or infected with the escape mutant; ((i) and (j)) and ((k), (l)), cells cotransfected with phCMV.F + phCMV.H.D332G/T380I or phCMV.H.D332G, respectively.

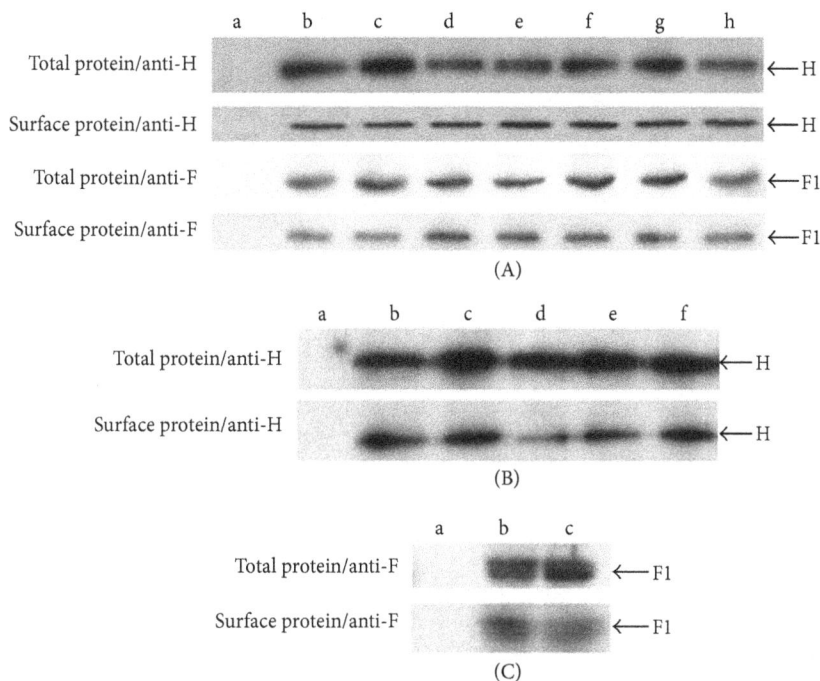

FIGURE 2: Determination of cell surface expression of escape mutant glycoproteins. Vero-SLAM cells infected with the escape mutant viruses or transfected with phCMV plasmids expressing the mutated H or mutated F proteins in the presence, when necessary, of the antifusion tripeptide FIP were lysed 1 day after infection (or after transfection) with or without previous biotinylation to extract the surface proteins or the total proteins, respectively. Western blot analysis was then done to detect the H and F proteins using anti-H mAb BH195 and anti-F mAb CB. (A) Cells infected with escape mutant viruses: (a) noninfected cells, (b) normal Halle, (c) Halle-CD46 1.2, (d) Halle-CD46 1.3, (e) Halle-CD46 3.3, (f) Halle-SLAM 1.2, (g) Halle-SLAM 2.3, and (h) Halle-SLAM 5.1. (B) Cells transfected with phCMV plasmids expressing H proteins of escape mutants: (a) normal Halle-F (control), (b) normal Halle-H, (c) Halle-CD46 1.3 H, (d) Halle-CD46 3.3 H, (e) Halle-SLAM 1.2 H, and (f) Halle-SLAM 2.3 H. (C) Western blot to detect the expression of Halle-SLAM 5.1 F; (a) normal Halle-H, (b) normal Halle-F, and (c) Halle-SLAM 5.1 F.

obtained for the wt virus (Figure 2(A), f). A similar result was obtained for the mutated H protein expressed from plasmid phCMV (Figure 2(B), e).

These results suggest that a mutation at distance, but still within the same domain of the H protein, can be more important for changing the conformation of an epitope—and thus changing its interaction with an antibody—than a mutation that touches the epitope directly.

3.5. Evidence That Immune Escape Can Result from Mutations in the MeV-H Stalk Region and Globular Head Acting in Concert

3.5.1. Escape Mutant Halle-CD46 1.3.
This CD46-dependant escape mutant has two mutations in the H protein (T177I and R533G), two in the M protein (I69T and I196V), but none in the F protein. The H protein mutations occur in separate domains: the stalk (T177I) and the globular head (R533G). It should be noted that this latter mutation occurs in a residue important for the SLAM-binding site [19, 25]. Indeed, the cytometric analysis for Vero-SLAM cells infected with this escape mutant shows a reduced interaction with the SSPE serum and with mAb 55 (63%; $P < 0.001$ and 33%; $P < 0.05$ resp.) and while there is a reduction in mAb BH129, it is not statistically significant (Figures 3(a)

and 3(b)). Very similar results were obtained with cells transfected with phCMV.H.T177I/R533G (Figures 3(c) and 3(d)) that suggest that the mutated H protein alone is responsible for the immune escape. However, the observed reduction in the relative fluorescence intensity with SSPE serum was more important with the escape mutant virus than with its H protein (63% in comparison with 40%, resp.; P for the difference: < 0.05). We thus transfected Vero-SLAM cells with either phCMV.H.T177I or phCMV.H.R533G. We found that with the former construction there was no reduction in the interaction with SSPE serum or mAb 55 (data not shown). Importantly, although in the latter case the mutation R533G largely reduced the interaction with mAb 55 (59%), there was no concomitant negative shift with the SSPE serum (see Figures 3(e) and 3(f)). Our interpretation of these results is that although the R533G mutation greatly perturbs the mAb 55 epitope—residue R533 is an important component of the SLAM binding-site [19, 25]—the presence of both mutations perturbs other epitopes in the H protein, which is reflected by the negative shift with the (anti-MeV polyclonal) SSPE serum. As the negative shift observed with phCMV.H.T177I/R533G + phCMV.F + phCMV.M.I69T/I196V was similar to that obtained with phCMV.H.T177I/R533G and with phCMV.H.T177I/R533G + phCMV.F and no such shift was observed with phCMV.H + phCMV.F + phCMV.M.I69T/I196V (data not shown), we

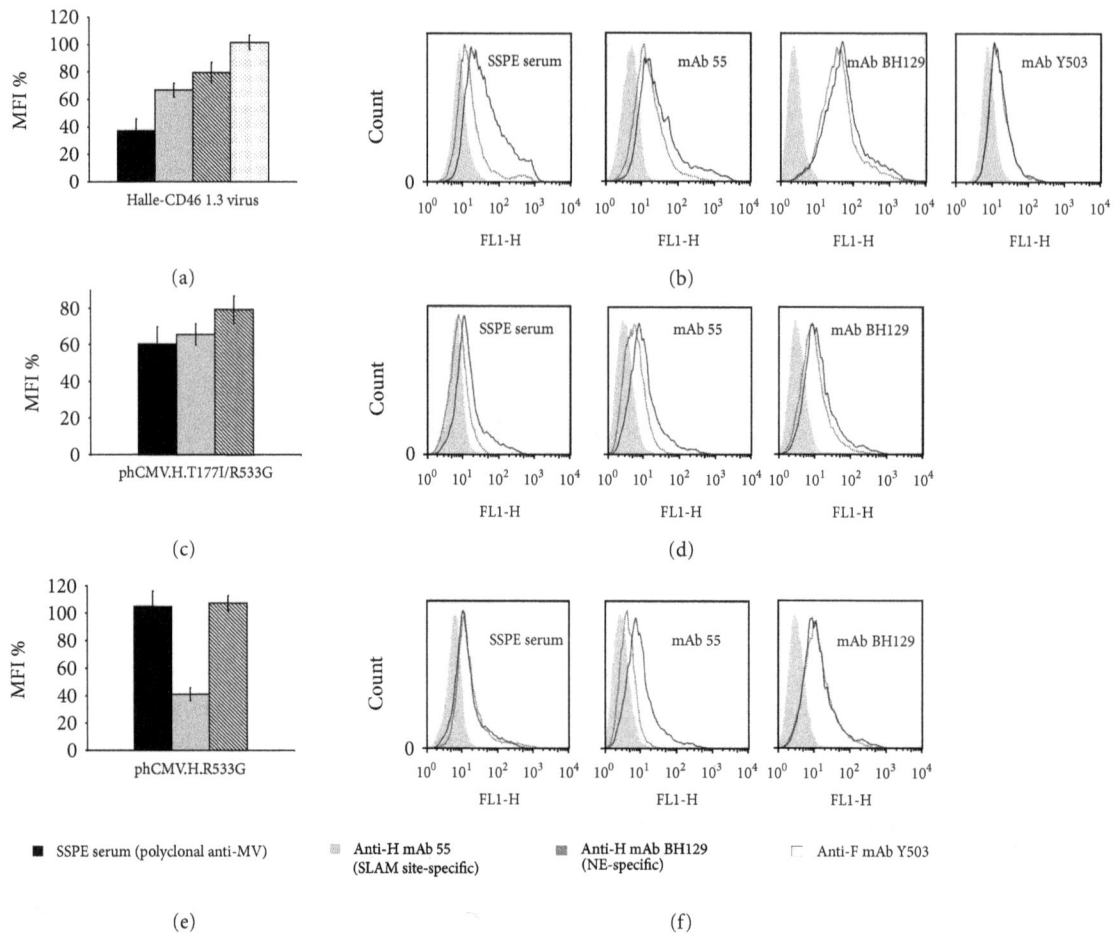

FIGURE 3: Flow cytometry analysis of escape mutant Halle-CD46 1.3 (H: T177I, R533G; F: no change; M: I69T, I196V). Vero-SLAM cells infected with the escape mutant virus or transfected with phCMV plasmid expressing the mutated H, in the presence of antifusion tripeptide FIP when necessary, were analysed 1 day after infection (or after transfection) using the indicated antibodies. Each overlay histogram represents one experiment; the MFI histogram data represent the mean percentages ± standard deviations for three experiments. These values were obtained by using nonmutated Halle virus or expressed nonmutated Halle-H protein as controls. (a) and (b)) Escape mutant Halle-CD46 1.3-infected cells; ((c) and (d)) and ((e) and (f)) cells transfected with phCMV.H.T177I/R533G and phCMV.H.R533G, respectively.

concluded that the mutations present in the M protein of this escape mutant play no role in its escape. Moreover, the biotinylation/western blot assay demonstrated that the mutated H protein and nonmutated F protein in this escape mutant are well expressed at the cell surface (Figure 2(A), d). A similar result was obtained for the mutated H protein expressed from plasmid phCMV (Figure 2(B), c). It should be noted that the mutation T177I is present in a SSPE strain, UK85/56 (accession number: AF399850).

Importantly, the results for this escape mutant suggest that mutations in one domain of the H protein, the stalk, can cause conformational changes in another domain, the globular head, that result in a loss of epitope recognition.

3.6. Evidence That a Single Mutation in the MeV-H Stalk Region Can Be Sufficient to Allow Immune Escape

3.6.1. Escape Mutant Halle-SLAM 2.3. This SLAM-dependant escape mutant has one mutation in the H protein

(A158V) that localizes to the stalk region, one mutation in the M protein (M239I), and no mutations in the F protein. For the Vero-SLAM cells infected with this escape mutant, the cytometric analysis revealed important reductions in the interaction with anti-H mAbs 55 and BH129 (50% and 39%, resp.; $P < 0.001$) and a moderate reduction with SSPE serum (32%; $P < 0.05$), but no reduction occurred in the relative fluorescence intensity with anti-F mAb Y503 (Figures 4(a) and 4(b)). We then transfected Vero-SLAM cells with phCMV.H.A158V to determine whether this profile could be reproduced by the expression of the H protein alone and found that this was indeed the case (Figures 4(c) and 4(d)). In addition, co-transfection of plasmids phCMV.H.A158V + phCMV.F gave a similar profile as well as plasmids phCMV.H.A158V + phCMV.F + phCMV.M.M239I. (data not shown). Moreover, as the co-transfection of plasmids phCMV.H + phCMV.F + phCMV.M.M239I did not result in a negative shift with any of the antibodies, we concluded that the M protein mutation does not play

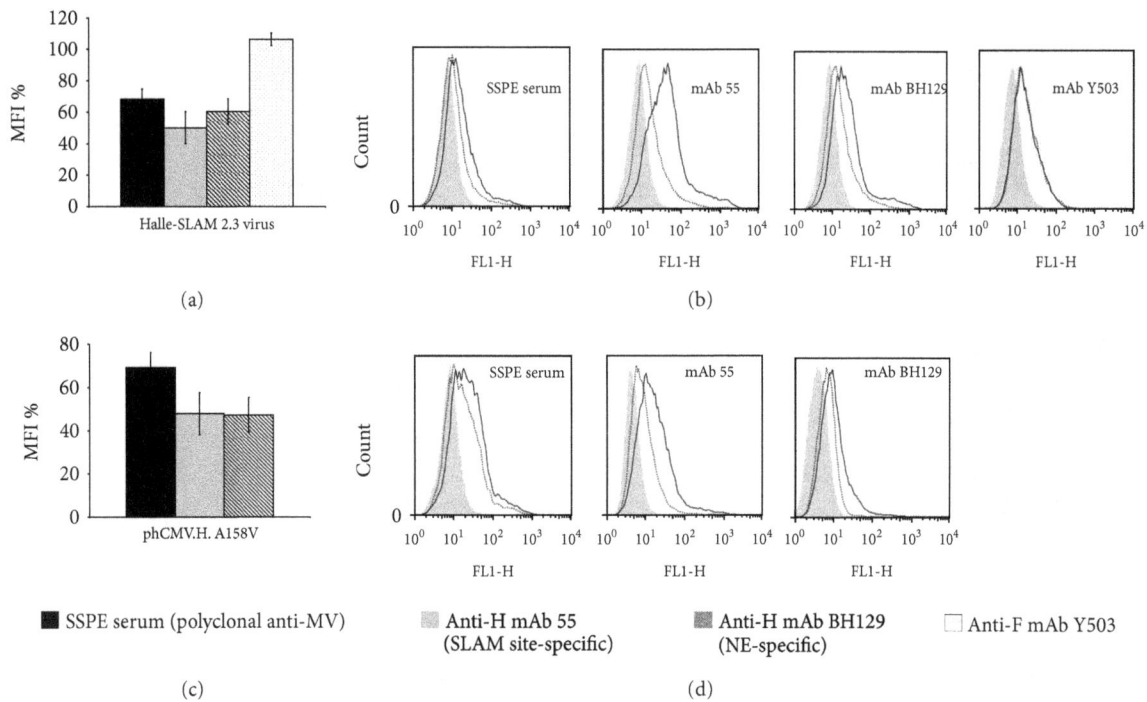

FIGURE 4: Flow cytometry analysis of escape mutant Halle-SLAM 2.3 (H: A158V; F: no change; M: M239I). Vero-SLAM cells infected with the escape mutant virus or transfected with phCMV expressing the mutated H, in the presence of antifusion tripeptide FIP when necessary, were analysed 1 day after infection or after transfection using the four antibodies indicated. Each overlay histogram represents one experiment; the MFI histogram data represent the mean percentages ± standard deviations for three experiments. These values were obtained by using nonmutated Halle virus or expressed nonmutated Halle-H protein as controls. (a) and (b)) Escape mutant Halle-SLAM 2.3-infected cells; ((c) and (d)) cells transfected with phCMV.H.A158V.

a role in the immune escape (data not shown). Importantly, the biotinylation/western blot assay demonstrated that the mutated H protein and nonmutated F protein in this escape mutant are well expressed at the cell surface (Figure 2(A), g). A similar result was obtained for the mutated H protein expressed from plasmid phCMV (Figure 2(B), f).

The results for the Halle-SLAM 2.3 mutant suggest that a single mutation (A158V) in the stalk of the H protein can be sufficient to change the oligomeric conformational state of the H protein, the globular head, thereby perturbing the recognition of major epitopes and allowing immune escape.

3.7. Evidence of Immune Escape Resulting from a Single Mutation in the Globular Head of the F Protein Indirectly Affecting H Protein Epitopes

3.7.1. Escape Mutant Halle-CD46 1.2.
This CD46-dependant escape mutant has two mutations in the F protein (M4I, V283A) but none in the H protein or the M protein. Cytometric analysis of Vero-SLAM cells infected with this escape mutant revealed important reductions in the relative fluorescence intensity for the interaction with the polyclonal anti-MeV SSPE serum and anti-H mAb 55 (49% and 44%, resp.; $P < 0.025$) and in particular for the anti-H mAb BH129, specific for the NE epitope [20] and anti-F mAb Y503 (61% and 60%, resp.; $P < 0.001$) (Figures 5(a)

and 5(b)). To determine whether these results could be reproduced when the F protein was expressed in the absence of the H protein, Vero-SLAM cells were transfected with phCMV.F.M4I/V283A. Surprisingly, the cytometric analysis showed no reduction in the relative fluorescence intensity for the interaction between the mutated F protein and the SSPE serum or the anti-F mAb Y503 (Figures 5(c) and 5(d)). However, when we cotransfected Vero-SLAM cells with phCMV.H + phCMV.F.M4I/V283A, we obtained a result similar to that obtained with the escape mutant virus with similar reductions for the interaction with all four antibodies (Figures 5(e) and 5(f)). As the protein F from this escape mutant has two mutations (M4I and V283A), we studied them separately to determine the role of each. We found that only the combination phCMV.H + phCMV.F.V283A could replicate the result obtained with the double mutant (Figures 5(g) and 5(h)). Moreover, the biotinylation/western blot assay demonstrated that the mutated F protein and nonmutated H protein in this escape mutant are well expressed at the cell surface (Figure 2(A)-c).

These results can be explained in terms of the reciprocal interaction between the H and F proteins [1]. We speculate that, as a consequence of the physical association between the two glycoproteins, the V283A mutation in the F protein modifies the 3D conformation of the H protein (and thereby its epitopes) which in turn modifies the 3D conformation of the F protein (and mAb Y503's epitope).

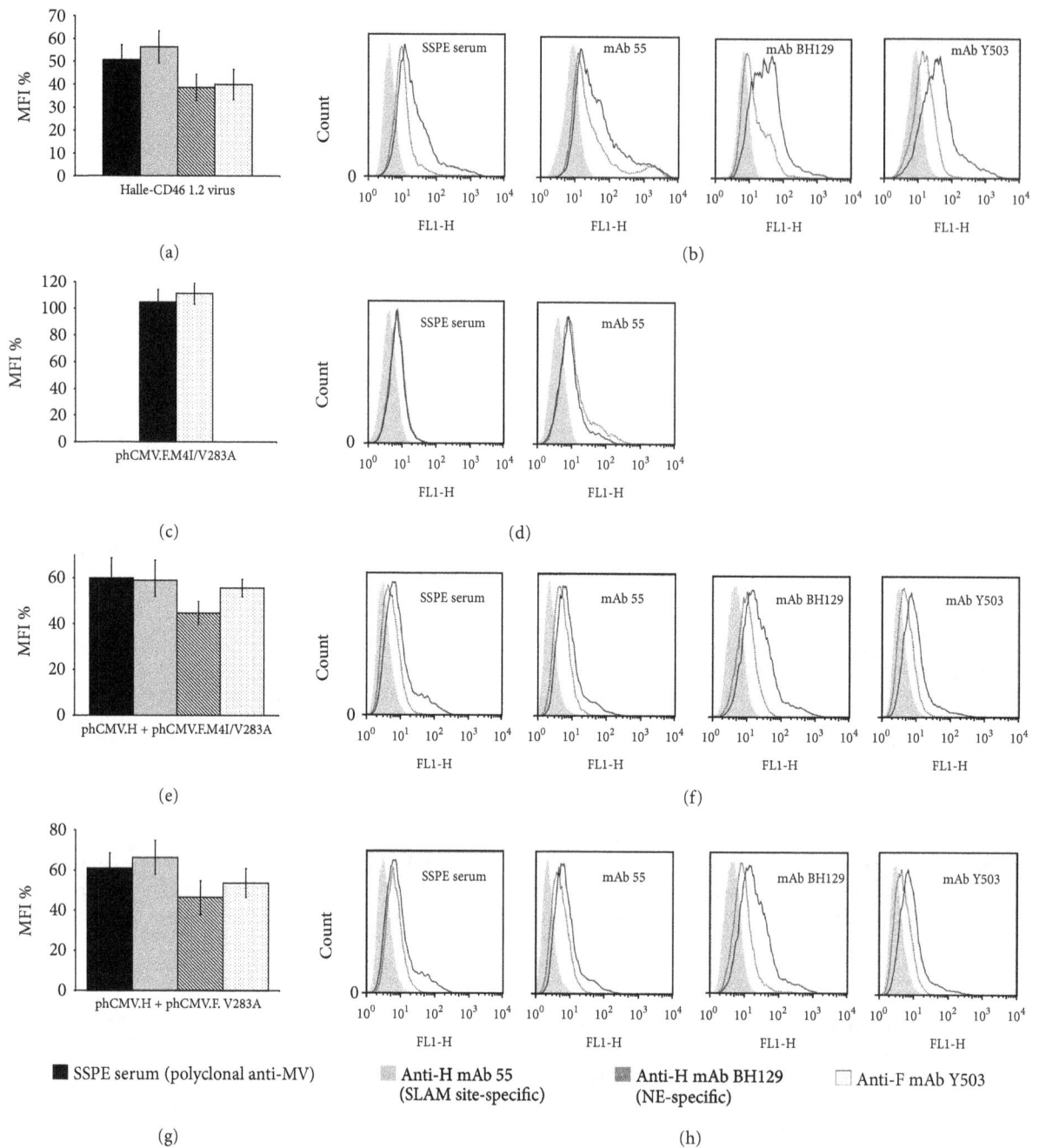

(a)

(b)

(c)

(d)

(e)

(f)

(g)

(h)

FIGURE 5: Flow cytometry analysis of escape mutant Halle-CD46 1.2 (H: no change; F: M4I, V283A; M: No change). Vero-SLAM cells infected with the escape mutant virus or transfected with phCMV plasmid expressing the mutated F protein, either alone or with a phCMV plasmid expressing the nonmutated H protein, in the presence of antifusion tri-peptide FIP, were analysed 1 day afterinfection (or after transfection) using the indicated antibodies. Each overlay histogram represents one experiment; the MFI histogram data represent the mean percentages ± standard deviations for three experiments. These values were obtained by using nonmutated Halle virus or expressed nonmutated Halle proteins as controls. (a) and (b)) Escape mutant Halle-CD46 1.2-infected cells; (c) and (d)) cells transfected with phCMV.F.M4I/V283A; ((e) and (f)) and ((g) and (h)) cells transfected with phCMV.H + phCMV.F.M4I/V283A or phCMV.F.V283A, respectively.

3.8. Evidence That Mutations in the MeV-M Protein in Combination with Mutations in Either the H Protein Or the F Protein Can Allow Immune Escape

3.8.1. Escape Mutant Halle-CD46 3.3. This CD46-dependant escape mutant has one mutation in the H protein (L136P)

that localizes to the stalk region of the protein, two in the M protein (K243E, S312P), but none in the F protein. For the Vero-SLAM cells infected by this escape mutant, the cytometric analysis revealed important reductions in the interaction with all four sera used, in particular for the anti-F mAb Y503 (52%, 60%, 55%, 67% for SSPE serum, mAB

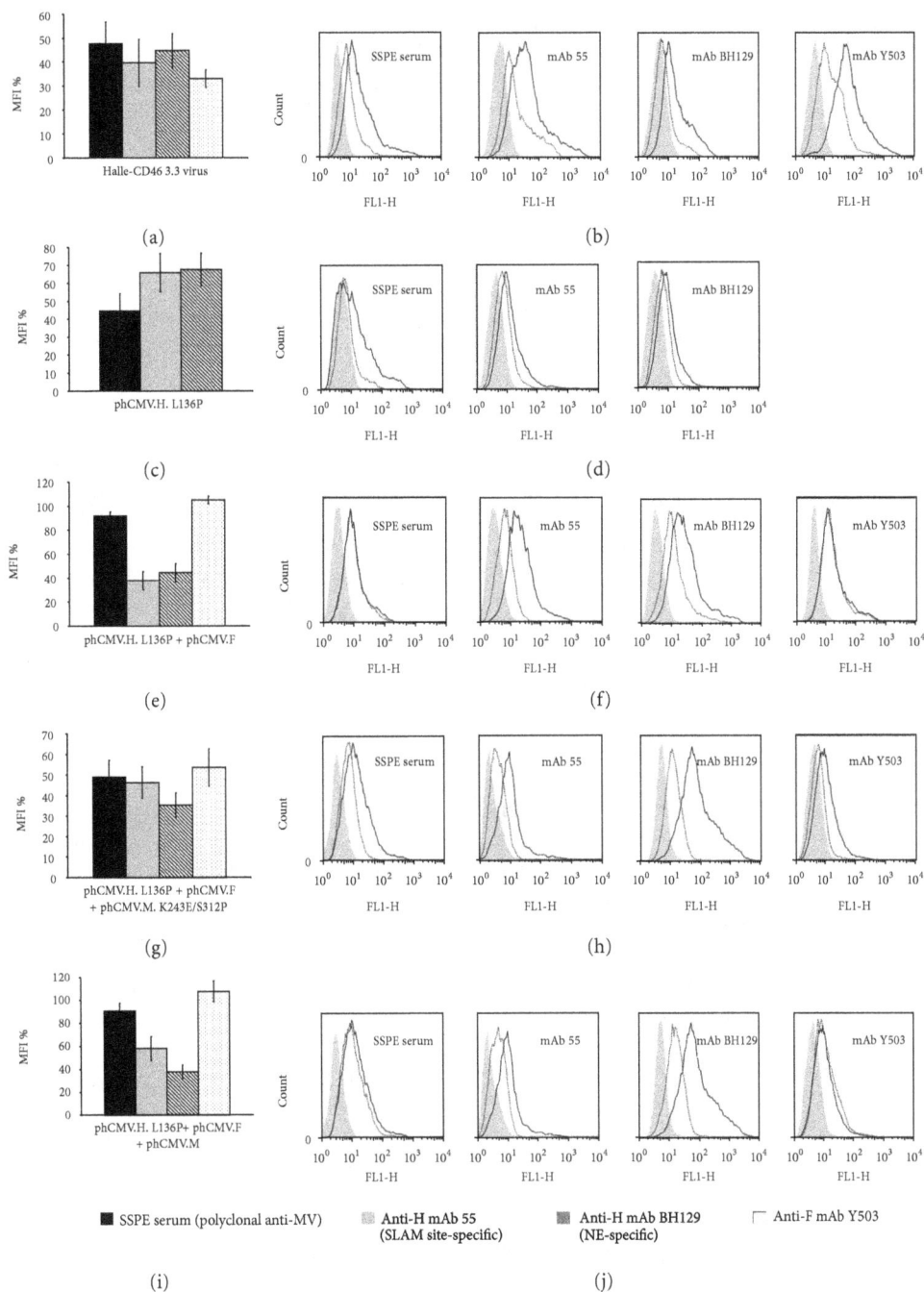

FIGURE 6: Flow cytometry analysis of escape mutant Halle-CD46 3.3 (H: L136P; F: no change; M: K243E, S312P). Vero-SLAM cells infected with the escape mutant virus or transfected with various phCMV plasmids in the presence, when necessary, of the antifusion tri-peptide FIP, were analysed 1 day after infection (or after transfection) using the indicated antibodies. Each overlay histogram represents one experiment; the MFI histogram data represent the mean percentages ± standard deviations for three experiments. These values were obtained by using nonmutated Halle virus or expressed nonmutated Halle proteins as controls. ((a) and (b)) Escape mutant Halle-CD46 3.3-infected cells; ((c) and (d)) cells transfected with phCMV.H.L136P; ((e) and (f)) cells cotransfected with phCMV.H.L136P + phCMV.F; ((g) and (h)) and ((i) and (j)) cells cotransfected with phCMV.H.L136P + phCMV.F + phCMV.M.K243E/S312P or phCMV.M, respectively.

55, mAb BH129, and mAb Y503, resp.; $P < 0.001$; Figures 6(a) and 6(b)). When the H protein from this escape mutant was expressed (phCMV.H.L136P), there was a significant decrease in the interaction with the SSPE serum (55%; $P < 0.001$; Figures 6(c) and 6(d)), but the reduction for the

interaction with anti-H mAbs 55 and BH129 was less for phCMV.H.L136P (34% and 32%, resp.; $P < 0.05$) than for the escape mutant virus. Interestingly, a large reduction in the relative fluorescence intensity was observed with the anti-F mAb Y503 despite no mutations being present in the F protein. Moreover, when we cotransfected phCMV plasmids expressing the mutated H with the F protein, the cytometric analysis revealed the absence of a shift with the anti-F mAb Y503 but a large reduction in the relative fluorescence intensity with both anti-H mAbs 55 and BH129 (62% and 55%, resp.; Figures 6(e) and 6(f)). Furthermore, the important negative shift previously observed with the interaction of the mutated H protein with SSPE serum (Figures 6(c) and 6(d)) was not reproduced (Figures 6(e) and 6(f)).

To investigate whether the mutations in the M protein were playing a role, we cotransfected cells with phCMV plasmids expressing the mutated H protein + the F protein + the mutated M protein. The cytometric analysis of this co-transfection revealed important negative shifts with all four sera (51%, 54%, 65%, and 46% for the SSPE serum, mAb 55, mAb BH129, and mAb Y503, resp.; Figures 6(g) and 6(h)). However, when we cotransfected cells with phCMV plasmids expressing the mutated H protein + the F protein + the (nonmutated) M protein, there was no longer a shift with mAb Y503 (Figures 6(i) and 6(j)). Moreover, no such shifts were obtained when cells were cotransfected with phCMV plasmids expressing the H protein + the F protein + the mutated M protein (data not shown). Taken together, these results strongly suggest that the shift observed with Y503 is due to the L136P mutation in the stem of the H protein acting in concert with the mutations in the M protein. As there are two mutations in the M protein (K243E and S312P), we next studied their individual contribution. These cotransfection studies revealed that a shift for mAb Y503 was not obtained when the mutations were present separately in the M protein (data not shown) indicating that the two M protein mutations act in concert to modify the interaction H-F-M with the result that the native conformation of both glycoproteins and hence their epitopes is perturbed. Interestingly, the M protein mutation S312P is present in a SSPE case (UK87/69; accession number: AF503526).

The biotinylation/western blot assay demonstrated that the nonmutated F protein and the mutated H protein in this escape mutant are well expressed at the cell surface (Figure 2(A)-e). A similar result was obtained for the mutated H protein expressed from plasmid phCMV (Figure 2(B)-d). Confocal microscopy confirmed that the mutated M protein, expressed from plasmid phCMV, localizes to the inner plasma membrane beneath them: in Figure 8(b), blue peaks corresponding to the mutated M protein can be seen to colocalize with the green and red peaks corresponding, respectively, to the H and F glycoproteins.

3.8.2. Escape Mutant Halle-SLAM 5.1. This SLAM-dependant escape mutant has one mutation in the F protein (I329N), one in the M protein (T172I), but none in the H protein. The cytometric analysis of cells infected with this escape mutant revealed important reductions in the interaction with all four sera used (64%, 60%, 58%, and 55% for the SSPE serum, mAb 55, mAb BH129, and mAb Y503, resp.; $P < 0.001$; Figures 7(a) and 7(b)), the largest reduction being with the SSPE serum. Transfection of Vero-SLAM cells with phCMV.F.I329N gave similar results for the SSPE serum and Y503 (59% and 46%,

resp.; $P < 0.001$) indicating that the single mutation in F is responsible for these negative shifts (Figures 7(c) and 7(d)). To determine why there is a shift with anti-H mAbs 55 and BH129 when there are no mutations in this protein, we transfected phCMV.H with phCMV.F.T172I. This time, the cytometric analysis showed a shift only with anti-F mAb Y503 (47%; Figures 7(e) and 7(f)). As this implicates the mutated M protein, we transfected cells with phCMV.H + phCMV.F.I329N + phCMV.MT173I. This time, the cytometric analysis revealed important negative shifts with all four sera in particular for the SSPE serum (66%, 43%, 50%, and 55% for the SSPE serum, mAb 55, mAb BH129, and mAb Y503, resp.; $P < 0.001$; Figures 7(g) and 7(h)). To determine whether the mutation in the M protein indeed acts in concert with the mutation in the F protein, we cotransfected cells with phCMV plasmids expressing the H protein + the mutated F protein + the (nonmutated) M protein rather than the mutated M protein. This time the only important negative shift was with mAb Y503 (50%; Figure 7(i); compare with Figure 7(g)). Additionally, we cotransfected cells with phCMV plasmids expressing the H protein + the F protein + the mutated M protein. Cytometric analysis of this co-transfection did not give such shifts with any of the four antibodies (data not shown).

The biotinylation/western blot assay demonstrated that the mutated F protein and nonmutated H protein in this escape mutant are well expressed at the cell surface (Figure 2(A)-h). A similar result was obtained for the mutated F protein expressed from plasmid phCMV (Figure 2(C)-c). Confocal microscopy was then made to confirm that the mutated M protein expressed from plasmid phCMV localizes to the inner plasma membrane beneath them: blue peaks corresponding to the mutated M protein can be seen to colocalize with the green and red peaks corresponding, respectively, to the H and F glycoproteins (Figure 8(c)).

We interpret these results in terms of the I329N mutation in the F protein acting in concert with the M protein's T172I mutation to induce epitope-perturbing conformational changes in the F protein, that, due to the glycoproteins' physical association, affect in turn the native conformation of the H protein and its epitopes, thereby allowing immune escape. Importantly, our results suggest that due to the M protein's interaction with the cytoplasmic tails of the glyco-proteins, mutations in this protein can potentially influence the conformational state of both the H and F proteins. It should be noted that the protein M mutation T172I is present in two of our escape mutants. Moreover, T172S is present in a SSPE case (Zagreb.CRO/47.02; accession number: DQ227318).

4. Discussion

We, and others, have made previous studies examining neutralizing antibody escape in the MV-H protein [10, 12, 19, 26]. Our current results support the finding that the anti-MV humoral response is primarily directed against this glycoprotein [9] but reveal that mutations allowing MV to

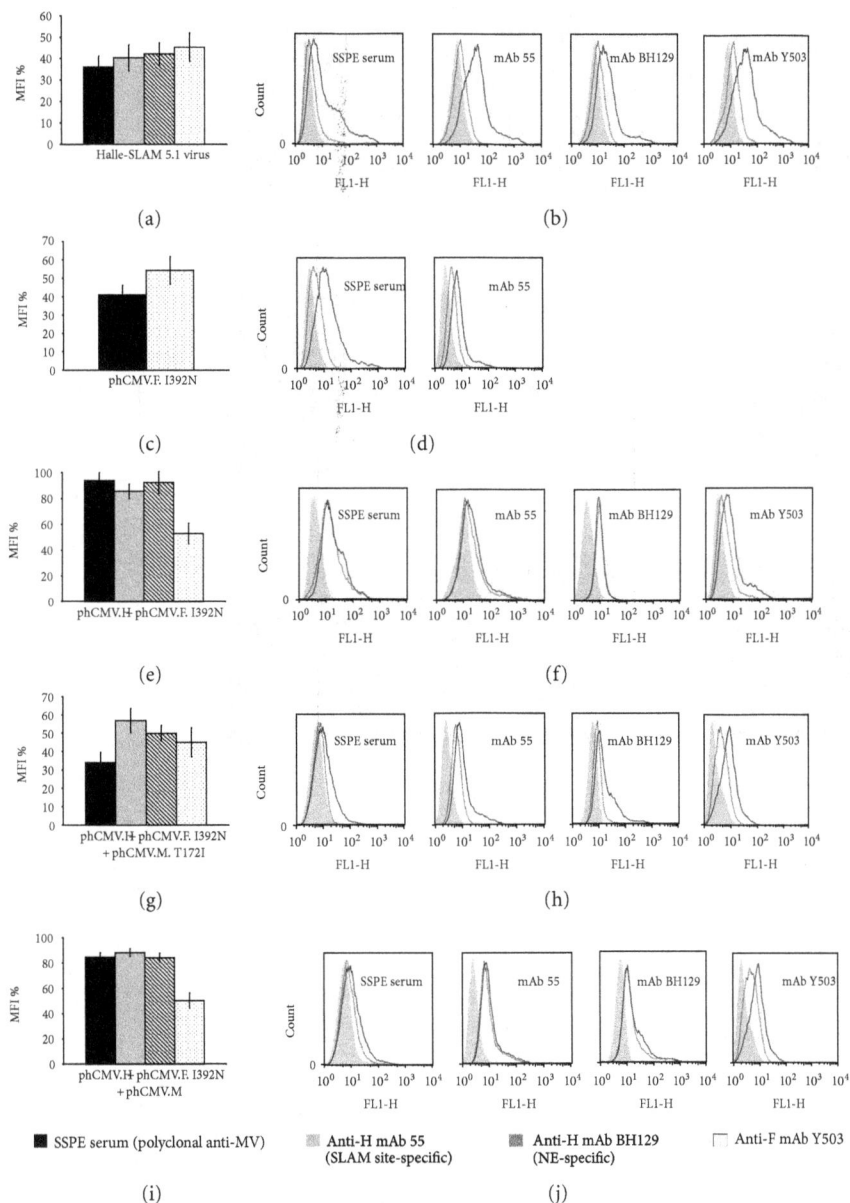

FIGURE 7: Flow cytometry analysis of escape mutant Halle-SLAM 5.1 (H: no change; F: I329N; M: T172I). Vero-SLAM cells infected with the escape mutant virus or transfected with various phCMV plasmids in the presence, when necessary, of the antifusion tri-peptide FIP were analysed 1 day after infection (or after transfection) using the indicated antibodies. Each overlay histogram represents one experiment; the MFI histogram data represent the mean percentages ± standard deviations for three experiments. These values were obtained by using nonmutated Halle virus or expressed nonmutated Halle proteins as controls. (a) and (b) Cells infected with escape mutant Halle-SLAM 5.1; (c) and (d) cells transfected with phCMV.F.I329N; ((e) and (f)) cells cotransfected with phCMV.F.I329N + phCMV.H; ((g) and (h)) and ((i) and (j)), cells cotransfected with phCMV.F.I329N + phCMV.H + phCMV.M.T172I or phCMV.M, respectively.

resist neutralization by anti-MV antibodies are not necessarily located in the major MV-H epitopes. As the analysis of our "minimal" mutants has shown, escape appears to be essentially dependant on conformational changes occurring in the H protein which compromise the recognition of its major epitopes, NE and Noose. Moreover, our results show that there are many ways in which epitopes on this protein can be affected. Most of the mutations we have identified have their effect at distance: mutations affecting the conformation of the H globular head, and hence its epitopes have been

found not only in the globular head itself but also in the stalk of the H protein, in the F protein, and even in the M protein (summarized in Figure 9).

That mutations in other domains of the H protein can have an effect on the conformation of this protein's major epitopes and hence their recognition by antibodies is not surprising. Indeed, previously we have shown that the addition of an extra glycosylation site—created by mutating a residue outside of the two major MV-H epitopes—increases escape from polyclonal MV-positive human serum [10]. In addition,

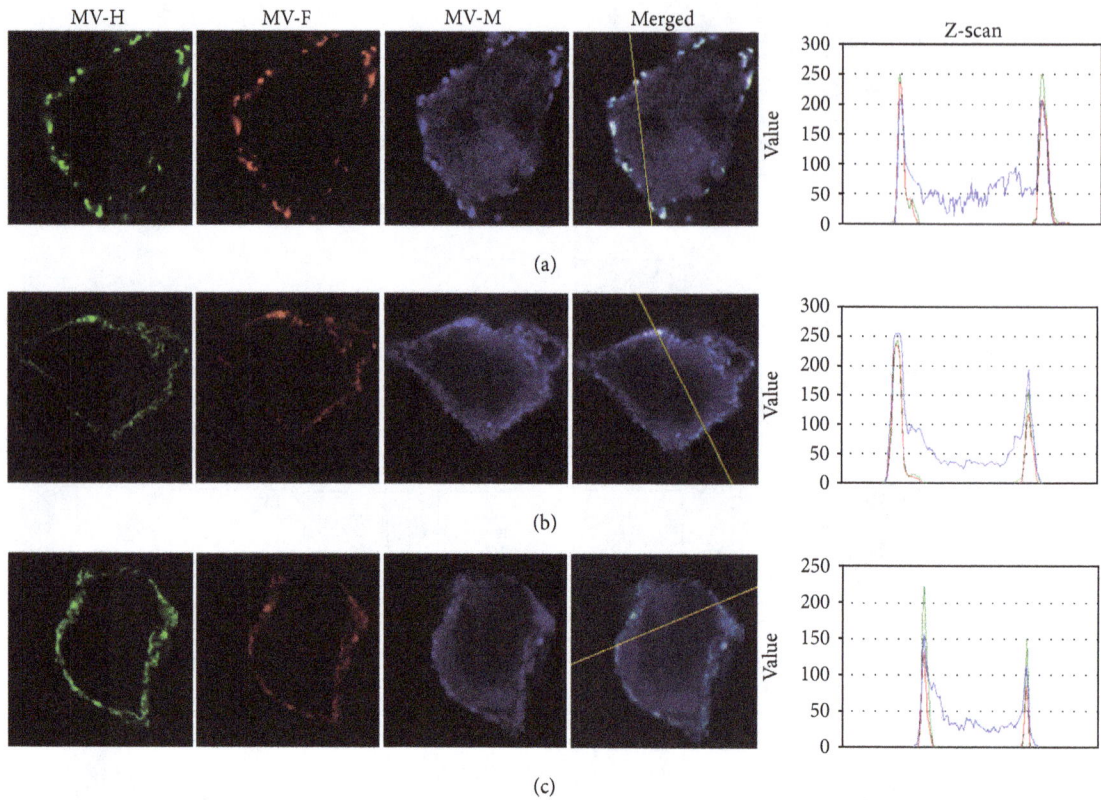

FIGURE 8: Confocal study of localization of H, F, and M proteins of escape mutants Halle-CD46 3.3 and Halle-SLAM 5.1: (a) phCMV.H + phCMV.F + phCMV.M (control); (b) phCMV.H.L136P + phCMV.F + phCMV.M.K243E/S312P; (c) phCMV.H + phCMV.F.I329N + phCMV.M.T172I. The H, F, and M proteins are labeled green, red, and blue, respectively. Magnification × 630.

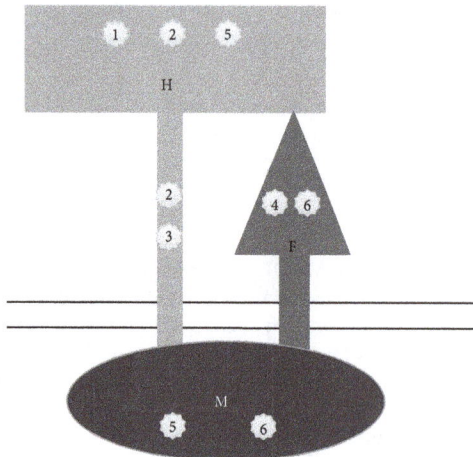

FIGURE 9: Schematic representation of the MV hemagglutinin (H), fusion (F), and matrix (M) proteins. The localization of the mutations responsible for the immune escape for each of the mutants (numbers1–6) dissected is indicated by the mutant's numbers.

we also found that incubation of MV with a cocktail of known MV-H-specific mAbs resulted in several mutations in the H protein that allowed MV to escape neutralization by the mAbs present in the selection cocktail but, importantly, not

neutralization by human serum from MV-vaccinated healthy donors [12].

That mutations in the F protein can have an effect on the conformation of the H protein and its epitopes and vice versa can be explained on the basis of the physical interaction between the two glycoproteins. It is believed that, for fusion to occur, structural changes, occurring in the MV-H globular head following receptor attachment, are transmitted to the F protein via the H's stalk domain, which is believed to interact directly with the F protein [27, 28]. Presumably, the F can also influence the H's conformation—again via the stalk domain.

That mutations in the M protein can have an "inside-out" type of effect on the conformation of the glycoprotein ectodomains would appear at first to be surprising but previous studies found that alterations in one glycoprotein's cytoplasmic tail gave advanced cell-cell fusion while double-tail mutants gave even more cell-cell fusion [2]. Moreover, a recombinant matrix-less MV gave extensive cell fusion [3]. On the basis of their results, these authors proposed that, by interacting with the cytoplasmic tails of the glycoproteins, the M protein controls MV fusogenicity—presumably by holding the two glycoproteins in a particular conformational state. Our results thus reinforce the idea of a dynamic physical interaction between these three proteins.

Although generated *in vitro* rather than *in vivo*, it is interesting that both our wt and vaccine strain mutants have similar mutations to those found in SSPE. This introduces the possibility that vaccine strains of MV do not cause SSPE simply because they are less virulent than wt MV strains. SSPE, first described by Dawson in 1933 [29], is a rare fatal sequela of wild-type MV infection where the virus is found persisting in the human brain 8–10 years after an apparently banal acute infection. The mechanism responsible for the pathogenesis of SSPE is still unknown but it has been proposed that anti-MV antibody plays a role in the persistence of the virus [30, 31]. Our results appear to support this hypothesis. It would perhaps be interesting to make similar dissections of the mutations found in SSPE cases to determine whether they contain particular mutations in the H, F, or M proteins of MV that have an effect at distance on the conformation of MV-H epitopes.

It is of interest that cases of SSPE have been reported in vaccinated individuals and that this is due to subsequent infection by a wt virus rather than the vaccine strain [32]. If vaccinated individuals nominally protected by anti-MV antibody are susceptible to wt MV strains, this raises concerns not only for neurological complications of MV but also for its global eradication. That wild-type MV can also accept mutations that do not compromise receptor recognition but allow immune escape underlines the importance of maintaining the monitoring of new emerging strains of the virus.

Acknowledgments

The authors thank Christian Buchholz (Paul Ehrlich Institute, Langen, Germany) for the blottable anti-F antibody, Yusuke Yanagi for CHO-SLAM cells, Olivier Duc and Christophe Chamot (USB/UMS3444 Platim Platform, ENSL), Olivier Reynaud for advice on confocal microscopy, Sébastien Dussurgey and Thibault Andrieu (USB/UMS3444 Cytometry Platform) for advice on flow cytometry, Didier Décimo for advice on transfection, Thierry Defrance (INSERM U851) for much good advice on flow cytometry, and Denis Gerlier for anti-F mAb 503 and useful discussions. This work was supported financially by INSERM. Hasan Kweder has a grant from Syria; Evelyne Manet and Robin Buckland are CNRS scientists.

References

[1] R. Buckland and T. F. Wild, "Functional aspects of envelope-associated measles virus proteins," in Current Topics in Microbiology and Immunology, V. ter Meulen and M. Billeter, Eds., vol. 191, pp. 51 64, Springer, Heidelberg, Germany, 1995.

[2] T. Cathomen, H. Y. Naim, and R. Cattaneo, "Measles viruses with altered envelope protein cytoplasmic tails gain cell fusion competence," Journal of Virology, vol. 72, no. 2, pp. 1224–1234, 1998.

[3] T. Cathomen, B. Mrkic, D. Spehner et al., "A matrix-less measles virus is infectious and elicits extensive cell fusion: consequences for propagation in the brain," The EMBO Journal, vol. 17, no. 14, pp. 3899–3908, 1998.

[4] Y. Yanagi, M. Takeda, S. Ohno, and T. Hashiguchi, "Measles virus receptors," Current Topics in Microbiology and Immunology, vol. 329, pp. 13–30, 2009.

[5] M. D. Mühlebach, M. Mateo, P. L. Sinn et al., "Adherens junction protein nectin-4 is the epithelial receptor for measles virus," Nature, vol. 480, no. 7378, pp. 530–533, 2011.

[6] R. S. Noyce, D. G. Bondre, M. N. Ha et al., "Tumor cell marker pvrl4 (nectin 4) is an epithelial cell receptor for measles virus," PLoS Pathogens, vol. 7, no. 8, Article ID e1002240, 2011.

[7] J. Klingele, H. K. Hartter, F. Adu, W. Ammerlaan, W. Ikusika, and C.P. Muller, "Resistance of recent measles virus wild-type isolates to antibody-mediated neutralization by vaccinees with antibody," Journal of Medical Virology, vol. 62, pp. 91–98, 2000.

[8] C. J. Hickman, T. B. Hyde, S. B. Sowers et al., "Laboratory characterization of measles virus infection in previously vaccinated and unvaccinated individuals," The Journal of Infectious Diseases, vol. 204, supplement 1, pp. S549–S558, 2011.

[9] R. L. de Swart, S. Yüksel, and A. D. M. E. Osterhaus, "Relative contributions of measles virus hemagglutinin- and fusion protein-specific serum antibodies to virus neutralization," Journal of Virology, vol. 79, no. 17, pp. 11547–11551, 2005.

[10] C. Lévy, F. Amirache, C. Costa et al., "Lentiviral vectors displaying modified measles gp overcome pre-existing immunity in in vivo-like transduction of human T and B cells," Molecular Therapy, vol. 20, no. 9, pp. 1699–1712, 2012.

[11] S. Santibanez, S. Niewiesk, A. Heider et al., "Probing neutralizing-antibody responses against emerging measles viruses (MVs): immune selection of MV by H protein-specific antibodies?" Journal of General Virology, vol. 86, no. 2, pp. 365–374, 2005.

[12] P. J. Lech, G. J. Tobin, R. Bushnell et al., "Epitope dampening montypic measles virus hemagglutinin glycoprotein results in resistance to cocktail of monoclonal antibodies," PLoS ONE, vol. 8, no. 1, Article ID e52306, 2013.

[13] J. P. Moore and D. D. Ho, "Antibodies to discontinuous or conformationally sensitive epitopes on the gp120 glycoprotein of human immunodeficiency virus type 1 are highly prevalent in sera of infected humans," Journal of Virology, vol. 67, no. 2, pp. 863–875, 1993.

[14] J. Gutierrez, R. S. Issacson, and B. S. Koppel, "Subacute sclerosing panencephalitis: an update," Developmental Medicine and Child Neurology, vol. 52, no. 10, pp. 901–907, 2010.

[15] W. R. Kiessling, W. W. Hall, L. L. Yung, and V. Ter Meulen, "Measles virus specific immunoglobulin M response in subacute sclerosing panencephalitis," The Lancet, vol. 1, no. 8007, pp. 324–327, 1977.

[16] T. F. Wild, E. Malvoisin, and R. Buckland, "Measles virus: both the haemagglutinin and fusion glycoproteins are required for fusion," Journal of General Virology, vol. 72, no. 2, pp. 439–442, 1991.

[17] R. K. Plemper, A. L. Hammond, and R. Cattaneo, "Measles virus envelope glycoproteins hetero-oligomerize in the endoplasmic reticulum," Journal of Biological Chemistry, vol. 276, no. 47, pp. 44239–44246, 2001.

[18] R. A. Lamb and T. S. Jardetzky, "Structural basis of viral invasion: lessons from paramyxovirus F," Current Opinion in Structural Biology, vol. 17, no. 4, pp. 427–436, 2007.

[19] N. Massé, M. Ainouze, B. Néel, T. F. Wild, R. Buckland, and J. P. M. Langedijk, "Measles virus (MV) hemagglutinin: evidence

that attachment sites for MV receptors SLAM and CD46 overlap on the globular head," *Journal of Virology*, vol. 78, no. 17, pp. 9051–9063, 2004.

[20] P. Fournier, N. H. C. Brons, G. A. M. Berbers et al., "Antibodies to a new linear site at the topographical or functional interface between the haemagglutinin and fusion proteins protect against measles encephalitis," *Journal of General Virology*, vol. 78, no. 6, pp. 1295–1302, 1997.

[21] D. Ziegler, P. Fournier, G. A. H. Berbers et al., "Protection against measles virus encephalitis by monoclonal antibodies binding to a cystine loop domain of the H protein mimicked by peptides which are not recognized by maternal antibodies," *Journal of General Virology*, vol. 77, no. 10, pp. 2479–2489, 1996.

[22] V. Guillaume, H. Aslan, M. Ainouze et al., "Evidence of a potential receptor-binding site on the Nipah virus G protein (NiV-G): identification of globular head residues with a role in fusion promotion and their localization on an NiV-G structural model," *Journal of Virology*, vol. 80, no. 15, pp. 7546–7554, 2006.

[23] C. D. Richardson, A. Scheid, and P. W. Choppin, "Specific inhibition of paramyxovirus and myxovirus replication by oligopeptides with amino acid sequences similar to those at the N-termini of the F1 or HA2 viral polypeptides," *Virology*, vol. 105, no. 1, pp. 205–222, 1980.

[24] R. K. Plemper, A. L. Hammond, D. Gerlier, A. K. Fielding, and R. Cattaneo, "Strength of envelope protein interaction modulates cytopathicity of measles virus," *Journal of Virology*, vol. 76, no. 10, pp. 5051–5061, 2002.

[25] S. Vongpunsawad, N. Oezgun, W. Braun, and R. Cattaneo, "Selectively receptor-blind measles viruses: identification of residues necessary for SLAM- or CD46-induced fusion and their localization on a new hemagglutinin structural model," *Journal of Virology*, vol. 78, no. 1, pp. 302–313, 2004.

[26] M. Tahara, Y. Ito, M. A. Brindley et al., "Functional and structural characterization of neutralizing epitopes of measles virus hemagglutinin protein," *Journal of Virology*, vol. 87, pp. 666–675, 2013.

[27] E. O. Saphire and M. B. A. Oldstone, "Measles virus fusion shifts into gear," *Nature Structural and Molecular Biology*, vol. 18, no. 2, pp. 115–116, 2011.

[28] E. A. Corey and R. M. Iorio, "Mutations in the stalk of the measles virus hemagglutinin protein decrease fusion but do not interfere with virus-specific interaction with the homologous fusion protein," *Journal of Virology*, vol. 81, no. 18, pp. 9900–9910, 2007.

[29] J. R. Dawson, "Cellular inclusions in cerebral lesions of lethargic encephalitis," *The American Journal of Pathology*, vol. 9, pp. 7–16, 1933.

[30] K. W. Rammohan, H. F. McFarland, and D. E. McFarlin, "Induction of subacute murine measles encephalitis by monoclonal antibody to virus haemagglutinin," *Nature*, vol. 290, no. 5807, pp. 588–589, 1981.

[31] K. W. Rammohan, H. F. McFarland, and D. E. McFarlin, "Subacute sclerosing panencephalitis after passive immunization and natural measles infection: role of antibody in persistence of measles virus," *Neurology*, vol. 32, no. 4, pp. 390–394, 1982.

[32] H. Campbell, N. Andrews, K. E. Brown, and E. Miller, "Review of the effect of measles vaccination on the epidemiology of SSPE," *International Journal of Epidemiology*, vol. 36, no. 6, pp. 1334–1348, 2007.

12

Phage-Displayed Peptides Selected to Bind Envelope Glycoprotein Show Antiviral Activity against Dengue Virus Serotype 2

Carolina de la Guardia,[1,2] Mario Quijada,[1] and Ricardo Lleonart[1]

[1]*Center of Cellular and Molecular Biology of Diseases, Instituto de Investigaciones Científicas y Servicios de Alta Tecnología (INDICASAT AIP), Building 219, Ciudad del Saber, Apartado 0843-01103, Panamá, Panama*
[2]*Department of Biotechnology, Acharya Nagarjuna University, Guntur, India*

Correspondence should be addressed to Ricardo Lleonart; rlleonart@indicasat.org.pa

Academic Editor: Gary S. Hayward

Dengue virus is a growing public health threat that affects hundreds of million peoples every year and leave huge economic and social damage. The virus is transmitted by mosquitoes and the incidence of the disease is increasing, among other causes, due to the geographical expansion of the vector's range and the lack of effectiveness in public health interventions in most prevalent countries. So far, no highly effective vaccine or antiviral has been developed for this virus. Here we employed phage display technology to identify peptides able to block the DENV2. A random peptide library presented in M13 phages was screened with recombinant dengue envelope and its fragment domain III. After four rounds of panning, several binding peptides were identified, synthesized, and tested against the virus. Three peptides were able to block the infectivity of the virus while not being toxic to the target cells. Blind docking simulations were done to investigate the possible mode of binding, showing that all peptides appear to bind domain III of the protein and may be mostly stabilized by hydrophobic interactions. These results are relevant to the development of novel therapeutics against this important virus.

1. Introduction

Dengue virus is a growing public health problem worldwide as about 390 million people get infected annually and almost 96 million people develop clinical manifestations of the disease [1]. Other authors estimate that about 3.9 billion people from 128 countries share the risk of infection with this virus [2]. Dengue virus can produce a wide spectrum of clinical presentations, from asymptomatic or mild manifestation to a more severe life threating manifestation, known as dengue hemorrhagic fever (DHF) and dengue shock syndrome (DSS). Severe dengue can be deadly, especially in children, due to plasma leaking, respiratory distress, edema, severe bleeding, and organ impairment [3].

This virus is a *Flavivirus* that is transmitted by mosquitoes of the genus *Aedes*. The incidence of dengue infections has increased due to the spread of these mosquitoes into new regions [4]. DENV are present as 4 serotypes (DENV1, DENV2, DENV3, and DENV4) circulating in most endemic countries [5]. Dengue fever is considered the most rapidly growing mosquito-transmitted disease worldwide [6].

The dengue virion has an icosahedral symmetry, with diameter between 500 Å and 600 Å [7]. The viral genome consists of 11 Kbp single positive stranded RNA coding for a single polyprotein. The polyprotein is cleaved in the cytoplasm into several structural and nonstructural polypeptides [8]. The structural proteins include the capsid (C), premembrane (PrM)/membrane (M), and envelope glycoprotein (E), which are involved in the formation of the viral particle. The nonstructural proteins (NS1, NS2a, NS2b, NS3, NS4a, NS4b, and NS5) are responsible for the viral replication, assembly, and immune response escape [8].

E glycoprotein is the most important molecule during the viral entry process as it appears to be responsible for receptor recognition and attachment to the cell surface, triggering the clathrin mediated endocytosis and the subsequent fusion of viral and cellular membranes. This protein can interact with diverse cellular molecules; therefore, it is an ideal target for the development of new antivirals [9–12]. E glycoprotein is formed by three domains, plus a membrane proximal stem and a transmembrane anchor [13–15]. Domain I is formed by eight β-strand barrels containing two insertion loops. Domain II contains hydrophobic sequences that are conserved among all flaviviruses. These hydrophobic sequences, also known as fusion peptides, are responsible for the insertion of the rearranged E trimer into the cellular membrane during fusion [9, 14–16]. Domain III (DIII) is an immunoglobulin-like carboxyterminal domain, responsible for the initial cellular receptor binding [17]. Additionally, the DIII is the main target for neutralizing antibodies [18].

Due to the complexities of the immune response and the pathology generated by this virus, particularly during subsequent infections, the development of vaccines has been slow and there is only one vaccine that has been registered, which is still in the early stages of testing in several territories [19, 20].

Currently, there is no approved antiviral for the treatment of dengue infection. It is known that the duration of the viremia is short in dengue patients. However, since high viremia has been related to a severe onset of the disease, the use of antivirals at early stages of the disease may block the progression of the disease and accelerate the recovery of patients. Here, we proposed the use of dengue virus envelope (E) glycoprotein as a target to search for peptides that could interfere with the first step of the infection process. We selected the DENV-2 E glycoprotein and the DIII as targets for the screening random peptides displayed on M13 bacteriophages.

Phage display technology has been used in past years to identify peptides with specific binding activities to a variety of targets. This technique has been very useful to identify mimotopes, novel antivirals [21], peptidomimetic drugs [22], and many other applications.

Here we show that the DENV2 E glycoprotein, as well as domain III (DIII), is useful target for the identification of phage-displayed peptides with potential as novel antivirals. Here we describe three peptides that inhibit the viral infectivity and are not cytotoxic to the permissive cells. Fully blind docking simulations with these peptides suggest binding sites at domain III of the envelope protein, stabilized by predominant hydrophobic interactions. These findings open new possibilities to optimize and refine the design of new peptidic inhibitors of infection by this virus.

2. Materials and Methods

2.1. Cells, Viruses, and Peptides. Vero cells (CCL-81, ATCC, USA) were used for dengue 2 virus propagation, plaque formation assays, and cytotoxicity assays. These cells were grown in MEM with 10% FCS and 50 μg/ml gentamicin, at 37°C and 5% CO_2. Dengue virus, serotype 2, was a kind gift from Department of Research, Instituto Conmemorativo

Gorgas de Estudios de la Salud (ICEGS), Panamá. Heptapeptide random library was obtained from New England Biolabs (USA). Synthetic peptides were obtained from Genscript (USA). Full DENV2 E glycoprotein (Cat. number Den-034) was obtained from ProSpec-Tany TechnoGene (Israel). Tissue culture, molecular biology grade, and general reagents were from Nunc and Sigma-Aldrich (USA).

2.2. Recombinant Protein Expression. Sequence coding for DENV2 E DIII (aa 289 to 405) was obtained from GenBank (complete genome NCBI Reference Sequence: NC_001474.2) and optimized for expression in *Escherichia coli* using an online tool (GeneOptimizer, Geneart). The optimized sequence was synthesized (IDT) and subcloned into NdeI-XhoI restriction sites of the expression plasmid pET-30b(+) (Novagen) using standard recombinant DNA techniques. This plasmid directs the inducible production of the DIII plus the amino acids Leu and Glu and a 6xHis tail at the C-terminal. This fragment of the E protein contains 125 amino acids and an approximate molecular weight of 14.2 kDa. The expression plasmid was verified by sequencing and transformed into *E. coli* strain BL21 (DE3) for expression. Single colonies were cultured in LB medium containing antibiotic at 37°C and 250 rpm and induced with 0.5 mM IPTG when OD_{600} reached 0.6. After 4 hours of induction cells were collected by centrifugation and kept at −20°C until purification. As expression of the recombinant protein was mostly at the insoluble fraction, further purification was done including inclusion body isolation, solubilization, and affinity purification in denaturing conditions, refolding, and dialysis.

2.3. Purification of Recombinant DIII. Frozen pellet from induced culture was suspended in cold lysis buffer (10 mM Tris-HCl pH 7.5, 5 mM benzamidine-HCl, 5 mM EDTA, 5 mM DTT, 0.3 mg/ml lysozyme, and 1 mM PMSF) and lysed by sonication on ice. The inclusion bodies were recovered by centrifugation, washed with 50 mM phosphate buffer, 5 mM EDTA, 200 mM NaCl, 0.5 M urea, and 1% Triton X-100, recovered as before, washed with 50 mM phosphate buffer, 1 mM EDTA, and 1 M NaCl, and solubilized in denaturing buffer (10 mM Tris-HCl pH 8.0, 100 mM NaH_2PO_4, 100 mM NaCl, and 6 M GuHCl). The solubilized inclusion bodies were purified using a Ni-NTA superflow cartridge (Qiagen). Following loading and washing with buffer C (6 M GuHCl, 100 mM NaH_2PO_4, and 100 mM HCl, pH 6.3), protein was eluted with buffer E (6 M GuHCl, 100 mM NaH_2PO_4, and 100 mM HCl, pH 4.5). Attempts to dialyze against PBS resulted in protein precipitation; therefore we tested several refolding conditions using a Protein Refolding kit (Pierce, USA). Finally, refolding was done by dilution into refolding buffer RB7 (1 mM GSH, 1 mM GSSH, 1 mM EDTA, 1.1 M GuHCl, 55 mM Tris-HCl, pH 8.0, 21 mM NaCl, and 0.88 mM KCl). The refolded protein was dialyzed against TBS, filter sterilized, and quantitated by BCA method.

2.4. In Vitro Virus Blocking Assay to Test Protein Refolding. Aliquots of refolded recombinant DIII, obtained from several refolding conditions, were first visually checked for aggregation. Those not showing aggregation were then tested in virus infection blocking experiments, to check for their ability to bind cellular receptors and inhibit these first steps in viral infection. Vero cells were plated in 96-well plates at 10^4 cells per well in complete medium (MEM, 10% FCS, and 50 μg/ml gentamicin). On the next day, cells were washed with PBS and incubated with the refolded protein (100 μl, 50 μg/ml) in maintenance medium (MEM, 1% FCS, and gentamicin) for 30 min at 37°C. Then cells were washed and incubated with DENV2 (MOI = 3, 50 μl), containing the refolded protein at the same concentration, for 1 h at 37°C in maintenance medium. Cells were again washed, replenished with maintenance medium, and incubated at 37°C, 5% CO_2. After 5 days, cytopathic effect was quantified using chemiluminescent based ATP detection (CellTiter-Glo® cell viability assay, Promega).

2.5. Biopanning. Phage display was performed using a Ph.D.-7 Phage Display Peptide Library according to the manufacturer's instructions. Four rounds of panning were done using two targets, the full DENV2 envelope, or the recombinant DIII obtained as described above. For each panning step, several wells in ninety-six-well plates were coated overnight at 4°C and then blocked with 400 μl of 0.1 M NaHCO$_3$ (pH 8.6) and 5 mg/ml BSA for 1 h at 4°C. After six washes with TBST (TBS, 0.1% Tween 20), 100 μl of phages in TBST, containing 10^{11} pfu were allowed to bind for 1 h at RT. Then wells were washed ten times with TBST and phages eluted with 0.2 M glycine-HCl pH 2.2, 1 mg/ml BSA. After neutralization with Tris pH 9.1, phages were titered and bulk amplified in *E. coli* for next round. In order to increase the stringency of the selection, successive rounds were done increasing the concentration of Tween-20 during washes and reducing the amount of target fixed to the solid phase. After the fourth round, eluted phages were plated, picked, and propagated for subsequent phage-ELISA and sequencing.

2.6. Phage-ELISA. DENV2 E, DIII, and BSA were used to coat 96-well microtiter plate (MaxiSorp, Nunc) overnight at 4°C. Plates were blocked with TBS-BSA, washed with TBST, and incubated with 10^{12} pfu of individual phages. After washing, HRP-conjugated anti-M13 monoclonal antibody (GE Healthcare, USA) was added, incubated, and washed. Color was developed by adding TMB substrate solution for 10 min and stopped with 1 N HCl. Plates were read at 405 nm. Phage clones with the best target-to-background signal ratio were selected for DNA sequencing. The DNA insert was determined by cycle sequencing using a 96-gIII primer as recommended by manufacturer. Sequence logos and residue coloring was done as implemented in JalView v.2.10.1 [23]. Coloring was following Zappo scheme, where residues are colored according to their physicochemical properties as follows: pink, aliphatic/hydrophobic; orange, aromatic; red, positive; green, negative; blue, hydrophilic; magenta, proline/glycine; and yellow, cysteine.

2.7. Viral Plaque Reduction Assay. Vero cells were seeded (2 × 10^5 cells per well) in 6-well plates and incubated

overnight. The peptides were prepared at several dilutions and mixed with 100 pfu of DENV2, for a final volume of 150 μl. Mixtures were incubated at RT for 1 h and then added to cell monolayers and incubated 1 h at 37°C for adsorption. Then inoculum was removed and cells were washed and overlaid with maintenance medium with 1% methylcellulose and incubated for 5 days. Then medium was aspirated; cells were fixed with 10% formaldehyde and stained with 1% crystal violet. The viral plaques were visually counted and values from experimental groups compared to virus alone, nontreated controls.

2.8. Peptide Cytotoxicity Assay. Peptides were incubated with cells exactly as described in the viral plaque reduction assay (1 h for 37°C) and then washed away and cells refed with maintenance medium and incubated at 37°C for 24 h. Then cytotoxicity was estimated by measuring cellular ATP using luminescence as described by the manufacturer of the CellTiter-Glo® cell viability assay. Viability was estimated by comparing readings of peptide-treated cells with those of nontreated cells.

2.9. Peptide-Protein Docking Simulations. The proteins structures used for the peptide-docking analysis were either obtained from the database of protein structures (in the case of the DENV envelope protein, PDB 1OAN) or obtained in our lab by homology modeling (for DIII). Sequence corresponding to the expressed protein was used to obtain the structure by homology modeling using I-TASSER web server [24]. The best model predicted by I-TASSER presented good predictive value (*C*-score = 0.77) that was structurally closely related to several viral envelope glycoproteins in the PDB database. To study the probable binding mode of the active peptides, we performed computational docking analysis using CABS-dock web server with the default parameters except for the increase of simulation cycles to 200 [25]. The best poses, ranked according to trajectory characteristics, were further analyzed to describe intermolecular interactions using LigPlot$^+$ [26].

2.10. Statistical Analysis. Where indicated, medians were compared between the nontreated control and all other groups using Kruskal-Wallis with Dunn's multiple comparison test, as implemented in GraphPad Prism software.

3. Results and Discussion

3.1. Expression, Purification, and Refolding of Recombinant DENV2 Envelope Domain III. The dengue virus envelope is a class II virus fusion protein. This protein is a β strand rich, elongated molecule with three ectodomains, the centrally located DI, the apical DII which bears the fusion loop, and the Ig-like DIII, which is connected to the short, stem, and transmembrane C-terminus [14, 27]. DIII undergoes drastic repositioning during transition to the fusogenic conformation of the E protein, leading to rearrangement from dimeric to trimeric form [15, 28]. DENV E domain III forms an Ig-like β-barrel structure that is stabilized by one disulfide bridge.

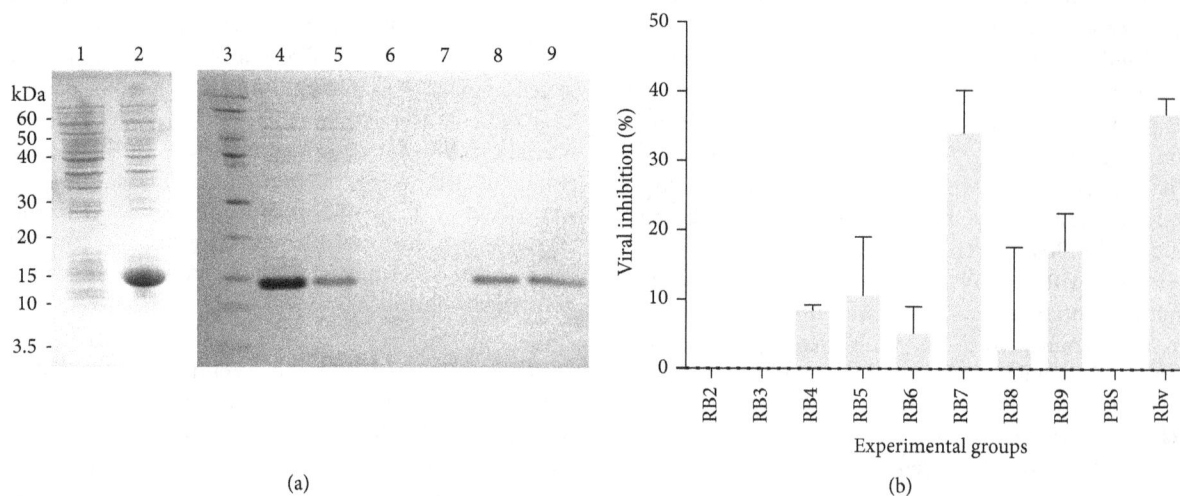

(a)

(b)

FIGURE 1: Cloning, expression in *E. coli*, and purification of codon optimized, dengue 2 envelope domain III. (a) SDS-PAGE showing several steps of protein expression and purification. 1, total lysate of cells before induction; 2, total lysate of cell after IPTG induction; 3, protein molecular weight standards; 4, solubilized inclusion bodies before loading into Ni-NTA column; 5, column flow-through; 6-7, two successive washes; 8-9, two successive elution. (b) Activity of purified and refolded recombinant domain III against DENV2 in Vero cells. Protein was refolded using several refolding conditions (RB1 to RB9) before dialysis and filtration for testing against the virus, in parallel with virus alone and ribavirin as a control. Data is presented as the mean and SD of two independent experiments. Condition RB1 was not used in this experiment as protein consistently precipitated.

The DIII is a key part of the envelope, as it is the receptor binding domain [13, 29]. It is also very antigenic, and antibodies to this domain are able to efficiently neutralize virus infection [18, 30]. Taking into account the important role of the envelope protein, as well as its domain III, we decided to search for peptides with binding activity to this protein, in an attempt to find novel molecules able to impair the infection process. For this purpose, we selected phage display as a robust methodology that allows the rapid screening and selection of millions of peptidic variants for binding capacity to a variety of targets. As the full envelope glycoprotein is a large protein that may present many possible nonproductive binding sites, we chose also to screen the random peptide library with domain III obtained in our lab.

The region coding for domain III of the envelope protein of DENV2 was obtained as a synthetic fragment from a commercial source after codon optimization for *E. coli* expression. This fragment was subcloned into the *E. coli* expression vector pET-30b under the control of a T7 promoter and fused to a 6xHis tail at the C-terminal. After IPTG induction, *E. coli* cultures showed high level of expression of the recombinant protein (Figure 1(a)), accounting for more than 50% of total protein, as judged from densitometric analysis of Coomassie stained SDS-PAGE gels (data not shown). As the recombinant protein was mainly present in insoluble aggregates, the purification protocol included an isolation and wash of inclusion bodies, followed by an affinity purification by affinity Ni-NTA chromatography under denaturing conditions. The affinity purification procedure implemented here allowed a high purity preparation of DIII (Figure 1(a)).

Several attempts to refold this protein by dilution into PBS failed due to precipitation; therefore the refolding conditions for this protein were further explored using several experimental conditions. These tests included varying concentrations of refolding agents and additives, such as L-arginine, reduced and oxidized glutathione, and/or polyethylene glycol [31]. It has been shown that the dengue envelope DIII may act as a dominant negative inhibitor of the infection process [10, 32–35]; therefore we decided to use this criterion to assess the functionality of the refolded DIII. After the refolding tests, each protein preparation was dialyzed against TBS, filter sterilized, and tested in a DENV2 virus binding blocking assay in Vero cells. We found that recombinant protein refolded in buffer RB7 (1.1 M guanidine, 55 mM Tris-HCl, pH 8.2, 21 mM NaCl, 0.88 mM KCl, 1 mM GSH, and 1 mM EDTA) showed a significant inhibition of the infection (Figure 1(b)) while there were no signs of cytotoxicity (data not shown), suggesting that the protein was folded and functional. This procedure was then used at a higher scale to isolate the recombinant EDIII for the biopanning and further steps.

3.2. Selection of Phage-Displayed Envelope Binding Peptides. In order to select for peptidic binders to both targets, we used a random library of heptapeptides presented in the M13 phage pIII protein. These phages would contain 1 to 5 copies of the peptides per capsid, theoretically allowing for the selection of higher affinity binders. As baits for the panning procedure, two sources of proteins were used, (1) a recombinant, insect-expressed DENV2 envelope and (2) recombinant, *E. coli* expressed domain III. The panning procedure involved four rounds of phage selection on solid phase-bound protein

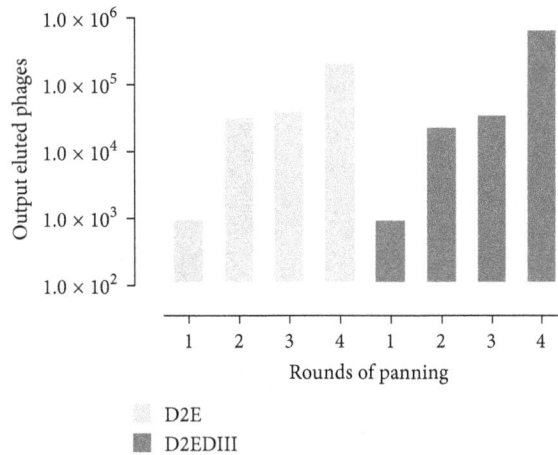

FIGURE 2: Eluted phages were enriched in successive rounds or panning. During each panning, 10^{11} input phages were incubated with the target proteins (full DENV2 envelope or domain III) and eluted as described in Materials and Methods. The eluted phages were neutralized and titer was estimated. Data is presented as the absolute number of eluted phages at each step.

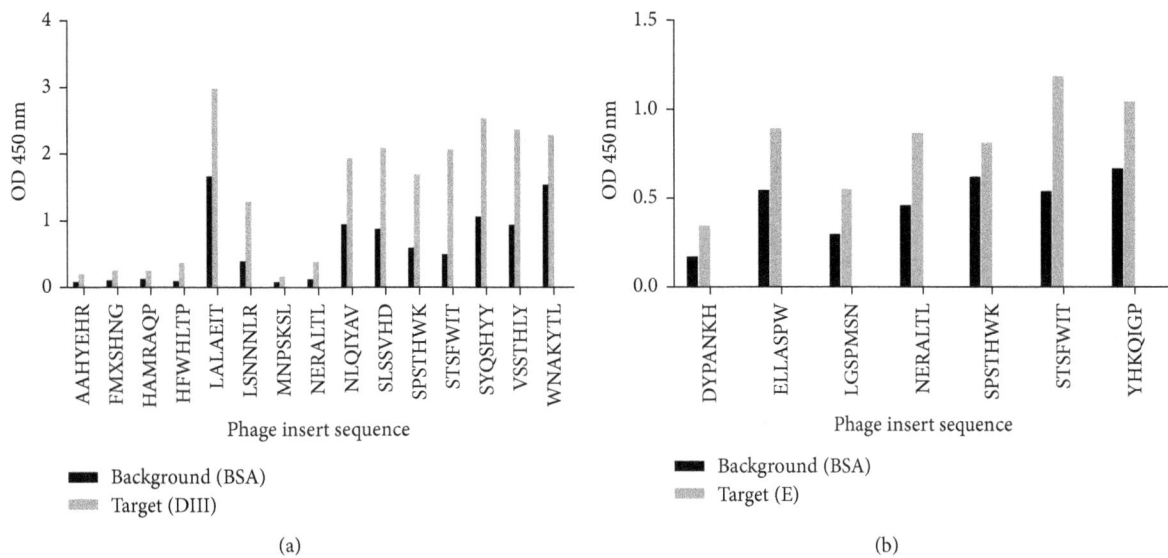

(a)

(b)

FIGURE 3: In vitro binding of selected clones as shown by phage-ELISA. Absorbance values by ELISA when phages are incubated in the presence of solid phase-bound specific ligand (target) or unspecific ligand (bovine serum albumin, as background). The binding of phages was revealed by anti-M13 antibody. (a) Results of phage-ELISA using E domain III as target. (b) Results obtained when using full E glycoprotein as target.

and subsequent elution and amplification. During panning progression against both proteins, the amounts of eluted phages increase stepwise (Figure 2), suggesting an effective enrichment of particular phages. After final round, 24 phages were randomly selected from each panning scheme, propagated, and purified to test their binding ability to the respective bait by phage-ELISA.

The analysis of eluted phages by ELISA showed that many appear to bind in a specific manner, indicated by absorbance values higher for the target as compared to those against the background (BSA) (Figure 3). Out of 24 randomly picked phages from each panning, 12 were confirmed for the full E glycoprotein and 21 were positive for the E domain III fragment. This behavior was shown by phages bearing

the peptides STSFWIT, NERALTL, ELLASPW, SPSTHWK, LALAEIT, NLQIYAV, and SLSSVHD. Some other clones did not show good target-to-background signal ratio and were not used for sequencing or subsequent analyses. This result is not unexpected, as panning procedure may yield artifact binding phages which later do not bind well in the context of the ELISA.

The phage-ELISA positive phages, 33 in total, were submitted to DNA sequencing of the randomized region. DNA sequencing revealed that binding phages carried peptides with some common features, with similar consensus sequences for the envelope glycoprotein (SxSAxxx) and for the domain III protein (SxSxHTL) (Table 1, Figure 4). The corresponding peptides had abundance of negatively charged

FIGURE 4: Peptide sequences resulting from binder phages confirmed by phage-ELISA. (a) Peptides resulting from panning on full DENV2 envelope. (b) Peptides resulting from panning on recombinant DENV2 envelope domain III. Sequence logos and consensus sequence, as calculated using JalView, are depicted below each list of peptides. Coloring of residues and sequence logos as per Zappo scheme, where residues are colored according to their physicochemical properties as follows: pink, aliphatic/hydrophobic; orange, aromatic; red, positive; green, negative; blue, hydrophilic; magenta, proline/glycine; and yellow, cysteine.

residues, intercalated with positively charged, hydrophilic, and aliphatic/hydrophobic residues (Figure 4). The resulting peptide sequences are consistent with one of the proteins (DIII) being a smaller portion of the other (E), as the consensus sequences share some resemblance, and the consensus for peptides selected against DIII appears to be more defined. Interestingly, several phages bearing the same peptide were selected with both proteins and appeared to be repeated in the final round elution (Table 1). The fact that some phages were observed several times is also congruent with a successful enrichment of specific binders during the panning steps.

Some of the binding phages shown here carried peptide sequences already reported at the biopanning data bank, a repository that, at the moment of writing this paper, had more than 23,700 sequences appearing in biopanning data [36]. The peptides SPSTHWK and WNAKYTL were also previously reported to bind crystalline Ni3B nanoparticles [37]. Also, the peptide NERALTL appeared twice in the database, with binding activity to epoxy covered surfaces [38] and to the fusion protein of the infectious salmon anemia virus [39]. Additionally, the peptide LSNNNLR was previously reported to bind poly(dimethylsiloxane) [40].

3.3. Activity of Selected Peptides and Possible Mode of Binding. Peptides from ELISA verified binding phages were synthesized containing the additional Gly-Gly-Gly-Ser spacer at the C-terminal and amidation of the carboxylate to block the negative charge, as suggested by the manufacturer to mimic

TABLE 1: Peptide sequences at the randomized region and frequency of appearance of bearing phages.

Target: envelope		Target: domain III	
Peptide	Frequency (%)	Peptide	Frequency (%)
DYPANKH	10	AAHYEHR	5
ELLASPW	20	FMXSHNG	5
LGSPMSN	20	HAMRAQP	5
NERALTL	20	HFWHLTP	5
SPSTHWK	10	LALAEIT	5
STSFWIT	10	LSNNNLR	10
YHKQIGP	10	MNPSKSL	5
		NERALTL	15
		NLQIYAV	5
		SLSSVHD	5
		SPSTHWK	15
		STSFWIT	5
		SYQSHYY	5
		VSSTHLY	5
		WNAKYTL	5

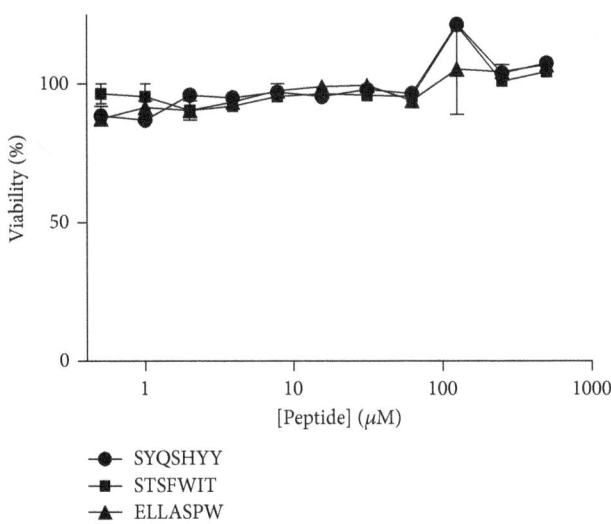

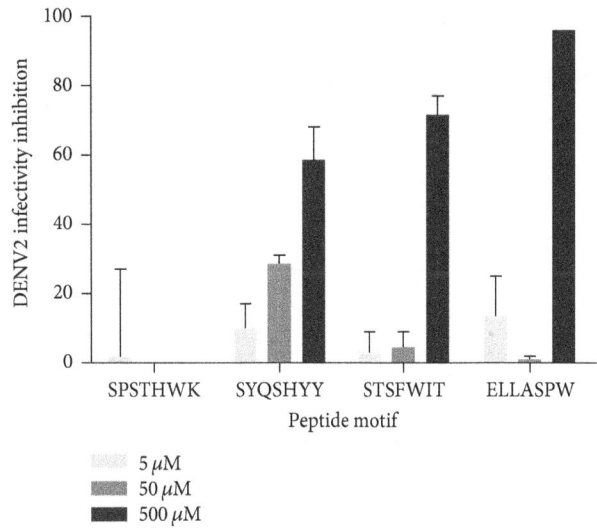

(a)   (b)

FIGURE 5: Biological activities of active peptides in Vero cells. (a) Cytotoxicity of active peptides. Cells were incubated with increasing concentrations of the peptide for 1 h and washed away and cells incubated in maintenance medium for 24. Then cytotoxicity was estimated from ATP levels using luminescence. The viability was estimated by comparing with nontreated cells. Data is presented as means and standard deviations of two independent experiments. Medians were compared between the nontreated control and all other groups using Kruskal-Wallis with Dunn's multiple comparison test. (b) DENV2 infectivity inhibition by peptides, as shown in a plaque reduction assay. The reduction in the number of plaques was estimated by comparison with virus-infected cells alone. The peptide SPSTHWK is presented as negative control since it showed no consistent inhibitory activity against the virus. Data is presented as means and standard deviations of two independent experiments.

the context of presentation in the M13 pIII protein. Then synthetic peptides were tested to check their ability to impair the infection of DENV2 in Vero cells. The synthetic peptides ELLASPW, SYQSHYY, and STSFWIT showed inhibitory activity against DENV2 infection in a plaque reduction test (Figure 5(b)). Some peptides that showed good in vitro binding to their target proteins did not show consistent inhibitory activity against the virus, that is, the peptide

SPSTHWK, and were considered negative for viral inhibition (Figure 5(b)). Since the peptides were incubated only during the adsorption period (1 h at 37°C), it may be possible that the observed inhibition is due to interference in this initial step of the virus life cycle or due to virucidal activity. More experiments are required to elucidate the exact mechanism of action of these peptides. When the active peptides were incubated alone with Vero cells in the same conditions as

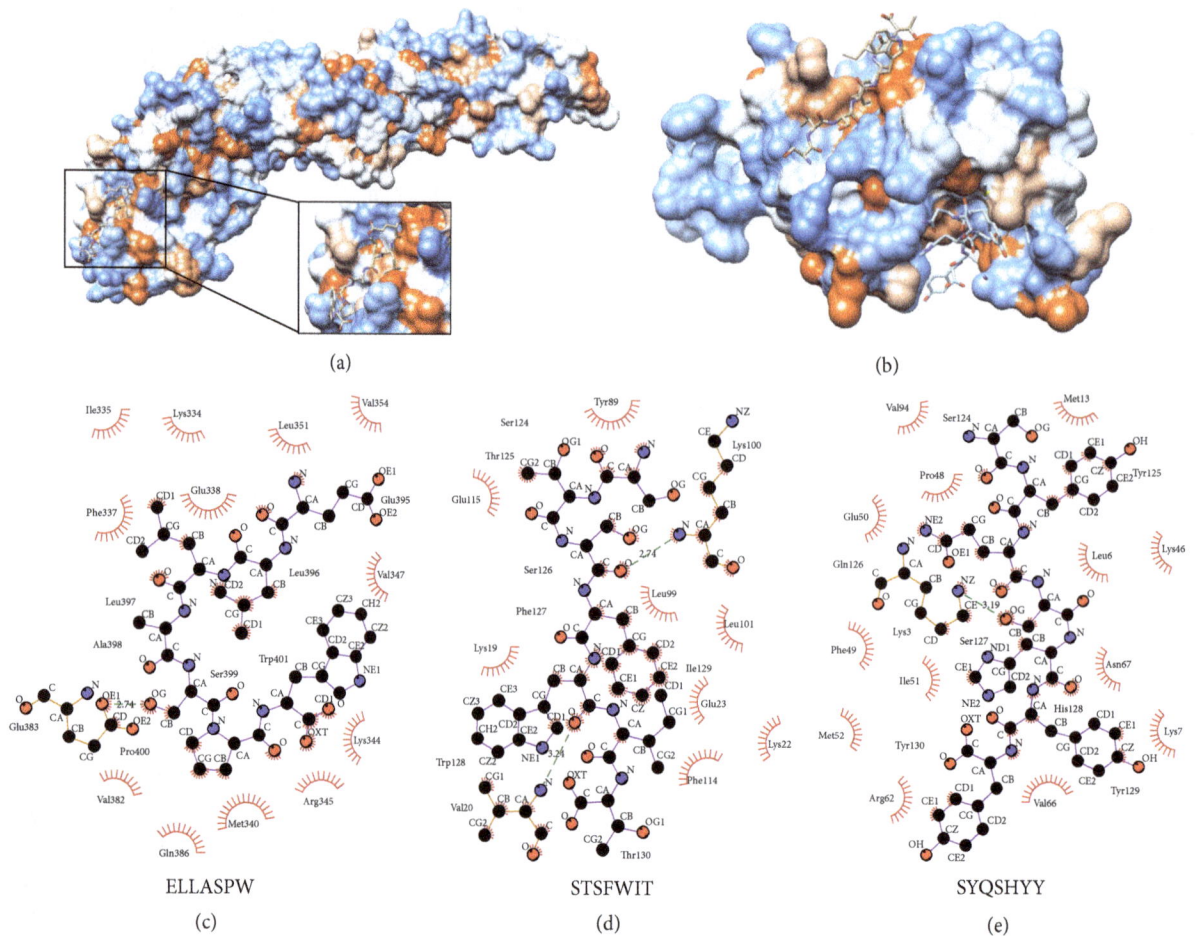

FIGURE 6: Probable binding modes of active peptides on their targets, as explored by molecular docking simulation. (a) Most favorable pose of peptide ELLASPW on the DENV2 envelope. Inset shows more details of putative binding site. (b) Most favorable poses of peptides STSFWIT and SYQSHYY on domain III. Peptide STSFWIT occupies the upper binding site of the molecule in this diagram. Proteins are depicted in their hydrophobicity surfaces, while peptides are in stick model. (c) Interaction diagram for ELLASPW-envelope. (d) Interaction diagram for STSFWIT-domain III. (e) Interaction diagram for SYQSHYY-domain III. Interaction diagrams were done in LigPlot+. Intermolecular hydrogen bonds are indicated by green dashed lines, while hydrophobic interactions are represented by arcs and radiating spokes. Pose figures were prepared using UCSF Chimera.

those of the viral infection, no sign of cytotoxicity was observed (Figure 5(a)).

The fact that some other peptides with good binding activity at the panning and ELISA tests did not inhibit the DENV2 infection is not surprising, as some of these small peptides might be binding to target regions that are not critical for their functions during binding and entry. It may be also possible that some of these peptides interact differently with their targets in their synthetic form compared with the case when they were presented in the context of the phage protein pIII.

In order to explore possible modes of binding of these peptides to their targets, we performed computational docking analysis using CABS-dock online server [25]. The algorithm pipeline followed by this software allows flexible, fully blind protein-peptide docking, finishing with all atom reconstruction and optimization of the best poses [41, 42]. The structures of the proteins used for the docking were either

obtained from the database of protein structures (in the case of the DENV2 envelope protein, PDB 1OAN) or obtained in our lab by homology modeling (in the case of the DENV2 EDIII).

The docking simulations allowed prediction of possible poses for the binding of the active peptides to their target proteins. The best models predicted indicated a predominance of hydrophobic interactions over hydrogen bonds (Figure 6). Interestingly, the best pose predicted for the envelope glycoprotein and the corresponding peptide ELLASPW proposed a binding site also at the domain III of the protein (Figure 6). The suggested pose with the lowest interaction energy for the peptide ELLASPW on the envelope glycoprotein indicates hydrophobic interactions of the peptide with the protein residues I335, K334, L351, V354, F337, E338, V347, K344, R345, M340, Q386, and V382 and a hydrogen bond with E383. The suggested interactions of peptide STSFWIT with DIII include hydrophobic links with Y377, E403, L387, L389,

K307, E311, K310, and F402 and hydrogen bonds with K388 and V308. Similarly, the best pose proposed for peptide SYQSHYY includes hydrophobic interactions with V382, M301, P336, E338, L294, K334, F337, I339, N355, M340, K295, V354, and R350 and hydrogen bond with K291.

Domain III of the dengue envelope has been shown to be involved in the initial steps of the binding of the virion to the receptor/attachment factors, particularly the amino acids in the sequence 380-IGVEPGQLKL-389 [13, 43, 44]. Interestingly, the best poses proposed by the simulation indicate some overlap with this region. The peptide STSFWIT appears to bind a region of DIII overlapping the 380–389 sequence, with three of its amino acids interacting with the peptide: L387, K388, and L389. The peptide ELLASPW does not appear to block or interfere with the receptor binding sequence, although two of its residues do participate in hydrophobic interactions with the peptide, E383 and Q386. The peptide SYQSHYY does not bind overlapping the receptor binding region, although one of its amino acids does interact with the peptide (V382). Interestingly, this last peptide also appears to bind very close to the sequence DKLQLK, interacting with the residues K291, Q293, and L294. This binding site may be relevant to the entry process as lysine 291 and lysine 295 were shown to be key for binding to the cellular GAGs [45]. Although best poses proposed here for the interactions between these peptides and their targets show favorable interaction energy profiles (data not shown), it should be also noted that, due to the complexities of modeling small peptides, there may be other valid binding sites. Further studies may be required to validate the proposed binding modes of the peptides and perform a rational, structure directed improvement of their biological activities.

The phage display technique has been used by several authors to search for interacting peptides, with the ultimate objective of developing new drugs against several viruses. Some examples include peptides described with activity against HIV [21, 46], Puumala virus [47], West Nile virus [48], Newcastle virus [49, 50], hepatitis C virus [51, 52], porcine reproductive and respiratory syndrome virus [53, 54], infectious bronchitis virus [55], hepatitis B virus [56], Mink enteritis virus [57], Japanese encephalitis virus [58], classical swine fever virus [59], and influenza virus [60].

Regarding peptidic inhibitors of dengue virus, three main strategies have been followed: (1) in silico designed peptides, (2) peptides mimicking the viral protein sequences, and (3) peptides obtained by panning against viral proteins. While the development of antiviral peptides against DENV has been dominated by reports using the first two approaches [61–64], a few attempts have been made to make use of the versatile phage display technology to find new antivirals. The peptides identified using random peptide libraries include those reported by Panya et al., 2014 [65], which target the envelope hydrophobic pocket, and those reported by Chew et al., 2015 [66], which target the full DENV2 virion. The peptides reported here do not share homology with the ones found by these authors, although results may not be fully comparable as different targets and libraries were used.

4. Conclusions

In conclusion, we have used a novel strategy to identify inhibitors of the DENV2 infectivity and we have found three peptides out of a random peptide library, which are able to specifically bind viral envelope protein and inhibit infectivity in vitro, without showing toxicity to the cells. The binding modes suggested by blind docking simulations indicate that the interactions between these active peptides and their targets may be stabilized by hydrophobic interactions, providing information relevant to the future improvement of new antivirals against this important virus.

Authors' Contributions

Carolina de la Guardia participated in experimental work and drafting of the manuscript. Mario Quijada participated in the execution of the experimental work. Ricardo Lleonart participated in the design, experimental work, and manuscript preparation.

Acknowledgments

This work was partially supported by grants from Banco Interamericano de Desarrollo (Grant no. IND-JAL-02-DENGUE), Secretaría Nacional de Ciencia, Tecnología e Innovación de Panamá (SENACYT), and Sistema Nacional de Investigación de Panamá (SNI). Authors are also grateful to Alejandro Llanes for revision of the manuscript and to researchers from ICGES, Panamá, for donation of viral DENV2 strain.

References

[1] S. Bhatt, P. W. Gething, O. J. Brady et al., "The global distribution and burden of dengue," Nature, vol. 496, no. 7446, pp. 504–507, 2013.

[2] O. J. Brady, P. W. Gething, S. Bhatt et al., "Refining the global spatial limits of dengue virus transmission by evidence-based consensus," PLoS Neglected Tropical Diseases, vol. 6, no. 8, Article ID e1760, 2012.

[3] WHO, Dengue and severe dengue, World Health organization, 2017, http://www.who.int/mediacentre/factsheets/fs117/en/.

[4] J. S. Mackenzie, D. J. Gubler, and L. R. Petersen, "Emerging flaviviruses: the spread and resurgence of Japanese encephalitis, West Nile and dengue viruses," Nature Medicine, vol. 10, no. 12, pp. S98–S109, 2004.

[5] J. P. Messina, O. J. Brady, T. W. Scott et al., "Global spread of dengue virus types: mapping the 70 year history," Trends in Microbiology, vol. 22, no. 3, pp. 138–146, 2014.

[6] D. A. Shroyer, "Aedes albopictus and arboviruses: a concise review of the literature," Journal of the American Mosquito Control Association, vol. 2, no. 4, pp. 424–428, 1986.

[7] R. J. Kuhn, W. Zhang, M. G. Rossmann, S. V Pletnev et al., "Structure of dengue virus: implications for flavivirus organization, maturation, and fusion," *Cell*, vol. 108, no. 5, pp. 717–725, 2002.

[8] B. D. Lindenbach and C. M. Rice, "Flaviviridae: the viruses and their replication," *Fields Virol*, pp. 1101–1151, 2007.

[9] S. Bressanelli, K. Stiasny, S. L. Allison et al., "Structure of a flavivirus envelope glycoprotein in its low-pH-induced membrane fusion conformation," *The EMBO Journal*, vol. 23, no. 4, pp. 728–738, 2004.

[10] M. Liao and M. Kielian, "Domain III from class II fusion proteins functions as a dominant-negative inhibitor of virus membrane fusion," *Journal of Cell Biology*, vol. 171, no. 1, pp. 111–120, 2005.

[11] D. Kato, S. Era, I. Watanabe et al., "Antiviral activity of chondroitin sulphate E targeting dengue virus envelope protein," *Antiviral Research*, vol. 88, no. 2, pp. 236–243, 2010.

[12] C. O. Nicholson, J. M. Costin, D. K. Rowe et al., "Viral entry inhibitors block dengue antibody-dependent enhancement in vitro," *Antiviral Research*, vol. 89, no. 1, pp. 71–74, 2011.

[13] F. A. Rey, F. X. Heinz, C. Mandl, C. Kunz, and S. C. Harrison, "The envelope glycoprotein from tick-borne encephalitis virus at 2 Å resolution," *Nature*, vol. 375, no. 6529, pp. 291–298, 1995.

[14] Y. Modis, S. Ogata, D. Clements, and S. C. Harrison, "A ligand-binding pocket in the dengue virus envelope glycoprotein," *Proceedings of the National Academy of Sciences of the United States of America*, vol. 100, no. 12, pp. 6986–6991, 2003.

[15] Y. Modis, S. Ogata, D. Clements, and S. C. Harrison, "Structure of the dengue virus envelope protein after membrane fusion," *Nature*, vol. 427, no. 6972, pp. 313–319, 2004.

[16] S. L. Allison, K. Stiasny, K. Stadler, C. W. Mandl, and F. X. Heinz, "Mapping of functional elements in the stem-anchor region of tick-borne encephalitis virus envelope protein E," *Journal of Virology*, vol. 73, no. 7, pp. 5605–5612, 1999.

[17] J. Hung, M. Hsieh, M. Young, C. Kao, C. King, and W. Chang, "An external loop region of domain III of dengue virus type 2 envelope protein is involved in serotype-specific binding to mosquito but not mammalian cells," *Journal of Virology*, vol. 78, no. 1, pp. 378–388, 2004.

[18] W. D. Crill and J. T. Roehrig, "Monoclonal antibodies that bind to domain III of dengue virus E glycoprotein are the most efficient blockers of virus adsorption to vero cells," *Journal of Virology*, vol. 75, no. 16, pp. 7769–7773, 2001.

[19] A. R. Precioso, R. Palacios, B. Thomé, G. Mondini, P. Braga, and J. Kalil, "Clinical evaluation strategies for a live attenuated tetravalent dengue vaccine," *Vaccine*, vol. 33, no. 50, pp. 7121–7125, 2015.

[20] "Weekly Epidemiological Record (WER)," vol. 91, no. 30, 349–364, 2016, http://www.who.int/wer/.

[21] M. Ferrer and S. C. Harrison, "Peptide ligands to human immunodeficiency virus type 1 gp120 identified from phage display libraries," *Journal of Virology*, vol. 73, no. 7, pp. 5795–5802, 1999.

[22] E. Koivunen, W. Arap, D. Rajotte, J. Lahdenranta, and R. Pasqualini, "Identification of receptor ligands with phage display peptide libraries," *Journal of Nuclear Medicine*, vol. 40, no. 10, pp. 883–888, 1999.

[23] A. M. Waterhouse, J. B. Procter, D. M. A. Martin, M. Clamp, and G. J. Barton, "Jalview Version 2-A multiple sequence alignment editor and analysis workbench," *Bioinformatics*, vol. 25, no. 9, pp. 1189–1191, 2009.

[24] J. Yang, R. Yan, A. Roy, D. Xu, J. Poisson, and Y. Zhang, "The I-TASSER Suite: protein structure and function prediction," *Nature Methods*, vol. 12, no. 1, pp. 7-8, 2015.

[25] M. Kurcinski, M. Jamroz, M. Blaszczyk, A. Kolinski, and S. Kmiecik, "CABS-dock web server for the flexible docking of peptides to proteins without prior knowledge of the binding site," *Nucleic Acids Research*, vol. 43, no. 1, pp. W419–W424, 2015.

[26] R. A. Laskowski and M. B. Swindells, "LigPlot+: multiple ligand-protein interaction diagrams for drug discovery," *Journal of Chemical Information and Modeling*, vol. 51, no. 10, pp. 2778–2786, 2011.

[27] Y. Modis, S. Ogata, D. Clements, and S. C. Harrison, "Variable surface epitopes in the crystal structure of dengue virus type 3 envelope glycoprotein," *Journal of Virology*, vol. 79, no. 2, pp. 1223–1231, 2005.

[28] Y. Zhang, W. Zhang, S. Ogata et al., "Conformational changes of the flavivirus E glycoprotein," *Structure*, vol. 12, no. 9, pp. 1607–1618, 2004.

[29] J. T. Roehrig, R. A. Bolin, and R. G. Kelly, "Monoclonal antibody mapping of the envelope glycoprotein of the dengue 2 virus, Jamaica," *Virology*, vol. 246, no. 2, pp. 317–328, 1998.

[30] P. Thullier, P. Lafaye, F. Mégret, V. Deubel, A. Jouan, and J. C. Mazié, "A recombinant Fab neutralizes dengue virus in vitro," *Journal of Biotechnology*, vol. 69, no. 2-3, pp. 183–190, 1999.

[31] A. P. J. Middelberg, "Preparative protein refolding," *Trends in Biotechnology*, vol. 20, no. 10, pp. 437–443, 2002.

[32] M.-W. Chiu and Y.-L. Yang, "Blocking the dengue virus 2 infections on BHK-21 cells with purified recombinant dengue virus 2 E protein expressed in Escherichia coli," *Biochemical and Biophysical Research Communications*, vol. 309, no. 3, pp. 672–678, 2003.

[33] J. J. H. Chu, R. Rajamanonmani, J. Li, R. Bhuvananakantham, J. Lescar, and M.-L. Ng, "Inhibition of West Nile virus entry by using a recombinant domain III from the envelope glycoprotein," *Journal of General Virology*, vol. 86, part 2, pp. 405–412, 2005.

[34] Z.-S. Zhang, Y.-S. Yan, Y.-W. Weng et al., "High-level expression of recombinant dengue virus type 2 envelope domain III protein and induction of neutralizing antibodies in BALB/C mice," *Journal of Virological Methods*, vol. 143, no. 2, pp. 125–131, 2007.

[35] J. F. L. Chin, J. J. H. Chu, and M. L. Ng, "The envelope glycoprotein domain III of dengue virus serotypes 1 and 2 inhibit virus entry," *Microbes and Infection*, vol. 9, no. 1, pp. 1–6, 2007.

[36] B. He, G. Chai, Y. Duan et al., "BDB: biopanning data bank," *Nucleic Acids Research*, vol. 44, pp. D1127–D1132, 2016.

[37] M. Ploss, S. J. Facey, C. Bruhn et al., "Selection of peptides binding to metallic borides by screening M13 phage display libraries," *BMC Biotechnology*, vol. 14, article no. 12, 2014.

[38] S. Swaminathan and Y. Cui, "Recognition of epoxy with phage displayed peptides," *Materials Science and Engineering C*, vol. 33, no. 5, pp. 3082–3084, 2013.

[39] N. Ojeda, C. Cárdenas, F. Guzmán, and S. H. Marshall, "Chemical synthesis and in vitro evaluation of a phage display-derived peptide active against infectious salmon anemia virus," *Applied and Environmental Microbiology*, vol. 82, no. 8, pp. 2563–2571, 2016.

[40] S. Swaminathan and Y. Cui, "Recognition of poly(dimethylsiloxane) with phage displayed peptides," *RSC Advances*, vol. 2, no. 33, pp. 12724–12727, 2012.

[41] M. Blaszczyk, M. Kurcinski, M. Kouza et al., "Modeling of protein-peptide interactions using the CABS-dock web server for binding site search and flexible docking," *Methods*, vol. 93, pp. 72–83, 2016.

[42] M. P. Ciemny, M. Kurcinski, K. J. Kozak, A. Kolinski, and S. Kmiecik, "Highly flexible protein-peptide docking using CABS-dock," in *Modeling Peptide-Protein Interactions*, vol. 1561 of *Methods in Molecular Biology*, pp. 69–94, 2017.

[43] S. Mukhopadhyay, R. J. Kuhn, and M. G. Rossmann, "A structural perspective of the Flavivirus life cycle," *Nature Reviews Microbiology*, vol. 3, no. 1, pp. 13–22, 2005.

[44] S. M. Erb, S. Butrapet, K. J. Moss et al., "Domain-III FG loop of the dengue virus type 2 envelope protein is important for infection of mammalian cells and *Aedes aegypti* mosquitoes," *Virology*, vol. 406, no. 2, pp. 328–335, 2010.

[45] D. Watterson, B. Kobe, and P. R. Young, "Residues in domain III of the dengue virus envelope glycoprotein involved in cell-surface glycosaminoglycan binding," *Journal of General Virology*, vol. 93, no. 1, pp. 72–82, 2012.

[46] B. D. Welch, J. N. Francis, J. S. Redman et al., "Design of a potent D-peptide HIV-1 entry inhibitor with a strong barrier to resistance," *Journal of Virology*, vol. 84, no. 21, pp. 11235–11244, 2010.

[47] T. Heiskanen, A. Lundkvist, A. Vaheri, and H. Lankinen, "Phage-displayed peptide targeting on the Puumala hantavirus neutralization site," *Journal of Virology*, vol. 71, no. 5, pp. 3879–3885, 1997.

[48] F. Bai, T. Town, D. Pradhan et al., "Antiviral peptides targeting the West Nile virus envelope protein," *Journal of Virology*, vol. 81, no. 4, pp. 2047–2055, 2007.

[49] M. Ozawa, K. Ohashi, and M. Onuma, "Identification and characterization of peptides binding to Newcastle disease virus by phage display," *Journal of Veterinary Medical Science*, vol. 67, no. 12, pp. 1237–1241, 2005.

[50] S. L. Chia, W. S. Tan, K. Shaari, N. Abdul Rahman, K. Yusoff, and S. D. Satyanarayanajois, "Structural analysis of peptides that interact with Newcastle disease virus," *Peptides*, vol. 27, no. 6, pp. 1217–1225, 2006.

[51] H. W. Hong, S. W. Lee, and H. Myung, "Selection of peptides binding to HCV E2 and inhibiting viral infectivity," *Journal of Microbiology and Biotechnology*, vol. 20, no. 12, pp. 1769–1771, 2010.

[52] X. Lü, M. Yao, J.-M. Zhang et al., "Identification of peptides that bind hepatitis C virus envelope protein E2 and inhibit viral cellular entry from a phage-display peptide library," *International Journal of Molecular Medicine*, vol. 33, no. 5, pp. 1312–1318, 2014.

[53] K. Liu, X. Feng, Z. Ma et al., "Antiviral activity of phage display selected peptides against Porcine reproductive and respiratory syndrome virus in vitro," *Virology*, vol. 432, no. 1, pp. 73–80, 2012.

[54] H. Wang, R. Liu, J. Cui et al., "Characterization and utility of phages bearing peptides with affinity to porcine reproductive and respiratory syndrome virus nsp7 protein," *Journal of Virological Methods*, vol. 222, pp. 231–241, 2015.

[55] B. Peng, H. Chen, Y. Tan, M. Jin, H. Chen, and A. Guo, "Identification of one peptide which inhibited infectivity of avian infectious bronchitis virus in vitro," *Science in China, Series C: Life Sciences*, vol. 49, no. 2, pp. 158–163, 2006.

[56] K. L. Ho, K. Yusoff, H. F. Seow, and W. S. Tan, "Selection of high affinity ligands to hepatitis B core antigen from a phage-displayed cyclic peptide library," *Journal of Medical Virology*, vol. 69, no. 1, pp. 27–32, 2003.

[57] Q. Zhang, Y. Wang, Q. Ji et al., "Selection of antiviral peptides against mink enteritis virus using a phage display peptide library," *Current Microbiology*, vol. 66, no. 4, pp. 379–384, 2013.

[58] X. Zu, Y. Liu, S. Wang et al., "Peptide inhibitor of Japanese encephalitis virus infection targeting envelope protein domain III," *Antiviral Research*, vol. 104, no. 1, pp. 7–14, 2014.

[59] L. Yin, Y. Luo, B. Liang et al., "Specific ligands for classical swine fever virus screened from landscape phage display library," *Antiviral Research*, vol. 109, no. 1, pp. 68–71, 2014.

[60] T. Matsubara, A. Onishi, D. Yamaguchi, and T. Sato, "Heptapeptide ligands against receptor-binding sites of influenza hemagglutinin toward anti-influenza therapy," *Bioorganic and Medicinal Chemistry*, vol. 24, no. 5, pp. 1106–1114, 2016.

[61] Y. M. Hrobowski, R. F. Garry, and S. F. Michael, "Peptide inhibitors of dengue virus and West Nile virus infectivity," *Virology Journal*, vol. 2, article 49, 2005.

[62] A. G. Schmidt, P. L. Yang, and S. C. Harrison, "Peptide inhibitors of dengue-virus entry target a late-stage fusion intermediate," *PLOS Pathogens*, vol. 6, no. 4, Article ID e1000851, 2010.

[63] A. G. Schmidt, P. L. Yang, and S. C. Harrison, "Peptide inhibitors of flavivirus entry derived from the E protein stem," *Journal of Virology*, vol. 84, no. 24, pp. 12549–12554, 2010.

[64] A. Panya, N. Sawasdee, M. Junking, C. Srisawat, K. Choowongkomon, and P.-T. Yenchitsomanus, "A Peptide Inhibitor Derived from the Conserved Ectodomain Region of DENV Membrane (M) Protein with Activity Against Dengue Virus Infection," *Chemical Biology and Drug Design*, vol. 86, no. 5, pp. 1093–1104, 2015.

[65] A. Panya, K. Bangphoomi, K. Choowongkomon, and P.-T. Yenchitsomanus, "Peptide inhibitors against dengue virus infection," *Chemical Biology and Drug Design*, vol. 84, no. 2, pp. 148–157, 2014.

[66] M.-F. Chew, H.-W. Tham, M. Rajik, and S. H. Sharifah, "Anti-dengue virus serotype 2 activity and mode of action of a novel peptide," *Journal of Applied Microbiology*, vol. 119, no. 4, pp. 1170–1180, 2015.

13

Mixed Viral Infections Circulating in Hospitalized Patients with Respiratory Tract Infections in Kuwait

Sahar Essa,[1] Abdullah Owayed,[2] Haya Altawalah,[3] Mousa Khadadah,[4] Nasser Behbehani,[4] and Widad Al-Nakib[1]

[1]*Department of Microbiology, Faculty of Medicine, Kuwait University, 24923 Safat, Kuwait*
[2]*Department of Pediatrics, Faculty of Medicine, Kuwait University, 24923 Safat, Kuwait*
[3]*Virology Unit, Mubarak Hospital, Ministry of Health, 24923 Safat, Kuwait*
[4]*Department of Medicine, Faculty of Medicine, Kuwait University, 24923 Safat, Kuwait*

Correspondence should be addressed to Sahar Essa; sahar@hsc.edu.kw

Academic Editor: Trudy Morrison

The aim of this study was to determine the frequency of viral mixed detection in hospitalized patients with respiratory tract infections and to evaluate the correlation between viral mixed detection and clinical severity. Hospitalized patients with respiratory tract infections (RTI) were investigated for 15 respiratory viruses by using sensitive molecular techniques. In total, 850 hospitalized patients aged between 3 days and 80 years were screened from September 2010 to April 2014. Among the 351 (47.8%) patients diagnosed with viral infections, viral mixed detection was identified in 49 patients (14%), with human rhinovirus (HRV) being the most common virus associated with viral mixed detection (7.1%), followed by adenovirus (AdV) (4%) and human coronavirus-OC43 (HCoV-OC43) (3.7%). The highest combination of viral mixed detection was identified with HRV and AdV (2%), followed by HRV and HCoV-OC43 (1.4%). Pneumonia and bronchiolitis were the most frequent reason for hospitalization with viral mixed detection (9.1%). There were statistical significance differences between mixed and single detection in patients diagnosed with bronchiolitis ($P = 0.002$) and pneumonia ($P = 0.019$). Our findings might indicate a significant association between respiratory virus mixed detection and the possibility of developing more severe LRTI such as bronchiolitis and pneumonia when compared with single detection.

1. Introduction

The progress of molecular techniques for the identification of respiratory viruses allows for quick and specific diagnosis which is vital for the management of patients with respiratory tract infections (RTI) [1, 2]. Polymerase chain reaction (PCR) technology has been determined as an adequate tool for the identification of respiratory viruses as shown in many studies [3–5]. Respiratory virus infections represent a major public health problem because of their worldwide occurrence, ease of spread in the community, and considerable morbidity and mortality [6, 7]. The frequency of mixed respiratory viral detection varies from 10% to 30% in hospitalized children [8–11]. Several studies suggested an association between mixed detection and increase in the disease and/or clinical severity

[9, 12–15]. Others propose the absence of a relationship between mixed respiratory detection and increase in the disease and/or clinical severity [16, 17].

Respiratory viruses such as human rhinovirus (HRV), respiratory syncytial virus (RSV), influenza A virus (FluA), influenza B virus (FluB), parainfluenza virus-1 (PIV-1), parainfluenza virus-2 (PIV-2), parainfluenza virus-3 (PIV-3), human coronavirus-OC43 (HCoV-OC43), human coronavirus-229E (HCoV-229E), and adenoviruses (AdV) have been recognized as causative agents of RTI [3, 18]. The panel of viruses determined responsible for RTI has been extended more by including more viruses such as human metapneumovirus (hMPV) [19, 20], HCoV-NL63 [21, 22], human bocavirus (Boca) [23, 24], human polyomavirus KI (KIV), and human polyomavirus WU (WUV) [25, 26].

The aim of this study was to determine the frequency of viral mixed detection in hospitalized patients with RTI and to evaluate the correlation between viral mixed detection and clinical severity.

2. Methods and Materials

2.1. Study Population. The study included 850 hospitalized children and adult patients with upper respiratory tract infections (URTI) or lower respiratory tract infections (LRTI) in Mubarak Al-Kabir Hospital, Kuwait. All patients hospitalized with RTI were screened during the period from September 2010 to April 2014. Specimens were collected in the hospital and stored at −70°C until processed in the Virology Unit, Faculty of Medicine, Kuwait University, to detect the presence of viral nucleic acids using PCR techniques. The age of the patients ranged between 3 days and 80 years. The majority of samples were collected during autumn and winter. Autumn in Kuwait is between September and November, and winter is between December and March.

2.2. Clinical Samples. Nasopharyngeal swab specimens for URTI and bronchoalveolar lavage for LRTI were collected after obtaining written informed consent from the hospitalized patients. Ethical permission to perform this research study was granted by the Health Science Center and Kuwait Institute for Medical Specialization (KIMS) Joint Committee of the Protection of Human Subjects in Research. Clinical data were collected from medical record using a uniform data collection form.

2.3. Molecular Detection of Respiratory Viruses

2.3.1. Extraction Method. The nucleic acid extraction was done using the automated nucleic acid extraction method, MagNA Pure LC 2.0 (Roche Diagnostics Ltd., Rotkreuz, Switzerland). All 850 respiratory samples were extracted using the MagNA Pure LC Total Nucleic Acid Isolation Kit (Roche Applied Science, Mannheim, Germany) according to the manufacturer's instruction. The extraction resulting in 60 μL eluates of viral nucleic acid was stored at −70°C until processing.

Determination of the presence of viral nucleic acids from respiratory viruses was performed as described before [4]. Briefly, a single PCR was used to detect adenovirus and parainfluenza virus-2 (PIV-2); duplex PCR was carried out to detect influenza A and B viruses; triplex PCR was carried out to detect respiratory syncytial virus (RSV) and parainfluenza viruses (PIV) 1 and 3, and another triplex PCR was performed to detect human rhinovirus and human coronavirus-229E and coronavirus-OC43.

2.3.2. PCR and RT-PCR for the Detection of Newly Discovered Respiratory Viruses. Upon extraction of nucleic acids from clinical specimens, determination of the presence of hMPV RNA was performed using primers described by Ordás et al. [27]. HCoV-NL63 RNA was detected using primers described by Moës et al. [28], and Boca DNA was detected using primers described by Allander et al. [23]. The primers

used to detect KIV and WUV DNA were previously described by Allander et al. and Bialasiewicz et al. [25, 26].

2.3.3. PCR and RT-PCR Conditions. The RT step was performed for 60 min at 37°C in a 10 μL reaction volume containing 1X GeneAmp RNA PCR buffer, 5 μL of 25 mM MgCl$_2$, 1 mM of each deoxynucleoside triphosphate, 2.5 μM of random hexamers, 0.5 μL RNase inhibitor, 3 μL of viral nucleic acid, and 2.5 U/μL reverse transcriptase enzyme (GeneAmp RNA Core Kit; Applied Biosystems, Chicago, IL). Following heat inactivation of the reverse transcriptase at 90°C for 5 min, the entire reaction mixture was used for PCR in a total volume of 50 μL. The reaction mixture composition was 2 mM MgCl$_2$ solution, 1X PCR buffer containing 0.02 pg of each forward and reverse primer, and 0.05 μL of 5 U/μL Ampli Taq DNA Polymerase (Thermo Fisher Scientific, Pittsburgh, USA). PCR was performed as follows: 94°C denaturation for 1 min, followed by 40 cycles of denaturation at 94°C for 30 sec, annealing at 50°C for 30 sec and elongation at 72°C for 30 sec, and a final extension at 72°C for 7 min. Water was used instead of nucleic acids as a negative control. The specificity of the PCR was established for each PCR format using a panel of ATCC reference viruses to check for cross-reactivity to old respiratory viruses. DNA templates (110–140 bp, Thermo Scientific) encompassing the annealing sites of the primers and probes were used as positive controls for the detection of nucleic acid from HCoV-NL63, hMPV, Boca, WUV, and KIV.

2.4. Statistical Analysis. Data analysis was performed using the Statistical Package for the Social Sciences (SPSS version 20.0, IBM Corp, Armonk, NY, USA). The descriptive statistics of the continuous variables were compared using a nonparametric Mann-Whitney U test or Kruskal-Wallis test. For the categorical variables, a Chi-square or Fisher's exact test or Z-test was applied to test the difference between proportions or to assess whether any association existed between the proportions. The two-tailed probability value $P < 0.05$ was considered statistically significant.

3. Results

From the overall number of 850 hospitalized patients three hundred fifty one patients (47.8%) were diagnosed with viral respiratory infections, 210 (59.8%) of them were males and 141 (40.2%) were females. Results show that from the 351 patients 408 viruses were detected. Table 1 shows that HRV was the most detected virus in clinical respiratory specimens of patients with respiratory symptoms (41.6%), followed by FluA (15.1%), RSV (13.1%), and HCoV-OC43 (12.3%). Among the 351 hospitalized patients viral mixed detection was detected in 49 patients (14%). HRV was the most common virus associated with mixed detection (7.1%), followed by AdV (4%), HCoV-OC43 (3.7%), RSV (3.1%), and FluA (2.8%) (Table 1).

It was interesting to note that four patients had triple viral mixed detection. The first patient was infected with Boca, HCoV-OC43, and HRV, the second patient was diagnosed with WUV, Boca, and HCoV-229E, the third one was infected

TABLE 1: Clinical manifestations of patients with mixed and single viral detection for twelve respiratory viruses.

Viruses	URTI		Bronchitis		Bronchiolitis		Pneumonia		LRTI		Total		Total (%)** Viral detection
	Mixed detection	Single detection	Mixed detection	Single detection	Mixed detection	Single detection	Mixed detection	Single detection	Mixed detection	Single detection	Mixed detection	Single detection	
WUV	2	0	0	3	2	3	2	1	4	7	6 (1.7)	7 (2)	13 (3.7)
KIV	1	1	1	1	1	0	1	0	3	1	4 (1.1)	2 (0.6)	6 (1.7)
Boca	0	5	2	3	3	5	1	0	6	8	6 (1.5)	13 (3.7)	19 (5.4)
HCoV-OC43	1	9	1	13	4	3	7	5	12	21	13 (3.7)	30 (8.5)	43 (12.3)
HCoV-229E	0	1	0	5	2	5	1	0	3	10	3 (0.9)	11 (3.1)	14 (4)
AdV	2	4	3	6	4	3	5	3	12	12	14 (4)	16 (4.6)	30 (8.5)
FluA	4	5	2	11	3	8	1	19	6	38	10 (2.8)	43 (12.3)	53 (15.1)
RSV	3	9	2	0	3	19	3	7	8	26	11 (3.1)	35 (10)	46 (13.1)
PIV-3	1	2	0	2	1	2	0	2	1	6	2 (0.6)	8 (2.3)	10 (2.8)
HRV	2	5	2	39	8	45	13	32	23	116	25 (7.1)	121 (33.9)	146 (41.6)
PIV-1	0	1	0	1	1	1	1	1	2	3	2 (0.6)	4 (1.1)	6 (1.7)
hMPV	4	5	1	8	0	3	1	0	2	11	6 (1.7)	16 (4.6)	22 (6.3)
Total (%)†	20 (19.6)	47 (15.4)	14 (13.7)	92 (30.1)	32 (31.4)*	97 (31.7)	36 (35.3)*	70 (22.9)	82 (80.3)	259 (84.6)	102 (25)	306 (75)	408

†Number in parentheses represents the number of detection incidences (single or mixed) in relation to number of total (single or mixed) detection incidences in percent.
**Number in parentheses represents the number of detection incidences in relation to the total number of patients ($n = 351$) in percent.
*Statistical significance differences between single and mixed detection in patients diagnosed with bronchiolitis and pneumonia $P < 0.05$.

TABLE 2: Frequency of viral mixed detection for twelve respiratory viruses.

Viruses	WUV	KIV	Boca	HCoV-OC43	HCoV-E229	AdV	FluA	RSV	PIV-3	HRV	P1V-1	hMPV
WUV												
KIV	—											
Boca	1	—										
HCoV-OC43	—	1	1									
HCoV-229E	1	—	1	—								
AdV	—	1	1	3	—							
FluA	1	1	—	—	—	1						
RSV	1	1	—	3	—	—	2					
PIV-3	—	—	—	—	—	—	—	1				
HRV	1	—	2	5	1	7	4	3	—			
P1V-1	—	—	—	—	—	—	—	—	—	1		
hMPV	1	—	—	—	—	1	1	—	1	1	1	1
Total*	6	4	6	13	3	14	10	11	2	25	2	6

Data are number of samples positive for viral mixed detection.
*Total number of mixed detection for each virus.

TABLE 3: Distribution of patients with undetected, single, and mixed detection in relation to age, gender, and hospital admission.

	Undetected	Single detection	Mixed detection
	Patient number 499 (%)*	Patient number 302 (%)**	Patient number 49 (%)**
Age (yrs)			
<1	90 (18)	148 (42.2)	20 (5.7)
1–14	144 (29)	66 (18.8)	17 (4.8)
≥15	265 (53.1)	88 (25.1)	12 (3.4)
Gender			
Male	276 (55.3)	177 (50.4)	33 (9.4)
Female	223 (44.7)	125 (35.6)	16 (4.6)
Hospital Unit			
ICU	184 (36.9)	36 (10.3)	6 (1.7)
PICU	175 (35.1)	91 (30)	11 (3.1)
Ward	139 (27.9)	175 (50)	32 (9.1)

*Number in parentheses represents the number of undetected viral infections in relation to the total number of patients ($n = 499$) in percent.
**Number in parentheses represents the number of detected viral infections in relation to the total number of patients ($n = 351$) in percent.
ICU: Intensive care unit.
PICU: Pediatric intensive care unit.

with HCoV-OC43, FluA, and HRV, and the fourth patient was infected with KIV, RSV, and hMPV.

Table 2 shows the frequency of viral mixed detection among the 49 patients. The highest combination of viral mixed detection was identified with HRV and AdV in 7 patients (2%), followed by HRV and HCoV-OC43 in 5 patients (1.4%), and HRV and FluA in 4 patients (1.1%).

From the 49 (14%) patients with mixed detection, 33 (9.4%) of them were males (32 patients (9.1%) with double detection and one patient (0.3%) with triple detection), and 16 (4.6%) were females (14 patients (4%) with double detection and 2 patients (0.6%) with triple detection) (Table 3).

In total, 20 of the 49 (5.7%) patients with viral mixed detection were aged <1 years (18 patients (5.1%) with double detection and 2 patients (0.6%) with triple detection), 17 patients (4.8%) were 1–14 years (16 patients (4.6%) with double detection and one patient (0.3%) with triple detection),

and 12 patients (3.4%) were ≥15 with double viral detection (Table 3). Overall, the majority of viral mixed detection, reaching 8.5% ($n = 30$), was among children ≤5 years of age. Table 4 shows the distribution of median age, range, and IQ of patients with mixed detection for each virus. The median age was <1 years for Boca, HCoV-OC43, and RSV whereas for WU, AdV, FluA, PIV-3, HRV, and hMPV it was 1–11.5 years. Furthermore, for the rest of the respiratory viruses KI, HCoV-229E, and PIV-1 it was ≥15 years of age.

Mixed viral detection was identified in 17 patients (4.8%) with pneumonia (15 patients (4.3%) with double viral detection and 2 patients (0.6%) with triple detection), 15 patients (4.8%) with bronchiolitis (14 patients (4%) with double viral detection and one patient (0.3%) with triple detection), 10 patients (2.8%) with URTI all suffered from double viral detection, and 7 patients (2%) with respiratory distress (RD) all suffered from double viral mixed detection (Table 3).

TABLE 4: Age distribution of respiratory virus found in mixed detection.

Virus	Median age*	Range	IQ
WUV	11.5	3–68	4–35
KIV	19.5	1–45	1–43
Boca	0	0–3	0–2
HCoV-OC43	0.5	0–38	0–7
HCoV-229E	50	3–80	3–65
AdV	4	0–70	0–16
FluA	4.5	0–60	0–45
RSV	0	0–24	0-1
PIV-3	2	0–80	0–24
HRV	1	0–80	0–19
PIV-1	36	33–39	33–36
hMPV	1	0–68	0–21

*Median age = 0 (zero was coded for <1 year).
IQ: Interquartile range.

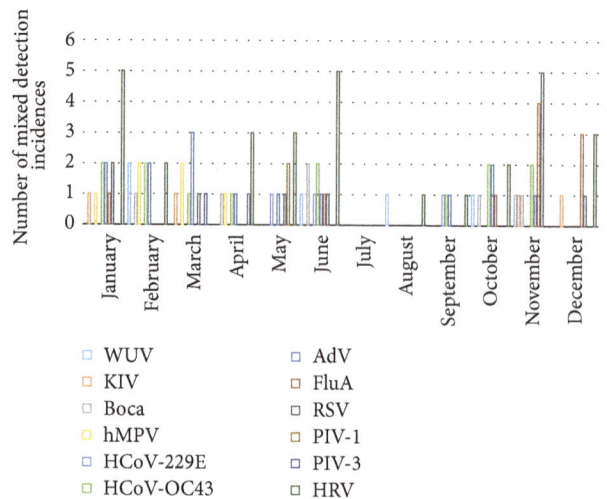

FIGURE 1: Monthly distribution of viral mixed detection.

the majority of infections by the investigated respiratory viruses affected the lower respiratory tract (39 patients or 11.1%) rather than the upper respiratory tract (10 patients, or 2.8%). Pneumonia and bronchiolitis were the most frequent reason for hospitalization with viral mixed detection (32 patients or 9.1%). Table 1 compares the clinical manifestation of patients with mixed and single viral detection. There were statistical significance differences between mixed and single detection in patients diagnosed with bronchiolitis (P = 0.002) and pneumonia (P = 0.019).

The majority (32 patients or 9.1%) of hospitalized patients were admitted to wards, followed by pediatric intensive care unit (PICU) (11 patients or 3.1%), and intensive care unit (ICU) (6 patients or 1.7%) (Table 3).

The peak incidence of viral mixed detection was identified during the month of November (15 incidences of detection or 14.7%) followed by January and June (14 incidences of detection each or 13.7%). The lowest incidence was detected during the month of August (2 incidences of detection or 2%) and no viral mixed detection was identified during the month of July (Figure 1).

4. Discussion

The usage of molecular techniques for viral infections has improved the identification of mixed viral detection in a single sample [29]. In this study, we assessed the incidence of viral mixed detection in Kuwait during three and a half consecutive years, September 2010 to April 2014 by PCR techniques in hospitalized children and adults with URTI and LRTI. The overall prevalence of viral mixed detection in Kuwait among hospitalized patients with RTI was 14%. The frequency of mixed viral detection was approximately 8% higher in LRTI than in URTI. From the published studies that use molecular diagnostics to report respiratory viral mixed detection, no other studies match our study population (children and adults) or clinical presentation (URTI and LRTI). A community-based study in Jinan, China, of 720 samples from

inpatient and outpatients with RTI during a one-year period identified viral mixed detection in 95 samples (13.19%). Also, in this study the virus positive rate was approximately 20% higher in LRTIs than in URTIs [9]. In a recent study 48% (140/292) of the samples from hospitalized children and adults with acute LRTI, viral mixed detection, were observed in 8% (22/292) of the samples [10]. In another recent study of 131 samples from children aged 0–8 with acute RTI, 19 (14.5%) were identified with mixed viral detection [11].

The three principal pathogens involved in mixed viral detection were HRV, AdV, and HCoV-OC43. Similar results were reported, where they identified HRV, AdV, and HCoV-OC43 as the leading viruses involved in mixed detection [30]. Other studies reported different groupings of leading viruses involved in mixed detection. Recent studies reported RSV, HRV, and AdV as the leading viruses involved in mixed detection among children [8, 31] and among children/adults [10]. In another study the most prevalent viruses involved in mixed detection among children with RTI were HRV, PIV, and Flu viruses [9]. These differences may be attributed to the panel of respiratory viruses tested, regional or environmental variability and the difference of the virus detection techniques.

Out of the 49 virally coinfected patients, 45 (12.8%) suffer from double viral detection and 4 (1.1%) triple viral detection. In an epidemiological study from Korea the mixed viral detection analysis showed 17.1% of double detection and 1.8% of triple detection, which is higher than our result probably due to the fact that they tested larger sample size and we tested a larger panel of viruses [30]. Another study also reported double 20.3% and triple 3.9% viral detection among children with RSV infection [8]. The most frequently detected combinations were HRV/AdV, HRV/HCoV-OC43, and HRV/FluA. The combination of HRV/AdV is the leading combination; this finding is directly comparable with those from previous reports [8, 30]. In this study, the majority of viral mixed detection was among children <1 years of age (20 patients or 5.7%). This is comparable with other recent studies [8–10, 31]. This may be due to an immature immune system of the infants and the absence of earlier exposure to respiratory viruses which could increase their susceptibility to a mixed infection [12].

In this study virus mixed detection was not identified between RSV and hMPV although a number of studies have found hMPV and RSV coinfection rates of approximately ~5–14% [20, 32, 33]. However, in a study conducted in Netherlands in hospitalized children with LRTI, no virus coinfection between RSV and hMPV was detected [34].

As shown in Table 2, HCoV-OC43 positive patients were most commonly coinfected with HRV and RSV. In a study conducted in China from 2006 to 2009 aimed to assess the overall prevalence of 10 respiratory viruses in children with acute LRTI, coronaviruses-positive samples were most commonly coinfected with HRV and RSV [35]. Similar data describing a high rate of mixed detection of coronaviruses with RSV has also been previously described [36, 37].

Since the first identification of KIV and WUV, their viral sequences have been identified globally in respiratory samples from patients with RTI [38–41]. However WUV and KIV were found at similar rates in control individuals without respiratory diseases so the association between these polyomaviruses and respiratory diseases remains hypothetical [38, 40, 42]. A mixed detection rate of 74% has been identified for KIV and rates stretching from 68 to 79% for WUV [39–41]. In this study, hospitalized patients with a single WUV detection were diagnosed with bronchitis, bronchiolitis, and pneumonia (Table 1). In a study in Southern China, hospitalized children with a single WUV detection presented with cough, moderate fever, and wheezing and they were also diagnosed with pneumonia, bronchiolitis, URTI, and bronchitis. These findings suggest that polyomavirus can cause URTI and LRTI [43]. In another study assessing the incidence and viral load of WUV and KIV in respiratory samples from immunocompromised and immunocompetent children revealed that the prevalence of WUV and KIV is similar in immunocompromised patients compared with that of the immunocompetent population [44]. Nevertheless these data have to be confirmed in further studies.

Several studies have shown that Boca detection tends to be associated with other respiratory viruses such as HRV, AdV, and RSV [23, 35, 45]. In this study Boca virus mixed detection was identified with HRV, HCoV-OC43, HCoV-229E, and AdV (Table 2). Persistent viral shedding and high frequency of mixed detection have led to an argument over its role as a true pathogen [42, 46]. Other studies confirmed that Boca virus is most probably the cause of RTI if the patient has a single detection and high viral load in clinical samples [45, 47]. In this study, our patients who were diagnosed with a single Boca virus detection suffered from both URTI and LRTI (Table 1). Nevertheless, despite this debate it is becoming increasingly obvious that Boca virus is an important respiratory virus [48].

Our findings might indicate an association between respiratory virus mixed detection and the possibility of developing more severe LRTI such as bronchiolitis ($P = 0.002$) and pneumonia ($P = 0.019$) when compared with single detection. The relationship between mixed viral detection and disease/clinical severity is debatable. Earlier studies have reported that mixed detection with respiratory viruses increased the risk of hospitalization and pneumonia [8, 9, 13, 14], while other studies reported no association between mixed detection and disease/clinical severity [16, 49]. However, despite the availability of sensitive molecular assays, reports are still controversial concerning the role of mixed detection in the disease/clinical severity in comparison to single detection. A number of theories have been proposed to explain the association between mixed respiratory virus detection and RTI severity; these theories include alteration of immune responses after the primary infection [50, 51] and host vulnerability to multiple viruses [15].

The seasonal incidence of mixed viral detection was detectable throughout the year except for the month of July, with the peak incidence during the months of January, June, and November (43 incidences of detection or 42.1%).

In summary, our findings may indicate that viral mixed detection in patients with RTI is not uncommon and that mixed detection may increase the clinical severity of patients with pneumonia or bronchiolitis. Further investigations are necessary to investigate the determinants of disease severity in viral mixed detection in RTI.

Although this study has several limitations like the lack of study controls (matched hospitalizations without RTI necessary to estimate attributable disease), difference in RT-PCR sensitivity/specificity among targeted pathogens, and lack of systematic testing for potential bacterial pathogens, viral loads were not detected but these data provide representative results of mixed respiratory viral detection in Kuwait.

Acknowledgments

The authors are thankful for the continuous support of this research (MI 03/08) by the Research Sector, Kuwait University. Special thanks are due to Dr. Wassim Chehadeh (Associate Professor, Microbiology Department, Faculty of Medicine, Kuwait University) for his excellent technical assistance in analyzing viruses by PCR. Special thanks are due to Dr. Prem Sharma (Technical advisor, Vice President Office, Kuwait University) for his outstanding help in the statistical analysis of the data.

References

[1] L. Ivaska, J. Niemelä, T. Heikkinen, T. Vuorinen, and V. Peltola, "Identification of respiratory viruses with a novel point-of-care multianalyte antigen detection test in children with acute respiratory tract infection," *Journal of Clinical Virology*, vol. 57, no. 2, pp. 136–140, 2013.

[2] N. Lévêque, F. Renois, and L. Andréoletti, "The microarray technology: facts and controversies," *Clinical Microbiology and Infection*, vol. 19, no. 1, pp. 10–14, 2013.

[3] S. Bierbaum, J. Forster, R. Berner et al., "Detection of respiratory viruses using a multiplex real-time PCR assay in Germany, 2009/10," *Archives of Virology*, vol. 159, no. 4, pp. 669–676, 2014.

[4] M. Khadadah, S. Essa, Z. Higazi, N. Behbehani, and W. Al-Nakib, "Respiratory syncytial virus and human rhinoviruses are the major causes of severe lower respiratory tract infections in Kuwait," *Journal of Medical Virology*, vol. 82, no. 8, pp. 1462–1467, 2010.

[5] J. Li, S. Qi, C. Zhang et al., "A two-tube multiplex reverse transcription PCR assay for simultaneous detection of sixteen human respiratory virus types/subtypes," *BioMed Research International*, vol. 2013, Article ID 327620, 8 pages, 2013.

[6] F. W. Denny Jr., "The clinical impact of human respiratory virus infections," *The American Journal of Respiratory and Critical Care Medicine*, vol. 152, no. 4, pp. S4–S12, 1995.

[7] G. M. Karaivanova, "Viral respiratory infections and their role as public health problem in tropical countries (review)," *African Journal of Medicine and Medical Sciences*, vol. 24, no. 1, pp. 1–7, 1995.

[8] Y. Harada, F. Kinoshita, L. M. Yoshida et al., "Does respiratory virus coinfection increases the clinical severity of acute respiratory infection among children infected with respiratory syncytial virus?" *The Pediatric Infectious Disease Journal*, vol. 32, no. 5, pp. 441–445, 2013.

[9] Y. Lu, S. Wang, L. Zhang et al., "Epidemiology of human respiratory viruses in children with acute respiratory tract infections in Jinan, China," *Clinical and Developmental Immunology*, vol. 2013, Article ID 210490, 8 pages, 2013.

[10] A.-C. Sentilhes, K. Choumlivong, O. Celhay et al., "Respiratory virus infections in hospitalized children and adults in Lao PDR," *Influenza and other Respiratory Viruses*, vol. 7, no. 6, pp. 1070–1078, 2013.

[11] C. Tecu, M. E. Mihai, V. I. Alexandrescu et al., "Single and multipathogen viral infections in hospitalized children with acute respiratory infections," *Roumanian Archives of Microbiology and Immunology*, vol. 72, no. 4, pp. 242–249, 2013.

[12] A. L. Drews, R. L. Atmar, W. P. Glezen, B. D. Baxter, P. A. Piedra, and S. B. Greenberg, "Dual respiratory virus infections," *Clinical Infectious Diseases*, vol. 25, no. 6, pp. 1421–1429, 1997.

[13] A. Franz, O. Adams, R. Willems et al., "Correlation of viral load of respiratory pathogens and co-infections with disease severity in children hospitalized for lower respiratory tract infection," *Journal of Clinical Virology*, vol. 48, no. 4, pp. 239–245, 2010.

[14] V. Foulongne, G. Guyon, M. Rodière, and M. Segondy, "Human metapneumovirus infection in young children hospitalized with respiratory tract disease," *Pediatric Infectious Disease Journal*, vol. 25, no. 4, pp. 354–359, 2006.

[15] M. G. Semple, A. Cowell, W. Dove et al., "Dual infection of infants by human metapneumovirus and human respiratory syncytial virus is strongly associated with severe bronchiolitis," *Journal of Infectious Diseases*, vol. 191, no. 3, pp. 382–386, 2005.

[16] E. T. Martin, J. Kuypers, A. Wald, and J. A. Englund, "Multiple versus single virus respiratory infections: viral load and clinical disease severity in hospitalized children," *Influenza and Other Respiratory Viruses*, vol. 6, no. 1, pp. 71–77, 2012.

[17] Y. Wang, W. Ji, Z. Chen, Y. D. Yan, X. Shao, and J. Xu, "Comparison of severe pneumonia caused by human metapneumovirus and respiratory syncytial virus in hospitalized children," *Indian Journal of Pathology and Microbiology*, vol. 57, no. 3, pp. 413–417, 2014.

[18] E. Karadag-Oncel, M. A. Ciblak, Y. Ozsurekci, S. Badur, and M. Ceyhan, "Viral etiology of influenza-like illnesses during the influenza season between December 2011 and April 2012," *Journal of Medical Virology*, vol. 86, no. 5, pp. 865–871, 2014.

[19] B. G. van den Hoogen, J. C. de Jong, J. Groen et al., "A newly discovered human pneumovirus isolated from young children with respiratory tract disease," *Nature Medicine*, vol. 7, no. 6, pp. 719–724, 2001.

[20] J. V. Williams, P. A. Harris, S. J. Tollefson et al., "Human metapneumovirus and lower respiratory tract disease," *The New England Journal of Medicine*, vol. 350, no. 5, pp. 443–450, 2004.

[21] K. Pyrc, B. Berkhout, and L. van der Hoek, "The novel human coronaviruses NL63 and HKU1," *Journal of Virology*, vol. 81, no. 7, pp. 3051–3057, 2007.

[22] L. van der Hoek, K. Pyrc, M. F. Jebbink et al., "Identification of a new human coronavirus," *Nature Medicine*, vol. 10, no. 4, pp. 368–373, 2004.

[23] T. Allander, T. Jartti, S. Gupta et al., "Human bocavirus and acute wheezing in children," *Clinical Infectious Diseases*, vol. 44, no. 7, pp. 904–910, 2007.

[24] T. Allander, M. T. Tammi, M. Eriksson, A. Bjerkner, A. Tiveljung-Lindell, and B. Andersson, "Cloning of a human parvovirus by molecular screening of respiratory tract samples," *Proceedings of the National Academy of Sciences of the United States of America*, vol. 102, no. 36, pp. 12891–12896, 2005.

[25] T. Allander, K. Andreasson, S. Gupta et al., "Identification of a third human polyomavirus," *Journal of Virology*, vol. 81, no. 8, pp. 4130–4136, 2007.

[26] S. Bialasiewicz, D. M. Whiley, S. B. Lambert, A. Gould, M. D. Nissen, and T. P. Sloots, "Development and evaluation of real-time PCR assays for the detection of the newly identified KI and WU polyomaviruses," *Journal of Clinical Virology*, vol. 40, no. 1, pp. 9–14, 2007.

[27] J. Ordás, J. A. Boga, M. Alvarez-Argüelles et al., "Role of metapneumovirus in viral respiratory infections in young children," *Journal of Clinical Microbiology*, vol. 44, no. 8, pp. 2739–2742, 2006.

[28] E. Moës, L. Vijgen, E. Keyaerts et al., "A novel pancoronavirus RT-PCR assay: frequent detection of human coronavirus NL63 in children hospitalized with respiratory tract infections in Belgium," *BMC infectious diseases*, vol. 5, article 6, 2005.

[29] J. Kuypers, N. Wright, J. Ferrenberg et al., "Comparison of real-time PCR assays with fluorescent-antibody assays for diagnosis of respiratory virus infections in children," *Journal of Clinical Microbiology*, vol. 44, no. 7, pp. 2382–2388, 2006.

[30] J. K. Kim, J.-S. Jeon, J. W. Kim, and I. Rheem, "Epidemiology of respiratory viral infection using multiplex RT-PCR in Cheonan, Korea (2006–2010)," *Journal of Microbiology and Biotechnology*, vol. 23, no. 2, pp. 267–273, 2013.

[31] A. Cantais, O. Mory, S. Pillet et al., "Epidemiology and microbiological investigations of community-acquired pneumonia in children admitted at the emergency department of a university hospital," *Journal of Clinical Virology*, vol. 60, no. 4, pp. 402–407, 2014.

[32] F. Canducci, M. Debiaggi, M. Sampaolo et al., "Two-year prospective study of single infections and co-infections by respiratory syncytial virus and viruses identified recently in infants with acute respiratory disease," *Journal of Medical Virology*, vol. 80, no. 4, pp. 716–723, 2008.

[33] P. Xepapadaki, S. Psarras, A. Bossios et al., "Human metapneumovirus as a causative agent of acute bronchiolitis in infants," *Journal of Clinical Virology*, vol. 30, no. 3, pp. 267–270, 2004.

[34] J. B. M. van Woensel, A. P. Bos, R. Lutter, J. W. A. Rossen, and R. Schuurman, "Absence of human metapneumovirus co-infection in cases of severe respiratory syncytial virus infection," *Pediatric Pulmonology*, vol. 41, no. 9, pp. 872–874, 2006.

[35] Y. Jin, R.-F. Zhang, Z.-P. Xie et al., "Newly identified respiratory viruses associated with acute lower respiratory tract infections in children in Lanzou, China, from 2006 to 2009," *Clinical Microbiology and Infection*, vol. 18, no. 1, pp. 74–80, 2012.

[36] J. Kuypers, E. T. Martin, J. Heugel, N. Wright, R. Morrow, and J. A. Englund, "Clinical disease in children associated with newly described coronavirus subtypes," *Pediatrics*, vol. 119, no. 1, pp. e70–e76, 2007.

[37] E. R. Gaunt, A. Hardie, E. C. J. Claas, P. Simmonds, and K. E. Templeton, "Epidemiology and clinical presentations of the four human coronaviruses 229E, HKU1, NL63, and OC43 detected over 3 years using a novel multiplex real-time PCR method," *Journal of Clinical Microbiology*, vol. 48, no. 8, pp. 2940–2947, 2010.

[38] Y. Abed, D. Wang, and G. Boivin, "WU polyomavirus in children, Canada," *Emerging Infectious Diseases*, vol. 13, no. 12, pp. 1939–1941, 2007.

[39] S. Bialasiewicz, D. M. Whiley, S. B. Lambert et al., "Presence of the newly discovered human polyomaviruses KI and WU in Australian patients with acute respiratory tract infection," *Journal of Clinical Virology*, vol. 41, no. 2, pp. 63–68, 2008.

[40] T. H. Han, J. Y. Chung, J. W. Koo, S. W. Kim, and E.-S. Hwang, "WU polyomavirus in children with acute lower respiratory tract infections, South Korea," *Emerging Infectious Diseases*, vol. 13, no. 11, pp. 1766–1768, 2007.

[41] B. M. Le, L. M. Demertzis, G. Wu et al., "Clinical and epidemiologic characterization of WU polyomavirus infection, St. Louis, Missouri," *Emerging Infectious Diseases*, vol. 13, no. 12, pp. 1936–1938, 2007.

[42] M. Debiaggi, F. Canducci, E. R. Ceresola, and M. Clementi, "The role of infections and coinfections with newly identified and emerging respiratory viruses in children," *Virology Journal*, vol. 9, article 247, 2012.

[43] W. L. Zhuang, X. D. Lu, G. Y. Lin et al., "WU polyomavirus infection among children in South China," *Journal of Medical Virology*, vol. 83, no. 8, pp. 1440–1445, 2011.

[44] S. Rao, R. L. Garcea, C. C. Robinson, and E. A. F. Simões, "WU and KI polyomavirus infections in pediatric hematology/oncology patients with acute respiratory tract illness," *Journal of Clinical Virology*, vol. 52, no. 1, pp. 28–32, 2011.

[45] A. Christensen, S. A. Nordbø, S. Krokstad, A. G. W. Rognlien, and H. Døllner, "Human bocavirus commonly involved in multiple viral airway infections," *Journal of Clinical Virology*, vol. 41, no. 1, pp. 34–37, 2008.

[46] O. Schildgen, A. Müller, T. Allander et al., "Human bocavirus: passenger or pathogen in acute respiratory tract infections?" *Clinical Microbiology Reviews*, vol. 21, no. 2, pp. 291–304, 2008.

[47] M. Söderlund-Venermo, A. Lahtinen, T. Jartti et al., "Clinical assessment and improved diagnosis of bocavirus-induced wheezing in children, Finland," *Emerging Infectious Diseases*, vol. 15, no. 9, pp. 1423–1430, 2009.

[48] T. Jartti, K. Hedman, L. Jartti, O. Ruuskanen, T. Allander, and M. Söderlund-Venermo, "Human bocavirus—the first 5 years," *Reviews in Medical Virology*, vol. 22, no. 1, pp. 46–64, 2012.

[49] M. de Paulis, A. E. Gilio, A. A. Ferraro et al., "Severity of viral coinfection in hospitalized infants with respiratory syncytial virus infection," *Jornal de Pediatria*, vol. 87, no. 4, pp. 307–313, 2011.

[50] J. H. Aberle, S. W. Aberle, E. Pracher, H. P. Hutter, M. Kundi, and T. Popow-Kraupp, "Single versus dual respiratory virus infections in hospitalized infants: impact on clinical course of disease and interferon-γ response," *Pediatric Infectious Disease Journal*, vol. 24, no. 7, pp. 605–610, 2005.

[51] K. M. Spann, K. C. Tran, B. Chi et al., "Suppression of the induction of alpha, beta, and lambda interferons by the NS1 and NS2 proteins of human respiratory syncytial virus in human epithelial cells and macrophages," *Journal of Virology*, vol. 78, no. 8, pp. 4363–4369, 2004.

14

Seroprevalence of Herpes Simplex Virus Infection in HIV Coinfected Individuals in Eastern India with Risk Factor Analysis

Soumyabrata Nag,[1] **Soma Sarkar,**[2] **Debprasad Chattopadhyay,**[3] **Sanjoy Bhattacharya,**[4] **Rahul Biswas,**[5] **and Manideepa SenGupta**[2]

[1]*Department of Microbiology, IIMSAR & BCRH, Haldia, West Bengal, India*
[2]*Department of Microbiology, Medical College Kolkata, 88 College Street, Kolkata, West Bengal, India*
[3]*ICMR Virus Unit, I.D. and B.G. Hospital, GB-4, 1st Floor, 57 Dr. S. C. Banerjee Road, Beliaghata, Kolkata, India*
[4]*Department of Medicine, Medical College Kolkata, 88 College Street, Kolkata, West Bengal, India*
[5]*Department of Community Medicine, A.I.I.H. & P.H., Kolkata, West Bengal, India*

Correspondence should be addressed to Manideepa SenGupta; manideepa.sengupta2305@gmail.com

Academic Editor: Jay C. Brown

Herpes simplex virus type 2 (HSV-2) is the cause of most genital herpes while HSV-1 is responsible for orolabial and facial lesions. In immunocompromised individuals, like HIV patients, impaired immunity leads to more frequent symptomatic and asymptomatic HSV infection. Fifty-two blood samples from HIV patients with clinically diagnosed HSV infection were taken as cases, while 45 blood samples each from HIV-infected (HIV control) and noninfected patients without any herpetic lesion (non-HIV control) were taken as control. Serum was tested for IgM and IgG antibodies of both HSV-1 and HSV-2 by ELISA. The seroprevalence was compared among the three groups of study population, considering the demographic and socioeconomic parameters. The HSV-2 IgM was significantly higher ($p < 0.005$) in the HIV patient group (34.6%) than the HIV control (2.2%) and non-HIV control (2.2%) groups, whereas HSV-2 IgG seroprevalence was higher in both HIV patient (61.5%) and HIV control (57.8%) groups than the non-HIV control group (17.8%). The prevalence of HSV-2 was significantly higher in persons with multiple partners and in the reproductive age group. The overall seroprevalence of HSV-1 IgM was too low (<5%), whereas it was too high (about 90%) with HSV-1 IgG in all three study groups.

1. Introduction

Most Herpes simplex virus type 1 (HSV-1) and type 2 (HSV-2) infections are subclinical. However, in symptomatic infections, the clinical manifestations are characterized by recurrent orolabial and facial lesions in HSV-1 and recurrent vesicular, ulcerative genital or anal lesions in HSV-2 [1–3].

HSV is a life-long infection and serological testing provides the best method to estimate its prevalence. Since 1976, the CDC has monitored the HSV-2 seroprevalence in the United States through the National Health and Nutrition Examination Survey (NHANES). Reports indicate that HSV-2 prevalence was increased to 31% between 1976 and 1980 (NHANES II) and was decreased to 21.0% in 1988–1994 (NHANES III) and 17.0% in 1999–2004. In 2005–2008 it was 16.2%, which was statistically same with the seroprevalence in 1999–2004 [4].

Classically, HSV-1 is acquired in childhood through contact, whereas HSV-2 is transmitted sexually. After initial infection, the virus persists for life in a latent form in neurons of the host, periodically reactivate to cause recurrent episodes. Daily suppressive therapy with acyclovir, famciclovir, and valacyclovir decreases HSV shedding dramatically and thereby decreases transmission along with decreased HIV viral loads. Vaccines, interleukins, interferons, therapeutic proteins, antibodies, immunomodulators, small-molecule drugs, and inhibitors of the HSV helicase-primase are in the developmental stages. It is increasingly evident that HSV-2 facilitates HIV transmission [5, 6] which strengthens the importance of the implementation of available HSV control methods [7–9]. The majority of HSV infections are asymptomatic or silent and thus, the infected people shedding the virus are potentially infectious. Therefore, seroepidemiological studies are critical to understand the pattern and distribution of infection within populations [9].

Till date, a limited amount of data on the HSV prevalence and its association with HIV infection are available in Eastern India, particularly in West Bengal. Hence, the aim of this study was to find out the prevalence of HSV infection in HIV patients attending the HIV Clinic of Medical College and Hospital, Kolkata. Specifically, we sought to know the prevalence of HSV-1 and HSV-2 antibodies (both IgM and IgG) in HIV patients with herpetic blister and/or ulcer (HIV group), compared to that in both HIV and non-HIV patients without any herpetic blister and/or ulcer (HIV control) and non-HIV control group. Moreover, the associations, if any, with various demographic, socioeconomic, and behavioral factors were correlated.

2. Methods

After obtaining the institutional ethical clearance, 52 blood samples were collected from patients of both sexes of 18–55 years of age attending the HIV Clinic of Medical College and Hospital, Kolkata, with oral or genital blisters (clinically diagnosed as Herpes simplex lesions) from April 2012 to March 2013. The HIV control group consisted of 45 blood samples, collected from age- and sex-matched HIV seropositive individuals of the clinic, while HIV seronegative blood collected from the Surgery and Gynaecology OPD served as a non-HIV control. Informed consent was obtained from each individual prior to collection of blood. The personal, demographic, and clinical data of all the patients were obtained by a pretest questionnaire containing name, age, sex, socioeconomic status, occupation, marital status, contact history, medical history, sexual behaviour, risk factors, knowledge of STDs (particularly HSV and HIV/AIDS), and clinical symptoms. Patients below 18 years or above 55 years of age, suffering from critical or deteriorating diseases, or with a history of receiving antiviral therapy aside from ART were excluded from the study.

Serum separated from blood samples collected by venipuncture was tested for HSV-1 and HSV-2 (IgG and IgM) antibodies, using commercial ELISA kits (SERION ELISA classic; Manufacturer Fabricant; Institut Virion/Serion GmbH, Germany) that distinguished the type-specific antibody response of both viruses. Microtitre plates of SERION ELISA classic HSV-1 and HSV-2 IgG were coated with recombinant glycoprotein gG1 or gG2, respectively. The use of envelope proteins gG1 in HSV-1 IgG and gG2 in HSV-2 IgG allowed differentiation of type-specific antibody response to HSV-1 and HSV-2. Microtitre plates of SERION ELISA classic HSV-1 IgM and HSV-2 IgM were coated with the corresponding whole virus antigen to ensure immediate and sensitive

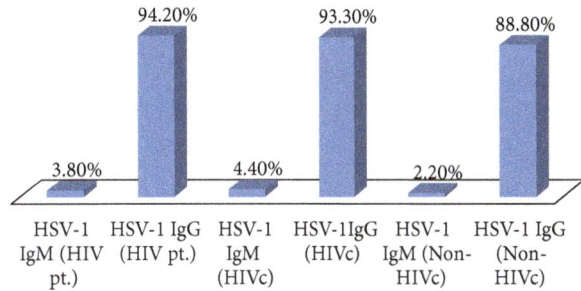

FIGURE 1: Prevalence of HSV-1 IgM and IgG in HIV pt., HIVc, and non-HIVc group.

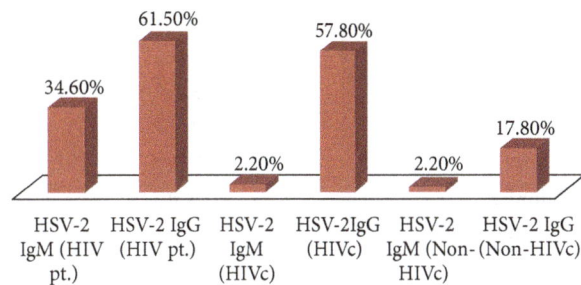

FIGURE 2: Prevalence of HSV-2 IgM and IgG in HIV pt., HIVc, and non-HIVc group.

detection of acute infections. All the tests were done according to the manufacturer's instructions. To fix the cut-off ranges, the mean absorbance value of the supplied standard serum (STD) was multiplied with the numerical data quality control provided by the manufacturer, for example, OD = 0.502 × MW (STD) with upper cut-off and OD = 0.352 × MW (STD) with lower cut-off.

If the measured mean absorbance value of the supplied standard serum is 0.64, the range of the cut-off is in between 0.225 and 0.321.

Statistical analysis was done by a statistician using standard statistical software (SPSS). Data were entered into Microsoft Excel (2007) and further exported to SPSS version 16.0 for analysis. Pearson's chi-square test was performed at 95% confidence interval and significant level was accepted at $p < 0.05$. p values were calculated to observe any statistically significant difference among the seroprevalence of HSV-1 and 2 antibodies (IgM and IgG) in different study groups.

3. Results

Most of the patients participating in the three study groups were males, between 26–45 years, married, literate, and belonged to the upper lower or lower middle class background while most HIV seropositive patients had multiple partners (Table 1). The seroprevalence of HSV-1 IgM was found to be very low (<5%), while HSV-1 IgG was very high (≈90%) in all three groups (Figures 1 and 2). This suggest a high prevalence of HSV-1 in the general population irrespective of age, sex, literacy, and socioeconomic status (Table 2).

TABLE 1: Distribution of study population according to various parameters.

Overall demographic profile studied	HIV patients with HSV blister/ulcer (HIV patient group) $n = 52$	HIV patients without HSV blister/ulcer (HIV control group) $n = 45$	Non-HIV patients without HSV blister/ulcer (non-HIV control group) $n = 45$
Age			
18–25 years	8 (15.38%)	8 (17.78%)	9 (20%)
26–35 years	14 (26.92%)	17 (37.78%)	8 (17.78%)
36–45 years	23 (44.23%)	15 (33.33%)	21 (46.67%)
46–55 years	7 (13.46%)	5 (11.11%)	7 (15.55%)
Gender			
Male	30 (57.7%)	23 (51.11%)	25 (55.56%)
Female	22 (42.3%)	22 (48.89%)	20 (44.44%)
Marital status			
Never married	2 (3.85%)	5 (11.11%)	11 (24.44%)
Married	44 (84.62%)	31 (68.89%)	30 (66.67%)
Separated	1 (1.92%)	1 (2.22%)	2 (4.44%)
Widowed	5 (9.61%)	8 (17.78%)	2 (4.44%)
Socioeconomic status			
Lower (L)	4 (7.69%)	3 (6.67%)	2 (4.44%)
Upper lower (UL)	30 (57.69%)	18 (40%)	15 (33.33%)
Lower middle (LM)	17 (32.69%)	18 (40%)	18 (40%)
Upper middle (UM)	1 (1.92%)	3 (6.67%)	7 (15.56%)
Upper (U)	0	3 (6.67%)	3 (6.67%)
Literacy level			
Illiterate (Ill)	7 (13.46%)	11 (24.44%)	8 (17.78%)
Up to primary (P)	29 (55.77%)	3 (6.67%)	4 (8.89%)
Up to middle (M)	9 (17.31%)	17 (37.78%)	10 (22.22%)
Secondary (S)	4 (7.69%)	10 (22.22%)	12 (26.67%)
HS and above (H)	3 (5.77%)	4 (8.89%)	11 (24.44%)
Number of partners			
0	0	9 (20%)	8 (17.78%)
1	28 (53.85%)	9 (20%)	31 (68.89%)
>1	24 (46.15%)	27 (60%)	6 (13.33%)

Different socioeconomic classes as per modified (for 2012) Prasad's Scale of socioeconomic status are based on per capita income in Rupees/month.
Lower = <585; upper lower = 585–1169; lower middle = 1170–1949; upper middle = 1950–3899; upper = ≥3900.

On the other hand, HSV-2 IgM seroprevalence was significantly higher (p value < 0.005 by χ^2 test) in the HIV patient group (34.6%) than the HIV control (2.2%) and non-HIV control (2.2%) group (Tables 3 and 4). In comparison to the non-HIV control group, sera from HIV patients were 23 times more reactive for HSV-2 IgM (odds ratio −23.294; 95% confidence interval 2.961–183.278). Further, the seroprevalence was found to be higher in males of 18–25 years' having more than one partner, literate, and in the upper lower socioeconomic class (Table 3).

HSV-2 IgG seroprevalence was higher in both HIV patient (61.53%) and HIV control (57.78%) groups than the non-HIV control group (17.78%). When compared with the non-HIV control group, the HIV patient group was 29 times (odds ratio −29.421; 95% confidence interval −6.331–136.720) and the HIV control group was 34 times (odds ratio −34.400;

95% confidence interval −7.495–157.895) more likely to be seropositive for HSV-2 IgG, significant at 5% level (p value < 0.005). However, HSV-2 IgG did not vary significantly among patients of different age groups, sex, socioeconomic strata, and literacy levels but varied significantly with the number of partners among the patients of the HIV-patient and the non-HIV control group (p value < 0.05).

4. Discussion

HSV infection is highly prevalent worldwide and varies between regions and populations.

In this study, it was found that the overall seroprevalence of HSV-2 IgG was 42.3%. While it was 59.79% in HIV-infected patients (61.53% in case and 57.7% in control), and only 17.78% in the non-HIV group. However, higher rates of

TABLE 2: Seroprevalence of HSV-1 IgM and HSV-1 IgG antibody.

	Number of pts.	HIV patient group				HIV control group				Non-HIV control group			
		HSV-1 IgM		HSV-1 IgG		HSV-1 IgM		HSV-1 IgG		HSV-1 IgM		HSV-1 IgG	
Overall	$n = 52$	2 (3.8%)	$n = 52$	49 (94.2%)	$n = 45$	2 (4.4%)	$n = 45$	42 (93.3%)	$n = 45$	1 (2.2%)		40 (88.9%)	
According to gender													
Male	$n = 30$	3.3%	$n = 30$	93.3%	$n = 23$	4.3%	$n = 23$	95.7%	$n = 25$	4%	$n = 25$	92%	
Female	$n = 22$	4.5%	$n = 22$	95.5%	$n = 22$	4.5%	$n = 22$	90.0%	$n = 20$	0	$n = 20$	85%	
According to age (in yrs)													
18–25	$n = 8$	12.5%	$n = 8$	100%	$n = 8$	0	$n = 8$	100%	$n = 9$	0	$n = 9$	100%	
26–35	$n = 14$	0	$n = 14$	100%	$n = 17$	5.9%	$n = 17$	94.1%	$n = 8$	0	$n = 8$	87.5%	
36–45	$n = 23$	4.3%	$n = 23$	87%	$n = 15$	6.7%	$n = 15$	86.7%	$n = 21$	4.8%	$n = 21$	85.7%	
46–55	$n = 7$	0	$n = 7$	100%	$n = 5$	0	$n = 5$	100%	$n = 7$	0	$n = 7$	85.7%	
According to number of partners													
0	$n = 0$	0	$n = 0$	0	$n = 9$	0	$n = 9$	100%	$n = 8$	0	$n = 8$	7.5%	
1	$n = 28$	0	$n = 28$	100%	$n = 9$	0	$n = 9$	7.7%	$n = 31$	3.3%	$n = 31$	90.3%	
>1	$n = 24$	8.3%	$n = 24$	87.5%	$n = 27$	7.4%	$n = 27$	9.6%	$n = 6$	0	$n = 6$	100%	
According to Income groups													
L	$n = 4$	25%	$n = 4$	75%	$n = 3$	0	$n = 3$	100%	$n = 2$	0	$n = 2$	100%	
UL	$n = 30$	3.3%	$n = 30$	93.3%	$n = 18$	5.5%	$n = 18$	100%	$n = 15$	0	$n = 15$	93.3%	
LM	$n = 17$	0	$n = 17$	100%	$n = 18$	5.5%	$n = 18$	88.9%	$n = 18$	5.5%	$n = 18$	88.9%	
UM	$n = 1$	0	$n = 1$	100%	$n = 3$	0	$n = 3$	100%	$n = 7$	0	$n = 7$	71.4%	
U	$n = 0$	0	$n = 0$	0	$n = 3$	0	$n = 3$	66.7%	$n = 3$	0	$n = 3$	100%	
According to literacy status													
Ill	$n = 7$	0	$n = 7$	100%	$n = 11$	9.1%	$n = 11$	81.8%	$n = 8$	0	$n = 8$	100%	
P	$n = 29$	3.4%	$n = 29$	93.1%	$n = 3$	33.3%	$n = 3$	100%	$n = 4$	0	$n = 4$	100%	
M	$n = 9$	11.1%	$n = 9$	100%	$n = 17$	0	$n = 17$	100%	$n = 10$	0	$n = 10$	90%	
S	$n = 4$	0	$n = 4$	75%	$n = 10$	0	$n = 10$	100%	$n = 12$	8.3%	$n = 12$	83.3%	
H	$n = 3$	0	$n = 3$	100%	$n = 4$	0	$n = 4$	75%	$n = 11$	0	$n = 11$	81.8%	

L, lower; UL, upper lower; LM, lower middle; UM, upper middle; U, upper; Ill, illiterates; P, primary; M, middle; S, secondary; H, HS and above.

coinfection with HIV and HSV-2 ranging from 62.7–100% [10–12], 88% and 91% [13] have been reported in the US, Haiti, and Central African Republic, respectively, which was similar to the control group of this study. Another study on hospitalized patients and blood donors in Germany revealed overall 12.8% HSV-2 seropositivity, including 15% females and 10.5% males, but the prevalence in non-HIV control group was 17.78% (20% in males and 15% in females) [14].

There are several possible biological mechanisms where HSV-2 acts as a cofactor in HIV acquisition or transmission. First, the HSV-2 reactivations result in mucosal or epithelial disruption, creating a portal of exit or entry for HIV, to which the activated CD4 cells are recruited [14]. There also appear to be several cellular interactions that promote the establishment of HIV infection and its coinfection with HSV-2 which may lead to the creation of "pseudotypes" (i.e., HSV-2 particles containing the HIV genome enveloped with HSV surface glycoprotein). This allows HSV to infect the cells that could not be infected by HIV alone [11]. The HSV-2 infection may also promote the increased expression of the HIV target cells (i.e., the CCR5+ CD4 cells and the immature dendritic cells) [12].

In our study, the seroprevalence of HSV-1 IgG was found to be very high in all the study groups (overall 92.3%), showing a good correlation with the German study, where the prevalence of HSV-I antibodies showed a steady increase with age and reached high levels (88%) among patients aged 40 years or older [15]. In the German study, the seropositivity of HSV-1 (91.1%) and HSV-2 (47.9%) in HIV-infected populations supports our observation of 93.81% and 59.79% in the present study. However, the higher seropositivity of HSV-2 in males in this study was probably due to limited sample size. Higher prevalence of HSV-1 antibodies (73.3%) among 168 HIV-antibody negative and 132 HIV-antibody positive men, with no difference between HIV seronegative and seropositive men (p value = 0.48), while about 20% of HIV seronegative and 61% of seropositive men showed antibodies to HSV-2 ($p < 0.0001$). Similarly, 83.5% and 63.4% seroprevalence of HSV-1 and HSV-2 among patients at higher risk for HIV reported by Lupi [17], is similar to the findings in this study.

The present cross-sectional study on seroprevalence of HSV-2 corroborated the prospective observational study of Patel et al. [18]. Similar results on the seroprevalence of HSV-2 in adult HIV-infected patients and blood donors were also

TABLE 3: Seroprevalence of HSV-2 IgM and HSV-2 IgG antibody.

	Number of pts.	HIV patient group				HIV control group				Non-HIV control group			
		HSV-2 IgM		HSV-2 IgG		HSV-2 IgM		HSV-2 IgG		HSV-2 IgM		HSV-2 IgG	
Overall	$n = 52$	18 (34.6%)	$n = 52$	32 (61.5%)	$n = 45$	1 (2.2%)	$n = 45$	26 (57.8%)	$n = 45$	1 (2.2%)	$n = 45$	8 (17.8%)	
According to gender													
Male	$n = 30$	40%	$n = 30$	63.3%	$n = 23$	0	$n = 23$	60.9%	$n = 25$	4%	$n = 25$	20%	
Female	$n = 22$	27%	$n = 22$	59.1%	$n = 22$	4.5%	$n = 22$	54.5%	$n = 20$	0	$n = 20$	15%	
According to age (in yrs)													
18–25	$n = 8$	25%	$n = 8$	62.5%	$n = 8$	0	$n = 8$	75%	$n = 9$	11.1%	$n = 9$	11.1%	
26–35	$n = 14$	35.7%	$n = 14$	85.7%	$n = 17$	5.9%	$n = 17$	47.1%	$n = 8$	0	$n = 8$	12.5%	
36–45	$n = 23$	34.8%	$n = 23$	52.2%	$n = 15$	0	$n = 15$	60%	$n = 21$	0	$n = 21$	23.8%	
46–55	$n = 7$	42.9%	$n = 7$	42.9%	$n = 5$	0	$n = 5$	60%	$n = 7$	0	$n = 7$	14.3%	
According to number of partners													
0	$n = 0$	0	$n = 0$	0	$n = 9$	0	$n = 9$	55.5%	$n = 8$	0	$n = 8$	0	
1	$n = 28$	35.7%	$n = 28$	42.8%	$n = 9$	0	$n = 9$	55.5%	$n = 31$	0	$n = 31$	12.9%	
>1	$n = 24$	33.3%	$n = 24$	83.3%	$n = 27$	3.7%	$n = 27$	59.2%	$n = 6$	16.6%	$n = 6$	66.7%	
According to income groups													
L	$n = 4$	25%	$n = 4$	75%	$n = 3$	0	$n = 3$	66.7%	$n = 2$	0	$n = 2$	0	
UL	$n = 30$	43.3%	$n = 30$	63.3%	$n = 18$	0	$n = 18$	61.1%	$n = 15$	0	$n = 15$	26.7%	
LM	$n = 17$	23.5%	$n = 17$	58.8%	$n = 18$	0	$n = 18$	55.5%	$n = 18$	5.5%	$n = 18$	16.7%	
UM	$n = 1$	0	$n = 1$	0	$n = 3$	33.3%	$n = 3$	0	$n = 7$	0	$n = 7$	14.2%	
U	$n = 0$	0	$n = 0$	0	$n = 3$	0	$n = 3$	100%	$n = 3$	0	$n = 3$	0	
According to literacy status													
Ill	$n = 7$	28.5%	$n = 7$	71.4%	$n = 11$	0	$n = 11$	45.4%	$n = 8$	0	$n = 8$	25%	
P	$n = 29$	41.3%	$n = 29$	55.1%	$n = 3$	0	$n = 3$	66.7%	$n = 4$	0	$n = 4$	25%	
M	$n = 9$	11.1%	$n = 9$	66.7%	$n = 17$	0	$n = 17$	58.7%	$n = 10$	0	$n = 10$	20%	
S	$n = 4$	50%	$n = 4$	100%	$n = 10$	10%	$n = 10$	70%	$n = 12$	8.3%	$n = 12$	16.6%	
H	$n = 3$	33.3%	$n = 3$	33.3%	$n = 4$	0	$n = 4$	50%	$n = 11$	0	$n = 11$	9.1%	

L, lower; UL, upper lower; LM, lower middle; UM, upper middle; U, upper; Ill, illiterates; P, primary; M, middle; S, secondary; H, HS and above.

TABLE 4: Seroprevalence of HSV-1 and 2 antibodies (IgM and IgG) in different study groups.

	HIV patient group			HIV control group			Non-HIV control group			Total			Chi square test (p value)
	R	NR	T	R	NR	T	R	NR	T	R	NR	T	
HSV1 IgM	2 (3.8%)	50 (96.2%)	52 (100%)	2 (4.4%)	43 (95.6%)	45 (100%)	1 (2.2%)	44 (97.8%)	45 (100%)	5 (3.5%)	137 (96.5%)	142 (100%)	0.838 (not statistically significant)
HSV1 IgG	49 (94.2%)	3 (5.8%)	52 (100%)	42 (93.3%)	3 (6.7%)	45 (100%)	40 (88.9%)	5 (11.1%)	45 (100%)	131 (92.3%)	11 (7.7%)	142 (100%)	0.585 (not statistically significant)
HSV2 IgM	18 (34.6%)	34 (65.4%)	52 (100%)	1 (2.2%)	44 (97.8%)	45 (100%)	1 (2.2%)	44 (97.8%)	45 (100%)	20 (14.1%)	122 (85.9%)	142 (100%)	0.000 (statistically significant)
HSV2 IgG	32 (61.5%)	20 (38.5%)	52 (100%)	26 (57.8%)	19 (42.2%)	45 (100%)	8 (17.78%)	37 (82.22%)	45 (100%)	66 (46.48%)	76 (53.52%)	142 (100%)	0.000 (statistically significant)

reported by Rode et al. [19] from Croatia. Agabi et al. [20] reported a very high prevalence of HSV-2 (87%) among the patients attending STD Clinic in Jos, Nigeria, which was significantly higher than the present finding, probably due to the differences in sexual behavior and higher prevalence of HIV in Nigeria. Moreover, in this study, genital Herpes (genital blisters: 13, genital ulcers: 35) obtained from 62.3% HSV-2 seropositive HIV subjects indicates that about 37.7% of HIV patients were unaware of their HSV-2 infection.

Strengths and Limitations

(1) Limited amounts of data are available on the seroprevalence of HSV and its association with HIV infection in Eastern India. Hence, this study was conducted to find out the prevalence of HSV-1 and HSV-2 antibodies (both IgM and IgG) in an HIV patient group, compared with the seroprevalence in an HIV control and non-HIV control group(s) with their demographic, socioeconomic, and behavioural factors.

(2) The increased number of HSV seropositivity among HIV positive samples corroborate the fact that there is a synergistic relationship between HIV and HSV infection.

(3) Promoting awareness on HSV, its silent epidemic potential, and the role of HSV-2 treatment to decrease HIV transmission and disease progression may have substantial public health benefits.

However, small sample size was the limitation of the study.

5. Conclusions

The HSV seropositivity was found to be higher in HIV positive patient samples (HSV-2 and HSV-1 were 59.79 and 93.81%) when compared to non-HIV population (HSV-2 and HSV-1 were 17.78 and 88.88%). Thus, it was found that HSV-2 was more common in HIV-infected than in non-HIV-infected individuals. The increased number of HSV seropositivity among HIV positive samples indicates that there is a synergistic relationship between HIV and HSV infection. Moreover, genital herpes (blisters: 13, ulcers: 35) presented by 62.3% of the HSV-2 seropositive HIV subjects indicates that 37.7% of HIV patients were unaware of their HSV-2 infection, suggesting that the awareness of HSV-2 treatment to decrease HIV transmission and disease progression may have substantial public health benefits.

References

[1] L. Corey, "Herpes simplex virus," in *Principles of Harrison's Internal Medicine*, A. S. Fauci, E. Braunwald, D. L. Kasper, and S. L. Hauser, Eds., vol. 1, pp. 1048–1056, McGraw-Hill, New York, NY, USA, 17th edition, 2008.

[2] J. I. Cohen, J. T. Schiffer, and L. Corey, "Introduction to herpesviridiae: herpes simplex virus," in *Principles and Practice of Infectious Disease*, G. L. Mandell, J. E. Bennett, and R. Dolin, Eds., pp. 1937–1962, Churchill Livingstone, Philadelphia, Pa, USA, 7th edition, 2010.

[3] D. W. Kimberlin and D. J. Rouse, "Clinical practice: genital herpes," *The New England Journal of Medicine*, vol. 350, no. 19, pp. 1970–1977, 2004.

[4] F. Xu, M. R. Sternberg, B. J. Kottiri et al., "Trends in herpes simplex virus type 1 and type 2 seroprevalence in the United States," *The Journal of the American Medical Association*, vol. 296, no. 8, pp. 964–973, 2006.

[5] J. N. Wasserheit, "Epidemiological synergy: interrelationships between human immunodeficiency virus infection and other sexually transmitted diseases," *Sexually Transmitted Diseases*, vol. 19, no. 2, pp. 61–77, 1992.

[6] A. Wald and K. Link, "Risk of human immunodeficiency virus infection in herpes simplex virus type 2-seropositive persons: a meta-analysis," *Journal of Infectious Diseases*, vol. 185, no. 1, pp. 45–52, 2002.

[7] L. Corey and H. H. Handsfield, "Genital herpes and public health—addressing a global problem," *The Journal of the American Medical Association*, vol. 283, no. 6, pp. 791–794, 2000.

[8] WHO and UNAIDS, "Herpes simplex virus type 2: programmatic and research priorities in developing countries," Report of a WHO/UNAIDS/LSHTM Workshop HIV_AIDS/2001.05, WHO, UNAIDS, London, UK, 2001.

[9] F. M. Cowan, R. S. French, P. Mayaud et al., "Seroepidemiological study of herpes simplex virus types 1 and 2 in Brazil, Estonia, India, Morocco, and Sri Lanka," *Sexually Transmitted Infections*, vol. 79, no. 4, pp. 286–290, 2003.

[10] S. Safrin, R. Ashley, C. Houlihan, P. S. Cusick, and J. Mills, "Clinical and serologic features of herpes simplex virus infection in patients with AIDS," *AIDS*, vol. 5, no. 9, pp. 1107–1110, 1991.

[11] E. W. Hook III, R. O. Cannon, A. J. Nahmias et al., "Herpes simplex virus infection as a risk factor for human immunodeficiency virus infection in heterosexuals," *Journal of Infectious Diseases*, vol. 165, no. 2, pp. 251–255, 1992.

[12] R. Boulos, A. J. Ruff, A. Nahmias et al., "Herpes simplex virus type 2 infection, syphilis, and hepatitis B virus infection in haitian women with human immunodeficiency virus type 1 and human T lymphotropic virus type I infections," *Journal of Infectious Diseases*, vol. 166, no. 2, pp. 418–420, 1992.

[13] F.-X. Mbopi-Kéou, G. Grésenguet, P. Mayaud et al., "Interactions between herpes simplex virus type 2 and human immunodeficiency virus type 1 infection in African women: opportunities for intervention," *Journal of Infectious Diseases*, vol. 182, no. 4, pp. 1090–1096, 2000.

[14] B. A. S. Shameem, S. Lakshmi, K. Kaveri, and S. Jayakumar, "Correlation of serology, tissue culture and PCR in identification of herpes simplex type-2 infection among HIV patients," *Journal of Clinical and Diagnostic Research*, vol. 5, no. 6, pp. 1190–1194, 2012.

[15] P. Wutzler, H. W. Doerr, I. Färber et al., "Seroprevalence of herpes simplex virus type 1 and type 2 in selected German populations - Relevance for the incidence of genital herpes," *Journal of Medical Virology*, vol. 61, no. 2, pp. 201–207, 2000.

[16] D. B. Russell, S. N. Tabrizi, J. M. Russell, and S. M. Garland, "Seroprevalence of herpes simplex virus types 1 and 2 in HIV-Infected and uninfected homosexual men in a primary care

setting," *Journal of Clinical Virology*, vol. 22, no. 3, pp. 305–313, 2001.

[17] O. Lupi, "Prevalence and risk factors for herpes simplex infection among patients at high risk for HIV infection in Brazil," *International Journal of Dermatology*, vol. 50, no. 6, pp. 709–713, 2011.

[18] P. Patel, T. Bush, K. H. Mayer et al., "Prevalence and risk factors associated with herpes simplex virus-2 infection in a contemporary cohort of HIV-infected persons in the united states," *Sexually Transmitted Diseases*, vol. 39, no. 2, pp. 154–160, 2012.

[19] O. D. Rode, S. Ž. Lepej, and J. Begovac, "Seroprevalence of herpes simplex virus type 2 in adult HIV-infected patients and blood donors in Croatia," *Collegium Antropologicum*, vol. 32, no. 3, pp. 693–695, 2008.

[20] Y. A. Agabi, E. B. Banwat, J. D. Mawak et al., "Seroprevalence of herpes simplex virus type-2 among patients attending the sexually transmitted infections clinic in Jos, Nigeria," *Journal of Infection in Developing Countries*, vol. 4, no. 9, pp. 572–575, 2010.

15

Transmitted Drug Resistance among People Living with HIV/Aids at Major Cities of Sao Paulo State, Brazil

Joao Leandro Paula Ferreira,[1] Rosangela Rodrigues,[1] Andre Minhoto Lança,[1] Valeria Correia de Almeida,[2] Simone Queiroz Rocha,[3] Taisa Grotta Ragazzo,[2] Denise Lotufo Estevam,[3] and Luis Fernando de Macedo Brigido[1]

[1] Laboratório de Retrovírus, Centro de Virologia, Instituto Adolfo Lutz, Avenue Dr. Arnaldo 355, 01246-902 São Paulo, SP, Brazil
[2] Centro de Referência em DST/Aids, 13013-051 Campinas, SP, Brazil
[3] Centro de Referência e Treinamento em DST/Aids, 04121-000 São Paulo, SP, Brazil

Correspondence should be addressed to Luis Fernando de Macedo Brigido; lubrigido@gmail.com

Academic Editor: Michael Bukrinsky

Human immunodeficiency virus type 1 (HIV-1) transmitted drug resistance (TDR) is an important public health issue. In Brazil, low to intermediate resistance levels have been described. We assessed 225 HIV-1 infected, antiretroviral naïve individuals, from HIV Reference Centers at two major metropolitan areas of Sao Paulo (Sao Paulo and Campinas), the state that concentrates most of the Brazilian Aids cases. TDR was analyzed by Stanford Calibrated Population Resistance criteria (CPR), and mutations were observed in 17 individuals (7.6%, 95% CI: 4.5%–11.9%). Seventy-six percent of genomes (13/17) with TDR carried a nonnucleoside reverse transcriptase inhibitor (NNRTI) resistance mutation, mostly K103N/S (9/13, 69%), potentially compromising the preferential first-line therapy suggested by the Brazilian HIV Treatment Guideline that recommends efavirenz-based combinations. Moreover, 6/17 (35%) had multiple mutations associated with resistance to one or more classes. HIV-1 B was the prevalent subtype (80%); other subtypes include HIV-1 F and C, mosaics BC, BF, and single cases of subtype A1 and CRF02_AG. The HIV Reference Center of Campinas presented more cases with TDR, with a significant association of TDR with clade B infection ($P < 0.05$).

1. Introduction

Access to free antiretroviral therapy (ART) is part of the Brazilian response to the Aids epidemic and transmitted drug resistance (TDR); it has been a concern since the introduction of highly active antiretroviral therapy (HAART) in the late 1990s [1]. TDR surveillance is an important strategy to monitor the emergence of genetic resistance as it may impact ART efficacy [2]. This issue was especially sensible in Brazil that deployed a free ARV program in the late 90s amidst a suboptimal health care system. This initiative could boost the emergence of transmitted drug resistance variants and jeopardize Human immunodeficiency virus (HIV-1) treatment [3]. However, most studies in Brazil have shown TDR prevalence similar to that observed among developed countries. Two recent Brazilian national surveys had accessed this issue [4, 5] but included a small representation of São Paulo metropolitan areas. We and others have analyzed mutations in treatment-naive individuals [6–12]; but to trace trends for TDR prevalence, continual monitoring is necessary.

We analyzed ARV naïve individuals living with HIV/Aids, recruited at HIV Reference Centers from the two major metropolitan areas of Sao Paulo state to investigate the TDR prevalence. These metropolitan areas concentrate 44% of notified Aids cases of Sao Paulo state and about one third of the Brazilians living with HIV/Aids.

2. Materials and Methods

People living with HIV, asymptomatic and naïve to ART were recruited at outpatient Clinic or Voluntary Counseling and Testing (VCT) at two metropolitan areas, the HIV State Reference Center at the city of Sao Paulo and the Municipal HIV Reference Center at the city of Campinas. Volunteers' selection was conducted by their primary physicians' or by VCT counseling personnel, with additional revision to confirm ARV exposure history. Individuals that agreed to participate were interviewed by clinical staff to access risk, review of potential previous exposures to ART (e.g., MTCT or postexposure prophylaxis), and document the knowledge of a partnership (sexual or sharing of drug paraphernalia) with individuals using ARV. Blood samples were collected from May 2008 to November 2009. Briefly, HIV-1 RNA was extracted with QIAamp viral RNA mini kit (Qiagen, Germany) and reverse transcribed with random primers and Superscript III enzyme (Invitrogen, USA). In samples of low HIV-1 RNA viral load or negative plasma detection, DNA from peripheral blood mononuclear cells (PBMCs) was extracted (Qiagen, Germany). Nested PCR products were sequenced using Big Dye terminators at an ABI 3100 Genetic Analyzer (ABI, USA) to evaluate protease (PR, codons 1 to 99) and partial reverse transcriptase (RT, codons 1 to 235) genes as previously described [6]. Sequences were manually edited using Sequencher 4.7 software (Gene Codes, USA). Ambiguous DNA bases (mixtures) at resistance associated codons were considered at sequence edition. HIV genotyping resistance test results were reported to the HIV Reference Centers. TDR was defined according to the Calibrated Population Resistance Version 6.0 (CPR, Stanford Database, SDRM 2009), an algorithm specifically designed for the epidemiologic surveillance of HIV-1 transmitted drug resistance mutations (DRMs) [13]. International Antiviral Society (IAS) 2011 resistance list [14] was additionally considered to evaluate the impact of resistance in ART response, which considers all mutations that impact ARV susceptibility. To contribute to HIV molecular epidemiology surveillance, HIV-1 subtyping was performed at NCBI genotyping and REGA HIV subtyping tools and confirmed by phylogenetic methods (PAUP* 4.10b), using evolution model selected by ModelTest3.7. Sequences are available at GenBank with accession numbers: HM533970 to HM534205; HQ015155 to HQ015157.

The Institutional Review Boards of the participating HIV Reference Centers and Adolfo Lutz Institute approved this study.

3. Results

Of the 243 HIV-1 infected individuals enrolled in the study, partial HIV-1 *pol* sequences from 230 (95%) individuals were successfully sequenced (96% from plasma and 4% from PBMC). The inclusion criterion of no previous exposure to ART was met by 225 individuals and included in the analysis. Among the study individuals, six females had previous exposure to MTCT prophylaxis, documented at the interview. Most of the sequences analyzed from these women (3/4) exhibited one or more DRM but were not included in TDR prevalence estimate that was generated from the 225 ART naive individuals. Also, one case with unknown information of exposure to ART, but with history of undetectable viremia in previous years, had several resistance mutations and was

excluded from analysis. Unprotected exposure among men who have sex with men (MSM) was the most frequent transmission route. Table 1 depicts demographic and laboratorial data from study volunteers at Campinas and São Paulo sites.

According to Stanford CPR, TDR mutations were detected in 17 sequences (7.6%, 95% CI: 4.5%–11.9%). The prevalence of TDR at Campinas was 9.6% (15/156) and at Sao Paulo 2.9% was (2/69) ($P = 0.13$). The two individuals from Sao Paulo site had more extensive TDR patterns, both with mutations to protease inhibitor (PI) and NNRTI and one with additional resistance mutations to nucleoside reverse transcriptase inhibitor (NRTI) (Table 2). At Campinas site, most TDR were NNRTI-resistance mutations (73%, 11/15), followed by PI and NRTI (20%, 3/15 each), and 2 sequences were resistant to two ARV classes. Overall, 76% (13/17) of sequences bearing at least one TDR had an NNRTI resistance, 71% (12/17) impairing susceptibility to efavirenz (EFV). K103N/S (53%, 9/17) was the most frequent mutation observed. No association of TDR to transmission risk group or gender was observed. Also, individuals referring sexual partner on ART have a similar number of TDR ($P = 0.8$). Campinas tended to have more cases with TDR, with a significant association of TDR to HIV-1 B infection using either CPR or IAS list ($P < 0.05$).

To evaluate the impact of mutations or polymorphisms associated to resistance, we additionally used the 2011 IAS updated mutation list, and 32 sequences (14.2%) had at least one IAS major mutation (Table 2). Both IAS and CPR criteria were used to assess the relevance of mutations on successful therapy in patients where, according the Brazilian treatment guidelines, ART was recommended. The genotyping test reports were available to the HIV Reference Centers, and the ART was selected taking into account the test report. Virological responses of patients with TDR were assessed until 2012 (Table 2). Out of the 33 individuals with drug resistance mutation, 18 initiated ART, and 16 had clinical and laboratory information, some cases with followup of 37 months. This followup showed that all but one individual (BR09CA175), who had documented adherence issues, were virally suppressed at last observation (Table 2).

4. Discussion

Among this population of antiretroviral naïve HIV-1-infected individuals attending the Campinas and São Paulo Reference Centers, transmitted drug resistance was detected overall in 7.6% of individuals. This TDR prevalence was similar to that reported in other regions of Brazil that used Stanford CPR criteria for these estimates, [4–12, 15]. The reasons for this stabilization, or even decrease of TDR prevalence in most surveys are unclear, but there may well be a plateau where the circulation of mutated isolates may come to some equilibrium, depending on multiple factors as the ARV therapy usage, therapy combinations, adherence, and social networking, among others. The followup of the small but consistent increase in TDR in Africa after the introduction of treatment programs [2] will allow verifying this hypothesis.

TABLE 1: Demographic and laboratorial characteristics of the study individuals according to transmitted drug resistance (TDR).

	TDR ($n = 33$)	Without TDR ($n = 192$)	Total ($n = 225$)
Age (year old)	32 (29–40)	34 (29–40)	34 (29–40)
Gender (male)	23 (69.7%)	138 (71.9%)	161 (71.6%)
CD4+ T cells (cells/mm^3)	461 (322–631)	475 (353–623)	463 (344–626)
Viral load (Log$_{10}$)	4.1 (3.6–4.8)	4.3 (3.6–4.7)	4.3 (3.6–4.7)
HIV exposure			
MSM	20 (61%)	100 (51%)	120 (52%)
Heterosexual	11 (33%)	76 (41%)	87 (40%)
IDU	1 (33%)	1 (0.5%)	2 (0.4%)
HIV-1 subtype (partial *pol*)			
A1	0	1 (0.5%)	1 (0.4%)
B	31 (94%)	149 (78%)	180 (76%)
F	0	14 (7%)	14 (6%)
C	1 (33%)	15 (8%)	16 (7%)
CRF02_AG	0	1 (0.5%)	1 (0.4%)
Recombinant mosaic	1 (33%)	12 (6%)	13 (6%)
Surveillance site			
Campinas	26 (79%)	130 (68%)	156 (69%)
São Paulo	7 (21%)	62 (32%)	69 (31%)

In this study, most individuals had high CD4+ T cell counts at collection, with a median of 463 cells/mm^3. However, 25% had CD4+ T cell counts below 350 cells/mm^3, CD4 counts that were indicative of ARV treatment by Brazilian Guidelines at the time. Currently, the recommendation is starting therapy when CD4+ T cells counts drops below 500 cells/mm^3. Although not the focus of this study, a small number of women exposed to MTCT prevention were recruited but excluded in TDR estimates. These women had significant resistance profile and constitute a population segment that should have access to pretreatment genotype test.

In Brazil, over 200,000 patients are currently using ART, with 80,000 in São Paulo state. However, some of these patients are not virally suppressed, representing a potential source of transmitted drug resistance. On the other hand, a large number of untreated individuals harbor wild-type variants (estimated to be at least twice the number of treated individuals), another source of HIV transmission that may play an important role in the low prevalence of documented transmitted resistance, consistent across most studies. Additionally, low fitness or other biological limitations in the transmissibility potential of these variants might also have a role in the observed prevalence of resistant strains. More important than point estimates of transmitted resistance would be to trace its tendency in time. Trends in TDR based on sequential assess of sentinel sites are probably one of the best ways to monitor TDR. The Brazilian AIDS Program have conducted a large TDR survey [4], but differences in methodology, design, and participating sites at a previous study [8] limit the comparability of these evaluations. Other recent national study [5], conducted almost at the same time, indicate similar numbers, with estimates at 5–15%, applying the HIV Threshold Survey methodology from WHO. For

sites in the Sao Paulo state, the estimates generated at these studies ranged from 0 to 15% [4, 5]. However, these studies had small samples size at sites in the state, ranging from 7 to 34 individuals, which decreased the strength of these to estimates. These differences probably also reflect issues as sampling effects, study design, stringency in the confirmation of ART naïve status, and criteria for defining TDR. Additionally, one of these studies [4] found an association of partnership of individuals using ART with the presence of TDR. This issue was also evaluated here, and the number of TDR was similar among cases reporting a partner using ART.

As expected for the subtype prevalence at this region of Brazil, HIV-1 subtype B was the most common in both sites. However, the proportions of HIV-1 C and BC mosaics were higher than previously reported in the area. This is further detailed elsewhere [16]. Also, a single CRF02_AG was identified, and although it is the most disseminated HIV recombinant form worldwide, it is rather uncommon at South America, with a report of its presence in Brazil [17]. In this study, we found an association within Campinas sequences of transmitted resistance to subtype B ($P = 0.05$) using either CPR or IAS criteria, in agreement with Sprinz et al. [4], where TDR was predominantly found among subtype B infected individuals.

The impact on treatment efficacy is the central problem of TDR. The high prevalence of NNRTIs resistance among cases with TDR is a potential problem as the Brazilian HIV Treatment Guideline recommends NNRTI, especially efavirenz-based HAART, as a preferential first-line therapy. Considering the low genetic barrier of NNRTI, which a single DRM may confer high level of resistance to the entire class, these patients may actually be initiating ARV with a functional dual therapy, leading to a partial viral suppression, a favoring environment for the emergence of

TABLE 2: Followup information from individuals with *TDR* associated with ARV resistance.

ID[a]	HIV exposure[b]	Age[c]/gender[d]	ART regimen[e]	CD4+ T cells (cells/mm³) Before ART[f]	Last test[g]	Viral load (Log10) Before ART[h]	Last test[i]	Time to last VL/CD4 test[j]	Resistance mutations (CPR)[k]	Resistance mutations (IAS)[l]	Subtype[m]
BR08SP441	MSM	25/M	TDF/3TC/EFV	312	311	4.40	4.78	N.A		E138A	B
BR09SP005	MSM	41/M	AZT/3TC/LPV	388	1025	4.77	<1.70	13 months		E138A	B
BR09CA065	MSM	37/M	TDF/3TC/EFV	248	641	4.43	<1.70	8 months		V82I	B
BR09CA067	MSM	38/M	AZT/3TC/EFV	244	817	4.58	<1.70	31 months		V82I	B
BR09CA071	MSM	29/M	—	487	562	3.58	2.61	—	K103N	K103N	B
BR09CA075	WSM	66/F	AZT/3TC/EFV	224	378	5.70	<1.70	18 months		E138A	B
BR09CA091	MSM	30/M	TDF/3TC/LPV/r	260	384	4.94	<1.70	26 months	K103N	K103N	B
BR09CA097	WSM	40/F	—	567	567	4.18	4.18	—	G190A	G190A	B
BR09CA175	MSM	29/M	AZT/3TC/ATV/r	283	164	3.76	3.58	No supression	K103N	K103N	B
BR09CA190	WSM	32/F	AZT/3TC/EFV	207	609	4.47	<1.70	37 months		E138A	B
BR09CA192	MSM	29/M	—	751	686	3.74	4.25	—		K103N	B
BR09CA194	IDU+MSM	34/M	N.A.	230	230	—	<1.70	N.A.		Q58E	B
BR09CA210	MSM	22/M	—	741	636	5.19	4.90	—	M41L, T215E	M41L	B
BR09CA262	IGN	29/F	N.A.	217	209	5.45	5.24	—		V82I/E138A	C
BR09CA264	MSM	39/M	AZT/3TC/EFV	243	629	4.78	<1.70	37 months		V82I	B
BR09CA271	MSM	29/M	TDF/3TC/EFV	197	557	4.50	<1.70	26 months	M41L	M41L	B
BR09CA296	MSM	26/M	TDF/3TC/LPV/r	221	723	3.95	<1.70	29 months	K103N	K103N	B
BR09CA298	MSW	26/M	—	731	668	4.00	4.65	—	M230L	M230L	B
BR09CA300	MSM	21/M	—	1062	729	3.29	3.20	—		V82I	B
BR09SP310	MSM	27/M	TDF/3TC/RAL	108	955	5.55	<1.70	15 months		E138A	B
BR09CA343	MSM	34/M	AZT/3TC/EFV	335	388	4.41	<1.70	35 months		E138A	B
BR09CA344	WSM	39/F	—	381	454	3.81	4.42	—	K101P, K103S	K101P, K103S	B
BR09CA355	WSM	53/F	TDF/3TC/EFV	318	787	3.52	<1.70	27 months	L90M	L90M	B
BR09CA369	MSW	58/M	TDF/3TC/EFV	348	348	4.76	—	N.A	E138A	E138A	BF
BR09SP588	MSM	33/M	—	517	441	4.49	4.07	—	M46I, L90M/K101P, G190A; M46I, G73S, I84V,	M46I, Q58E, L90M/K101P, G190A	B
BR09SP003	pARV+MSM	41/M	TDF/3TC/DRV/r/RAL	312	490	4.86	<1.70	3 months	L90M/M41L, V75M, F77L, T215D, Y188L	M46I, I54V, I84V, L90M/M41L, F77L, Y188L	B
BR09CA078	pARV+WSM	40/F	AZT/3TC/EFV	297	492	3.57	<1.70	14 months		A72V	B
BR09CA095	pARV+MSM	32/M	—	631	658	2.68	3.07	—	K103N, L90M	K103N, L90M	B
BR09CA187	pARV+WSM	25/F	—	656	592	3.93	4.24	—		V82I	B
BR09CA280	pARV	28/M	—	632	395	3.61	4.19	—	K103N	K103N	B

TABLE 2: Continued.

ID[a]	HIV exposure[b]	Age[c]/gender[d]	ART regimen[e]	CD4+ T cells (cells/mm^3)		Viral load (Log$_{10}$)		Time to last VL/CD4 test[j]	Resistance mutations (CPR)[k]	Resistance mutations (IAS)[l]	Subtype[m]
				Before ART[f]	Last test[g]	Before ART[h]	Last test[i]				
BR09CA372	pARV+MSM	29/M	TDF/3TC/ATV/r	273	**774**	4.73	**<1.70**	37 months	K103N	K103N	B
BR09CA197	pARV+WSM	47/F	—	N.A.	N.A.	N.A.	N.A.	—	T215S, I85V		B
BR09SP253	pARV+WSM	47/F	—	1142	1334	2.06	2.70			V82I	B

[a] Samples ID (BR from Brazil, 09 for year of collection, and CA or SP for clinical site and samples number).

[b] Exposure as MSM men that refers to sex with other men, WSM for women that refers to sex with men, IDU for intravenous drug use, and pARV for patients referring to sex with one or more partners using antiretroviral therapy.

[c] Age in years, [d] gender as M for male, and F for female.

[e] ART treatment as first regimen or no treatment.

[f] CD4+ T cells prior to treatment start (pre-ART) and [g] last CD4 available (bold for CD4 values higher than pre-ART, when ARV was taken);

[h] Viral load prior to ART (before ART) and [i] last VL available. (bold for undetectable viral copies/mL, <1.70 log$_{10}$, when ARV was taken).

[j] Time since treatment initiation until last CD4/VL determination in months.

[k,l] Resistance mutations observed (Stanford CPR and 2011 IAS list).

[m] Viral subtype at *pol* region.

N.A. for data not available.

further resistance mutations. Moreover, no decrease in viral fitness is expected for mutations to this class, and they tend to persist longer and be more readily detectable than others that interfere with viral fitness, such as M184V. Persistence as a detectable mutation in population sequencing does not mean the same as persistence throughout the viral quasispecies, and it is conceivable that mutations detected in population sequencing may be considered "sentinel mutations." These could indicate the presence of additional resistance mutations, below detection limits of this sequencing approach, which could interfere with treatment success [18]. If those detectable mutations signs that other occult mutations exist, one would expect treatment compromise even if physicians have access to pretreatment genotype. Although the number of cases is small, our study documented relatively long-term treatment outcome. As the physicians received the genotype test results, cases initiating HAART after the test result had a therapy influenced by the genotyping test. In most cases a combination therapy was chosen to include drugs not predicted to be affected by the observed DRM. This would have circumvented the direct effect of the detected mutation but not of the occult, additional resistances supposedly present in minority variants. However, the presence of TDR in these individuals did not have an impact in viremia control, and the only treated case without suppression had documented adherence issues.

5. Conclusion

We observed low to intermediate levels of transmitted drug resistance mutations (7.6%), with most of them impairing susceptibility to efavirenz, a preferential ARV in first-line therapy at Brazil, confirming previous findings in the country. It is important both to monitor transmitted resistance trends and to define algorithms that might subsidize treatment alternatives where access to genotype test prior to therapy initiation to all individuals may be unrealistic. On the other hand, targeting genotypic test to population segments most susceptible to TDR may be cost effective. This would have both the potential benefit in HIV treatment response and an impact in reducing the drug resistance transmission of to the overall population.

Acknowledgments

This work was supported by FAPESP, Grant nos. 2006/61311-0 and 2011/21958-2. The authors thank Andrade RB, Siqueira AFAC, Batista, JG, and Silva AJ for their contribution. They are grateful to the volunteers and the staff at the HIV Reference Centers involved in the study. The authors thank the São Paulo HIV Salvage Workgroup: Almeida RAMB, Vazquez CMP, Ferreira DM, Jamal L, Silva IO, Braga PE, Pereira LC, and Hornke L.

References

[1] M. Petrella, B. Brenner, H. Loemba, and M. A. Wainberg, "HIV drug resistance and implications for the introduction of antiretroviral therapy in resource-poor countries," *Drug Resistance Updates*, vol. 4, no. 6, pp. 339–346, 2001.

[2] WHO HIV Drug Resistance Report, 2012, http://apps.who.int/iris/bitstream/10665/75183/1/9789241503938_eng.pdf.

[3] M. Wadman, "Experts clash over likely impact of cheap AIDS drugs in africa," *Nature*, vol. 410, no. 6829, pp. 615–616, 2001.

[4] E. Sprinz, E. M. Netto, M. Patelli et al., "Primary antiretroviral drug resistance among HIV type 1-infected individuals in Brazil," *AIDS Research and Human Retroviruses*, vol. 25, pp. 861–867, 2009.

[5] L. A. Inocencio, A. A. Pereira, M. C. Sucupira et al., "Brazilian Network for HIV Drug Resistance Surveillance: a survey of individuals recently diagnosed with HIV," *Journal of the International AIDS Society*, vol. 12, p. 20, 2009.

[6] J. L. De Paula Ferreira, M. Thomaz, R. Rodrigues et al., "Molecular characterisation of newly identified HIV-1 infections in Curitiba, Brazil: preponderance of clade C among males with recent infections," *Memorias do Instituto Oswaldo Cruz*, vol. 103, no. 8, pp. 800–808, 2008.

[7] R. Rodrigues, L. C. Scherer, C. M. Oliveira et al., "Low prevalence of primary antiretroviral resistance mutations and predominance of HIV-1 clade C at polymerase gene in newly diagnosed individuals from south Brazil," *Virus Research*, vol. 116, no. 1-2, pp. 201–207, 2006.

[8] R. M. Brindeiro, R. S. Diaz, E. C. Sabino et al., "Brazilian Network for HIV Drug Resistance Surveillance (HIV-BResNet): a survey of chronically infected individuals," *AIDS*, vol. 17, no. 7, pp. 1063–1069, 2003.

[9] B. C. Carvalho, L. P. V. Cardoso, S. Damasceno, and M. M. A. Stefani, "Moderate prevalence of transmitted drug resistan and interiorization of HIV type 1 subtype C in the Inland North State of Tocantins, Brazil," *AIDS Research and Human Retroviruses*, vol. 27, pp. 1081–1087, 2011.

[10] E. Arruda, L. Simões, C. Sucupira et al., "Short communication: intermediate prevalence of HIV type 1 primary antiretroviral resistance in Ceará State, Northeast Brazil," *AIDS Research and Human Retroviruses*, vol. 27, no. 2, pp. 153–156, 2011.

[11] R. M. de Medeiros, D. M. Junqueira, M. C. Matte, N. T. Barcellos, J. A. Chies, and S. E. M. Almeida, "Co-circulation HIV-1 subtypes B, C, and CRF31_BC in a drug-naïve population from Southernmost Brazil: analysis of primary resistance mutations," *Journal of Medical Virology*, vol. 83, pp. 1682–1688, 2001.

[12] T. Gräf, C. P. Passaes, L. G. Ferreira et al., "HIV-1 genetic diversity and drug resistance among treatment naïve patients from Southern Brazil: an association of HIV-1 subtypes with exposure categories," *Journal of Clinical Virology*, vol. 51, pp. 186–191, 2011.

[13] R. W. Shafer, S. Y. Rhee, D. Pillay et al., "HIV-1 protease and reverse transcriptase mutations for drug resistance surveillance," *AIDS*, vol. 21, no. 2, pp. 215–223, 2007.

[14] V. A. Johnson, V. Calvez, H. F. Günthard et al., "2011 update of the drug resistance mutations in HIV-1," *Topics in Antiviral Medicine*, vol. 19, pp. 156–164, 2011.

[15] A. M. Lança, J. K. B. Colares, J. L. P. Ferreira et al., "HIV-1 tropism and CD4 T lymphocyte recovery in a prospective cohort of patients initiating HAART in Ribeirão Preto, Brazil," *Memórias do Instituto Oswaldo Cruz*, vol. 107, pp. 96–101, 2012.

[16] L. F. M. Brígido, J. L. P. Ferreira, V. C. Almeida et al., "Southern Brazil HIV type 1 C expansion into the State of São Paulo, Brazil," *AIDS Research and Human Retroviruses*, vol. 27, pp. 339–344, 2011.

[17] W. A. Eyer-Silva and M. G. Morgado, "Autochthonous horizontal transmission of a CRF02_AG strain revealed by a human immunodeficiency virus type 1 diversity survey in a small city in inner state of Rio de Janeiro, Southeast Brazil," *Memorias do Instituto Oswaldo Cruz*, vol. 102, no. 7, pp. 809–815, 2007.

[18] M. Pingen, M. Nijhuis, J. A. de Bruijn, C. A. B. Boucher, and A. M. J. Wensing, "Evolutionary pathways of transmitted drug-resistant HIV-1," *Journal of Antimicrobial Chemotherapy*, vol. 66, no. 7, pp. 1467–1480, 2011.

16

Viral Etiology of Chronic Obstructive Pulmonary Disease Exacerbations during the A/H1N1pdm09 Pandemic and Postpandemic Period

Ivan Sanz,[1,2] Sonia Tamames,[3] Silvia Rojo,[1,2] Mar Justel,[2] José Eugenio Lozano,[3] Carlos Disdier,[4] Tomás Vega,[3] and Raúl Ortiz de Lejarazu[1,2]

[1]*Valladolid National Influenza Centre, Avenida Ramón y Cajal No. 7, 47005 Valladolid, Spain*
[2]*Microbiology and Immunology Service, University Clinic Hospital of Valladolid, Avenida Ramón y Cajal s/n, 47005 Valladolid, Spain*
[3]*Consejería de Sanidad, Junta de Castilla y León, Paseo de Zorrilla No. 1, 47007 Valladolid, Spain*
[4]*Pulmonology Service, University Clinic Hospital of Valladolid, Avenida Ramón y Cajal s/n, 47005 Valladolid, Spain*

Correspondence should be addressed to Ivan Sanz; isanzm@saludcastillayleon.es

Academic Editor: R. C. Gallo

Viral infections are one of the main causes of acute exacerbations of chronic obstructive pulmonary disease (AE-COPD). Emergence of A/H1N1pdm influenza virus in the 2009 pandemic changed the viral etiology of exacerbations that were reported before the pandemic. The aim of this study was to describe the etiology of respiratory viruses in 195 Spanish patients affected by AE-COPD from the pandemic until the 2011-12 influenza epidemic. During the study period (2009–2012), respiratory viruses were identified in 48.7% of samples, and the proportion of viral detections in AE-COPD was higher in patients aged 30–64 years than ≥65 years. Influenza A viruses were the pathogens most often detected during the pandemic and the following two influenza epidemics in contradistinction to human rhino/enteroviruses that were the main viruses causing AE-COPD before the pandemic. The probability of influenza virus detection was 2.78-fold higher in patients who are 30–64 years old than those ≥65. Most respiratory samples were obtained during the pandemic, but the influenza detection rate was higher during the 2011-12 epidemic. There is a need for more accurate AE-COPD diagnosis, emphasizing the role of respiratory viruses. Furthermore, diagnosis requires increased attention to patient age and the characteristics of each influenza epidemic.

1. Introduction

Chronic obstructive pulmonary disease (COPD) is a slowly progressive and largely irreversible clinical condition characterized by airflow limitation [1]. In Spain, COPD affects over 10% of the population between 40 and 80 years of age [2, 3]. Acute exacerbations of COPD (AE-COPD) play a crucial role in the course of the disease, having a negative impact on morbidity, mortality, healthcare costs, and health-related quality of life [4, 5]. Patients with moderate and severe COPD are prone to exacerbations, and the frequency of these episodes increases with the severity of disease [6]. One of the key points in COPD management programs is prevention and treatment of exacerbations [7]. Results of follow-up studies show that patients who suffer a high number of exacerbations during a given period of time will continue to suffer frequent exacerbations in the future [8]. Therefore, the frequency of exacerbations will depend on the patient's underlying severity of lung disease and number of prior exacerbations [9].

The etiology of AE-COPD is diverse. Most AE-COPD cases are attributed to bacterial or viral respiratory infections [10] and to both types of microorganisms together [11]. However to a minor extent, exacerbations are also associated with pollution, tobacco consumption, temperature changes, allergens, and other comorbidities such as heart failure and pulmonary thromboembolism [8, 12]. Respiratory viral infections have been associated with more frequent and severe AE-COPD and also with longer recovery times than episodes caused by other factors including bacteria [13, 14]. Studies conducted before emergence of the pandemic H1N1pdm09 strain showed that half of all AE-COPD cases were associated with viral infections and that picornaviruses (especially human rhinovirus and enterovirus (HREV)) were the dominant viral pathogens diagnosed in these patients [15, 16]. HREVs are the main viruses responsible for the common cold, with high prevalence throughout the whole year and without an established epidemic circulation period. Currently they are the most important trigger of COPD exacerbations [17]. Related with that, exacerbations treated with antibiotics could lead to the emergence of resistances in cases with other etiologies, which constitutes a problem particularly in southern Mediterranean European countries where antibiotics are widely used for these kinds of patients [18]. Improvement of clinical diagnosis and correct identification of respiratory viruses may help reduce the use of these antibiotics. It is important to find clinical and analytical parameters to guide identification of the etiology of new AE-COPD cases, especially considering the laborious techniques currently used for diagnosis [19].

In 2009 the world experienced the first pandemic of influenza A virus in 40 years, and this pathogen is now known as the H1N1pdm09 pandemic virus [20]. During 2009-10, this virus spread worldwide, causing high infection rates but low mortality compared with previous pandemics (Spanish Flu, Asian Flu, and Hong Kong Flu) [21]. The pandemic resulted in a high rate of screening for respiratory viruses in patients with respiratory clinical manifestations, including those with COPD. During this pandemic and following influenza epidemics, Valladolid National Influenza Centre (Valladolid NIC, Spain) and the Microbiology & Immunology Service of Clinic University Hospital of Valladolid worked closely on several topics related to influenza A/H1N1pdm09 [22, 23]. Consequently, we received a large number of respiratory samples from the Hospital Network of Castile and León (2.5 million habitants) and currently we serve as a reference center for viral diagnostics for influenza-suspect cases and for other respiratory pathologies, including AE-COPD.

The aim of this study is to describe the etiological characteristics of respiratory viruses linked to COPD exacerbations after a singular pandemic period caused by a new influenza virus. We have placed special emphasis on the differences of viral etiology of AE-COPD between the pandemic and the following postpandemic period.

2. Materials and Methods

2.1. Study Design. This retrospective observational study was done at the Microbiology & Immunology Service of the Clinic University Hospital of Valladolid, Valladolid, Spain. Respiratory samples from 195 AE-COPD patients hospitalized in Castile and León Hospital Network (Spain) were sent to the microbiology laboratory for viral molecular diagnosis between October 2009 and September 2012. This work was exempt from Ethical Committee approval and from the need for informed consent following Spanish laws regarding the use of routine clinical samples in research studies.

TABLE 1: Inclusive study periods.

Period	Description	Duration
PAN	2009 pandemic	Week 35, 2009–week 32, 2010
INEP1	Interepidemic 2010	Week 33, 2010–week 39, 2010
FLUEP1	Influenza epidemic 2010-11	Week 40, 2010–week 20, 2011
INEP2	Interepidemic 2011	Week 21, 2011–week 39, 2011
FLUEP2	Influenza epidemic 2011-12	Week 40, 2011–week 20, 2012
INEP3	Interepidemic 2012	Week 21, 2012–week 39, 2012

2.2. Case Definition. Sample recruitment was done reviewing microbiology laboratory order slips. Only cases in which the microbiology order slip specifically showed COPD exacerbation causing respiratory sample submission were included. This inclusion criterion was established by the chief pulmonologist of our hospital network following the Anthonisen criteria [24] in which one of the following symptoms were present: cough, dyspnea, or sputum increasing in volume or purulence. Before the sample was included in the study, the clinical chart of each of the potentially included patients was checked for the presence of at least one of the cited symptoms. Demographic data such as age and sex were also obtained from the order slips. Clinical samples from AE-COPD patients without Anthonisen symptoms or who lacked clinical or demographic information were excluded from this study. Because COPD is a disease that affects only adults [3], only patients ≥30 years old were included in this work.

2.3. Epidemic Information. The infection rates and prevalence of the different respiratory viruses diagnosed in this work were obtained from local epidemiological surveillance data. This free information is provided weekly by the public health authorities through the Influenza Sentinel Surveillance Network (ISSN) of Castile and León. We defined six study periods (Table 1) following the World Health Organization guidelines [25].

2.4. Viral Analysis. Viral detection was done by means of a set of molecular diagnostic techniques implemented in the lab routine during the 2009 pandemic. Briefly, genetic material was extracted from the respiratory samples by using an *EasyMag* (*Biomerieux, Craponne, France*) automatic extractor, and the eluted final volume of 50 μL was used for multiple molecular diagnostic assays. Primary screening was by multiplex real-time polymerase chain reaction (RT-PCR) for 17 different respiratory viral targets (influenza virus A/H3, A/H1, A/H1N1pdm09 and B; respiratory syncytial viruses A and B (RSV A and RSV B); HREV; coronavirus OC43, 229E, HKU1, and NL63; metapneumovirus; parainfluenza 1, 2, 3, and 4; adenovirus; and bocavirus) using *Luminex 200* platform (Luminex, Austin, TX, USA) and *Respiratory Viral Panel-XTAG RVP* (Abbott, Chicago, IL, USA). Influenza A viruses not subtyped by this technique were identified by means of real-time RT-PCR *Roche 2.0* platform (Roche, Basel, Switzerland) using *Influenza A/H1N1 Detection Set*

TABLE 2: Number and percentage of positives, negatives, gender distribution, average age, and pathogens affecting AE-COPD patients during the entire period and in each separate influenza period included in the study.

	Whole period studied	Pandemic	INEP1 2010	FLUEP1 2010-11	INEP2 2011	FLUEP2 2011-12	INEP3 2012
AE-COPD cases	195	124	0	40	4	25	2
Mean Age (SD)	63.9 (13.1)	62.7 (13.1)	0 (0)	61.1 (14.2)	66.5 (2.1)	67.9 (12.6)	69.5 (2.1)
Males (%)	136 (69.7)	85 (68.6)	0 (0)	25 (62.5)	2 (50)	22 (88.0)	2 (100)
Negatives (%)	100 (51.3)	65 (52.4)	0 (0)	20 (50)	4 (100)	9 (36.0)	2 (100)
Positives (%)	95 (48.7)	59 (47.6)	0 (0)	20 (50)	0 (0)	16 (64.0)	0 (0)
Pathogen most represented	H1N1pdm09	H1N1pdm09	N/A	H1N1pdm09/RSV	N/A	H3N2	N/A
H1N1pdm09 (%)	41 (21.0)	35 (28.2)	0 (0)	6 (15.0)	0 (0)	0 (0)	0 (0)
H3N2 (%)	11 (5.6)	0 (0)	0 (0)	0 (0)	0 (0)	11 (44.0)	0 (0)
Influenza B (%)	1 (0.5)	0 (0)	0 (0)	1 (2.5)	0 (0)	0 (0)	0 (0)
HREV (%)	24 (12.3)	14 (11.3)	0 (0)	5 (12.5)	0 (0)	5 (20.0)	0 (0)
RSV (%)	13 (6.7)	4 (3.2)	0 (0)	6 (15.0)	0 (0)	3 (12.0)	0 (0)
ORP (%)	12 (6.2)	6 (4.8)	0 (0)	3 (7.5)	0 (0)	3 (12.0)	0 (0)
Coinfections (%)	5 (2.6)	0 (0)	0 (0)	1 (2.5)	0 (0)	4 (16.0)	0 (0)

N/A: data not available; HREV: human rhino/enterovirus; ORP: other respiratory pathogens; I: influenza interepidemic period; E: influenza epidemic period.

(Roche) for the specific detection of A/H1N1pdm09 virus. Also, to specifically characterize 16 haemagglutinin and 9 neuraminidase types of non subtypable influenza A viruses, we used *Clondiag Array Mate* and *Influenza A Genotyping* reagents (Alere, Waltham, MA, USA). Influenza B lineages Victoria and Yamagata were identified using a real-time RT-PCR as previously described [26].

2.5. Data Analysis. This study included data from two groups of patients: those who were 30 to 64 years old and those ≥65 years. A descriptive analysis was conducted by calculating the appropriate summary measures for quantitative and qualitative variables. Means and standard deviations were calculated for continuous variables. Associations from basic clinical data and frequency of viral infections were analyzed by Student's t-test adjusted by age with 95% confidence interval ($\alpha = 0.01$). Detection probability of the different respiratory pathogens involved in this work was calculated using odds ratio (OR) adjusted by different demographic and epidemiological characteristics such as sex, age, and the influenza circulation periods included in the study. OR was analyzed using 95% confidence interval (CI95%) and $\alpha = 0.05\%$. The statistical package employed was SPSS 19.0.

3. Results

Of the 195 patients included in the study, 94 (48.2%) were between the ages of 30 and 64 years and 101 (51.8%) were ≥65 years. The average age of all patients was 63.9 ± 13.1 years old and 136 were males (69.7%). From September 2009 until September 2012, respiratory viruses were diagnoses in 95 AE-COPD samples (48.7%), and no pathogen was detected in 100 samples (51.3%). The most frequently detected respiratory virus during this period was influenza A/H1N1pdm09, present in 41 cases (21.0%), followed by HREV ($n = 24$; 12.3%), RSV ($n = 13$; 6.7%), and H3N2 influenza virus

($n = 11$; 5.6%). Twelve samples (6.2%) were positive for other respiratory pathogens included in the molecular diagnostic assays. Influenza B virus was detected in only one AE-COPD patient (0.5%), and 5 cases (2.6%) of coinfection were also detected.

Viral diagnostic findings in AE-COPD patients decreased with age. Viruses were found in 61 samples (62.2%) from patients in the 30–64-year age group and in 41 patients (38.0%) in the ≥65-year age group (Figure 1). There was no age difference between AE-COPD patients with viral infection (62.1 ± 1.5 years) and those that tested negative by molecular diagnostics (65.6 ± 1.4 years, $p = 0.11$). Influenza A/H1N1pdm09 was the most often detected virus in the 30–64-year age group ($n = 30$; 30.6%). On the other hand, most of the viruses had prevalences similar to one another in the ≥65-year age group: A/H1N1pdm09, 10.2%; HREV, 9.3%; other respiratory pathogens, 7.4%. Thus the absence of positive diagnostics was more common in this age group.

The highest number of AE-COPD episodes ($n = 124$; 63.6%) was recorded for the pandemic, followed by FLUEP1 ($n = 40$; 20.5%) and FLUEP2 ($n = 25$; 12.8%) (Table 2). Only 6 AE-COPD episodes occurred during any of the interepidemic periods, and none of them were diagnosed with respiratory viral infection. The average age of AE-COPD patients during the pandemic, 62.7 ± 13.1 years (Table 2), was not significantly different from the patient ages in the first or second epidemic. The proportion of viral findings in AE-COPD patients increased from the pandemic ($n = 59$; 47.6%) until the end of the study period (Figure 2). The maximum proportion occurred during FLUEP2 ($n = 16$; 64.0%) despite the low number of patients recruited in this period. In contrast, for the general Spanish population, the maximum incidence of influenza occurred during the pandemic, 221.7 cases/100,000 inhabitants, rather than in the following two influenza epidemics, 195.9 cases/100,000 habitants during FLUE1 and 200.3 cases/100,000 habitants during FLUEP2 (Figure 3).

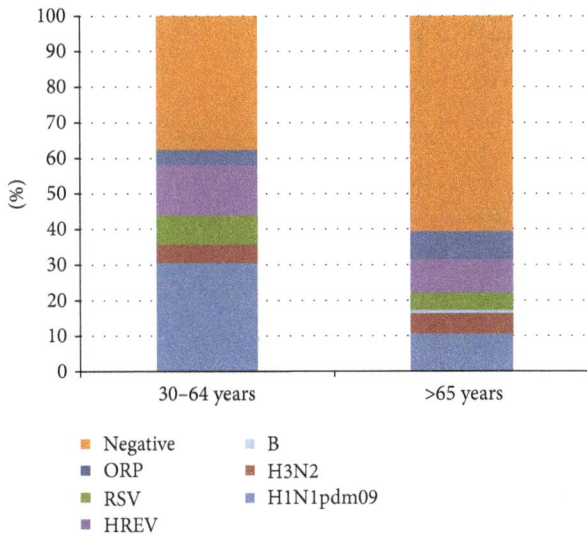

FIGURE 1: Cumulative percentage of respiratory virus prevalence causing AE-COPD in adults aged 30–64 years and elderly patients aged ≥65 years. The presence of viruses in AE-COPD declined with the age of individuals in the study. ORP: other respiratory pathogens; HREV: human rhino-enterovirus.

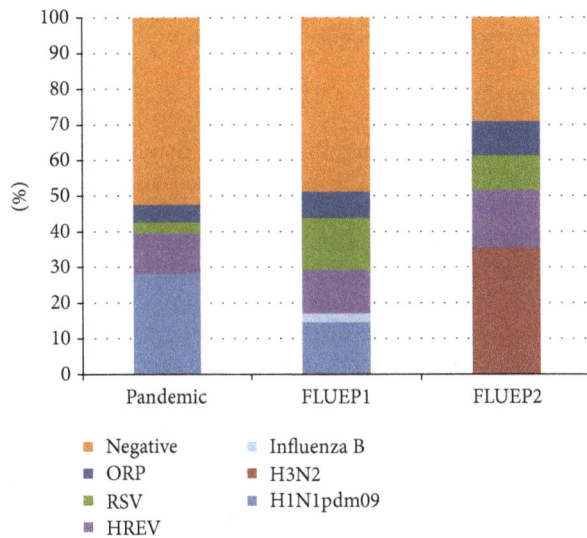

FIGURE 2: Cumulative percentage distribution of respiratory viral pathogens causing AE-COPD episodes during the 2009 pandemic and influenza epidemics described in the study. ORP: other respiratory pathogens; HREV: human rhino-enterovirus.

Influenza viruses were the most detected respiratory pathogens in AE-COPD patients during the pandemic and following epidemics. Specifically, influenza A/H1N1pdm09 was detected in 35 cases (28.2%) during the pandemic and in 6 cases (15.0%) during FLUEP1, while H3N2 influenza virus was detected in 11 cases (44.0%) during FLUEP2. Indeed, these two influenza strains represent together 54.7% of viruses diagnosed in AE-COPD episodes in this study. Meanwhile, HREV and RSV were the second and third most diagnosed viruses during the pandemic (n = 14; 11.3%

and n = 4; 3.2%, resp.). However, diagnosis of these two viruses constantly increased in the two following influenza epidemics. Also other respiratory pathogens increased in prevalence in AE-COPD cases from the pandemic (n = 6; 4.8%) until FLUEP2 (n = 3; 12.0%). Coinfections were more commonly detected during FLUEP2.

We used the OR to analyze the probability of detection of viral categories (ORP, HREV, any influenza virus, and RSV) as well as viral coinfections in AE-COPD patients among different demographic and epidemiological characteristics such as gender, age groups, and the different periods analyzed. There was no gender difference in the rate of detection of respiratory viral or coinfections. The OR for detecting influenza in the 30–64-year age group was 2.78-fold (CI95% = 1.44–5.38) greater than for the ≥65-year old group (p = 0.002). The probability for detecting RSV in the pandemic was significantly lower than detecting it in the first epidemic (OR = 0.19; CI95% = 0.05–0.71; p = 0.013). Additionally, the probability for detecting influenza virus in the first epidemic was significantly lower than in the second epidemic (OR = 0.27, CI95% = 0.09–0.84; p = 0.024). Finally, the probability for detecting coinfection during the pandemic was significantly lower than in the second epidemic (OR = 0.02; IC95% = 0.01–0.37; p = 0.009).

4. Discussion

Accurate detection of the causes of AE-COPD is important to develop and improve specific therapies and health care for patients that suffer this disease. In this way, clinicians and microbiology laboratories can be better prepared for the constant emergence of new respiratory pathogens such as avian influenza viruses and MERS-coronavirus. Our study has revealed differences between the 2009 pandemic and the following two influenza epidemics and other differences in the etiology dynamic of AE-COPD described in the scientific literature before 2009. Even though most of the AE-COPD respiratory samples were acquired during the pandemic, the viral etiology increased from 47.6% in the pandemic to 64.0% in the FLUEP2. Respiratory viruses affecting AE-COPD episodes have been communicated in epidemics prior to 2009, ranging from 40 to 60% in previous publications [27, 28]. Also in several studies, viruses were associated with higher frequencies of AE-COPD than bacterial infection or air pollution [29]. Furthermore, viral infections serve as causes of secondary bacterial infections that are associated with a rapid decline and severe respiratory symptoms [11, 30, 31]. Our findings are consistent with the global relevance of viruses in AE-COPD as previously described [11, 15, 16]. This suggests that, in addition to the independence of the viral epidemiologic characteristics, there exists a balance between bacterial and viral infections which promotes these exacerbations.

The distribution of AE-COPD patients within the 30–64 and the ≥65-year age groups was similar (48.2 and 51.8%, resp.). Within these age groups, there was a decrease of respiratory viruses in the AE-COPD episodes with increasing age. Thus viral infections were the etiology in 62.2% of

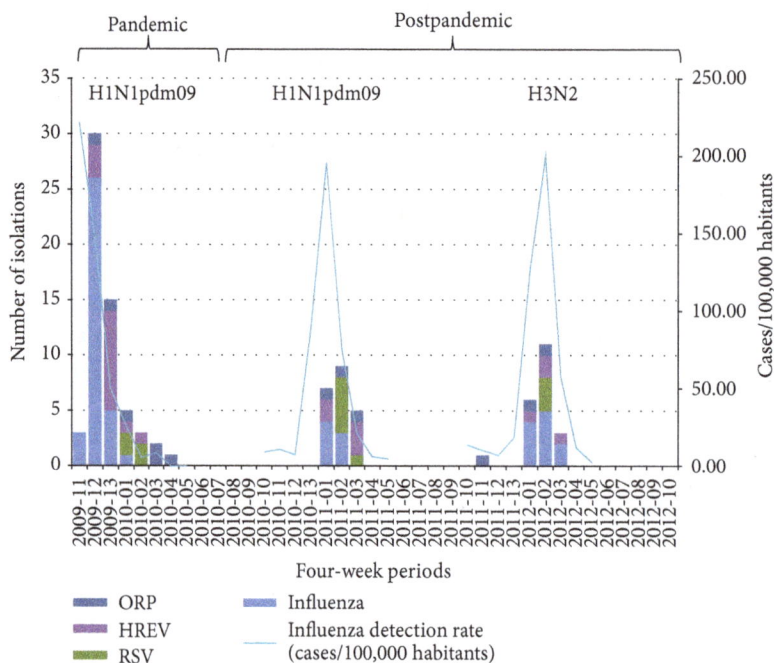

FIGURE 3: Isolations of respiratory viruses causing AE-COPD episodes and influenza detection rate during the pandemic and the following epidemics included in the study. For the Y-axis, the cumulative columns refer to the number of isolations and the lines refer to cases per 100,000 habitants. For the X-axis, the timeline represents merged four-week periods beginning with the first week of each year. The influenza viruses most often detected are shown at the top, corresponding with the pandemic and following postpandemic period.

AE-COPD episodes in the 30–64-year age group, while accounting for only 38.0% of AE-COPD cases in ≥65-year-old population. These data reveal a clear decrease in the role of respiratory viruses in AE-COPD episodes as the susceptible population ages. Thus, it is likely that bacterial infections or environmental conditions are the most frequent causes of exacerbations in elderly people. Despite the fact that age seems to be a factor for detection of respiratory viruses in AE-COPD episodes, we did not find any differences in the average age between people with respiratory infection and those with negative viral diagnostics. On the other hand, the average age of patients was also similar among the pandemic and influenza epidemics studied.

The virus most frequently detected in AE-COPD patients during the study period was influenza A/H1N1pdm09, followed by HREV and RSV. Analyzing each period, influenza A/H1N1pdm09 was the most frequent pathogen detected in the pandemic and during FLUEP1. RSV was also diagnosed in the same proportion as influenza A/H1N1pdm09 during FLUEP1. On the other hand, influenza A/H3N2 was the most frequently diagnosed pathogen in AE-COPD patients during FLUEP2. Before emergence of the new influenza strain A/H1N1pdm09 in 2009, several works cited HREV as the main etiological agent causing AE-COPD [28, 32]. The prevalence of HREV ranged from 10 to 12% of the total exacerbations, while other viruses like RSV and influenza A and B had prevalences ranging from 6 to 7% and 3 to 4%, respectively [28, 32]. Emergence of this new influenza virus seems to have changed the etiological viral pattern of AE-COPD episodes. It resulted in higher rates of AE-COPD of A influenza

viruses compared to other different viral families such as the Picornaviruses at least during the period studied. Despite that, HREV remained as the second most frequent cause of AE-COPD episodes in the patients included in this work.

We analyzed the probability of detection of the different respiratory viruses and coinfections within the demographic and epidemiological characteristics of the study population. Although COPD is a chronic disease that has been historically more associated with males, it has been continuously increasing in women in the recent years [33]. In our study, men represented more than 60% of population with AE-COPD; however, we did not find differences in the detection of any virus or coinfections between men and women.

We also studied the probability of detecting different viruses and coinfections between the two age groups. There were 2.78-fold more influenza virus detections in adults aged 30–64 than in the elderly patients aged ≥65. On the other hand, all of the other viruses were detected at the same proportion in both age groups. Most of the influenza viruses were subtyped as A/H1N1pdm09 influenza strain and were diagnosed during pandemic. This viral strain has strongly affected younger individuals since its emergence in 2009. The apparent tropism of this influenza strain for younger people could be due to a low level or even absence of immunological memory and cross-immunity compared to that present in older individuals, phenomenon well demonstrated as a protective factor in this age group [34].

We also studied the probability of detecting the different respiratory viruses and coinfections among the periods studied in this work. We found differences in detection of

RSV in the pandemic and FLUEP1, differences in coinfections in the pandemic and FLUEP2, and differences in influenza detection between the two epidemics. These data support the dynamic etiology of the respiratory viruses on AE-COPD episodes described in this study. Our study shows that the emergence of a new pandemic influenza virus completely changed the etiology of viral infections in AE-COPD patients. These changes were probably caused by the absence of immunity in a large part of the population. This change can be seen in the following years as fluctuations of the viruses causing these exacerbations. For this reason, it is necessary to continue studying this data series to know if the etiology of respiratory viruses can be absolutely changed in AE-COPD by the emergence of a new virus or, alternatively, if this new behavior occurs only for a few years after a pandemic event.

Influenza viruses have been associated with mortality and morbidity in chronic lung disease [31]. Recent studies have focused on the importance of influenza vaccination with emphasis on risk groups such as COPD patients and especially working adults (30–64 years old) with COPD who are not usually covered by vaccination [35, 36]. Also, the early use of antiviral drugs against influenza viruses in hospitalized patient results in better management of AE-COPD episodes [37], especially regarding the high proportion of patients that suffer an infection by influenza viruses causing AE-COPD described in this work. For this reason, it is important to design empiric and rapid laboratory diagnostic strategies to start treatment as soon as possible for these kinds of patients. In connection with that, our data offers highly valuable information based on demographic and epidemiological characteristics that can help clinicians with the diagnosis of AE-COPD patients. Thus the criteria can be adapted to the specific clinical characteristics of each patient and time of the year. Following these data, diagnostic suspicion may be supported on two different aspects: viral infections of AE-COPD patients are more likely in younger patients than in older ones and that detection of viral respiratory infection causing AE-COPD directly depends on influenza epidemic characteristics, at least in the years following a pandemic influenza emergence. Also, these data support the need for multiplex microbiological diagnostic techniques that allow detecting the most frequent viral targets involved in AE-COPD.

The low number of samples after the pandemic limited the performance of more complex statistical analysis. Specifically, the low number of recruited patients during the interepidemic periods impaired the ability to compare epidemics with periods without sustained influenza circulation. The lack of AE-COPD clinical information in some microbiological order slips handled in the laboratory also generated a loss of AE-COPD cases, which could not then be included in the study. However, recognition of this problem has generated a better dynamic on clinical requests completed by clinicians in their medical services. The high percentage of negative samples diagnosed showed the need for improved diagnostics to identify the role of bacterial infections in the AE-COPD in our sanitary area. It is important to continue this study during following influenza epidemics to check for changes in the etiology dynamic after A/H1N1pdm09 has become completely epidemic.

5. Conclusions

Emergence of the new influenza pandemic virus A/H1N1pdm09 caused influenza A viruses to be the main pathogens that affected COPD patients during the period studied. However A/H1N1pdm09 did not change the global role of respiratory viruses as the primary cause of COPD exacerbations during the pandemic and following two influenza epidemics. The presence of respiratory viruses in AE-COPD episodes that require hospitalization is related with several demographic and epidemiological factors, such as the age of the patient and characteristics of the influenza epidemic activity. These factors need to be used by clinicians to complete clinical and laboratory diagnostic guides that are focused on the role of the respiratory viruses in exacerbations. Vaccination and antiviral drug use is strongly recommended in these kinds of patients.

Acknowledgments

This work has been possible by the support of Consejería de Sanidad de la Junta de Castilla y León, Valladolid, Spain and Instituto de Estudios de Ciencias de la Salud de Castilla y León (IECSCYL), Soria, Spain.

References

[1] W. MacNee, "Pathogenesis of chronic obstructive pulmonary disease," *Proceedings of the American Thoracic Society*, vol. 2, no. 4, pp. 258–266, 2005.

[2] M. Miravitlles, J. B. Soriano, F. García-Río et al., "Prevalence of COPD in Spain: impact of undiagnosed COPD on quality of life and daily life activities," *Thorax*, vol. 64, no. 10, pp. 863–868, 2009.

[3] J. B. Soriano, J. Ancochea, M. Miravitlles et al., "Recent trends in COPD prevalence in Spain: a repeated cross-sectional survey 1997–2007," *European Respiratory Journal*, vol. 36, no. 4, pp. 758–765, 2010.

[4] A. Anzueto, S. Sethi, and F. J. Martinez, "Exacerbations of chronic obstructive pulmonary disease," *Proceedings of the American Thoracic Society*, vol. 4, no. 7, pp. 554–564, 2007.

[5] D. Proud and C.-W. Chow, "Role of viral infections in asthma and chronic obstructive pulmonary disease," *American Journal of Respiratory Cell and Molecular Biology*, vol. 35, no. 5, pp. 513–518, 2006.

[6] C. Fletcher and R. Peto, "The natural history of chronic airflow obstruction," *British Medical Journal*, vol. 1, no. 6077, pp. 1645–1648, 1977.

[7] F. P. Gómez and R. Rodriguez-Roisin, "Global Initiative for Chronic Obstructive Lung Disease (GOLD) guidelines for chronic obstructive pulmonary disease," *Current Opinion in Pulmonary Medicine*, vol. 8, no. 2, pp. 81–86, 2002.

[8] S. Gompertz, D. L. Bayley, S. L. Hill, and R. A. Stockley, "Relationship between airway inflammation and the frequency of exacerbations in patients with smoking related COPD," *Thorax*, vol. 56, no. 1, pp. 36–41, 2001.

[9] M. Miravitlles, T. Guerrero, C. Mayordomo, L. Sánchez-Agudo, F. Nicolau, and J. L. Segú, "Factors associated with increased risk of exacerbation and hospital admission in a cohort of ambulatory COPD patients: a multiple logistic regression analysis. The EOLO Study Group," *Respiration*, vol. 67, no. 5, pp. 495–501, 2000.

[10] S. Sethi and T. F. Murphy, "Bacterial infection in chronic obstructive pulmonary disease in 2000: a state-of-the-art review," *Clinical Microbiology Reviews*, vol. 14, no. 2, pp. 336–363, 2001.

[11] T. M. A. Wilkinson, J. R. Hurst, W. R. Perera, M. Wilks, G. C. Donaldson, and J. A. Wedzicha, "Effect of interactions between lower airway bacterial and rhinoviral infection in exacerbations of COPD," *Chest*, vol. 129, no. 2, pp. 317–324, 2006.

[12] A. F. Connors Jr., N. V. Dawson, C. Thomas et al., "Outcomes following acute exacerbation of severe chronic obstructive lung disease," *The American Journal of Respiratory and Critical Care Medicine*, vol. 154, no. 4 I, pp. 959–967, 1996.

[13] A. Mohan, S. Chandra, D. Agarwal et al., "Prevalence of viral infection detected by PCR and RT-PCR in patients with acute exacerbation of COPD: a systematic review," *Respirology*, vol. 15, no. 3, pp. 536–542, 2010.

[14] H. Frickmann, S. Jungblut, T. O. Hirche, U. Groß, M. Kuhns, and A. E. Zautner, "The influence of virus infections on the course of COPD," *European Journal of Microbiology and Immunology*, vol. 2, no. 3, pp. 176–185, 2012.

[15] A. F. Hutchinson, A. K. Ghimire, M. A. Thompson et al., "A community-based, time-matched, case-control study of respiratory viruses and exacerbations of COPD," *Respiratory Medicine*, vol. 101, no. 12, pp. 2472–2481, 2007.

[16] T. E. McManus, A.-M. Marley, N. Baxter et al., "Respiratory viral infection in exacerbations of COPD," *Respiratory Medicine*, vol. 102, no. 11, pp. 1575–1580, 2008.

[17] J. B. Varkey and B. Varkey, "Viral infections in patients with chronic obstructive pulmonary disease," *Current Opinion in Pulmonary Medicine*, vol. 14, no. 2, pp. 89–94, 2008.

[18] S. Nseir and F. Ader, "Prevalence and outcome of severe chronic obstructive pulmonary disease exacerbations caused by multidrug-resistant bacteria," *Current Opinion in Pulmonary Medicine*, vol. 14, no. 2, pp. 95–100, 2008.

[19] R. Boixeda, N. Rabella, G. Sauca et al., "Microbiological study of patients hospitalized for acute exacerbation of chronic obstructive pulmonary disease (AE-COPD) and the usefulness of analytical and clinical parameters in its identification (VIRAE study)," *International Journal of Chronic Obstructive Pulmonary Disease*, vol. 7, pp. 327–335, 2012.

[20] S. J. Sullivan, R. M. Jacobson, W. R. Dowdle, and G. A. Poland, "2009 H1N1 influenza," *Mayo Clinic Proceedings*, vol. 85, no. 1, pp. 64–76, 2010.

[21] L. Simonsen, P. Spreeuwenberg, R. Lustig et al., "Global mortality estimates for the 2009 influenza pandemic from the GLaMOR project: a modeling study," *PLoS Medicine*, vol. 10, no. 11, Article ID e1001558, 2013.

[22] S. G. Paquette, D. Banner, Z. Zhao et al., "Interleukin-6 is a potential biomarker for severe pandemic H1N1 influenza a infection," *PLoS ONE*, vol. 7, no. 6, Article ID e38214, 2012.

[23] R. Almansa, L. Socias, D. Andaluz-Ojeda et al., "Viral infection is associated with an increased proinflammatory response in chronic obstructive pulmonary disease," *Viral Immunology*, vol. 25, no. 4, pp. 249–253, 2012.

[24] N. R. Anthonisen, J. Manfreda, and C. P. W. Warren, "Antibiotic therapy in exacerbations of chronic obstructive pulmonary disease," *Annals of Internal Medicine*, vol. 106, no. 2, pp. 196–204, 1987.

[25] S. Al Hajjar and K. McIntosh, "The first influenza pandemic of the 21st century," *Annals of Saudi Medicine*, vol. 30, no. 1, pp. 1–10, 2010.

[26] B. Biere, B. Bauer, and B. Schweiger, "Differentiation of influenza b virus lineages yamagata and victoria by real-time PCR," *Journal of Clinical Microbiology*, vol. 48, no. 4, pp. 1425–1427, 2010.

[27] T. Seemungal, R. Harper-Owen, A. Bhowmik et al., "Respiratory viruses, symptoms, and inflammatory markers in acute exacerbations and stable chronic obstructive pulmonary disease," *American Journal of Respiratory and Critical Care Medicine*, vol. 164, no. 9, pp. 1618–1623, 2001.

[28] J. A. Wedzicha, "Role of viruses in exacerbations of chronic obstructive pulmonary disease," *Proceedings of the American Thoracic Society*, vol. 1, no. 2, pp. 115–120, 2004.

[29] M. Bafadhel, S. McKenna, S. Terry et al., "Acute exacerbations of chronic obstructive pulmonary disease: identification of biologic clusters and their biomarkers," *American Journal of Respiratory and Critical Care Medicine*, vol. 184, no. 6, pp. 662–671, 2011.

[30] P. Mallia, J. Footitt, R. Sotero et al., "Rhinovirus infection induces degradation of antimicrobial peptides and secondary bacterial infection in chronic obstructive pulmonary disease," *American Journal of Respiratory and Critical Care Medicine*, vol. 186, no. 11, pp. 1117–1124, 2012.

[31] S. A. Harper, J. S. Bradley, J. A. Englund et al., "Seasonal influenza in adults and children-diagnosis, treatment, chemoprophylaxis, and institutional outbreak management: clinical practice guidelines of the Infectious Diseases Society of America," *Clinical Infectious Diseases*, vol. 48, no. 8, pp. 1003–1032, 2009.

[32] S. B. Greenberg, M. Allen, J. Wilson, and R. L. Atmar, "Respiratory viral infections in adults with and without chronic obstructive pulmonary disease," *American Journal of Respiratory and Critical Care Medicine*, vol. 162, no. 1, pp. 167–173, 2000.

[33] J. B. Soriano, W. C. Maier, P. Egger et al., "Recent trends in physician diagnosed COPD in women and men in the UK," *Thorax*, vol. 55, no. 9, pp. 789–794, 2000.

[34] N. Verma, M. Dimitrova, D. M. Carter et al., "Influenza virus H1N1pdm09 infections in the young and old: evidence of greater antibody diversity and affinity for the hemagglutinin globular head domain (HA1 Domain) in the elderly than in young adults and children," *Journal of Virology*, vol. 86, no. 10, pp. 5515–5522, 2012.

[35] P. J. Poole, E. Chacko, R. W. Wood-Baker, and C. J. Cates, "Influenza vaccine for patients with chronic obstructive pulmonary disease," *The Cochrane Database of Systematic Reviews*, no. 4, Article ID CD002733, 2000.

[36] K. L. Nichol, J. D. Nordin, D. B. Nelson, J. P. Mullooly, and E. Hak, "Effectiveness of influenza vaccine in the community-dwelling elderly," *The New England Journal of Medicine*, vol. 357, no. 14, pp. 1373–1381, 2007.

[37] L. Kaiser, C. Wat, T. Mills, P. Mahoney, P. Ward, and F. Hayden, "Impact of oseltamivir treatment on influenza-related lower respiratory tract complications and hospitalizations," *Archives of Internal Medicine*, vol. 163, no. 14, pp. 1667–1672, 2003.

17

A Cross-Study Biomarker Signature of Human Bronchial Epithelial Cells Infected with Respiratory Syncytial Virus

Luiz Gustavo Gardinassi

Department of Biochemistry and Immunology, Ribeirão Preto Medical School, University of São Paulo, 14049-900 Ribeirão Preto, SP, Brazil

Correspondence should be addressed to Luiz Gustavo Gardinassi; gugard@gmail.com

Academic Editor: Jay C. Brown

Respiratory syncytial virus (RSV) is a major cause of lower respiratory tract infections in children, elderly, and immunocompromised individuals. Despite of advances in diagnosis and treatment, biomarkers of RSV infection are still unclear. To understand the host response and propose signatures of RSV infection, previous studies evaluated the transcriptional profile of the human bronchial epithelial cell line—BEAS-2B—infected with different strains of this virus. However, the evolution of statistical methods and functional analysis together with the large amount of expression data provide opportunities to uncover novel biomarkers of inflammation and infections. In view of those facts publicly available microarray datasets from RSV-infected BEAS-2B cells were analyzed with linear model-based statistics and the platform for functional analysis InnateDB. The results from those analyses argue for the reevaluation of previously reported transcription patterns and biological pathways in BEAS-2B cell lines infected with RSV. Importantly, this study revealed a biosignature constituted by genes such as *ABCC4, ARMC8, BCLAF1, EZH1, FAM118A, FAM208B, FUS, HSPH1, KAZN, MAP3K2, N6AMT1, PRMT2, S100PBP, SERPINA1, TLK2, ZNF322,* and *ZNF337* which should be considered in the development of new molecular diagnosis tools.

1. Introduction

Respiratory syncytial virus (RSV) is a major etiologic agent causing acute lower respiratory infections that can progress to bronchiolitis and pneumonia in children, elderly, and immunocompromised individuals [1, 2]. RSV outbreaks are influenced by virus diversity and evolution [3, 4], environmental factors [5], and host immunity [6].

The epithelium is the primary site for host-virus interface, where cells recognize pathogen-associated patterns on microbes through innate immunity receptors [7, 8]. Indeed, epithelial cells constitute an important line of defense against RSV and other airborne pathogens [9]. They form a physical barrier and produce mucus to inhibit microbes from entering the body. Moreover, they express molecules with antimicrobial properties, as lysozyme, lactoferrin, collectins, and antimicrobial peptides [10]. Two human cell lines have been extensively used to understand the interaction between host and RSV, the alveolar epithelial cell, A549, and one from proximal airways, the bronchial epithelial cell, BEAS-2B.

Genome-wide microarrays are powerful tools to investigate host transcriptional response during infections in the pulmonary epithelium, including those induced by RSV [11, 12]. Indeed, two studies evaluated the patterns of gene expression from BEAS-2B cell lines infected with RSV [10, 13]. However, it is intriguing that after 4 h of infection Huang and collaborators (2008) found that RSV-modulated genes were only associated with the neuroactive ligand-receptor interaction pathway [13]; in contrast, Mayer and collaborators (2007) identified that the same time of RSV infection of BEAS-2B cells induced transcriptional changes similar to those found for other respiratory pathogens as *Pseudomonas aeruginosa* [10]. In spite of differences, publicly available microarray data offers an interesting opportunity to reveal common features of RSV induced transcriptional profiles to understand the early response of BEAS-2B cell lines and extend the knowledge on biomarkers of acute infections with this virus. Therefore, those datasets were evaluated in a meta-analysis by fitting linear models for each array probe and Empirical Bayesian approach to detect transcriptional changes that revealed significant associations with unreported pathways. Of importance, this strategy also rendered a biomarker signature of BEAS-2B cell lines infected with RSV that can be useful for the design of molecular diagnosis tools.

2. Materials and Methods

The datasets GSE3397 and GSE6802 were obtained from GEO database (http://www.ncbi.nlm.nih.gov/), which compared BEAS-2B cells infected with RSV with control experiments. Only arrays in which cells were infected with RSV for 4 h were selected for further analysis. Raw data were processed using the R Language and Environment for Statistical Computing (R) 3.2.0 [14] and Bioconductor 3.1 [15]. The *affy* package for R [16] was used to perform quality control when applicable. Data was $\log_2$ transformed and quantile normalization was applied for dataset GSE3397 due the absence of CEL files. The dataset GSE6802 was already RMA normalized. Batch effects were corrected with Combat( ) function [17] of *sva* package for R [18]. Expression data were weighted with the arrayWeights( ) function from *limma* package for R [19]. Differential gene expression was also evaluated with *limma* package for R [19], whereby differentially expressed genes (DEGs) were identified by a false discovery rate (FDR) <0.05. Hierarchical clustering was performed with Euclidian distance for metric calculations and the complete linkage method, which were displayed as heatmaps drawn with *gplots* package for R [20]. Pathway analyses were performed with the online platform for functional analysis InnateDB [21] and significant pathway overrepresentation was computed with hypergeometrical distribution and Benjamini-Hochberg correction for multiple comparisons. Significantly enriched pathways were determined by a P value < 0.05 and FDR < 0.1.

3. Results and Discussion

3.1. Dataset Selection and Preprocessing Analysis. To define a robust transcriptional signature of BEAS-2B acutely infected with RSV, two publicly available datasets, GSE3397 and GSE6802, were used to conduct a meta-analysis from which data were extracted for BEAS-2B cells infected with RSV for 4 h and controls. First, background subtracted expression data from GSE3397 (Figure 1(a)) were preprocessed and normalized (Figure 1(b)). However, in a first attempt to conduct differential gene expression analysis using *limma* [19], there were no statistically significant differences in gene expression. Therefore, principal component analysis (PCA) was used to evaluate the expression profiles of each array and, except for arrays named here Control2 and RSV2, the consistent pattern of clustering in Figure 1(c) suggests a batch effect. After normalization, this effect was even more evident (Figure 1(d)), which led to the speculation that Huang and collaborators (2008) [13] analyzed only three microarray experiments from this dataset based on the assumption that differences found for those microarrays were due to failures in experimental procedures; however they did not consider or correct for

batch effects. In view of those facts, the datasets were adjusted with Combat function for R, which removed such effects from GSE3397 expression data (Figure 1(e)). Batch correction of GSE3397 did not change the profiles of arrays Control2 and RSV2; nevertheless, those arrays were included in further analysis because the variation observed in this experiment could have a substantial impact over the final result. Even adverse experimental variations that may change the overall expression patterns of a dataset could be useful to power up the identification of genes that are robustly modulated in BEAS-2B cells infected with RSV. The expression dataset GSE6802 (Figure 1(f)) was also included in the analysis. PCA from expression data extracted from GEO demonstrates that most of the variability between the arrays is explained (76.6%) by the infection with RSV, as the standardized PC1 separates RSV-infected from control arrays (Figure 1(g)), whereas standardized PC2 (11.4%) separates one pair of arrays (RSV_3 and ctrl2) and, although these arrays are supposedly from different batches, clustering features of this axis also suggested a batch effect (Figure 1(g)). $\log_2$ transformation of data impacted the profile of array RSV_1 however did not change the profiles from RSV_3 and ctrl_2 (Figure 1(h)). Combat( ) function was also applied to the expression dataset GSE6802; however, PCA shows that the adjustment did not to improve further clustering between specific arrays (Supplementary Figure 1; see Supplementary Material available online at http://dx.doi.org/10.1155/2016/3605302). In view of that, downstream analyses were carried out with normalized $\log_2$ transformed data.

3.2. Differential Gene Expression. Next, linear model-based statistical analyses with a FDR < 0.05 were conducted to identify differentially expressed genes (DEGs). The dataset GSE3397 exhibited ninety-four DEGs (Figure 2(a) and Table 1). Those genes are highly discordant from DEGs previously reported by Huang and collaborators (2008) [13], which identified 277 DEGs based on different statistical analysis and assumptions. Fifty genes were downregulated and forty-four were upregulated (Table 1). The differences found in this study might reflect the inclusion of all microarray experiments from controls and 4 h after RSV infection; exclusion of expression data from 24 h after RSV infection; distinct preprocessing approaches as normalizing method and batch effect correction; and the assessment of statistical significance with a linear model-based method and corrected P values. In contrast, 1965 DEGs were identified for the dataset GSE6802. The top hundred DEGs ranked by fold changes (Figure 2(b) and Table 2) included genes such as *JUNB, KLF4, CXCL1, CXCL2,* and *IL6,* which are in agreement with those reported by Mayer and collaborators (2007) [10]. Several factors should account for the notable differences in expression analysis from both datasets. First, different RSV strains were used to stimulate BEAS-2B cells. Second, experimental conditions of controls were also different, as control experiments from GSE3397 were incubated with vehicle (not specified) and those from GSE6802 were not stimulated. Third, despite both datasets being generated with affymetrix microarray platform, those include distinct versions, HU133 plus 2.0 for GSE3397 and HU133A 2.0 for GSE6802.

TABLE 1: Differentially expressed genes identified in dataset GSE3397.

ProbeID	Gene symbol	Gene name	$\log_2$ fold change	FDR
1560754_at	CMTM7	CKLF like MARVEL transmembrane domain containing 7	−1,54756	0,017104
239439_at	AFF4	AF4/FMR2 family member 4	−1,53581	0,023832
238929_at	SRSF8	Serine/arginine-rich splicing factor 8	−1,51887	0,018433
223142_s_at	UCK1	Uridine-cytidine kinase 1	−1,47939	0,017104
242636_at	PRCP	Prolylcarboxypeptidase	−1,45095	0,034358
228007_at	CEP85L	Centrosomal protein 85 kDa-like	−1,4103	0,017104
235573_at	HSPH1	Heat shock protein family H (Hsp110) member 1	−1,39959	0,0371
228391_at	CYP4V2	Cytochrome P450 family 4 subfamily V member 2	−1,38799	0,01671
219376_at	ZNF322	Zinc finger protein 322	−1,3491	0,046761
1553689_s_at	METTL6	Methyltransferase like 6	−1,34723	0,017104
242837_at	SRSF4	Serine/arginine-rich splicing factor 4	−1,34071	0,044693
237215_s_at	TFRC	Transferrin receptor	−1,32685	0,017104
208819_at	RAB8A	RAB8A, member RAS oncogene family	−1,32593	0,042264
236665_at	CCDC18	Coiled-coil domain containing 18	−1,31494	0,034201
206147_x_at	SCML2	Sex comb on midleg-like 2 (Drosophila)	−1,30586	0,016454
229325_at	ZZZ3	Zinc finger ZZ-type containing 3	−1,30495	0,017104
1565716_at	FUS	FUS RNA binding protein	−1,29415	0,049505
205062_x_at	ARID4A	AT-rich interaction domain 4A	−1,28877	0,033039
1552312_a_at	MFAP3	Microfibrillar associated protein 3	−1,28521	0,046511
223223_at	ARV1	ARV1 homolog, fatty acid homeostasis modulator	−1,27987	0,023832
232001_at	PRKCQ-AS1	PRKCQ antisense RNA 1	−1,27987	0,035983
233195_at	DNAI1	Dynein axonemal intermediate chain 1	−1,25963	0,047083
219094_at	ARMC8	Armadillo repeat containing 8	−1,25527	0,043392
235232_at	GMEB1	Glucocorticoid modulatory element binding protein 1	−1,2492	0,046511
218643_s_at	CRIPT	CXXC repeat containing interactor of PDZ3 domain	−1,24229	0,0371
1566851_at	TRIM42	Tripartite motif containing 42	−1,24057	0,042149
221821_s_at	KANSL2	KAT8 regulatory NSL complex subunit 2	−1,23799	0,017104
244115_at	FAM126A	Family with sequence similarity 126 member A	−1,23114	0,033039
215541_s_at	DIAPH1	Diaphanous related formin 1	−1,22774	0,033039
203196_at	ABCC4	ATP binding cassette subfamily C member 4	−1,22519	0,033039
225024_at	RPRD1B	Regulation of nuclear pre-mRNA domain containing 1B	−1,22264	0,043765
37860_at	ZNF337	Zinc finger protein 337	−1,22095	0,023832
212997_s_at	TLK2	Tousled like kinase 2	−1,21841	0,04814
225690_at	CDK12	Cyclin-dependent kinase 12	−1,21083	0,0371
232103_at	BPNT1	$3'(2')$, $5'$-Bisphosphate nucleotidase 1	−1,20748	0,0371
224848_at	CDK6	Cyclin-dependent kinase 6	−1,20247	0,0371
214962_s_at	NUP160	Nucleoporin 160 kDa	−1,20247	0,046319
219629_at	FAM118A	Family with sequence similarity 118 member A	−1,19831	0,028374
212290_at	SLC7A1	Solute carrier family 7 member 1	−1,19748	0,042264
227187_at	CBLL1	Cbl proto-oncogene like 1, E3 ubiquitin protein ligase	−1,19582	0,030047
233208_x_at	CPSF2	Cleavage and polyadenylation specific factor 2	−1,19334	0,046319
230566_at	MORC2-AS1	MORC2 antisense RNA 1	−1,17691	0,0371
238795_at	FAM208B	Family with sequence similarity 208 member B	−1,17609	0,0371
204980_at	CLOCK	Clock circadian regulator	−1,17283	0,0371
238653_at	LRIG2	Leucine-rich repeats and immunoglobulin like domains 2	−1,17202	0,048527
229939_at	ENDOV	Endonuclease V	−1,16878	0,041349
218185_s_at	ARMC1	Armadillo repeat containing 1	−1,16151	0,046319
201083_s_at	BCLAF1	BCL2 associated transcription factor 1	−1,15509	0,049505

TABLE 1: Continued.

ProbeID	Gene symbol	Gene name	$\log_2$ fold change	FDR
227840_at	C2orf76	Chromosome 2 open reading frame 76	−1,15109	0,042264
201686_x_at	API5	Apoptosis inhibitor 5	−1,14076	0,046761
221699_s_at	DDX50	DEAD-box helicase 50	1,140764	0,046511
1556178_x_at	TAF8	TATA-box binding protein associated factor 8	1,159096	0,034358
205623_at	ALDH3A1	Aldehyde dehydrogenase 3 family member A1	1,163927	0,049505
212495_at	KDM4B	Lysine demethylase 4B	1,193336	0,044693
1569057_s_at	MIA3	Melanoma inhibitory activity family member 3	1,193336	0,047866
222494_at	FOXN3	Forkhead box N3	1,19582	0,048527
223311_s_at	MTA3	Metastasis associated 1 family member 3	1,19582	0,041439
215424_s_at	SNW1	SNW domain containing 1	1,196649	0,049505
213478_at	KAZN	Kazrin, periplakin interacting protein	1,19914	0,025143
227864_s_at	MVB12A	Multivesicular body subunit 12A	1,201636	0,030287
228674_s_at	EML4	Echinoderm microtubule associated protein like 4	1,204137	0,040345
224196_x_at	DPH5	Diphthamide biosynthesis 5	1,205808	0,025143
224652_at	CCNY	Cyclin Y	1,207481	0,046761
212968_at	RFNG	RFNG O-fucosylpeptide 3-beta-N-acetylglucosaminyltransferase	1,211673	0,0371
1555486_a_at	PRR5L	Proline rich 5 like	1,212513	0,017104
232837_at	KIF13A	Kinesin family member 13A	1,214195	0,042264
224320_s_at	MCM8	Minichromosome maintenance 8 homologous recombination repair factor	1,217566	0,033039
230131_x_at	ARSD	Arylsulfatase D	1,221793	0,0371
218225_at	ECSIT	ECSIT signalling integrator	1,224336	0,034358
222610_s_at	S100PBP	S100P binding protein	1,226885	0,030047
32259_at	EZH1	Enhancer of zeste 1 polycomb repressive complex 2 subunit	1,229439	0,0371
203854_at	CFI	Complement factor I	1,232852	0,042264
221600_s_at	AAMDC	Adipogenesis associated, Mth938 domain containing	1,260503	0,0371
209558_s_at	HIP1R	Huntingtin interacting protein 1 related	1,263127	0,042264
224814_at	DPP7	Dipeptidyl peptidase 7	1,26488	0,016454
232280_at	SLC25A29	Solute carrier family 25 member 29	1,277214	0,030047
228424_at	NAALADL1	N-Acetylated alpha-linked acidic dipeptidase-like 1	1,286989	0,042264
203409_at	DDB2	Damage specific DNA binding protein 2	1,288775	0,023832
229975_at	BMPR1B	Bone morphogenetic protein receptor type 1B	1,297739	0,034358
227073_at	MAP3K2	Mitogen-activated protein kinase kinase kinase 2	1,297739	0,017104
225347_at	ARL8A	ADP ribosylation factor like GTPase 8A	1,298639	0,02672
221774_x_at	SUPT20H	SPT20 homolog, SAGA complex component	1,308578	0,016454
223679_at	CTNNB1	Catenin beta 1	1,318594	0,018043
227679_at	HDAC11	Histone deacetylase 11	1,328686	0,044693
220020_at	XPNPEP3	X-Prolyl aminopeptidase 3, mitochondrial	1,342573	0,031097
203199_s_at	MTRR	5-Methyltetrahydrofolate-homocysteine methyltransferase reductase	1,360371	0,017104
228722_at	PRMT2	Protein arginine methyltransferase 2	1,370783	0,016454
228951_at	SLC38A7	Solute carrier family 38 member 7	1,431969	0,016454
217529_at	ORAI2	ORAI calcium release-activated calcium modulator 2	1,453973	0,043775
220311_at	N6AMT1	N-6 adenine-specific DNA methyltransferase 1 (putative)	1,460032	0,017104
213402_at	ZNF787	Zinc finger protein 787	1,469169	0,017104
226055_at	ARRDC2	Arrestin domain containing 2	1,477338	0,017104
219756_s_at	POF1B	Premature ovarian failure, 1B	1,580083	0,016454
202833_s_at	SERPINA1	Serpin peptidase inhibitor, clade A (alpha-1 antiproteinase, antitrypsin), and member 1	2,488023	0,0371

TABLE 2: Top hundred differentially expressed genes identified in dataset GSE6802.

ProbeID	Gene symbol	Gene name	$\log_2$ fold change	FDR
212615_at	CHD9	Chromodomain helicase DNA binding protein 9	−3,69609	0,00131
221840_at	PTPRE	Protein tyrosine phosphatase, receptor type E	−3,56524	0,000195
220817_at	TRPC4	Transient receptor potential cation channel subfamily C member 4	−3,39168	0,001582
221703_at	BRIP1	BRCA1 interacting protein C-terminal helicase 1	−2,88786	0,021463
207012_at	MMP16	Matrix metallopeptidase 16	−2,82647	0,000119
219494_at	RAD54B	RAD54 homolog B (S. cerevisiae)	−2,81279	0,000177
207034_s_at	GLI2	GLI family zinc finger 2	−2,79723	0,005157
203518_at	LYST	Lysosomal trafficking regulator	−2,75872	$5,90E-05$
205282_at	LRP8	LDL receptor related protein 8	−2,7549	0,000311
214440_at	NAT1	N-Acetyltransferase 1 (arylamine N-acetyltransferase)	−2,68515	0,001777
219627_at	ZNF767P	Zinc finger family member 767, pseudogene	−2,67957	0,00024
218984_at	PUS7	Pseudouridylate synthase 7 (putative)	−2,67586	0,001308
206554_x_at	SETMAR	SET domain and mariner transposase fusion gene	−2,63536	0,002432
219779_at	ZFHX4	Zinc finger homeobox 4	−2,62624	0,001411
213103_at	STARD13	StAR related lipid transfer domain containing 13	−2,57219	0,002525
210138_at	RGS20	Regulator of G-protein signaling 20	−2,55974	0,000415
204291_at	ZNF518A	Zinc finger protein 518A	−2,54383	$9,70E-05$
204651_at	NRF1	Nuclear respiratory factor 1	−2,49147	0,003659
205408_at	MLLT10	Myeloid/lymphoid or mixed-lineage leukemia; translocated to, 10	−2,48975	$5,10E-05$
219581_at	TSEN2	tRNA splicing endonuclease subunit 2	−2,45377	0,001774
218242_s_at	SUV420H1	Lysine methyltransferase 5B	−2,44698	0,000754
203242_s_at	PDLIM5	PDZ and LIM domain 5	−2,43851	0,001699
203868_s_at	VCAM1	Vascular cell adhesion molecule 1	−2,43513	0,000761
220206_at	ZMYM1	Zinc finger MYM-type containing 1	−2,36362	0,008439
207616_s_at	TANK	TRAF family member associated NFKB activator	−2,34567	0,000424
218303_x_at	KRCC1	Lysine-rich coiled-coil 1	−2,34567	0,003187
218490_s_at	ZNF302	Zinc finger protein 302	−2,32785	0,001816
206876_at	SIM1	Single-minded family bHLH transcription factor 1	−2,32624	0,001681
219128_at	C2orf42	Chromosome 2 open reading frame 42	−2,28628	0,002926
212861_at	MFSD5	Major facilitator superfamily domain containing 5	−2,27048	0,000823
218653_at	SLC25A15	Solute carrier family 25 member 15	−2,25636	0,000562
206943_at	TGFBR1	Transforming growth factor beta receptor I	−2,24856	0,025349
201995_at	EXT1	Exostosin glycosyltransferase 1	−2,247	0,000421
221430_s_at	RNF146	Ring finger protein 146	−2,23457	0,001084
212286_at	ANKRD12	Ankyrin repeat domain 12	−2,2253	0,00029
219544_at	BORA	Bora, aurora kinase A activator	−2,21914	0,000333
210455_at	R3HCC1L	R3H domain and coiled-coil containing 1 like	−2,2176	0,0039
219459_at	POLR3B	Polymerase (RNA) III subunit B	−2,2176	0,000832
219078_at	GPATCH2	G-patch domain containing 2	−2,19923	0,000723
204547_at	RAB40B	RAB40B, member RAS oncogene family	−2,17648	0,001741
209760_at	KIAA0922	KIAA0922	−2,17347	0,001048
218791_s_at	KATNBL1	Katanin regulatory subunit B1 like 1	−2,17347	0,001187
205173_x_at	CD58	CD58 molecule	−2,17196	0,00022
204352_at	TRAF5	TNF receptor associated factor 5	−2,16895	0,002659
212441_at	KIAA0232	KIAA0232	−2,16595	0,006084
204236_at	FLI1	Fli-1 proto-oncogene, ETS transcription factor	−2,15397	0,005141
203072_at	MYO1E	Myosin IE	−2,15248	0,000154
219904_at	ZSCAN5A	Zinc finger and SCAN domain containing 5A	−2,14801	0,00144
219133_at	OXSM	3-Oxoacyl-ACP synthase, mitochondrial	−2,12285	0,002424
205798_at	IL7R	Interleukin 7 receptor	−2,11257	0,00506

Table 2: Continued.

ProbeID	Gene symbol	Gene name	$\log_2$ fold change	FDR
205476_at	CCL20	C-C motif chemokine ligand 20	4,613942	$9,50E-05$
213497_at	ABTB2	Ankyrin repeat and BTB domain containing 2	4,623547	$1,40E-05$
219179_at	DACT1	Dishevelled-binding antagonist of beta-catenin 1	4,642816	$9,00E-06$
219228_at	ZNF331	Zinc finger protein 331	4,723971	$6,00E-06$
213139_at	SNAI2	Snail family zinc finger 2	4,76673	$1,40E-05$
218177_at	CHMP1B	Charged multivesicular body protein 1B	4,806544	$1,00E-05$
203304_at	BAMBI	BMP and activin membrane-bound inhibitor	4,826576	$3,00E-06$
201631_s_at	IER3	Immediate early response 3	4,833271	$3,00E-06$
218559_s_at	MAFB	v-maf avian musculoaponeurotic fibrosarcoma oncogene homolog B	4,870264	0,000468
220266_s_at	KLF4	Kruppel-like factor 4 (gut)	4,890561	0,00022
209211_at	KLF5	Kruppel-like factor 5 (intestinal)	4,924578	0,002036
209681_at	SLC19A2	Solute carrier family 19 member 2	4,927992	$5,90E-05$
205266_at	LIF	Leukemia inhibitory factor	4,955395	$2,20E-05$
204790_at	SMAD7	SMAD family member 7	5,073566	0,000283
221667_s_at	HSPB8	Heat shock protein family B (small) member 8	5,422657	$2,90E-05$
212665_at	TIPARP	TCDD-inducible poly(ADP-ribose) polymerase	5,525098	$1,00E-05$
202935_s_at	SOX9	SRY-box 9	5,971114	$3,30E-05$
202023_at	EFNA1	Ephrin-A1	6,164569	$3,30E-05$
202393_s_at	KLF10	Kruppel-like factor 10	6,194552	0,000195
213146_at	KDM6B	Lysine demethylase 6B	6,203146	$1,90E-05$
205193_at	MAFF	v-maf avian musculoaponeurotic fibrosarcoma oncogene homolog F	6,2941	$2,00E-06$
209457_at	DUSP5	Dual specificity phosphatase 5	6,639157	$1,30E-05$
206029_at	ANKRD1	Ankyrin repeat domain 1	6,65759	0,008591
209283_at	CRYAB	Crystallin alpha B	6,703897	0,000118
201693_s_at	EGR1	Early growth response 1	7,056731	$4,10E-05$
212099_at	RHOB	ras homolog family member B	7,300524	0,000406
219682_s_at	TBX3	T-box 3	7,722136	$5,80E-05$
201473_at	JUNB	jun B proto-oncogene	8,322402	$7,00E-06$
200664_s_at	DNAJB1	DnaJ heat shock protein family (Hsp40) member B1	8,586082	$2,00E-05$
205828_at	MMP3	Matrix metallopeptidase 3	8,711976	$1,90E-05$
201169_s_at	BHLHE40	Basic helix-loop-helix family member e40	8,870405	0,00011
203665_at	HMOX1	Heme oxygenase 1	9,32433	0,000544
202643_s_at	TNFAIP3	TNF alpha induced protein 3	9,573192	$2,50E-05$
205207_at	IL6	Interleukin 6	10,18236	$3,00E-06$
202388_at	RGS2	Regulator of G-protein signaling 2	10,25318	$1,40E-05$
204472_at	GEM	GTP binding protein overexpressed in skeletal muscle	10,8003	$1,00E-06$
202149_at	NEDD9	Neural precursor cell expressed, developmentally down-regulated 9	11,06553	$2,50E-05$
219480_at	SNAI1	Snail family zinc finger 1	11,70457	$2,00E-06$
218839_at	HEY1	hes related family bHLH transcription factor with YRPW motif 1	12,07541	$6,00E-06$
206115_at	EGR3	Early growth response 3	14,19194	$1,20E-05$
204470_at	CXCL1	C-X-C motif chemokine ligand 1	17,61827	$2,00E-06$
204621_s_at	NR4A2	Nuclear receptor subfamily 4 group A member 2	18,77837	0
209774_x_at	CXCL2	C-X-C motif chemokine ligand 2	19,02731	$9,00E-06$
202859_x_at	CXCL8	C-X-C motif chemokine ligand 8	19,89039	$1,00E-06$
202340_x_at	NR4A1	Nuclear receptor subfamily 4 group A member 1	20,74943	$1,00E-06$
209189_at	FOS	FBJ murine osteosarcoma viral oncogene homolog	23,36051	$1,00E-06$
202672_s_at	ATF3	Activating transcription factor 3	24,18432	0
202768_at	FOSB	FBJ murine osteosarcoma viral oncogene homolog B	32,92245	0
207978_s_at	NR4A3	Nuclear receptor subfamily 4 group A member 3	43,80428	$1,00E-06$
117_at	HSPA6	Heat shock protein family A (Hsp70) member 6	90,82389	0

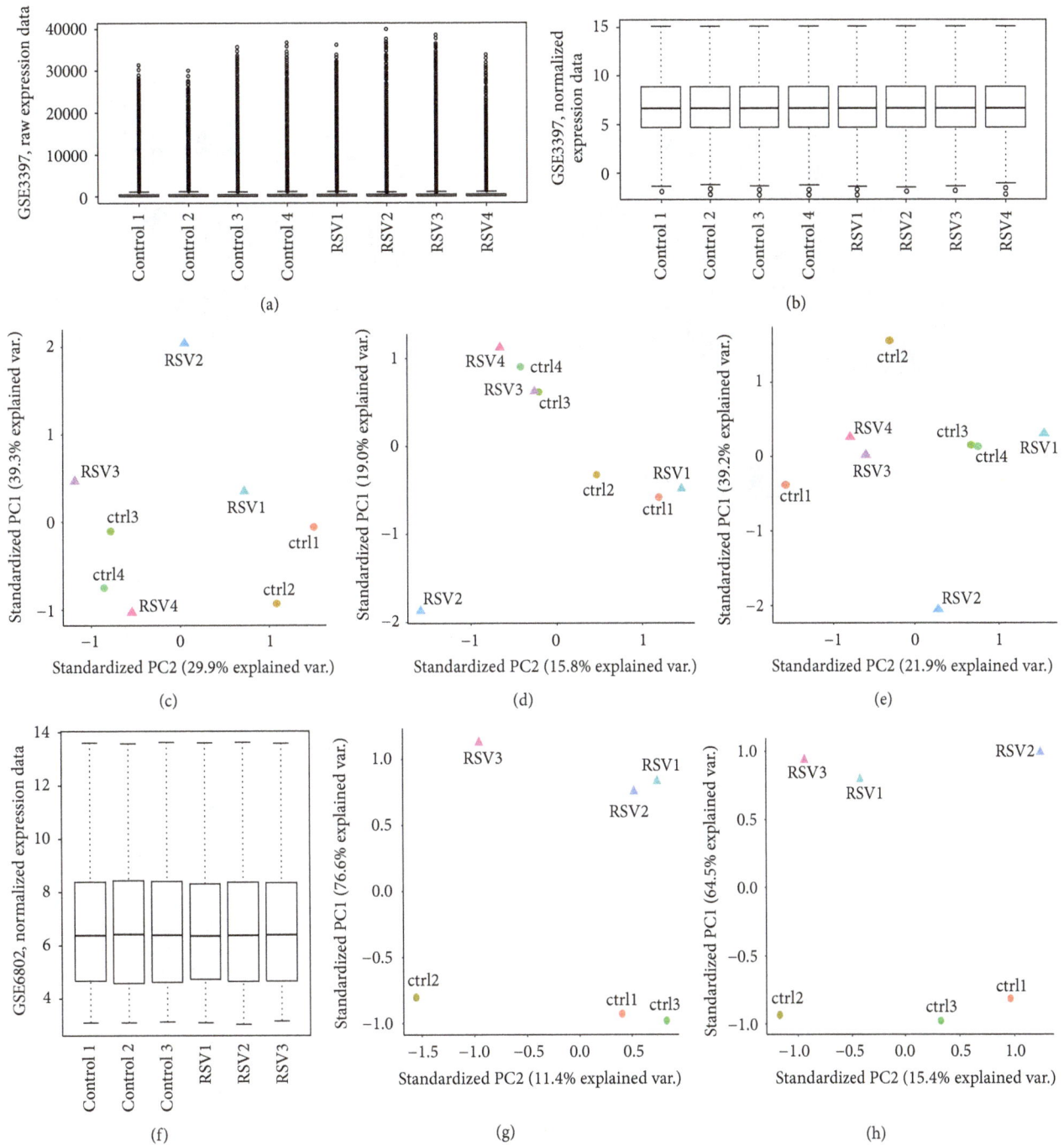

FIGURE 1: Preprocessing analysis of GEO datasets GSE3397 and GSE6802. (a) Boxplot of GSE3397, raw expression data. (b) Boxplot of GSE3397, normalized expression data. (c) Principal component analysis of GSE3397, raw expression data. (d) Principal component analysis of GSE3397, normalized expression data. (e) Principal component analysis of GSE3397, normalized and batch corrected expression data. (f) Boxplot of GSE6802, RMA normalized expression data. (g) Principal component analysis of GSE6802, normalized expression data. (h) Principal component analysis of GSE6802, $\log_2$ transformed RMA normalized expression data.

3.3. *Functional Analysis.* To obtain a biological interpretation of the transcriptional signature of RSV-infected BEAS-2B cells and compare with those reported by previous studies, enrichment analysis was performed with the online platform for functional analysis InnateDB [21]. Based on a FDR < 0.1, DEGs identified for GSE3397 were enriched in pathways related to Chromatin organization, histone acetylation, signaling by NOTCH, IL1, Integrin-linked kinase signaling,

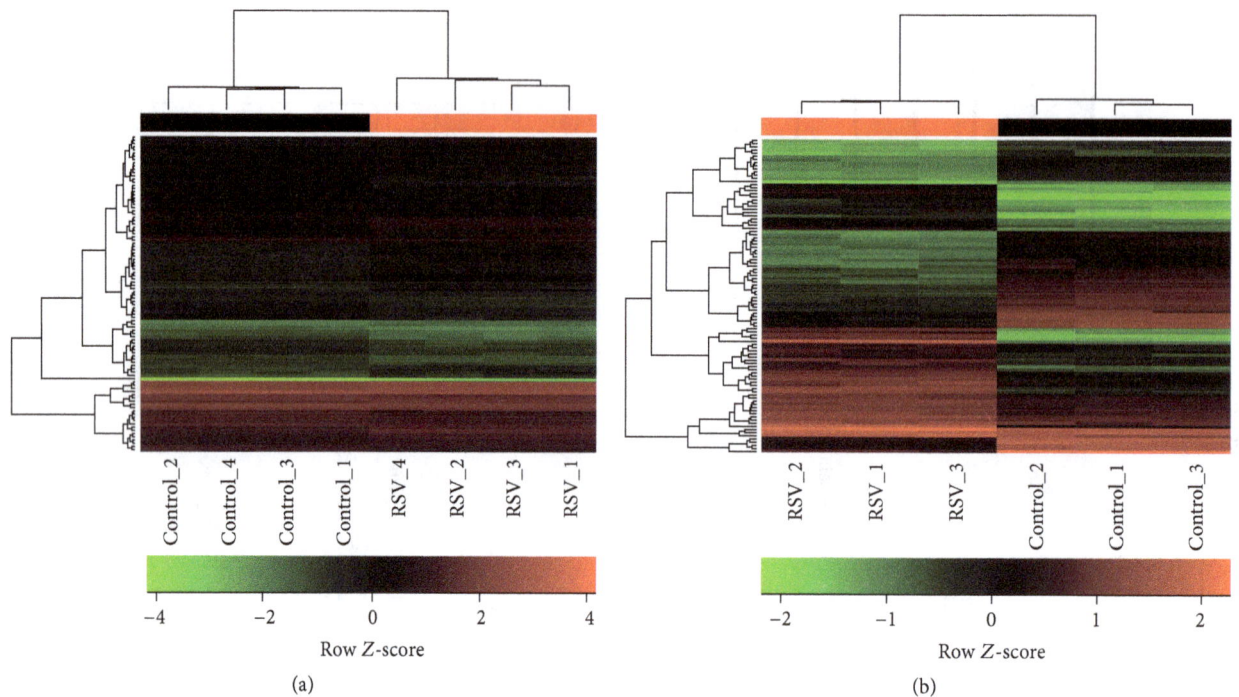

FIGURE 2: Transcriptional profiles of BEAS-2B cells infected with RSV for 4 h. (a) Hierarchical clustering of differentially expressed genes from dataset GSE3397. (b) Hierarchical clustering of differentially expressed genes from dataset GSE6802. Row Z-scores were calculated based on normalized expression data. The colors from green to red represent the transition of decreased to increased expression.

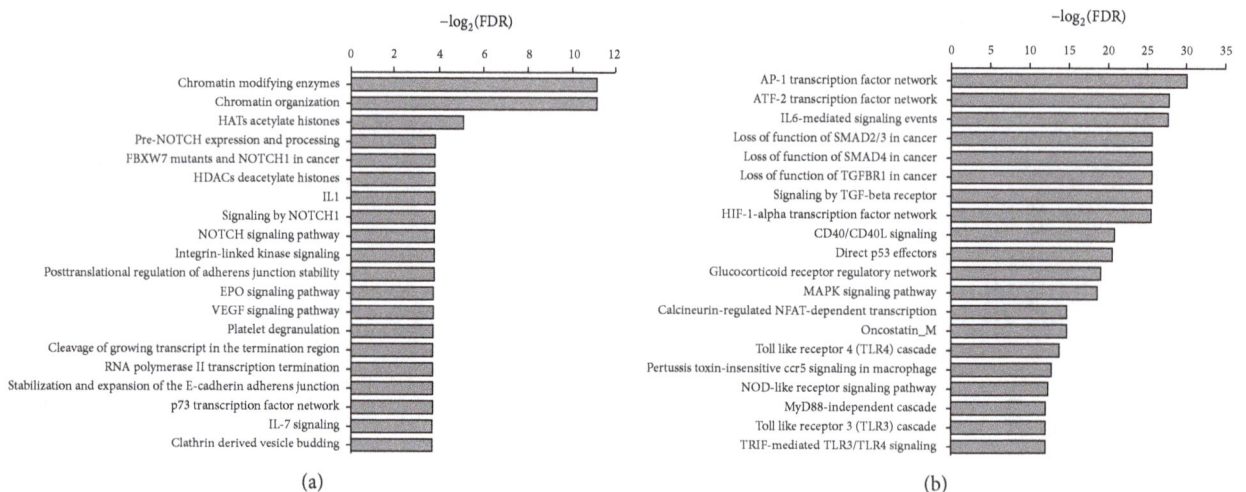

FIGURE 3: Pathway enrichment analysis with InnateDB. Differentially expressed genes from (a) GSE3397 or (b) GSE6802 were evaluated for overrepresentation in pathways annotated in databases as INOH, KEGG, NETPATH, PID NIC, and REACTOME.

EPO signaling pathway, VEGF signaling pathway, platelet degranulation, p73 transcription factor network, IL-7 signaling, p53 signaling pathway, and others (Figure 3(a) and Supplementary Data 1). Of interest, Huang and collaborators (2008) [13] reported gene overrepresentation within p53 signaling pathway, but only after 24 h following RSV infection of BEAS-2B cells. After 4 h following RSV infection, Huang and collaborators (2008) [13] only found a significant association with neuroactive ligand-receptor interaction pathway, which was not overrepresented in the present

analysis. In contrast, DEGs resultant from dataset GSE6802 were enriched in pathways related to AP-1 transcription factor, ATF-2 transcription factor, IL-6 signaling, SMAD function, signaling by TGFBR, HIF-1α transcription factor, signaling by CD40/CD40L, signaling by MAPK, signaling by innate immune receptors, and others (Figure 3(b) and Supplementary Data 1). Some of those pathways as CD40 signaling are indeed commonly induced by a variety of viral respiratory infections [22], whereas several of those pathways could indicate novel directions for studying the host response

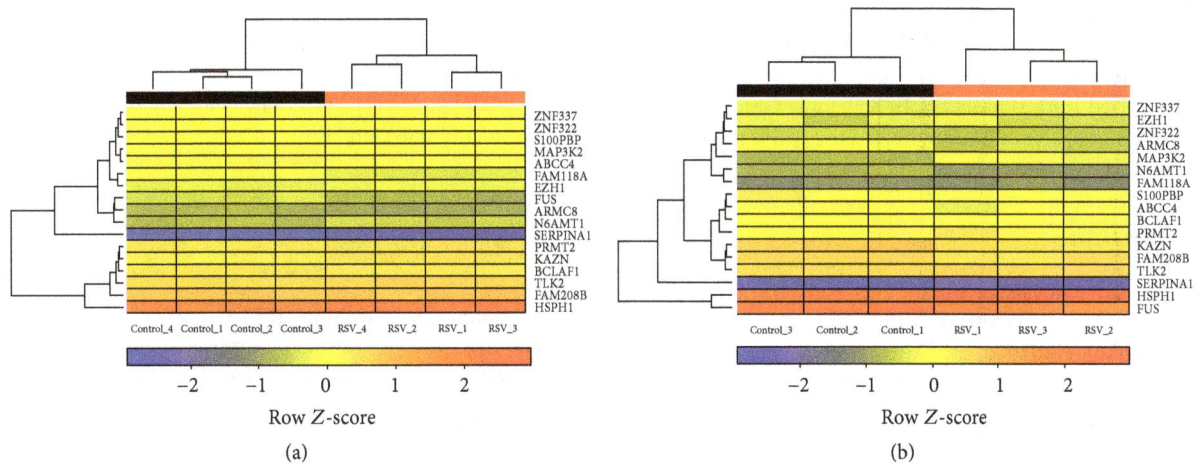

FIGURE 4: Biomarker signature of BEAS-2B cells infected with RSV for 4 h. Hierarchical clustering of expression data for *ABCC4*, *ARMC8*, *BCLAF1*, *EZH1*, *FAM118A*, *FAM208B*, *FUS*, *HSPH1*, *KAZN*, *MAP3K2*, *N6AMT1*, *PRMT2*, *S100PBP*, *SERPINA1*, *TLK2*, *ZNF322*, and *ZNF337* from (a) dataset GSE3397 and (b) dataset GSE6802. Row Z-scores were calculated based on normalized expression data. The colors from blue to red represent the transition of decreased to increased expression.

against RSV. Six pathways were enriched by DEGs from both datasets, the EPO signaling pathway, FBXW7 Mutants and NOTCH1 in Cancer, IL1, p53 signaling pathway, p73 transcription factor network, and signaling by NOTCH1. The erythropoietin (EPO) gene is a primary target of HIF-1α transcription factor, whereas binding of HIF-1α to the EPO enhancer promoter region induces transcriptional programs that influence inflammation and infection processes [23]. In addition, expression of Dll4, a major NOTCH ligand, is upregulated in dendritic cells infected with RSV, whereas blockage of Dll4 *in vivo* increased hyperreactivity of airways and mucus secretion that impacted the pathology of the disease, showing a key role of signaling by NOTCH in the regulation of immunity against RSV [24]. Moreover, besides modulations of the p53 signaling pathway by infection of RSV *in vitro* [10, 13], this pathway was found to be upregulated in whole blood of children with lower respiratory tract infection by RSV [25]. Taken together, those data point to key pathways which can impact infections of human bronchial epithelial cells with RSV.

3.4. Meta-Analysis Based Biomarker Signature of RSV-Infected BEAS-2B Cells. To determine a unique transcriptional signature of BEAS-2B cells induced by early infection with RSV, common DEGs for both datasets were further identified. The analysis retrieved a list of seventeen common genes: *ABCC4*, *ARMC8*, *BCLAF1*, *EZH1*, *FAM118A*, *FAM208B*, *FUS*, *HSPH1*, *KAZN*, *MAP3K2*, *N6AMT1*, *PRMT2*, *S100PBP*, *SER-PINA1*, *TLK2*, *ZNF322*, and *ZNF337* (Figure 4). Despite particular features in expression data from both datasets, unsupervised hierarchical clustering analysis based on this signature revealed the formation of robust clusters between RSV-infected or uninfected BEAS-2B cells (Figure 4). Of note, human airway epithelial cells were shown to express ABCC4/MRP4, a transporter for uric acid and cAMP [26].

Mucosal production of uric acid was recently linked to particulate matter-induced allergic sensitization [26]; therefore RSV infection could trigger such a response and contribute to the development and severity of allergic responses to particulate matter [27]. Moreover, both ABCC4 and SER-PINA1 are annotated into the platelet degranulation pathway (Figure 3(a)), suggesting a role in antiviral mechanisms from bronchial epithelial cells. After an initial encounter with RSV, the transcriptional activity of human bronchial epithelial cells is reprogrammed to counteract viruses and other pathogens [10], whereas *MAP3K2* and *ZNF322* are clearly involved on the activation and regulation of MAP kinase signaling pathway [28, 29]. Indeed, RSV infection leads to the activation of p38 MAPK [30] and c-JUN kinase pathway, which negatively regulates the production of TNF-α in human epithelial cells [31] and might contribute to virus evasion from an early immune response. Interestingly, the biosignature also included BCLAF1, a molecule involved in processes as apoptosis, transcription and processing of RNA, and export of mRNA from the nucleus [32]. However, this nuclear protein was also implicated as a viral restriction factor targeted to degradation by human cytomegalovirus [32]. Moreover, EZH1 was shown to be involved in the methylation of histone 3 at lysine 27 (H3K27) of the HIV provirus in resting cells [33] and could thus exert a significant function in infections with RSV, whereby other genes such as N6AMT1, FUS, and PRMT2 are also involved in protein methylation. Indeed, using coimmunoprecipitation and mass spectrometry, recent work demonstrated that RSV nucleoprotein (N) interacts with protein arginine N-methyltransferase 5 (PRMT5) [34], suggesting that PRMT2 could also interact with RSV proteins and play an important role during infections of human bronchial epithelial cells. Several of the genes identified in this study have been poorly studied in the context of RSV infection, whereby none of them was previously reported as a biomarker of infections by this virus. Of note, except for *FAM208B* and *KAZN*, analysis conducted by Smith and collaborators (2012) [22] which included both datasets (GSE3397 and GSE6802) also identified the significant modulation of the genes included in the biomarker signature identified herein.

4. Conclusions

The combined analysis of distinct datasets from BEAS-2B cells infected with RSV retrieved intriguing results, whereby using powerful statistical methods and assumptions this study identified a new set of biomarkers of early infection with RSV composed by seventeen genes: *ABCC4*, *ARMC8*, *BCLAF1*, *EZH1*, *FAM118A*, *FAM208B*, *FUS*, *HSPH1*, *KAZN*, *MAP3K2*, *N6AMT1*, *PRMT2*, *S100PBP*, *SERPINA1*, *TLK2*, *ZNF322*, and *ZNF337*. This transcriptional signature could be useful for the development of molecular diagnosis tools as well as future investigations of processes involved in host-pathogen interactions.

Competing Interests

The author declares that there are no competing interests regarding the publication of this paper.

Acknowledgments

The author is grateful to Dr. Fátima Pereira de Souza for critical comments on the paper. Luiz Gustavo Gardinassi was supported by scholarships from Fundação de Amparo à Pesquisa do Estado de São Paulo (FAPESP).

References

[1] L. J. Anderson, R. A. Parker, and R. L. Strikas, "Association between respiratory syncytial virus outbreaks and lower respiratory tract deaths of infants and young children," *The Journal of Infectious Diseases*, vol. 161, no. 4, pp. 640–646, 1990.

[2] A. R. Falsey and E. E. Walsh, "Respiratory syncytial virus infection in adults," *Clinical Microbiology Reviews*, vol. 13, no. 3, pp. 371–384, 2000.

[3] V. F. Botosso, P. M. D. Zanotto, M. Ueda et al., "Positive selection results in frequent reversible amino acid replacements in the G protein gene of human respiratory syncytial virus," *PLoS Pathogens*, vol. 5, no. 1, Article ID e1000254, 2009.

[4] L. G. A. Gardinassi, P. V. M. Simas, D. E. Gomes et al., "Diversity and adaptation of human respiratory syncytial virus genotypes circulating in two distinct communities: public hospital and day care center," *Viruses*, vol. 4, no. 11, pp. 2432–2447, 2012.

[5] L. G. Gardinassi, P. V. Marques Simas, J. B. Salomão et al., "Seasonality of viral respiratory infections in southeast of Brazil: the influence of temperature and air humidity," *Brazilian Journal of Microbiology*, vol. 43, no. 1, pp. 98–108, 2012.

[6] R. A. Tripp, "Respiratory syncytial virus (RSV) modulation at the virus-host interface affects immune outcome and disease pathogenesis," *Immune Network*, vol. 13, no. 5, pp. 163–167, 2013.

[7] T. H. Mogensen, "Pathogen recognition and inflammatory signaling in innate immune defenses," *Clinical Microbiology Reviews*, vol. 22, no. 2, pp. 240–273, 2009.

[8] G. Diamond, D. Legarda, and L. K. Ryan, "The innate immune response of the respiratory epithelium," *Immunological Reviews*, vol. 173, pp. 27–38, 2000.

[9] R. J. Boyton and P. J. Openshaw, "Pulmonary defences to acute respiratory infection," *British Medical Bulletin*, vol. 61, pp. 1–12, 2002.

[10] A. K. Mayer, M. Muehmer, J. Mages et al., "Differential recognition of TLR-dependent microbial ligands in human bronchial epithelial cells," *Journal of Immunology*, vol. 178, no. 5, pp. 3134–3142, 2007.

[11] I. Martínez, L. Lombardía, B. García-Barreno, O. Domínguez, and J. A. Melero, "Distinct gene subsets are induced at different time points after human respiratory syncytial virus infection of A549 cells," *Journal of General Virology*, vol. 88, no. 2, pp. 570–581, 2007.

[12] Y. Zhang, B. A. Luxon, A. Casola, R. P. Garofalo, M. Jamaluddin, and A. R. Brasier, "Expression of respiratory syncytial virus-induced chemokine gene networks in lower airway epithelial cells revealed by cDNA microarrays," *Journal of Virology*, vol. 75, no. 19, pp. 9044–9058, 2001.

[13] Y.-C. T. Huang, Z. Li, X. Hyseni et al., "Identification of gene biomarkers for respiratory syncytial virus infection in a bronchial epithelial cell line," *Genomic Medicine*, vol. 2, no. 3-4, pp. 113–125, 2008.

[14] R. Ihaka and R. Gentleman, "R: a language for data analysis and graphics," *Journal of Computational and Graphical Statistics*, vol. 5, no. 3, pp. 299–314, 1996.

[15] R. C. Gentleman, V. J. Carey, D. M. Bates et al., "Bioconductor: open software development for computational biology and bioinformatics," *Genome Biology*, vol. 5, no. 10, article R80, 2004.

[16] L. Gautier, L. Cope, B. M. Bolstad, and R. A. Irizarry, "Affy—analysis of Affymetrix GeneChip data at the probe level," *Bioinformatics*, vol. 20, no. 3, pp. 307–315, 2004.

[17] W. E. Johnson, C. Li, and A. Rabinovic, "Adjusting batch effects in microarray expression data using empirical Bayes methods," *Biostatistics*, vol. 8, no. 1, pp. 118–127, 2007.

[18] J. T. Leek, W. E. Johnson, H. S. Parker, A. E. Jaffe, and J. D. Storey, "The SVA package for removing batch effects and other unwanted variation in high-throughput experiments," *Bioinformatics*, vol. 28, no. 6, Article ID bts034, pp. 882–883, 2012.

[19] M. E. Ritchie, B. Phipson, D. Wu et al., "Limma powers differential expression analyses for RNA-sequencing and microarray studies," *Nucleic Acids Research*, vol. 43, article e47, 2015.

[20] G. R. Warnes, B. Bolker, L. Bonebakker et al., "gplots: various R programming tools for plotting data," *R Package Version*, vol. 2, no. 4, 2009.

[21] K. Breuer, A. K. Foroushani, M. R. Laird et al., "InnateDB: Systems biology of innate immunity and beyond—recent updates and continuing curation," *Nucleic Acids Research*, vol. 41, no. 1, pp. D1228–D1233, 2013.

[22] S. B. Smith, W. Dampier, A. Tozeren, J. R. Brown, and M. Magid-Slav, "Identification of common biological pathways and drug targets across multiple respiratory viruses based on human host gene expression analysis," *PLoS ONE*, vol. 7, no. 3, Article ID e33174, 2012.

[23] M. V. Sitkovsky, D. Lukashev, S. Apasov et al., "Physiological control of immune response and inflammatory tissue damage by hypoxia-inducible factors and adenosine A2A receptors," *Annual Review of Immunology*, vol. 22, pp. 657–682, 2004.

[24] M. A. Schaller, R. Neupane, B. D. Rudd et al., "Notch ligand Delta-like 4 regulates disease pathogenesis during respiratory viral infections by modulating Th2 cytokines," *The Journal of Experimental Medicine*, vol. 204, no. 12, pp. 2925–2934, 2007.

[25] A. Mejias, B. Dimo, N. M. Suarez et al., "Whole blood gene expression profiles to assess pathogenesis and disease severity in infants with respiratory syncytial virus infection," *PLoS Medicine*, vol. 10, no. 11, Article ID e1001549, 2013.

[26] M. J. Gold, P. R. Hiebert, H. Y. Park et al., "Mucosal production of uric acid by airway epithelial cells contributes to particulate matter-induced allergic sensitization," *Mucosal Immunology*, vol. 9, pp. 809–820, 2016.

[27] M. Barends, A. Boelen, L. de Rond et al., "Influence of respiratory syncytial virus infection on cytokine and inflammatory responses in allergic mice," *Clinical and Experimental Allergy*, vol. 32, no. 3, pp. 463–471, 2002.

[28] Y. Li, Y. Wang, C. Zhang et al., "ZNF322, a novel human C_2H_2 Krüppel-like zinc-finger protein, regulates transcriptional activation in MAPK signaling pathways," *Biochemical and Biophysical Research Communications*, vol. 325, no. 4, pp. 1383–1392, 2004.

[29] P. K. Mazur, N. Reynoird, P. Khatri et al., "SMYD3 links lysine methylation of MAP3K2 to Ras-driven cancer," *Nature*, vol. 510, no. 7504, pp. 283–287, 2014.

[30] T. R. Meusel and F. Imani, "Viral induction of inflammatory cytokines in human epithelial cells follows a p38 mitogen-activated protein kinase-dependent but NF-κB-independent pathway," *The Journal of Immunology*, vol. 171, no. 7, pp. 3768–3774, 2003.

[31] M. J. Stewart, S. B. Kulkarni, T. R. Meusel, and F. Imani, "c-Jun N-terminal kinase negatively regulates dsRNA and RSV induction of tumor necrosis factor-α transcription in human epithelial cells," *Journal of Interferon and Cytokine Research*, vol. 26, no. 8, pp. 521–533, 2006.

[32] S. H. Lee, R. F. Kalejta, J. Kerry et al., "BclAF1 restriction factor is neutralized by proteasomal degradation and microRNA repression during human cytomegalovirus infection," *Proceedings of the National Academy of Sciences of the United States of America*, vol. 109, no. 24, pp. 9575–9580, 2012.

[33] M. K. Tripathy, M. E. M. McManamy, B. D. Burch, N. M. Archin, and D. M. Margolis, "H3K27 demethylation at the proviral promoter sensitizes latent HIV to the effects of vorinostat in ex vivo cultures of resting $CD4^+$ T cells," *Journal of Virology*, vol. 89, no. 16, pp. 8392–8405, 2015.

[34] A. P. Oliveira, F. M. Simabuco, R. E. Tamura et al., "Human respiratory syncytial virus N, P and M protein interactions in HEK-293T cells," *Virus Research*, vol. 177, no. 1, pp. 108–112, 2013.

18

Detection of HIV-1 and Human Proteins in Urinary Extracellular Vesicles from HIV+ Patients

Samuel I. Anyanwu,[1] Akins Doherty,[1] Michael D. Powell,[1]
Chamberlain Obialo,[2] Ming B. Huang,[1] Alexander Quarshie (iD),[3]
Claudette Mitchell,[1] Khalid Bashir,[2] and Gale W. Newman (iD)[1]

[1]Department of Microbiology, Biochemistry and Immunology, Morehouse School of Medicine, Atlanta, GA, USA
[2]Department of Medicine, Morehouse School of Medicine, Atlanta, GA, USA
[3]Clinical Research Center, Morehouse School of Medicine, Atlanta, GA, USA

Correspondence should be addressed to Gale W. Newman; gnewman@msm.edu

Academic Editor: Jay C. Brown

Background. Extracellular vesicles (EVs) are membrane bound, secreted by cells, and detected in bodily fluids, including urine, and contain proteins, RNA, and DNA. Our goal was to identify HIV and human proteins (HPs) in urinary EVs from HIV+ patients and compare them to HIV− samples. *Methods.* Urine samples were collected from HIV+ ($n = 35$) and HIV− ($n = 12$) individuals. EVs were isolated by ultrafiltration and characterized using transmission electron microscopy, tandem mass spectrometry (LC/MS/MS), and nanoparticle tracking analysis (NTA). Western blots confirmed the presence of HIV proteins. Gene ontology (GO) analysis was performed using FunRich and HIV Human Interaction database (HHID). *Results.* EVs from urine were 30–400 nm in size. More EVs were in HIV+ patients, $P < 0.05$, by NTA. HIV+ samples had 14,475 HPs using LC/MS/MS, while only 111 were in HIV−. HPs in the EVs were of exosomal origin. LC/MS/MS showed all HIV+ samples contained at least one HIV protein. GO analysis showed differences in proteins between HIV+ and HIV− samples and more than 50% of the published HPs in the HHID interacted with EV HIV proteins. *Conclusion.* Differences in the proteomic profile of EVs from HIV+ versus HIV− samples were found. HIV and HPs in EVs could be used to detect infection and/or diagnose HIV disease syndromes.

1. Introduction

Extracellular vesicles (EVs) are membrane bound vesicles, between 30 nm and 1 μm in size, are secreted into blood, urine, saliva, semen, and other bodily fluids, and have been suggested as a potential source of biomarkers for disease progression [1, 2]. These EVs, microparticles and/or exosomes, are secreted by cells normally or while they are undergoing stress or apoptosis [3] and contain proteins, mRNA, and miRNA [4] that are involved in cell to cell communication, transfer of antigens to cells, and intracellular communication. EVs are described in cancer disease pathogenesis [5] in HIV infection [6], other viral infections [7], and other disease states such as cardiovascular, renal, liver, and metabolic disease [8–11].

EVs from urine are an attractive noninvasive source for biomarkers of diseases [12, 13]. In healthy individuals, protein only accounts for 0.01% of urine components; however, in certain disease states, the protein content and EV numbers can increase in urine [12–16]. The glomerular capsule filters blood that is passed into the renal tubule and accounts for thirty percent of the urinary protein content [14–16]. The remaining seventy percent of proteins in urine is derived from the kidney [17], and thus, urinary EVs are comprised of both renal and efferent components.

HIV proteins are detected in EVs of HIV+ patients and HIV Nef is the most prevalent protein found [18–21]. Other reports of HIV proteins in EVs are from *in vitro* transfected or HIV infected cultured cells and are not from HIV+ patient samples [6, 18, 19, 22, 23].

Biomarkers in urinary EVs are suggested for use in the diagnosis of many disease states [12, 13, 24–30]. The objectives of this study were to determine the differences in proteins from urinary EVs from HIV+ patients and HIV− individuals

TABLE 1: Patient demographics.

Characteristics	HIV-positive (N = 35)	HIV-negative (N = 12)
Age (median ± IQR)	41.5 ± 14.25	59 ± 18
Sex (n, %)		
Male	25 (71.4%)	7 (58.3%)
Female	10 (28.6%)	5 (41.7%)
Race (n, %)		
African American/Black	28 (80%)	12 (100%)
White	7 (20%)	-
Hispanic	-	-
Asian	-	-
Viral loads (copies/ml) (median ± IQR)	50 ± 0	-
CD4+ T cell (cells/μl) (median ± IQR)	66.5 ± 46.5	-
Antiretroviral therapy (n, %)	34 (97.1%)	-

using proteomics and mass spectrometry. The analysis of more patient samples could identify specific EV urinary proteins as biomarkers of HIV infection, treatment efficacy, and/or disease progression.

2. Methods

2.1. Sample Collection. Urine was collected from thirty-five (35) HIV+ patients and twelve (12) HIV− individuals in sterile collection cups. The subjects were recruited from clinics in the metropolitan Atlanta area, GA. Patient demographics are described in Table 1. The study was approved by the Institutional Review Board of Morehouse School of Medicine and written informed consent was obtained from all participants.

2.2. EV Isolation. Urine samples were centrifuged at 1000 ×g to remove cells and sediment then frozen at −80°C. Samples, 4 ml, were thawed and the EVs isolated followed by centrifugal filtration using Amicon Ultra-4 100 kDa centrifugal filter unit (Millipore, Billerica, MA), at 3000 ×g for 15 minutes at 4°C. The retentate, containing EVs, was collected from the top of the filter and resuspended in 200 μl phosphate buffered saline (PBS) for use in the transmission electron microscopy and tandem mass spectrometry (LC/MS/MS) analysis.

2.3. Transmission Electron Microscopy Analysis. Transmission electron microscopy (TEM) was used to identify EVs in two HIV-1 positive and two HIV-1 negative samples. Urinary EVs were fixed in 2.5% glutaraldehyde in 0.1 M cacodylate buffer for 2 hours at 4°C followed by 2 washes with 0.1 M cacodylate buffer, 5 minutes each. Samples were stained with 1% osmium tetroxide in 0.1 M cacodylate buffer for 1 hour at 4°C followed by 2 washes with the cacodylate buffer and 3 washes with deionized water, 5 minutes each. Samples were subsequently stained with 0.5% aqueous uranyl acetate for 2 hours at room temperature and subsequently viewed with

a JEOL 1200EX transmission electron microscope (JEOL, Peabody, MA).

2.4. Nanoparticle Tracking Analysis (NTA). Urine samples from HIV-negative (n = 8) and positive individuals (n = 11), 15 ml, were centrifuged at 300 ×g for 10 min at 4°C to remove cell debris. The supernatant was collected and centrifuged at 16,500 ×g for 20 min at 4°C and the supernatant collected and ultracentrifuged at 120,000 ×g at 4°C for 1.5 hr. The pellet was resuspended in 500 μl of PBS. The size and quantification of the EVs were analyzed using the NanoSight NS500 (NanoSight NTA 2.3 Nanoparticle Tracking and Analysis Release version build 0025). Particles were automatically tracked and sized based on Brownian motion and the diffusion coefficient. The NTA measurement conditions were temperature 21.0 +/− 0.5°C, viscosity 0.99 +/− 0.01 cP, frames per second 24.99−25, and measurement time 30 s. The detection threshold was similar in all samples. Two recordings were performed for each sample.

2.5. Mass Spectrometry Analysis. Thirty-five (35) HIV+ and twelve (12) HIV− EV samples were lysed and trypsinized and the sequence of peptides was determined by tandem mass spectrometry (LC/MS/MS), using an LTQ Ion Trap Mass Spectrometer (Thermo Fischer Scientific, Waltham, MA). Peptides were first reduced in DTT 10 mM at 56°C for at least 30 min and alkylated with 15 mM iodoacetic acid for 30 min at room temperature in the dark. Samples were then digested with mass spectrometry grade trypsin 20 ng/μl for 4 hours at 37°C. Just before analysis, the sample was acidified by the addition of formic acid to 0.1%. Peptides were separated by reverse phase HPLC (Agilent) on a 0.5 × 75 mm C-18 column (Michrom) at a flow rate of 500 nl/min using a linear gradient of acetonitrile (5−35%) over 100 min. Ions were directly introduced by nanospray and spectra were collected using Xcalibur 2.0 software using an intensity threshold of 200 counts. The resulting spectra were analyzed using Bioworks 1.1 software to search a hybrid Human-HIV database created from the complete nonredundant peptide database from NCBI. The threshold for inclusion in the search is a minimal S/N ratio of 3. False discovery rates were determined and set based on the control HIV− samples. An initial protein identification list was generated from matches with an Xcorr score versus charge state of 1.0 (+1) 1.5 (+2) and 1.7 (+3) and consensus scores greater than 10.0.

Bioinformatics techniques for analysis of HIV EV proteins were used on the LC/MS/MS detected proteins [31]. Functional enrichment analysis was performed using FunRich (Functional Enrichment analysis tool, http://funrich.org/index.html) [32] against a human database to detect proteins involved in biological processes, cellular components, sites of expression, and biological pathways. Only processes with a P value < 0.05, using the Benjamini-Hochberg False Discovery rate, were reported. The human proteins detected were compared to the top 100 EV proteins in ExoCarta (http://exocarta.org/exosome_markers_new) [33, 34], sixty EV proteins in the EV array [35], and proteins identified in EVs from HIV infected lymphocytic cells [36].

Pathway analysis comparing HIV+ samples with CD4+ T cells greater than 300 (n = 15) to those with less than 500 (n = 15) and HIV high VL, greater than 200 copies (n = 10), compared to HIV low viral loads, less than 200 copies (n = 10), was done using Pathway Studio version 11.4 Mammal Plus (Elsevier, Inc., Atlanta, GA). Gene Set Enrichment Analysis (GSEA) was used to identify the top 10 curated pathways for the proteins in the each of the patient groups. No comparisons were done between patients not on ART or undergoing ART because there was only one patient not on ART.

The HIV proteins, Nef, Vpr, Vpu, and Vif, were searched using the HIV-1 Human Interaction database (https://www.ncbi.nlm.nih.gov/genome/viruses/retroviruses/hiv-1/interactions/). This database contains all the known, published interactions of HIV-1 gene products with human proteins [37]. Proteins from the search were compared to the human proteins detected in the HIV EVs.

2.6. Western Blot Analysis. To validate the presence of HIV proteins in urinary EVs, western blot analysis (WB) was performed on twenty (20) randomly selected HIV+ and three (3) HIV− control urine samples. Recombinant HIV-1 Nef and HIV-1 p24 were used as positive controls, while HIV-negative urine and HIV-positive filtrate were used as negative controls. Samples were heated at 85°C for two minutes in a tris-glycine SDS sample buffer, were loaded into a 4–20% TGX gradient gel (Bio-Rad, Hercules, CA), and run for 50 mins at 200 V. A semidry transfer unit (Hoefer Scientific, Holliston, MA) was used to transfer the separated proteins onto a PVDF membrane (Bio-Rad) at 15 V for 50 mins. The filter was blocked for nonspecific binding using 5% nonfat dry milk in 1x tris buffered saline (TBS) with Tween 20. The membrane was incubated overnight in pooled plasma from twenty HIV+ patients as the primary antibody at a 1:500 dilution and rabbit anti-human IgG conjugated HRP antibody (1:1000, Bio-Rad, Hercules, CA) was used as secondary antibody. Super Signal West Femto (Thermo Fischer Scientific, Waltham, MA) was used as a chemiluminescent substrate for detection. The membrane was developed and imaged using the LAS 4000 biomolecular imager (GE Healthcare Life Sciences, Pittsburgh, PA). Recombinant HIV-1 Nef and p24 WB analyses were detected using anti-Nef and p24 monoclonal antibodies (1:500, EMD Millipore, Billerica, MA) and anti-mouse IgG conjugated HRP antibodies (1:1000, Bio-Rad, Hercules, CA) were used.

2.7. HIV p24 ELISA. Twenty-six (26) HIV+ and eleven (11) HIV− urine samples were tested for the presence of HIV p24 by ELISA (ImmunoDX, Woburn, MA).

3. Results

3.1. HIV Proteins Are Present in Urinary EVs of HIV-Positive Patients. LC/MS/MS mass spectrometry HIV EV protein results are presented in Table 2. Urinary EV proteins meeting the false discovery rate and Xcorr score criteria as HIV-1 proteins included Nef, Gag, Pol, Protease, gp120, gp160, gp41,

Rev, reverse transcriptase, Tat, Vif, Vpr, and Vpu. All HIV+ urine samples (n = 35) contained at least one HIV-1 protein in EVs, while no HIV proteins were found in the HIV− samples (n = 12) (Table 3). HIV-1 Nef was detected in twenty-six of thirty-five (26 of 35) (74.3%) HIV+ urine samples. Three (3) patients' urine samples, #173, #174, and #196, were tested 203, 311, and 35 days, respectively, after their first EV sample was analyzed. No difference in the HIV proteins detected in sample #196, 35 days after his previous sample, was found. #173's sample, tested 203 days after the first analysis, had a similar profile, except that Rev and Tat were not detected. In addition, #174's EVs examined 311 days after the first sampling found Rev and RT missing from the profile.

HIV p24 antigen was only detected in five of thirty-five (5 of the 35 patient) (14%) samples by LC/MS/MS, but of the twenty-six (26) HIV+ and eleven (11) HIV-negative samples tested by ELISA, no p24 was detected. There was no statistical correlation of the number of HIV proteins detected with CD4+ T cell counts, viral loads, or ART therapy.

Validation by WB analysis using polyclonal pooled patient serum and monoclonal antibodies against HIV Nef and HIV p24 indicated the presence of HIV proteins. Figure 1 is a WB using polyclonal pooled HIV+ serum used as the detection antibody. All the HIV+ patient samples contained HIV-1 proteins and the top panel shows patient samples reacting to anti-HIV Nef. HIV+ urine samples, 7 of 9 (77.7%), showed HIV-1 Nef bands at 27 kD.

3.2. TEM and NTA Analysis of EVs. TEM analysis of urine from HIV+ patients showed multiple EVs, ranging in size from 50 nm to 300 nm (Figure 2(a)), while two HIV-negative controls had fewer EVs present (Figure 2(b)). NTA analysis showed that there were significantly more EVs from HIV+ patients than healthy controls, 4.96 ± 0.0733 and 3.69 ± 0.075, respectively ($P < 0.05$). No significant differences were found in the size of the EVs, 110–227 nm for HIV-negative donors and 54–448 nm HIV+ samples. Representative Nanosight analyses for HIV-negative and HIV+ urine samples are shown in Figure 3.

3.3. Human Proteins in HIV+ and Negative EV Urine Samples. EV proteins from the HIV+ patients, 14,475, which entered into FunRich, functional enrichment analysis software, showed 29.44% or 1,932 proteins were associated with exosomes (Table 4). These EV identified proteins were compared to top 100 EV proteins in the ExoCarta database with 83% matching (http://exocarta.org/exosome_markers_new) [33], 22 EV proteins in the EV array [35] were similar, and 7 of 14 EV proteins identified in exosomes from HIV infected lymphocytes [36] were found and are highlighted in Table 4. Exosomal proteins found in the control samples are listed in Table 5.

The GO results of the FunRich analysis of the EVs from the HIV+ samples are summarized in Table 6 and Figure 4. The top five ($P < 0.01$) EV sites of expression were endothelial cells, plasma, liver, serum, and kidney and the most significant cellular components were lysosomes, exosomes, membranes, plasma membranes, the nucleus, and

TABLE 2: LC/MS/MS analysis of EV HIV proteins.

Accession	# AAs	MW [kDa]	Calc. pI	Description	Σ Coverage	Σ# peptides	Score A0	Coverage A0	# peptides A0
gi384917O5	192	22.7	10.1	Vif protein [human immunodeficiency virus 1]	13.54	12	9.22	13.54	4
gi73913089	104	11.7	10.1	Gag protein [human immunodeficiency virus 1]	14.42	15	6.27	14.42	4
gi58374258	869	98.1	8.8	Envelope glycoprotein [human immunodeficiency virus 1]	1.5	5	5.63	1.5	3
gi183197180	404	45.8	8.4	Pol protein [human immunodeficiency virus 1]	3.47	3	5.27	3.47	3
gi255984636	160	18.1	5.3	Reverse transcriptase [human immunodeficiency virus 1]	7.5	4	4.80	7.5	2
gi256012108	11?	13.5	5.7	Nef protein [human immunodeficiency virus 1]	14.04	3	4.53	14.04	2
gi9756252	524	60.4	8.7	Pol precursor [human immunodeficiency virus 1]	4.01	2	4.46	4.01	2
gi67082579	19?	22.3	9.4	Reverse transcriptase [human immunodeficiency virus 1]	10.47	2	4.43	10.47	2
gi2290009	852	96.7	8.5	Envelope glycoprotein [human immunodeficiency virus 1]	7.16	11	4.33	5.87	3
gi167886806	25	2.7	8.7	Rev protein [human immunodeficiency virus 1]	56	4	4.29	56	2
gi23344577	95	10.6	9.4	Protease [human immunodeficiency virus 1]	12.12	5	4.36	12.12	2
gi4324808	1437	161.9	8.3	Gag-pol polyprotein [human immunodeficiency virus 1]	2.51	7	4.05	1.6	2
gi222533599	73	8.0	9.1	Env C2V3 protein [human immunodeficiency virus 1]	23.29	4	3.85	23.29	2
gi71060450	206	23.7	6.3	Negative factor [human immunodeficiency virus 1]	4.85	4	3.84	4.85	2
gi37935985	85	10.3	4.8	Vpu protein [human immunodeficiency virus 1]	11.76	4	3.84	11.76	2
gi108860432	870	98.8	8.5	gp160 protein [human immunodeficiency virus 1]	3.33	2	3.74	3.33	2
gi114801226	209	24.4	10.1	Tat protein [human immunodeficiency virus 1]	5.26	2	3.70	5.26	2
gi22596451	341	38.6	8.0	Truncated envelope glycoprotein [human immunodeficiency virus 1]	3.81	4	3.68	3.81	2
gi183200570	342	38.7	9.2	Truncated pol protein [human immunodeficiency virus 1]	3.51	2	3.68	3.51	2
gi34786230	176	19.8	9.6	gp120 protein [human immunodeficiency virus 1]	10.8	3	3.60	10.8	3
gi1002239	104	11.5	8.9	Envelope glycoprotein, v3 region [human immunodeficiency virus 1]	18.27	2	3.54	18.27	2
gi77168129	95	11.3	7.6	Vpr protein [human immunodeficiency virus 1]	13.68	3	3.40	13.68	2
gi222532593	129	14.7	8.4	Gag p17 protein [human immunodeficiency virus 1]	8.53	3	2.61	8.53	1
gi219688191	132	14.9	10.2	Matrix protein [human immunodeficiency virus 1]	9.09	6	2.56	9.09	1
gi255687141	288	32.1	7.7	Integrase [human immunodeficiency virus 1]	9.03	1	2.52	9.03	1
gi222532161	132	15.0	9.5	Gag p17 protein [human immunodeficiency virus 1]	9.09	6	2.45	9.09	1
gi37934078	573	65.0	9.0	Gag-pol fusion polyprotein [human immunodeficiency virus 1]	2.79	2	2.42	2.79	2
gi405003	207	23.1	7.7	gp120 protein [human immunodeficiency virus 1]	12.08	1	2.41	12.08	1
gi54792352	213	23.7	5.6	Gag polyprotein [human immunodeficiency virus 1]	9.39	2	2.34	9.39	2
gi3885826	132	14.9	9.6	p17 matrix [human immunodeficiency virus 1]	11.36	2	2.31	11.36	1

Accession # = NCBI NR database, #AAs = total number of amino acids in the protein entry, MW = molecular weight of the protein in kDa, Description: description from the NCBI database, and Peptides = total number of unique peptides found.

TABLE 3: Presence of HIV-1 proteins in HIV+ patient urinary EVs.

ID	ART	AIDS	Viral load copies/ml	CD4 cells/ul	Nef	Gag	Pol	Protease	Rev	RT	Tat	Vif	p1	p24	p17	Poly	Vpu	Env	Vpr	Vif
22	No			224	X	X	X	X	X		X	X	X	X	X					
27	Yes	AIDS	<50	134	X															
28	Yes	AIDS	280100	22	X	X		X		X		X								
30	Yes	AIDS	>10000	<20	X	X	X	X		X	X	X		X	X	X				
41	Yes		29187	440	X	X	X													
46	Yes		<50	689		X	X		X	X										
45	Yes		400	345	X			X	X									X		X
48	Yes		4974	454	X				X		X	X								
51			NA	NA	X	X	X	X		X										
52	Yes		51	574	X	X	X		X		X						X		X	X
61			<50	655		X	X													
62	Yes	AIDS	<50	232	X															
63	No		2023	83	X															
65	No		NA	NA														X		
66	No		NA	NA			X													
67	Yes		75	509	X															
68	No		NA	NA														X		
69	Yes		<50	187	X															
70	Yes		<50	399	X															
71	Yes		<50	456														X		
74			NA	NA														X		
86	Yes		<75	1642	X	X	X	X												
103	Yes	AIDS	150	560	X	X	X													
104	Yes	AIDS	77	313	X	X														
108	Yes	AIDS	<50	653	X	X	X													
110	Yes		<50	379	X	X														
111	Yes	AIDS	<50	182		X	X													
112	Yes	AIDS	>200	581		X				X										
142	Yes		<50	487	X	X	X		X	X						X	X			X
173-1	Yes		<50	398	X	X	X	X	X	X	X	X		X				X		
173-2	Yes		<50	398	X	X	X	X		X		X								
174-1	Yes		48	315	X	X	X		X	X		X								
174-2	Yes		48	315	X	X	X					X								
196-1	Yes	AIDS	<50	113	X	X	X	X	X	X	X	X		X	X	X	X			X
196-2	Yes	AIDS	<50	113	X	X	X	X	X	X	X	X	X	X	X	X				X

An initial protein identification list was generated from matches with an Xcorr score versus charge state of 1.0 (+1), 1.5 (+2), and 1.7 (+3) and consensus scores greater than 10.0; NA = not available.

the cytoplasm ($P < 0.01$) (Figure 4). The top five ontologies (Table 6) were protein serine/threonine kinase activity, catalytic activity, GTPase activator activity, guanyl-nucleoside exchange factor activity, and cell adhesion molecule activity ($P < 0.0001$), the top biological process was regulation of nucleobase, nucleoside, and nucleic acid ($P < 0,0001$), and the most prominent biological pathway was integrin cell surface interactions ($P < 0.03$).

LC/MS/MS identified 15,571 proteins in EVs from HIV+ patients with CD4+ T cells greater than 300, 2,115 from CD4+ T cells less than 300, 15,028 proteins from patients with low VL, and 2486 from patients with high VLs. Pathway analysis was similar between EV proteins from patients with greater than 300 CD4+ T cells and low VLs and different between the low CD4+ T cells and high VLs (summarized in Table 7). The pathways found are detailed in Supplementary Material 1. Interleukin proteins detected were IL10, IL10RA, IL16, IL17RC, IL18, IL18BP, IL1RAP, IL1RL2, IL1RN, IL33, IL4I1, IL6, and IL6ST. Immunomodulatory molecules, HOXB4, CD81, CD9, TGF-β1, IDO, Notch1, ADAM17, Rab4, and HGF,

TABLE 4: Exosomal proteins found in urinary EVs from HIV+ patients.

	Genes in our analysis	Genes in the FunRich database	Percentage of genes	Fold enrichment
	1932	2001	29.44	2.11

Exosomal proteins

A1BG, *A2M*, AARS, ABCA7, ABCB1, ABCB6, ABCC1, ABCC11, ABCG2, ABHD8, ACAA2, ACAT1, ACAT2, ACE, ACE2, ACLY, ACO1, ACOT11, ACP2, ACSL2, ACSL3, ACSL4, ACSM1, *ACTA1*, **ACTA2**, *ACTB*, **ACTB12**, ACTC1, *ACTG1*, **ACTG16A, ACTN1, ACTN2, *ACTN4*, ACTR1A, ACTR1B, ACTR2, *ACTR3*,** ACY1, ACY3, ADAM10, ADAMTS3, ADCY1, ADH5, ADH6, ADK, ADSL, AEBP1, AGAP2, AGA2, AGR2, AGR3, AGRN, AGT, AHCTF1, AHCYL1, ACTC1, *ACTG1*, ACTL6A, AHSG, AK1, AK2, AKAP9, AKRIA1, AKR1B10, ALAD, *ALB*, ALCAM, ALDH1L1, ALDH2, *ALDOA*, ALDOB, ALDOC, ALK, ALOX12, ALPL, ALPP, ALYREF, AMBP, AHNAK, AHSA1, ALAD, ALDH3B1, ALDH1A1, AMN, ANGPT1, ANGPT1L, ANGPT2L4, ANKFY1, ANKRD11, ANO1, ANO6, *ANPEP*, ANXA1, ANXA2, ANXA3, ANXA4, ANXA6, ANXA7, AOX1, AP1M1, AP2A1, AP2A2, AP2M1, AP4M1, APAF1, APAF2, APOA1, APOA2, APOB, APOD, APOE, APOL1, APP, APPL1, APPL2, APRT, ARF5, ARF1P1, ARHGAP1, ARHGAP23, ARHGDIA, ARHGEF12, ARHGEF18, ARL15, ARL3, ARL8B, ARMC3, ARMC9, ARPC1A, ARPC1B, ARPC2, ARPC3, ARPC5, ARRDC1, ARSE, ARSF, ARVCF, ASAH1, ASB6, ASL, ASNA1, ASNS, ATAD2, ATIC, *ATP1A1*, ATP1A2, ATP1A3, ATP2B1, ATP2B4, ATP5A1, ATP5B, ATP5L, ATP6AP1, ATP6AP2, ATP6V0A1, ATP6V0A4, ATP6V0C, ATP6V0D2, ATP6V1A, ATP6V1C1, ATP6V1D, ATP6V1E1, ATP6V1H, ATRN, AUP1, AZGP1, AZU1, B2M, B3GAT3, B4GALT1, B4GALT3, BAIAP2, BAIAP2L1, BASP1, BAZ1B, BCAM, BCR, BDH2, BGN, BHLHB9, BHMT, BHMT2, BLMH, BLOC1S5, BLVRA, BLVRB, BMP3, BPI, BPIFB1, BPTF, BRI3BP, BROX, *BSG*, BTG2, BTN1A1, C11orf54, C16orf80, C19orf18, CIGALT1C1, C1orf16, CILQC, C1QTNF1, C1QTNF3, C1R, C2orf16, C3, C4BPA, C5, C9, CAB39L, CACNA2D1, CACYBP, CAD, CALM1, CALML3, CALR, CAMK4, CAMP, CAND1, CANX, CAP1, CAPN5, CAPN7, CAPNS1, CAPS, CAPZA2, CAPZB, CARD11, CASP9, CAV1, CBR3, CCDC105, CCDC32, CCL28, CCPG1, CCT2, CCT3, CCT4, CCT5, CCT6A, CCT7, CCT8, CD101, CD4, CD63L1, *CD9*, CD2, CD22, CD274, CD2AP, CD300A, CD36, **CD37**, CD40, CD44, CD53, CD55, CD58, CD59, **CD63**, CD70, CD74, CD79B, CD80, **CD81, CD9, CD97,** CDC42, CDC42BP, CDH1, CDH17, CDH2R2, CDHR5, CDK1, CDK5RAP2, CDKL1, CEACAM5, CELSR2, CEMIP, CEP250, CES2, CETP, CFD, CFH, CFI, *CFL1*, CHGB, CHID1, CHMP1A, CHMP4B, CHRDL2, CHST1, CHST14, CIB1, CKAP4, CKB, CLASP1, CLCA4, CLDN3, CLDN4, CLDN7, CLIC1, CLIC4, CLIC5, CLIC6, CLIP2, CLSTN1, *CLTC*, CLTCL1, CLU, CMPK1, CNDP2, CNKSR2, CNTLN, COASY, COBLL1, COL12A1, COL15A1, COL18A1, COL1A1, COL6A1, COL6A2, COL6A3, COLEC10, COLGALT1, COMT, COPA, COPB1, COPB2, COPS8, COROIA, COROIB, COX4I1, COX5B, CP, CPD, CPN2, CPNE1, CPNE3, CPNE8, CPVL, CR1, CR2, CRB2, CREB5, CRISPLD1, CRNN, CRTC2, CS, CSE1L, CSK, CSPG4, CST4, CSTB, CTDSPL, CTNNA1, CTNNB1, CTNND1, CTSB, CTSC, CTSG, CTTN, CUBN, CUL3, CUL4B, CUTA, CUX2, CXCR4, CYB5R1, CYBRD1, CYFIP1, CYFIP2, CYP2I2, DAAM2, DAG1, DAK, DARS, DBNL, DCD, DCTN2, DCXR, DDAH1, DDAH2, DDB1, DDC, DDR1, DDX1, DDX19A, DDX19B, DDX21, DDX23, DDX3X, DDX5, DERA, DHCR7, DHX36, DHX9, DIAPH2, DIP2A, DIP2B, DIP2C, DLD, DLG1, DMBT1, DNAH7, DNAH8, DNAJA1, DNAJA2, DNAJB1, DNAJB9, DNAJC13, DNAJC7, DNHD1, DNM2, DNPH1, DOCK10, DOCK2, DOPEY2, DPEP1, DPP3, DPP4, DPYS, DPYSL2, DSC2, DSG2, DSP, DSTN, DUOX2, DUSP26, DUT, DYNC1H1, DYNC2H1, DYSF, ECE1, ECH1, ECM1, EDIL3, EEA1, *EEF1A1*, EEF1A2, EEFID, EEFIG, *EEF2*, EFEMP1, EFEMP2, EGF, **EGFR**, EHD1, EHD2, EHD3, EIF2S1, EIF3B, EIF3E, EIF3L, EIF4A1, EIF4A2, EIF4A3, EIF4E, EIF4G1, *EHD4*, EIF3J, EIF5A, EML5, ENO1, ENO2, ENO3, ENPP1, ENPP4, ENPP6, ENTPD1, EPB41L2, **EPCAM**, EPHA2, EPHA5, EPHB1, EPHB2, EPHB3, EPHB4, EPHX2, EPN3, EPPK1, EPRS, EPS8, EPS8L1, EPS8L2, ERAP1, ERBB2, ERMN, EROIL, ERP44, ESD, ETFA, EVPL, EXOC4, EXOSC10, EXT2, EYS, *EZR*, F11, F11R, F5, F7, FABP1, FABP3, FAH, FAM129A, FAM129B, FAM151A, FAM208B, FAM209A, FAM20C, FAM49B, FAM65A, FAS, FASLG, FASN, FAT1, FAT2, FBL, FBP1, FBP2, FCGBP, FCN1, FCN2, FERMT3, FGA, FGB, FGG, FGL2, FGR, FH, FIGNL1, FKBP1A, FKBP4, FKBP5, *FLNA*, FLNB, FLNC, FLOT1, **FLOT1, FLOT2,** FMN1, FOLH1, FRK, FSCN1, FUCA1, FURIN, FUS, FUT1, FUT2, FUT3, FUT4, FUZ, G6PD, GAA, GABRB2, GAL3ST4, GALK1, GALM, GALNT3, GANAB, GARS, GART, GATSL3, GBEL, GBP6, GCNL1, GCNT2, GCNT3, GDF2, GDPD3, GEMIN4, GFPT1, GGCT, GGH, GHTM, GIPC1, GK, GK2, GLB1, GLDC, GLG1, GLIPR2, GLO1, GLUD1, GLUL, GNAI3, GNAI1, GNAI2, GNAQ, GNAS, GNB1, GNB2, GNB2L1, GNB3, GNB4, GNB5, GNG12, GNPDA1, GNPTG, GOLGA4, GOLGA8A, GOT1, GOT2, GPC1, GPC4, GPD1, GPI, GPM6A, GPR55, GPR64, GPR98, GPRASP1, GPRC5A, GPRC5B, GPT, GREB1, GRHPR, GRID1, GRN1, GRK4, GSN, GSR, GSS, GSTA3, GSTCD, GSTK1, GSTO1, GSTP1, GUSB, HIFOO, H2AFY, H2AFY2, HADHA, HAPLN3, HAUS5, HBB, HBD, HBS1L, HDHD2, HEBP1, HEBP2, HEPH, HGD, HGS, HINT1, HIRA, HIST1H1B, HIST1H2BA, HIST1H2BL, HIST1H2BM, HIST1H4B, HIST1H4H, HIST2H2AC, **HLA-A, HLA-B,** HLA-DPB1, HLA-DQB1, HLA-DRB1, **HLA-DRB1, HLA-DRB5,** HLAE, HNMT, HNRNPA1, HNRNPA2B1, HNRNPC, HNRNPF, HNRNPK, HNRNPL, HP, HPD, HPGD, HPR, HPRT1, HRG, HRNR, HSD11B2, HSD17B4, HSD17B10, HSPA12A, HSPA1A, HSPA1L, HSPA4, *HSPA5*, HSPA6, HSPA8, HSPA9, HSPB1, HSPD1, HNRNP, HNRNPH, HP, HPD, HPGD, HPR, HPRT1, HRG, HRNR, *IQCB1*, IQCG, IQGAP1, IQGAP2, IRF6, IST1, ITFG3, ITGA1, HSPG2, HSPH1, HTATIP2, HTRA1, HUWE1, HYOU1, IARS, **ICAM1,** ICAM3, IDH1, IFITM2, IFITM3, IGF2R, IGFALS, IGSF3, IGSF8, IKZF5, IMPDH2, INADL, INSR, *ITGA2, ITGA2B, ITGA3, ITGA4, ITGA6, ITGAL, ITGAV, **ITGB1,** ITGB2, ITGB3, ITGB4, ITGB7, ITGB8, ITIH2, ITIH4, ITM2C, ITSN1, ITSN2, IVL, JADE2, JUP, KALRN, KCNG2, KHK, KIAA1324, KIF12, KIF15, KIF8B, KIF3A, KIF3B, KIF9, KIFC3, KL, KNG1, **KPNB1,** KPRP, KRT1, KRT10, KRT12, KRT14, KRT15, KRT16A, KRT17, KRT18, KRT19, KRT2, KRT20, KRT24, KRT25, KRT27, KRT28, KRT3, KRT5, KRT6C, KRT73, KRT75, KRT76, KRT77, KRT78, KRT79, KRT8, KRT9, **LICAM,** *AD1*, LAMA3, LAMA4, LAMA5, LAMB2, LAMB3, LAMC1, LAMC2, LAMP1, **LAMP2,** LAMTOR3, LBP, LCK, LCP1, **LDHA, LDHB,** LEPRE1, LFNG, LGALS3, LGALS3BP, LGALS4, LIMA1, LIN7A, LIN7C, LMAN1, LMAN2, LOX14, LPO, LRP1, LRPIB, LRP2, LRP4, LRPPRC, LRRC15, LRRC16A, LRRC57, LRRK2, LRSAM1, LSP1, LSR, LTA4H, LTBP3, LTF, LUZP1, LYPLA2, MAGI3, MAL2, MAN1A1, MANIA2, MAN2A1, MAP4K4, MAP7, MARCKS1L, MARK3, MARS, MARVELD2, MASP1, MASP2, MBD5, MBLAC2, **MCAM,** MCFD2, MDH1, MDH2, MEGF8, MEPIA, MEST, METRNL, *MFGE8,* MGAM, MGAT1, MGAT4A, MIED, MIF, MINK1, MLT3, MLT4, MME, MMP24, MMP25, MMRN1, MMRN2, MNDA, MOB1A, MOB1B, MOGS, MPO, MPP5, MPP6, MS4A1, MSN, MSRA, MTA1, MTAP, MTCH2, MTHFD1, MTMR1, MTMR2, MUC13, MUC16, MUC4, MUM1L1, MVB12A, MVB12B, *MVP*, MX1, MXRA5, MXRA8, MYADM, MYH10, MYH11, MYH14, MYH3, MYH8, MYO15A, MYO1B, MYO1C, MYO1D, MYO1E, MYO1G, MYO5B, MYO6, MYOF, N4BP2L2, NAA16, NAA50, NACA, NAGLU, NAMPT, NAPIL4, NAPA, NAPG, NAPRT, NAP1A, NARS, NBR1, NCALD, NCCRP1, NCKAP1, NCL, NCOA3, NCSTN, NDRG1, NDRG2, NEB, NEBL, NEDD4, NEDD4L, NEU1, NID1, NIN, NIPBL, NIT2, NKXG1, NONO, NOTCH1, NOX3, NPC1, NPEPPS, NPHS1, NPHS2, **NPM1,** NPNT, NQO2, NT5C, NT5E, NUCB1, NUCB2, NUDT5, NUMA1, NXPE4, OLA1, OPTN, OR2A4, OS9, OSBPL1A, OXSR1, P2RX4, P4HB, PA2G4, PACSIN2, PACSIN3, PADI2, PAFAH1B1, PAFAH1B2, PAGE2, PAICS, PAM, PARD6B, PARP4, PBLD, PCBP1, PCDHGB5, PCK1, PCLO, PCNA, PCSK9, PCYOX1, **PDCD6,** PDCD5, **PDCD6, PDCD6IP,** PDDC1, PDE8A, PDIA2, PDIA4, PDIA6, PDLIM7, PDZK1, PEBP1, PEBP4, PECAM1, PEF1, PEPD, PEX1, PFAS, PFKL, PFKP, PGAM1, PGD, PHGDH1, PI4KA, PIGR, PIK3C2A, PIK3C2B, PILRA, PIR, PIP4K2C, PKD1, PKD1L3, PKD2, PKHD1, PKLR, PKM, PKN2, PKP3, PLAT, PLAU, PLCB1, PLCD1, PLCG2, PLD3, PLEC, PLEKHA1, PLEKHA7, PLEKHB2, PLG, PLIN2, PLOD1, PLOD2, PLS1, PLSCR1, PLTP, PLVAP, PLXNB2, PM20D1, PMEL, POM, PODXL, POFUT2, PON1, PON2, POTEE, POTEF, POTEI, POTEM, PPAL, PPARG, PPFIA2, *PPIA*, PPIB, PPL, PPM1L, PPP1CB, PPP2R1B, PPP2R2A, PPP2RIA, PPP2RIB, PRSS23, PRTN3, PSAP, PSAT1, PSMA2, PSMA3, PSMA5, PSMA7, PSMB1, PSMB3, PSMB4, PSMB5, PSMB6, PSMB8, PSMB9, PSMC2, PSMC4, PSMC6, PSMD1, PSMD12, PSMD13, PSMD2, PSMEL, PSME2, PSME3, PTBP1, PTER, *PTGFRN*, PTGR1, PTGSL, PTPN13, PTPN23, PTPRA, **PTPRC,** PTPRE, PTPRJ, PTPRO, PTRF, PTX3, PYGB, PYGL, QPRT, QPRT, QSOX1, RAB10, RAB13, RAB1A, RAB1B, RAB2A, RAB25, RAB39B, RAB34, RAB4B, RAB6A, RAB7A, *RAB8A*, RAB9A, **RAC1,** RACGAP1, RALA, RALB, RAPIA, RAPID5L, RAP2A, RAPGEF3, RAIRES1, RARS, RASAL3, RASSF9, RBL2, RCC2, REG4, RELN, RENBP, RFC1, RFTN1, RHEB, RHOB, RHOF, RIMS2, RLF, RNASE7, RNF213, RNH1, RNPEP, ROBO2, ROCK2, RP2, RPL10, RPL14, RPL15, RPL23, RPL3, RPL30, RPL34, RPL35A, RPL4, RPL5, RPL6, RPL8, RPLP2, RPN1, RPS16, RPS18, RPS2, RPS20, RPS21, RPS27A, RPS3A, RPS4X, RPS4Y2, RPS7, RPS9, RRAS, RREB1, RSU1, RTN4, RUSC2, RUVBL1, RUVBL2, RYR1, S100A11, S100A6, S100P, SAR1, SAFB2, SAMHD1, SCARB1, SCARB2, SCEL, SCIN, SCN10A, SCN11A, SCPEP1, SCRIB, SCYN2, **SDCBP, SDF4,** SEC31A, SELENBP1, **SELP,** SEMA3G, SEPP1, SEREP1, SERINC1, SERINC2, SERINC5, SERPINA1, SERPINA3, SERPINA5, SERPINB6, SERPING1, SERPINE2, SETD4, SFIL, SFN, SFRP1, SFT2D2, SH3BP4, SHMT1, SHMT2, SHROOM2, **SLA,** SIRPA, SIT1, SLAMF1, SLAMF6, SLC12A1, SLC12A9, SLC13A2, SLC16A1, SLC1A5, SLC20A2, SLC22A11, SLC22A12, SLC22A2, SLC22A5, SLC22A6, SLC23A1, SLC25A1, SLC25A4, SLC26A1, SLC26A9, SLC27A1, SLC2A3, SLC34A2, SLC35D1, SLC37A2, SLC38A1, SLC39A5, *SLC3A1*, *SLC3A2*, SLC44A1, SLC44A2, SLC46A3, SLC4A4, SLC4A1, SLC5A1, SLC5A10, SLC5A2, SLC5A6, SLC6A19, SLC7A5, SLC9A1, SLC9A3R1, SLC9A3R2, SLC0A4C1, SLIT2, SLK, SMC2, SMC3, SMIM22, SMIM24, SMO, SMPDL3B, SMURF1, SNCG, SND1, SNRNP200, SNX12, SNX18, SNX25, SNX3, SNX9, SOD1, SOGA1, SORD1, SORL1, SORT1, SPAG9, SPAST, SPEN, SPINK1, SPON2, SPRR3, SPTAN1, SPTBN1, SQSTM1, SRC, SRPR, SRSF7, ST13, ST3GAL1, ST3GAL6, STAMBP, STAU1, STIP1, STK10, STK11, STK24, *STOM*, STRIP1, STX3, STX4, STX7, STXBP1, STXBP2, STXBP3, TGFB1, TGFBR1, TGFBR3, TGM1, TGM2, TGM3, TGM4, *THBS1*, THBS2, THRAP3, THSD4, THY1, TIAM2, TINAGL1, TJP2, *TKT*, TLN1, TLR2, TM9SF2, TM9SF3, TM9SF4, TMBIM1, TMC6, TMC8, TMED2, TMED9, TMEM109, TMEM192, TMEM2, TMEM256, TMEM27, TMEM63A, TMPRSS11B, TMPRSS2, TNFAIP3, TNFRSF8, TNFSF10, TNFSF13, TNIK, TNKS1BP1, TNPO3, TOLLIP, TOM1, TOMI1L2, TOM1L2, TOMM70A, TORIA, TORIB, TOR3A, *TP11*, TPM3, TPP1, TPRGIL, TRAP1, TREH, TRIP10, TSNAX1P1, TSPAN1, TSPAN15, TSPAN3, TSSK3, TSTA3, TTC17, TTC18, TTLL3, TTN, TTR, *TUBA1B*, TUBA4A, TUBB3, TUBB4A, TUBB8, TUFM, TWF2, TXNDC16, TXNDC8, TXNRD1, TYK2, TYRP1, UACA, *UBA1*, UBAC1, UBASH3A, **UBE2N,** UBE2V2, UBL3, UBXN6, UEVLD, UGDH, UGGT1, UGP2, ULK3, UMOD, UPB1, UPK1A, UPK3A, UQCRC2, UTRN, UXS1, VAMP1, VAMP3, VAMP7, VAT1, VCL, *VCP*, VDAC3, VIL1, VIM, VMO1, VPSI3C, VPSI3D, VPS28, VPS36, VPS37B, VPS37C, VPS37D, VPS4A, VPS4B, VTA1, VWA2, VWF, WARS, WAS, WASF2, WASL, WDR1, WIZ, WNT5B, XDH, XPNPEP2, XPO1, XRCC5, XRCC6, YBX1, YES1, *YWHAE*, *YWHAH*, *YWHAQ*, *YWHAZ*, ZCCHC11, ZDHHC1, ZFYVE20, ZG16B, ZMPSTE24, ZNF14, ZNF486, ZNF571, and ZNHIT6.

GENE: ExoCarta (http://exocarta.org/exosome_markers_new) [33]; **GENE**: EV antibody array [35]; GENE: HIV exosomal proteins [36].

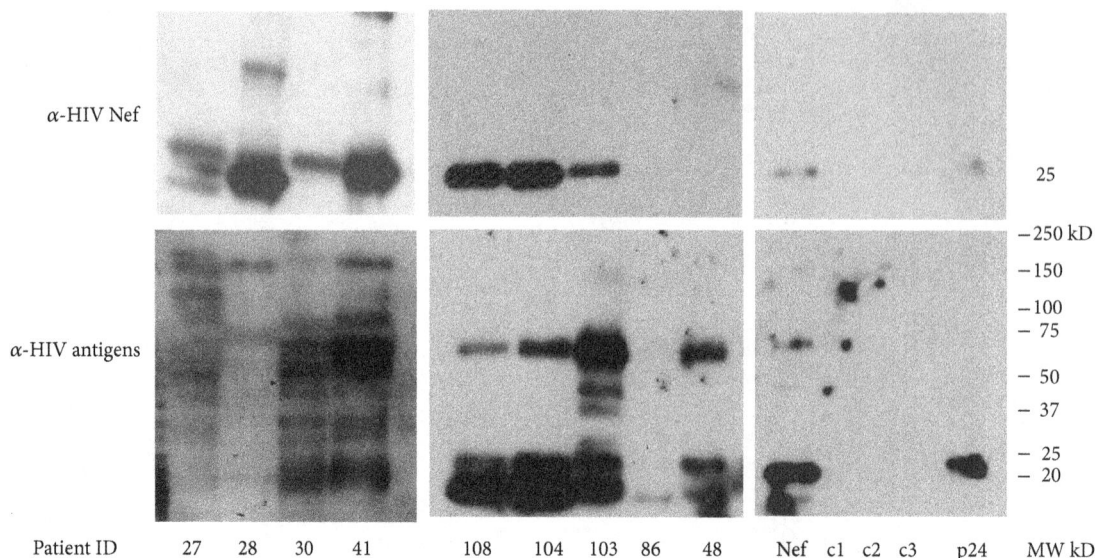

FIGURE 1: *Detection of HIV-1 proteins by western blot.* Extracellular vesicles were isolated from four ml of urine from HIV-1+ patients and HIV-1 negative individuals by Amicon ultrafiltration (MW cutoff = 100,000 kD). The western blot is representative of 9 HIV+ and 3 HIV-negative samples (c1, c2, and c3). Recombinant HIV Nef and p24 were added as positive controls (last panels on the right). Samples were isolated in a 4–20% gradient SDS gel and transferred to a PVDF membrane. The filter was incubated with the primary antibody, pooled HIV-1 positive plasma (bottom panels), or a monoclonal anti-HIV Nef (top panels). The secondary antibody, goat anti-mouse IgG for the anti-Nef blots or rabbit anti-human IgG for the anti-HIV antibodies, conjugated to horseradish peroxidase. Super Signal West Femto was used as chemiluminescent substrate for detection.

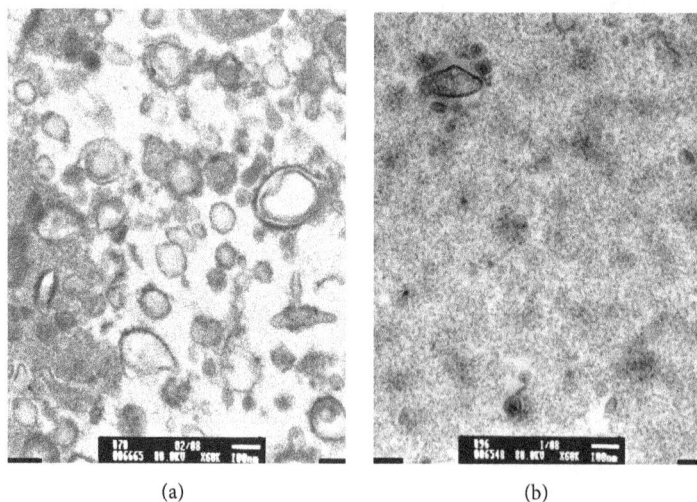

FIGURE 2: *Transmission electron microscopy of urinary extracellular vesicles.* Four mls of urine was used to isolate EVs by Amicon ultrafiltration (MW cutoff = 100,000 kD). EVs were fixed in 2.5% glutaraldehyde in 0.1 M cacodylate buffer. Samples were stained with 1% osmium tetroxide in 0.1 M cacodylate buffer and subsequently stained with 0.5% aqueous uranyl acetate. A JEOL 1200EX transmission electron microscope (JEOL, Peabody, MA) was used for observation and photography. **1A.** EVs from HIV-1 posi.

TABLE 5: Exosomal proteins found in urinary EVs from uninfected controls.

	Genes in our analysis	Genes in the FunRich database	Percentage of genes	Fold enrichment
Exosomal proteins	37	2001	72.54	5.26

A1BG, ACTA1, ACTA2, ACTB, ACTBL2, ACTC1, ACTG1, ACTG2, ALB, AMBP, APOA1, APOD, AZGP1, B2M, CDH1, CLU, CP, CRNN, DCTN2, EGF, HP, HPR, HSPB1, ITIH4, KNG1, LAMA3, LMAN2, POTEE, POTEF, POTEI, S100A8, SERPINA1, SERPING1, TF, TTR, UMOD, and VASN.

TABLE 6: Functional enrichment analysis of HIV+ EV proteins.

	Genes in the dataset	Genes in the Bkg. database	Percentage of genes	Fold enrichment	Corrected P value (BH FDR)
Molecular function					
Protein serine/threonine kinase activity	272	5,602	30	1.18	1.04^{-08}
Catalytic activity	456	827	4.9	1.1	1.12^{-05}
GTPase activator activity	131	836	4.7	1.2	8.14^{-05}
Guanyl-nucleoside exchange factor activity	105	614	3.6	1.2	8.54^{-05}
Cell adhesion molecule activity	307	531	3.3	1.1	0.0001
Biological process					
Regulation of nucleobase, nucleoside, and nucleic acid	2,236	4,658	24.8	1.05	3.24^{-05}
Biological pathway					
Integrin cell surface interactions	69	1,366	23.3	1.2	0.03

TABLE 7: Comparison of pathways between HIV+ groups from Pathway Studio 11.4.

HIV group	Pathway
CD4+ T cells greater than 300 $n = 15$	Natural killer cell inhibitor receptor signaling Intermediate filament polymerization Ca2+ flux regulation G1/S phase transition G2/M phase transition S/G2 phase transition Protein folding Golgi to endosome transport Endosomal recycling Kinetochore assembly
CD4+ T cells less than 300 $n = 15$	Neutrophil chemotaxis Vascular motility Platelet activation via GPCR signaling Insulin influence on protein synthesis mTOR signaling overview EDNRA/B → vascular motility Proplatelet maturation Natural killer cell activation through ITAM-containing receptors Taste sensor receptors activates mTOR signaling Natural killer cell activation
Low VLs $n = 14$	Intermediate filament polymerization Natural killer cell inhibitory receptor signaling golgi to endosome transport Ca+ flux regulation HRH1/3 → synaptic transmission Vascular motility Endosomal recycling G1/S phase transition Golgi transport G2/M phase transition
High VLs $n = 10$	Metaphase/anaphase phase transition S/G2 phase transition Spindle assembly Natural killer cell activation Histone ubiquitylation Eosinophil survival by cytokine signaling Protein folding G2/M phase transition

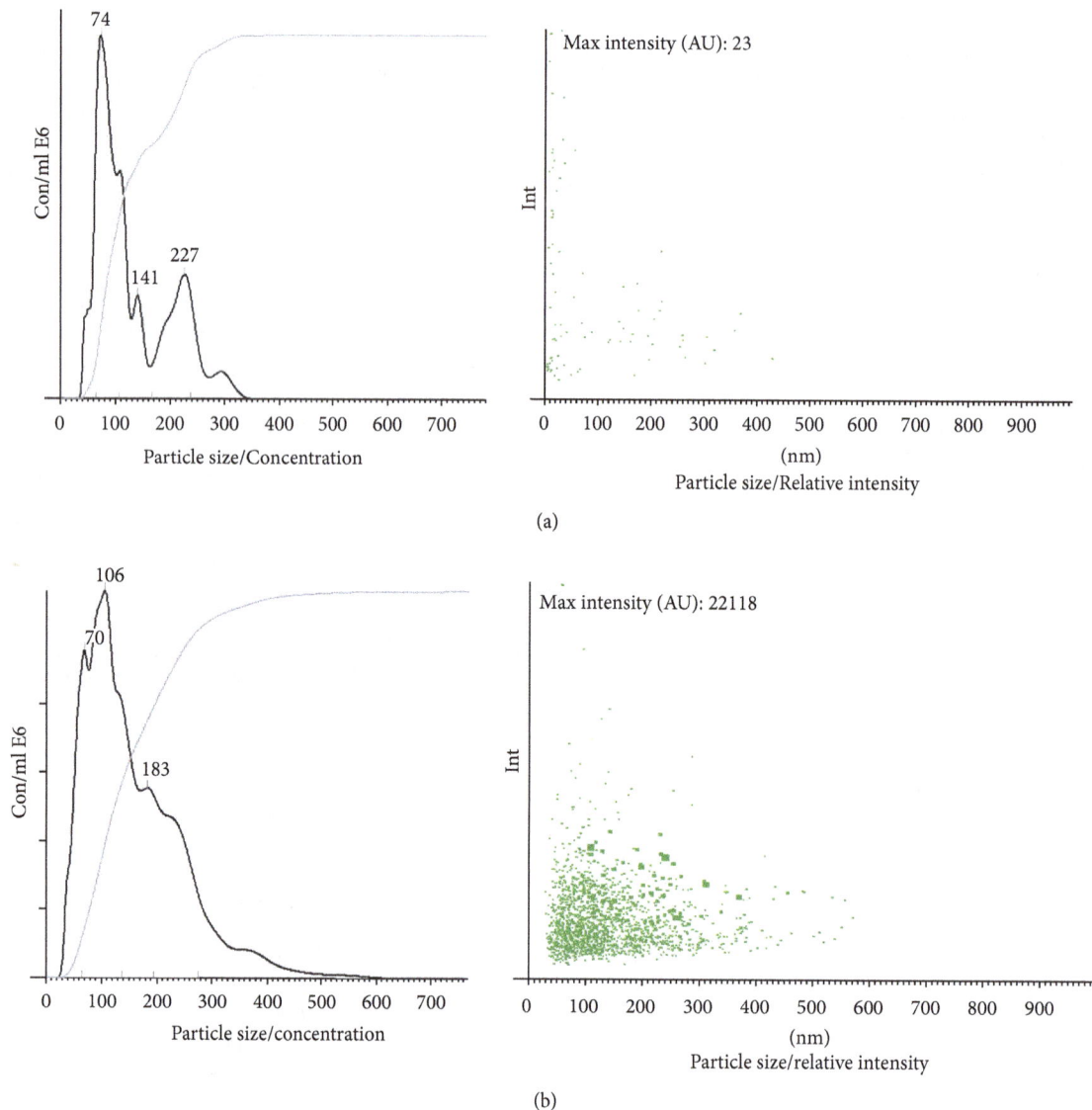

FIGURE 3: *Nanosight analysis* (representative analysis). (a) NTA analysis of an HIV-negative urine sample had 0.4×10^8 particles per ml (left panel) while (b) depicts an urine sample from a HIV+ patient that had 8.7×10^8 particles per ml and has a greater relative intensity profile (right panel (a) and (b)) when compared to the HIV-negative sample. The Rank Sum T test showed that HIV+ patient urine samples had more particles per ml than the negative control urine ($P < 0.05$).

were also found by LC/MS/MS in addition to MHC Class I and II antigens.

The HIV-1 Human Interaction database search found that HIV Nef interacted with 559 EV proteins of 770 total human proteins (72.6%); HIV Vpr interacted with 437 EV of 598 human (73.1%); HIV Vif interacted with 162 EV of 310 human (52.2%); and HIV Vpu interacted with 165 EV of 244 human proteins which were found in the HIV+ EVs (67.6%) (see Supplementary Material 2, including PMIDs for references).

Functional analysis of the control EVs are listed in Table 8. The major sites of expression were cervicovaginal fluid, neutrophils, and gastric juice ($P < 0.0001$). The most significant ontologies were molecular function of the proteins and defense/immunity protein activity and principal biological

processes were immune response, signal transduction, cell communication, and antigen presentation ($P < 0.0073$).

Only sixty-four (64) proteins overlapped between the HIV+ and control EV samples and are listed in Table 8. The top fourteen (14) GO ontologies for cellular components include extracellular exosome, extracellular region, extracellular space, hemoglobin complex, and blood microparticle ($P < 0.001$, Table 9), GO ontologies for molecular function were heparin binding, ion gated activity, and oxygen transporter activity, and the most significant biological processes found were response to yeast, defense response to fungus, macrophage chemotaxis, negative regulation of growth of symbiont in host, oxygen transport, and hydrogen peroxide catabolic process.

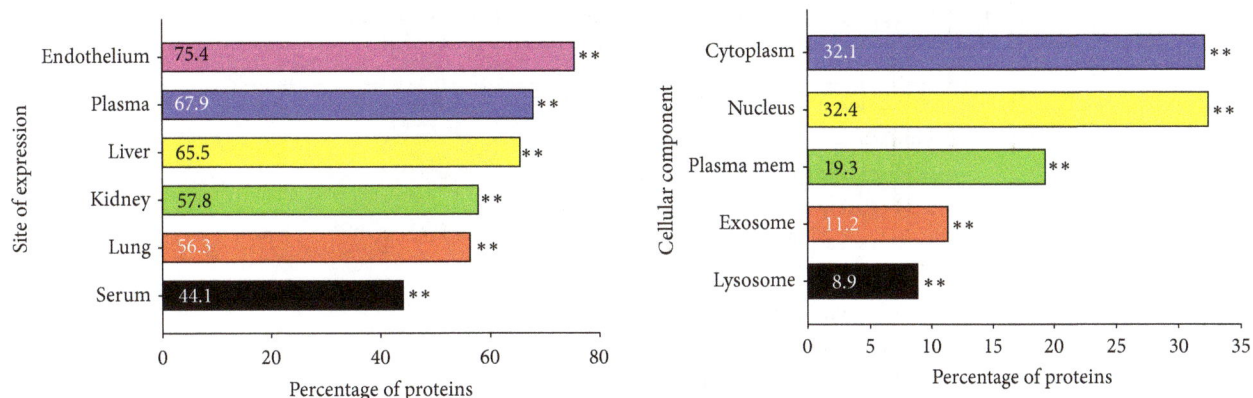

FIGURE 4: *Percentage of proteins found in HIV+ urinary EVs.* FunRich analysis of the LC/MS/MS proteins from HIV+ EVs determined the most likely tissue expressing the proteins, site of expression, and the cellular component from which the protein is derived. Data is graphed as the percentage of proteins found. ** denotes significance, $P < 0.01$.

TABLE 8: Functional enrichment analysis of control EV proteins.

	Genes in the database	Genes in the Bkg. database	Percentage of genes	Fold enrichment	Corrected P value (BH FDR)
Site of expression					
Cervicovaginal fluid	16	544	12.0	4.2	$2.59E - 06$
Neutrophils	13	392	9.7	4.8	$6.68E - 06$
Gastric juice	9	222	6.7	6.1	$4E - 05$
Molecular function					
Defense/immunity protein activity	5	52	3.7	15.7	$3.96E - 05$
Biological process					
Immune response	13	561	9.8	3.4	0.00026
Signal transduction	43	3907	32.5	1.5	0.0026
Cell communication	41	3687	31.1	1.5	0.0028
Antigen presentation	1	1	0.7	134.4	0.0073

4. Discussion

This is the first report of the detection of urinary EVs containing HIV and human proteins from HIV+ patients by mass spectrometry and western blot. EVs provide intercellular communication to cells through the delivery of their cargo, nucleic acids, miRNAs, and proteins, to recipient cells reviewed in [3]. Previous studies have found EVs in plasma of HIV+ patients but did not describe HIV or human proteins within them. Others have described EVs containing HIV proteins but these results were from *in vitro* HIV infected cell cultures and not from HIV+ patients [18, 20, 22, 23, 36, 38–47]. This study details both the HIV and human proteins found in urinary EVs from HIV+ patients.

According to the International Society for Extracellular Vesicles (ISEV), the minimal requirements for EVs or their presence in samples includes the simultaneous detection of transmembrane proteins and cytosolic proteins with membrane/receptor binding abilities, while major cell organelles are absent [48]. LC/MS/MS analysis identified these proteins and functional enrichment analysis determined a significant number which were of exosomal origin in both the EVs in

HIV+, 1,932, and HIV−, only 37. TEM analysis of HIV+ and HIV− urine showed pleiotropic membrane bound vesicles in both groups' urine samples and NTA analysis showed particles ranging in size from 50 nm to 300 nm in both groups, although the HIV+ samples had significantly more particles than uninfected samples. Other studies have found increased numbers of EVs in the plasma of HIV+ patients [43, 49]. Proteins from both the HIV+ and HIV− individuals were significantly associated with exosomal proteins, further substantiating our hypothesis that urine from HIV+ patients contains EVs (Table 10). The FunRich analysis of the sites of expression showed that a significant number of proteins were associated with the endothelium, plasma, serum, kidney, liver, and lung. These findings suggest that EVs from HIV+ patients may be filtered from these sites and concentrated in urine.

HIV has previously been detected in the urine of HIV+ patients; however, it was shown that HIV virions are associated with cell pellets and not in centrifuged urine [50, 51]. p24 is found in replicative HIV infectious virions but was not found in twenty-six of our HIV+ samples by ELISA and only five of thirty-five HIV+ EV urine samples had detectable p24 by LC/MS/MS analysis. p24 in urine pellets

TABLE 9: Overlapping EV proteins from HIV+ and HIV− samples, LC/MS/MS analysis.

Gene	
ABCB1	ATP-binding cassette, subfamily B (MDR/TAP), member 1
ANXA8	Annexin A8
ASIC1	Acid-sensing (proton-gated) ion channel 1
ASIC2	Acid-sensing (proton-gated) ion channel 2
AUTS2	Autism susceptibility candidate 2
AZU1	Azurocidin 1
BCAT1	Branched chain amino acid transaminase 1, cytosolic
BRD4	Bromodomain containing 4
CCL5	Chemokine (C-C motif) ligand 5
CEACAM8	Carcinoembryonic antigen-related cell adhesion molecule 8
CFH	Complement factor H
CHIT1	Chitinase 1 (chitotriosidase)
CLDN7	Claudin 7
COL16A1	Collagen, type XVI, alpha 1
CPB2	Carboxypeptidase B2 (plasma)
CRADD	CASP2 and RIPK1 domain containing adaptor with death domain
CTSG	Cathepsin G
CYP4A11	Cytochrome P450, family 4, subfamily A, polypeptide 11
DEFA1	Defensin, alpha 1
DNAH17	Dynein, axonemal, heavy chain 17
DUSP9	Dual specificity phosphatase 9
EIF4A1	Eukaryotic translation initiation factor 4A1
ELANE	Elastase, neutrophil expressed
	v-erb-b2 avian erythroblastic leukemia viral oncogene homolog 2
FARP1	FERM, RhoGEF (ARHGEF), and pleckstrin domain protein 1 (chondrocyte-derived)
GDF15	Growth differentiation factor 15
GNA12	Guanine nucleotide binding protein (G protein) alpha 12
GNL1	Guanine nucleotide binding protein-like 1
GRIN2A	Glutamate receptor, ionotropic, N-methyl D-aspartate 2A
HAAO	3-Hydroxyanthranilate 3,4-dioxygenase
HAL	Histidine ammonia-lyase
HBA1	Hemoglobin, alpha 1
HBB	Hemoglobin, beta
HBD	Hemoglobin, delta
IGKC	Immunoglobulin kappa constant
LGALS3	Lectin, galactoside-binding, soluble, 3
MEF2C	Myocyte enhancer factor 2C
MLLT4	Myeloid/lymphoid or mixed-lineage leukemia (trithorax homolog, Drosophila); translocated to, 4
MPO	Myeloperoxidase
MRC2	Mannose receptor, C type 2
MYBPC3	Myosin binding protein C, cardiac
NCAM1	Neural cell adhesion molecule 1
NKTR	Natural killer-tumor recognition sequence
NUP93	Nucleoporin 93 kDa
PDE1C	Phosphodiesterase 1C, calmodulin-dependent 70 kDa
PDLIM5	PDZ and LIM domain 5
PIK3R1	Phosphoinositide-3-kinase, regulatory subunit 1 (alpha)
RAB31	RAB31, member RAS oncogene family
RAP1GAP	RAP1 GTPase activating protein

TABLE 9: Continued.

Gene	
REG1A	Regenerating islet-derived 1 alpha
RNASE2	Ribonuclease, RNase A family, 2 (liver, eosinophil-derived neurotoxin)
RNASE3	Ribonuclease, RNase A family, 3
RPS14	Ribosomal protein S14
RUNX2	Runt-related transcription factor 2
SHBG	Sex hormone-binding globulin
SLC22A5	Solute carrier family 22 (organic cation/carnitine transporter), member 5
SLC6A6	Solute carrier family 6 (neurotransmitter transporter), member 6
TACC2	Transforming, acidic coiled-coil containing protein 2
TAF6L	TAF6-like RNA polymerase II, p300/CBP-associated factor (PCAF)-associated factor, 65 kDa
TNIK	TRAF2 and NCK interacting kinase
TRAPPC12	Trafficking protein particle complex 12
TRIM58	Tripartite motif containing 58
WNT2B	Wingless-type MMTV integration site family, member 2B
WNT6	Wingless-type MMTV integration site family, member 6

TABLE 10: Functional analysis of overlapping HIV+ and HIV− EV proteins.

	Genes in the data set	Genes in the Bkg. database	Percentage of genes	Fold enrichment	Corrected P value (Bonferroni method)
Site of expression					
Urine	31	3202	51.7	3.0	$6.85E-07$
Cervicovaginal fluid	12	544	20.0	7.2	$5.33E-05$
Neutrophils	9	392	15.0	7.7	$1.56E-03$
032403_BALF4_glypep	4	43	6.7	35.0	$3.53E-03$
Neutrophil	19	1979	31.7	3.0	$3.67E-03$
Monocyte	23	2786	38.3	2.6	$3.90E-03$
Cellular component					
Extracellular	22	1808	37.9	3.1	$4.61E-05$
Stored secretory granule	3	19	5.2	51.0	$3.27E-03$
Lysosome	17	1609	29.3	2.8	$7.73E-03$
Extracellular space	8	399	13.8	5.6	$1.05E-02$
Exosomes	19	2001	32.8	2.5	$1.14E-02$
Azurophil granule	2	6	3.4	108.9	$1.35E-02$

is derived from mononuclear cells but was found in only 3 of 80 analyzed samples [51]. This represents a low sensitivity, primarily because the HIV-1 p24 protein is not always present during advanced stages of HIV infection. To further confirm that these HIV proteins were from EVs, we tested the filtrate from ultracentrifugation (MW cutoff 100,000 kD) of HIV-positive urine, and no HIV proteins were present. We did not, however, perform an HIV infectivity assay, MAGI, on the isolated urinary EVs, and thus cannot be totally confident that HIV virions were not present in the EVs. HIV proteins in urinary EVs may be the result of a nonproductive HIV infection in the kidney [52–56] and/or EVs filtered from blood [21, 49, 57]. The type of HIV protein in the EVs remained relatively constant as demonstrated by the resampling of two patients,

203 and 311 days, after the first sample that had similar results. The identification of HIV proteins in urinary EVs may be a potential noninvasive diagnostic tool to monitor HIV disease states as well as treatment efficacy.

Different proteins and pathways were found in EVs from (1) CD4+ T cell > 300 versus <300 and (2) VLs < 200 versus >200 copies. It is interesting that EVs from HIV+ patients with low VLs and high CD4+ T cells, usually indicative of better health, had more proteins detected than EVs from high VLs and low CD4+ T cells (high VLs = 2486 vrs low = 15028; low CD4+ T cells = 2115 versus high CD4+ = 15761). These groups also had overlapping pathway results; however, proteins from high VLs and low CD4+ T cells did not have similar pathway results. Further comparison and analysis of the EV protein profile between the low VL/high CD4+ T cells and high VL/low CD4+ T cells may reveal more mechanisms involved in the evolving pathology of HIV infection.

Proteins contained in EVs can both enhance and inhibit host responses from innate, inflammatory, and adaptive reactions. Proteins from HIV+ patients showed a predominantly immunosuppressive profile. IL10 is a Th2 cytokine that downregulates macrophage function and inhibits T cell proliferation while IL6 can stimulate IL10 production and inhibit the effects of TNF-α and IL1. Both these cytokines were present in the EVs from HIV+ patients while TNF-α and IL1 were not detected suggesting an immunomodulatory effect may be elicited by the EVs. Other immune downregulating factors, IDO, HOXB4, HGF, and TGFβ1, were found. IDO [58], HLA-G [59], and HGF [60] can inhibit natural killer cell activation which was one of the top biological processes found in the pathway analysis of the EV proteins in patients with high CD4+ T cells and low VLs. TGFβ-1, an inhibitor of immune function, is induced by HIV Tat [61] and is a mediator of immune suppression in HIV infection [62–64]. These proteins were found in EVs from HIV+ patients while proinflammatory cytokines were not. New studies show that HIV+ nonprogressors have lower plasma TGFβ-1 and IL10 than patients with progressive disease [65] and it is possible that EVs may sequester TGFβ-1 and IL10 and remove them from circulation. The presence of over 16 different MHC Class I and II antigens in the EVs from HIV+ patients may support the hypothesis that this mechanism is used by intracellular pathogens to evade the immune response by decreasing cytotoxic T cell activity [66]. Herpes Simplex Virus-1 binds to HLA-DR inhibiting antigen presentation that leads to immune evasion [67]. Future studies should focus on the correlation of the concentration of these factors to HIV+ patients' clinical status.

In this study, we showed that structural, regulatory, and accessory HIV proteins could be detected in urinary EVs of HIV+ patients. Our WB analysis using polyclonal and monoclonal antibodies confirmed the presence of HIV proteins in the EVs from HIV+ patients. The most prevalent protein was HIV Nef. EVs from both *in vitro* and patient samples have been previously reviewed in [6]. HIV Nef induces an alternative pathway for TNF induction utilizing Notch-1, ADAM17, and Rab4+, all found in EVs from HIV+ patients, which leads to high plasma TNF levels [68]. Whether the isolation of these factors in EVs represents a diminishing or enhancement of TNF production remains to be examined.

The HIV Human Interaction database found significant interactions between HIV Nef, Vpr, Vif, and Vpu and human proteins. Serine/threonine protein kinases are important in T cell receptor signaling [69]. These kinases as well as CD4 and MHC antigens were found in EVs from the HIV+ samples; however, further studies are needed to determine the mechanisms involved with EV function in HIV infections. Cell adhesion molecules, ICAM, VCAM, and PECAM, were also found in the EVs from patients. Others have reported these molecules are present in HIV+ blood samples and may represent biomarkers from inflamed endothelium due to HIV infection [70].

One of the limitations of this study was a small sample size of specific HIV syndromes such as comorbidities, AIDS, HIV-associated nephropathy, and HIV-associated dementia as well as patients on or naïve to antiretroviral therapy. Increasing the numbers of HIV+ patients in these categories may allow us to determine whether specific HIV proteins as well as human proteins in urinary EVs could be associated with these conditions. Future studies will also quantify the amount of HIV proteins as well as human proteins to determine if a correlation exists between different HIV conditions and the amount of proteins detected.

HIV infection is usually detected by antibodies to HIV and can take up to three months to develop or by measuring VLs in blood whereas we can detect HIV-1 proteins in urinary EVs. In summary, urinary proteins in EVs from HIV+ patients may allow a noninvasive method to (1) rapidly screen for infection and identification of patients eligible for antiretroviral treatment (ART); (2) monitor ART treatment efficacy; and (3) diagnose HIV comorbidities.

Acknowledgments

The authors acknowledge Jane Chu and Mahfuz Khan of Morehouse School of Medicine for technical assistance and Dr. Douglas Paulsen for his support and editorial suggestions. This study was supported by the National Center for Advancing Translational Sciences of the National Institutes of Health under Award no. UL1TR000454. Other funds were received from the Minority Biomedical Research Support (MBRS) of the Research Initiative for Scientific Advancement (RISE) Program 5R25GM058268 funded by NIGMS and NIH Research Endowment S21MD000101 funded by the National Institute on Minority Health and Health Disparities (NIH/NIMHD). The authors also acknowledge the Research Centers in Minority Institutions (RCMI) G12 funded by the NIH/NIMHD, #8G12MD0076202. MEB Core facility was constructed with support from the Research Facilities Improvement Grant C06 RR18386 from NIH/NCRR. The newly renovated Core Resources space was funded by G20 RR031196 from NIH/NCRR. The R-CENTER was funded by Grant no. U54MD007588 from NIH/NIMHD and NIH/NCRR 5P20R R0111044 pilot for study support.

Supplementary Materials

Supplementary 1. Top 10 biological function pathways using Pathway Studio 11 Mammal Plus, Elsevier, Inc., for HIV+ EV proteins from HIV+ patients with (1) CD4+ T cells greater than 300, (2) CD4+ T cells less than 300, (3) viral loads less than 200 copies, and (4) viral loads greater than 200 copies.

Supplementary 2. EV HIV protein interactions (Nef, Vif, Vpr, and Vpu) with human proteins identified using HIV-1 Human Interaction database (https://www.ncbi.nlm.nih.gov/genome/viruses/retroviruses/hiv-1/interactions/). The file includes the gene symbol, human protein name, interaction keywords, protein accession ID, and PMID of references citing the interaction.

References

[1] J. S. Schorey, Y. Cheng, P. P. Singh, and V. L. Smith, "Exosomes and other extracellular vesicles in host-pathogen interactions," *EMBO Reports*, vol. 16, pp. 24–43, 2015.

[2] J. S. Schorey and C. V. Harding, "Extracellular vesicles and infectious diseases: New complexity to an old story," *The Journal of Clinical Investigation*, vol. 126, no. 4, pp. 1181–1189, 2016.

[3] F. Dreyer and A. Baur, "Biogenesis and functions of exosomes and extracellular vesicles," *Methods in Molecular Biology*, vol. 1448, pp. 201–216, 2016.

[4] B. Février and G. Raposo, "Exosomes: Endosomal-derived vesicles shipping extracellular messages," *Current Opinion in Cell Biology*, vol. 16, no. 4, pp. 415–421, 2004.

[5] C. Ciardiello, L. Cavallini, C. Spinelli et al., "Focus on extracellular vesicles: New frontiers of cell-to-cell communication in cancer," *International Journal of Molecular Sciences*, vol. 17, no. 2, 2016.

[6] J. H. Ellwanger, T. D. Veit, and J. A. B. Chies, "Exosomes in HIV infection: A review and critical look," *Infection, Genetics and Evolution*, vol. 53, pp. 146–154, 2017.

[7] H. S. Chahar, X. Bao, and A. Casola, "Exosomes and their role in the life cycle and pathogenesis of RNA viruses," *Viruses*, vol. 7, no. 6, pp. 3204–3225, 2015.

[8] F. Jansen, G. Nickenig, and N. Werner, "Extracellular vesicles in cardiovascular disease," *Circulation Research*, vol. 120, no. 10, pp. 1649–1657, 2017.

[9] D. Karpman, A.-L. Ståhl, and I. Arvidsson, "Extracellular vesicles in renal disease," *Nature Reviews Nephrology*, vol. 13, no. 9, pp. 545–562, 2017.

[10] M. C. Martínez and R. Andriantsitohaina, "Extracellular vesicles in metabolic syndrome," *Circulation Research*, vol. 120, no. 10, pp. 1674–1686, 2017.

[11] G. Szabo and F. Momen-Heravi, "Extracellular vesicles in liver disease and potential as biomarkers and therapeutic targets," *Nature Reviews Gastroenterology & Hepatology*, vol. 14, no. 8, pp. 455–466, 2017.

[12] M. Salih, R. Zietse, and E. J. Hoorn, "Urinary extracellular vesicles and the kidney: biomarkers and beyond," *American Journal of Physiology-Renal Physiology*, vol. 306, no. 11, pp. F1251–F1259, 2014.

[13] K. Barreiro and H. Holthofer, "Urinary extracellular vesicles. A promising shortcut to novel biomarker discoveries," *Cell and Tissue Research*, vol. 369, no. 1, pp. 217–227, 2017.

[14] V. Thongboonkerd and P. Malasit, "Renal and urinary proteomics: current applications and challenges," *Proteomics*, vol. 5, no. 4, pp. 1033–1042, 2005.

[15] V. Thongboonkerd, K. R. McLeish, J. M. Arthur, and J. B. Klein, "Proteomic analysis of normal human urinary proteins isolated by acetone precipitation or ultracentrifugation," *Kidney International*, vol. 62, no. 4, Article ID 4493245, pp. 1461–1469, 2002.

[16] R. Pieper, C. L. Gatlin, A. M. McGrath et al., "Characterization of the human urinary proteome: a method for high-resolution display of urinary proteins on two-dimensional electrophoresis gels with a yield of nearly 1400 distinct protein spots," *Proteomics*, vol. 4, no. 4, pp. 1159–1174, 2004.

[17] E. I. Christensen and H. Birn, "Megalin and cubilin: Multifunctional endocytic receptors," *Nature Reviews Molecular Cell Biology*, vol. 3, no. 4, pp. 258–268, 2002.

[18] M. Aqil, S. Mallik, S. Bandyopadhyay, U. Maulik, and S. Jameel, "Transcriptomic analysis of mRNAs in human monocytic cells expressing the HIV-1 nef protein and their exosomes," *BioMed Research International*, vol. 2015, Article ID 492395, 10 pages, 2015.

[19] M. Aqil, A. R. Naqvi, S. Mallik, S. Bandyopadhyay, U. Maulik, and S. Jameel, "The HIV Nef protein modulates cellular and exosomal miRNA profiles in human monocytic cells," *Journal of Extracellular Vesicles (JEV)*, vol. 3, Article ID 23129, 2014.

[20] M. Lenassi, G. Cagney, M. Liao et al., "HIV Nef is secreted in exosomes and triggers apoptosis in bystander CD4$^+$ T cells," *Traffic*, vol. 11, no. 1, pp. 110–122, 2010.

[21] M. B. Khan, M. J. Lang, M.-B. Huang et al., "Nef exosomes isolated from the plasma of individuals with HIV-associated dementia (HAD) can induce Aβ1–42 secretion in SH-SY5Y neural cells," *Journal of NeuroVirology*, vol. 22, no. 2, pp. 179–190, 2016.

[22] C. Arenaccio, S. Anticoli, F. Manfredi, C. Chiozzini, E. Olivetta, and M. Federico, "Latent HIV-1 is activated by exosomes from cells infected with either replication-competent or defective HIV-1," *Retrovirology*, vol. 12, article 87, 2015.

[23] C. Arenaccio, C. Chiozzini, S. Columba-Cabezas et al., "Exosomes from human immunodeficiency virus type 1 (HIV-1)-infected cells license quiescent CD4$^+$ T lymphocytes to replicate HIV-1 through a Nef- and ADAM17-dependent mechanism," *Journal of Virology*, vol. 88, no. 19, pp. 11529–11539, 2014.

[24] N. A. Kruh-Garcia, L. M. Wolfe, and K. M. Dobos, "Deciphering the role of exosomes in tuberculosis," *Tuberculosis*, vol. 95, no. 1, pp. 26–30, 2015.

[25] A. Marcilla, L. Martin-Jaular, M. Trelis et al., "Extracellular vesicles in parasitic diseases," *Journal of Extracellular Vesicles (JEV)*, vol. 3, Article ID 25040, 2014.

[26] D. Y. P. Fang, H. W. King, J. Y. Z. Li, and J. M. Gleadle, "Exosomes and the kidney: blaming the messenger," *Nephrology*, vol. 18, no. 1, pp. 1–10, 2013.

[27] K. Junker, J. Heinzelmann, C. Beckham, T. Ochiya, and G. Jenster, "Extracellular Vesicles and Their Role in Urologic Malignancies," *European Urology*, vol. 70, no. 2, pp. 323–331, 2016.

[28] M. Krause, A. Samoylenko, and S. J. Vainio, "Exosomes as renal inductive signals in health and disease, and their application as diagnostic markers and therapeutic agents," *Frontiers in Cell and Developmental Biology*, vol. 3, 2015.

[29] M. Nawaz, G. Camussi, H. Valadi et al., "The emerging role of extracellular vesicles as biomarkers for urogenital cancers," *Nature Reviews Urology*, vol. 11, no. 12, pp. 688–701, 2014.

[30] M. A. Pomatto, C. Gai, B. Bussolati, and G. Camussi, "Extracellular Vesicles in Renal Pathophysiology," *Frontiers in Molecular Biosciences*, vol. 4, 2017.

[31] M. Li and B. Ramratnam, "Proteomic characterization of exosomes from HIV-1-Infected cells," *Methods in Molecular Biology*, vol. 1354, pp. 311–326, 2016.

[32] M. Pathan, S. Keerthikumar, C.-S. Ang et al., "FunRich: An open access standalone functional enrichment and interaction network analysis tool," *Proteomics*, vol. 15, no. 15, pp. 2597–2601, 2015.

[33] R. J. Simpson, H. Kalra, and S. Mathivanan, "ExoCarta as a resource for exosomal research," *Journal of Extracellular Vesicles*, vol. 1, Article ID 18374, 2012.

[34] S. Mathivanan, C. J. Fahner, G. E. Reid, and R. J. Simpson, "Exocarta 2012: database of exosomal proteins, RNA and lipids," *Nucleic Acids Research*, vol. 40, no. 1, pp. D1241–D1244, 2012.

[35] M. M. Jørgensen, R. Bæk, and K. Varming, "Potentials and capabilities of the Extracellular Vesicle (EV) Array," *Journal of Extracellular Vesicles (JEV)*, vol. 4, Article ID 26048, 2015.

[36] M. Li, J. M. Aliotta, J. M. Asara et al., "Quantitative proteomic analysis of exosomes from HIV-1-infected lymphocytic cells," *Proteomics*, vol. 12, no. 13, pp. 2203–2211, 2012.

[37] D. Ako-Adjei, W. Fu, C. Wallin et al., "HIV-1, Human Interaction database: Current status and new features," *Nucleic Acids Research*, vol. 43, no. 1, pp. D566–D570, 2015.

[38] C. Arenaccio, C. Chiozzini, S. Columba-Cabezas, F. Manfredi, and M. Federico, "Cell activation and HIV-1 replication in unstimulated CD4$^+$ T lymphocytes ingesting exosomes from cells expressing defective HIV-1," *Retrovirology*, vol. 11, no. 1, article 46, 2014.

[39] A. M. Booth, Y. Fang, J. K. Fallon et al., "Exosomes and HIV Gag bud from endosome-like domains of the T cell plasma membrane," *The Journal of Cell Biology*, vol. 172, no. 6, pp. 923–935, 2006.

[40] E. Chertova, O. Chertov, L. V. Coren et al., "Proteomic and biochemical analysis of purified human immunodeficiency virus type 1 produced from infected monocyte-derived macrophages," *Journal of Virology*, vol. 80, no. 18, pp. 9039–9052, 2006.

[41] N. Izquierdo-Useros, M. Naranjo-Gómez, J. Archer et al., "Capture and transfer of HIV-1 particles by mature dendritic cells converges with the exosome-dissemination pathway," *Blood*, vol. 113, no. 12, pp. 2732–2741, 2009.

[42] I. Kadiu, P. Narayanasamy, P. K. Dash, W. Zhang, and H. E. Gendelman, "Biochemical and biologic characterization of exosomes and microvesicles as facilitators of HIV-1 infection in macrophages," *The Journal of Immunology*, vol. 189, no. 2, pp. 744–754, 2012.

[43] J.-H. Lee, S. Schierer, K. Blume et al., "HIV-Nef and ADAM17-Containing Plasma Extracellular Vesicles Induce and Correlate with Immune Pathogenesis in Chronic HIV Infection," *EBioMedicine*, vol. 6, pp. 103–113, 2016.

[44] I.-W. Park and J. J. He, "HIV-1 is budded from CD4+ T lymphocytes independently of exosomes," *Virology Journal*, vol. 7, article no. 234, 2010.

[45] P. Rahimian and J. J. He, "Exosome-associated release, uptake, and neurotoxicity of HIV-1 Tat protein," *Journal of NeuroVirology*, vol. 22, no. 6, pp. 774–788, 2016.

[46] W. Roth, M. Huang, K. Addae Konadu, M. Powell, and V. Bond, "Micro RNA in Exosomes from HIV-Infected Macrophages," *International Journal of Environmental Research and Public Health*, vol. 13, no. 12, p. 32, 2016.

[47] G. C. Sampey, M. Saifuddin, A. Schwab et al., "Exosomes from HIV-1-infected cells stimulate production of pro-inflammatory cytokines through trans-activating response (TAR) RNA," *The Journal of Biological Chemistry*, vol. 291, no. 3, pp. 1251–1266, 2016.

[48] J. Lötvall, A. F. Hill, F. Hochberg et al., "Minimal experimental requirements for definition of extracellular vesicles and their functions: a position statement from the International Society for Extracellular Vesicles," *Journal of Extracellular Vesicles (JEV)*, vol. 3, Article ID 26913, 2014.

[49] A. Hubert, C. Subra, M.-A. Jenabian et al., "Elevated abundance, size, and MicroRNA content of plasma extracellular vesicles in viremic HIV-1+ patients: Correlations with known markers of disease progression," *Journal of Acquired Immune Deficiency Syndromes*, vol. 70, no. 3, pp. 219–227, 2015.

[50] P. R. Skolnik, B. R. Kosloff, L. J. Bechtel et al., "Concise communications absence of infectious hiv-1 in the urine of seropositive viremic subjects," *The Journal of Infectious Diseases*, vol. 160, no. 6, pp. 1056–1060, 1989.

[51] J. J. Li, Y. Q. Huang, B. J. Poiesz, L. Zaumetzger-Abbot, and A. E. Friedman-Kien, "Detection of human immunodeficiency virus type 1 (HIV-1) in urine cell pellets from HIV-1-seropositive individuals," *Journal of Clinical Microbiology*, vol. 30, no. 5, pp. 1051–1055, 1992.

[52] A. K. Khatua, H. E. Taylor, J. E. K. Hildreth, and W. Popik, "Nonproductive HIV-1 infection of human glomerular and urinary podocytes," *Virology*, vol. 408, no. 1, pp. 119–127, 2010.

[53] D. Marras, L. A. Bruggeman, F. Gao et al., "Replication and compartmentalization of HIV-1 in kidney epithelium of patients with HIV-associated nephropathy," *Nature Medicine*, vol. 8, no. 5, pp. 522–526, 2002.

[54] L. A. Bruggeman, M. D. Ross, N. Tanji et al., "Renal epithelium is a previously unrecognized site of HIV-1 infection," *Journal of the American Society of Nephrology*, vol. 11, no. 11, pp. 2079–2087, 2000.

[55] N. Tanji, M. D. Ross, K. Tanji et al., "Detection and localization of HIV-1 DNA in renal tissues by in situ polymerase chain reaction," *Histology and Histopathology*, vol. 21, no. 4-6, pp. 393–401, 2006.

[56] G. Canaud, N. Dejucq-Rainsford, V. Avettand-Fenoël et al., "The kidney as a reservoir for HIV-1 after renal transplantation," *Journal of the American Society of Nephrology*, vol. 25, no. 2, pp. 407–419, 2014.

[57] A. D. Raymond, T. C. Campbell-Sims, M. Khan et al., "HIV type 1 Nef is released from infected cells in CD45$^+$ microvesicles and is present in the plasma of HIV-infected individuals," *AIDS Research and Human Retroviruses*, vol. 27, no. 2, pp. 167–178, 2011.

[58] S. Kai, S. Goto, K. Tahara, A. Sasaki, S. Tone, and S. Kitano, "Indoleamine 2,3-Dioxygenase is Necessary for Cytolytic Activity of Natural Killer Cells," *Scandinavian Journal of Immunology*, vol. 59, no. 2, pp. 177–182, 2004.

[59] N. Rouas-Freiss, P. Moreau, C. Menier, J. LeMaoult, and E. D. Carosella, "Expression of tolerogenic HLA-G molecules in cancer prevents antitumor responses," *Seminars in Cancer Biology*, vol. 17, no. 6, pp. 413–421, 2007.

[60] D. Wang, Y. Saga, N. Sato et al., "The hepatocyte growth factor antagonist NK4 inhibits indoleamine-2, 3-dioxygenase expression via the c-Met-phosphatidylinositol 3-kinase-AKT signaling pathway," *International Journal of Oncology*, vol. 48, no. 6, pp. 2303–2309, 2016.

[61] G. Zauli, B. R. Davis, M. C. Re, G. Visani, G. Furlini, and M. La Placa, "tat Protein stimulates production of transforming growth factor-β1 by marrow macrophages: A potential mechanism for human immunodeficiency virus- 1-induced hematopoietic suppression," *Blood*, vol. 80, no. 12, pp. 3036–3043, 1992.

[62] J. Kekow, W. Wachsman, J. A. McCutchan, M. Cronin, D. A. Carson, and M. Lotz, "Transforming growth factor β and noncytopathic mechanisms of immunodeficiency in human immunodeficiency virus infection," *Proceedings of the National*

Acadamy of Sciences of the United States of America, vol. 87, no. 21, pp. 8321–8325, 1990.

[63] J. Kekow, W. Wachsman, J. Allen McCutchan et al., "Transforming growth factor-β and suppression of humoral immune responses in HIV infection," *The Journal of Clinical Investigation*, vol. 87, no. 3, pp. 1010–1016, 1991.

[64] J. K. Lazdins, T. Klimkait, K. Woods-Cook et al., "In vitro effect of transforming growth factor-β on progression of HIV-1 infection in primary mononuclear phagocytes," *The Journal of Immunology*, vol. 147, no. 4, pp. 1201–1207, 1991.

[65] E. K. Maina, C. Z. Abana, E. A. Bukusi, M. Sedegah, M. Lartey, and W. K. Ampofo, "Plasma concentrations of transforming growth factor beta 1 in non-progressive HIV-1 infection correlates with markers of disease progression," *Cytokine*, vol. 81, pp. 109–116, 2016.

[66] S. A. Synowsky, S. L. Shirran, F. G. M. Cooke, A. N. Antoniou, C. H. Botting, and S. J. Powis, "The major histocompatibility complex class I immunopeptidome of extracellular vesicles," *The Journal of Biological Chemistry*, vol. 292, no. 41, pp. 17084–17092, 2017.

[67] J. Neumann, A. M. Eis-Hübinger, and N. Koch, "Herpes simplex virus type 1 targets the MHC class II processing pathway for immune evasion," *The Journal of Immunology*, vol. 171, no. 6, pp. 3075–3083, 2003.

[68] C. Ostalecki, S. Wittki, J.-H. Lee et al., "HIV Nef- and Notch1-dependent Endocytosis of ADAM17 Induces Vesicular TNF Secretion in Chronic HIV Infection," *EBioMedicine*, vol. 13, pp. 294–304, 2016.

[69] M. N. Navarro and D. A. Cantrell, "Serine-threonine kinases in TCR signaling," *Nature Immunology*, vol. 15, no. 9, pp. 808–814, 2014.

[70] K. De Gaetano Donati, R. Rabagliati, L. Iacoviello, and R. Cauda, "HIV infection, HAART, and endothelial adhesion molecules: Current perspectives," *The Lancet Infectious Diseases*, vol. 4, no. 4, pp. 213–222, 2004.

19

Apoptotic and Early Innate Immune Responses to PB1-F2 Protein of Influenza A Viruses Belonging to Different Subtypes in Human Lung Epithelial A549 Cells

Gunisha Pasricha ⓘⒹ, Sanjay Mukherjee, and Alok K. Chakrabarti ⓘⒹ

Microbial Containment Complex, National Institute of Virology, Sus Road, Pashan, Pune 411021, India

Correspondence should be addressed to Alok K. Chakrabarti; chakrabarti.alok@icmr.gov.in

Guest Editor: Binod Kumar

PB1-F2 is a multifunctional protein and contributes to the pathogenicity of influenza A viruses. PB1-F2 is known to have strain and cell specific functions. In this study we have investigated the apoptotic and inflammatory responses of PB1-F2 protein from influenza viruses of diverse pathogenicities in A549 lung epithelial cells. Overexpression of PB1-F2 resulted in apoptosis and heightened inflammatory response in A549 cells. Comparison revealed that the response varied with each subtype. PB1-F2 protein from highly pathogenic H5N1 virus induced least apoptosis but maximum inflammatory response. Results indicated that apoptosis was mediated through death receptor ligands TNFα and TRAIL via Caspase 8 activation. Significant induction of cytokines/chemokines CXCL10, CCL5, CCL2, IFNα, and IL-6 was noted in A549 cells transfected with PB1-F2 gene construct of 2008 West Bengal H5N1 virus (H5N1-WB). On the contrary, PB1-F2 construct from 2007 highly pathogenic H5N1 isolate (H5N1-M) with truncated N-terminal region did not evoke as exuberant inflammatory response as the other H5N1-WB with full length PB1-F2, signifying the importance of N-terminal region of PB1-F2. Sequence analysis revealed that PB1-F2 proteins derived from different influenza viruses varied at multiple amino acid positions. The secondary structure prediction showed each of the PB1-F2 proteins had distinct helix-loop-helix structure. Thus, our data substantiate the notion that the contribution of PB1-F2 to influenza pathogenicity is greatly strain specific and involves multiple host factors. This data demonstrates that PB1-F2 protein of influenza A virus, when expressed independently is minimally apoptotic and strongly influences the early host response in A549 cells.

1. Introduction

Influenza is pathogenic viral disease, causing the emergence of newer epidemics and pandemics in mammals [1, 2]. Since the beginning of 20th century, there have been four pandemics: Spanish influenza (H1N1) in 1918/1919, Asian influenza (H2N2) in 1957, Hong Kong influenza (H3N2) in 1968, and H1N1 influenza in 2009 [3, 4]. Influenza A virus (IAV) causes acute respiratory inflammation in humans and symptoms include high fever, body aches, and fatigue [4]. Symptoms of IAV infection is mostly mild in humans but may progress to fatal viral pneumonia if the virus spreads from the upper airways to the alveolar space in the lower respiratory tract [4].

IAV belongs to the Orthomyxoviridae family and its genome consists of eight negative strand RNA segments [4, 5]. PB1-F2 was the first accessory protein discovered in IAV [6]. Subsequently, more proteins like PB1-N40 and PA-N155 have been discovered [4, 5]. It is important to study these novel proteins in order to evaluate their role in the pathogenesis of IAV infection. PB1-F2 is the second protein encoded by the +1 alternate open reading frame within the PB1 gene. It is translated from the fourth initiation codon via a leaky ribosomal scanning mechanism wherein 43S ribosomal complex bypasses the PB1 start codon and two additional intervening AUG codons [6, 7]. PB1-N40 is the third protein translated from the fifth initiation codon of PB1 gene. Proteins encoded by PB1 gene have translational interdependence, which makes it difficult to understand their relative contributions to the replication cycle and pathogenicity of IAV [4, 5].

PB1-F2 has been shown to have proapoptotic activity when expressed either independently or during influenza virus infection [6]. The C-terminal region of PB1-F2 interacts with mitochondrial antiviral signaling protein (MAVS), resulting in decreased mitochondrial membrane potential and ultimately ensuing in apoptosis [8]. PB1-F2 has also been implicated in regulation of polymerase activity, exacerbating pathogenicity in animal models, causing susceptibility to secondary bacterial infection and regulation of innate immune response [9].

PB1-F2 has been discovered more than a decade ago and yet there is not much clarity about its pathogenicity determinants. The major deterrent in this is the fact that PB1-F2 protein is highly variable in many of the subtypes of IAV and truncated either from the C-terminal or N-terminal end, thus questioning its evolutionary utility in different subtypes and hosts [10]. Thus, it is not surprising that the effects of the PB1-F2 are strain, cell type, and host specific [11, 12]. Also, much of the research on PB1-F2 has been done using only a few important viral isolates, such as A/Puerto Rico/8/34 (PR8), A/WSN/1933 (WSN), 1918 Spanish flu belonging to H1N1 subtype and avian influenza A H5N1 isolates [13–16]. Thus, it is necessary to study and compare PB1-F2 protein from IAV isolates from different host species and have a complete understanding of how PB1-F2 proteins contribute to IAV replication or virulence.

In this study, a deeper understanding of PB1-F2 protein from various IAVs and their role in modulation of host cell response was sought. We used PB1-F2 gene of five different IAVs, belonging to four avian and one human isolate. Two isolates of highly pathogenic H5N1 viruses having PB1-F2 protein of different sizes were included in this study. The H5N1-M has an N-terminal truncation in PB1-F2 sequence and does not produce a complete PB1-F2 protein [17] while H5N1-WB produces a full length PB1-F2 protein [18]. We attempted to study the effects of PB1-F2 expression and understand its ability to initiate apoptosis and subvert or subdue the host immune response in A549 cells. We studied the PB1-F2 induced host responses in human lung epithelia since respiratory tract is the entry portals of influenza virus and plays a key role in the initial host response [19]. Most of *in vitro* studies on PB1-F2 have been carried out in monocyte/macrophage cell lines. Therefore, it seemed important to study the role of PB1-F2 in inducing apoptotic responses and cytokine production in epithelial cells of the respiratory mucosa.

2. Material and Methods

2.1. Sequence Analysis and Secondary Structure Prediction of PB1-F2 Proteins. PB1-F2 sequences were aligned and analyzed using ClustalX2 version 2.1 program. Secondary structure prediction of the PB1-F2 was performed using RaptorX program (http://raptorx.uchicago.edu/) [20].

2.2. Cell Line, Viruses and Reagents. Madin-Darby Canine Kidney (MDCK) and human lung epithelial (A549) cell lines were obtained from ATCC and were maintained in Dulbecco's modified Eagle's medium (DMEM; Gibco, San Diego, CA) with 10% fetal bovine serum (Gibco, San Diego, CA) in tissue culture flasks (Corning, USA) at 37°C in a CO_2 incubator. Plasmid constructs of PB1-F2 gene were made by amplifying a short region of PB1 gene from five strains of IAVs belonging to different subtypes. These included four avian influenza viruses A/chicken/India/WBNIV2653/2008 (H5N1-WB); A/chicken/India/NIV9743/2007 (H5N1-M); A/chicken/India/WB-NIV1057231/2010 (H9N2); and A/aquaticbird/India/NIV-17095/2007 (H11N1). PB1-F2 of human influenza A H1N1 (A/WSN/33) was amplified from a plasmid construct of PB1.

To test the transfection efficiency of the A549 cell line, cells were transfected with pmax- GFP vector (Lonza Group, Switzerland) and the fluorescence was detected using GFP antibody (FL), a rabbit polyclonal IgG (SantaCruz Biotechnology, CA). The other antibodies obtained from Santa Cruz Biotechnology, CA, USA, were β-Actin mouse monoclonal antibody (ACTBD11B7), goat anti-mouse IgG-Horse Radish Peroxidase (HRP), and mouse anti-rabbit IgG-HRP. Anti-V5 and anti-His (C-term) mouse monoclonal antibodies (Life technologies, CA, USA) were used to detect expression of PB1-F2 fusion protein from pcDNA™3.1/V5-His-TOPO plasmid (Life technologies, CA, USA). To detect the expression of PB1-F2 protein, at 8 h post transfection, proteasome inhibitor MG132 and calpain inhibitor were added alone and in combination at concentration of 10μM and 20μM respectively into the cell line.

2.3. RNA Extraction and RT-PCR. The viruses were grown and propagated in MDCK cells. The cell culture supernatant was used to extract viral RNA with QIAamp Viral RNA Mini Kit (Qiagen, Germany). Reverse transcription was carried out using Superscript II RT-PCR kit (Invitrogen, Carlsbad, CA, USA) and Uni12 primer [21]. PB1-F2 genes were amplified from the above-mentioned viral strains using AccuPrime *Taq* DNA polymerase system (Invitrogen, Carlsbad, CA, USA). The subtype specific primers which were used for the amplification are mentioned in the Table 1. The PCR program for all the amplifications was as follows: initial denaturation at 94°C for 2 mins followed by 30 cycles of 94°C for 30 secs, 55°C for 30 secs, and 68°C for 1 min. The final product was stored at 4°C.

2.4. Expression Plasmid Construction and Transient Transfection. Amplified products of the PB1-F2 gene from all the five cDNA were cloned into pcDNA 3.1/V5-His-TOPO expression vector (Invitrogen, Carlsbad, CA, USA). Competent *E. coli* DH5α cells were transformed with these constructs. The constructs containing PB1-F2 insert were isolated and purified using plasmid Midi kit (Qiagen, Germany). All these constructs were confirmed by restriction digestion and further reconfirmed by sequencing.

Plasmid constructs containing PB1-F2 gene from the different subtypes were transfected in A549 cells using Lipofectamine 2000 (Invitrogen, Carlsbad, CA, USA) according

TABLE 1: Primer sequences used in the study.

Primer	Sequences 5' to 3'	Size (bp)
Unil2	AGCRAAAGCAGG	
H1N1	F- CTTACAGCCATGGGACAGGAACAGG	270 bp
	R- CTTGTGTAAGCTTGTCCACTCGTGT	
H5N1	F- ACAGCCATGGAACAGGGACAGGATACA	270bp
	R- GACCTTGGGTGAGTTTATCCACTCTT	
H5N1-F2	F- ACCCAATTGATGGACCATTACCTGAG	156bp
	R- GACCTTGGGTGAGTTTATCCACTCTT	
H1lN1	F- CATACAGCCATGGAACAGGAACAGG	270bp
	R- CTTGGGTGAGTTTCTTGGGTGAGTTTGTCCACTCTTGT	

to the manufacturer's protocol. Transfection mixtures were then gently added to the respective wells. Twenty-four hours after transfection, total RNA and protein were extracted. The transfection efficiency of A549 cells was determined by transfecting the cell line with pmax-GFP vector (Lonza Group, Switzerland). Untransfected cells served as cell control and cells transfected with the empty expression vector pcDNA3.1 served as mock control.

2.5. PB1-F2 mRNA and Protein Expression in Transfected Cells. Twenty-four hours post transfection cells were harvested and RNA was extracted using TRIzol® Reagent (Invitrogen, Carlsbad, CA, USA). RNA was purified using RNeasy mini kit (Qiagen, Carlsbad, CA) and their yields were evaluated in a Nanodrop spectrophotometer (Nanodrop technologies, Wilmington, DE, USA) at 260 nm. To determine the mRNA expression of PB1-F2 gene in A549 cells, 500 ng of RNA was reverse transcribed using the Superscript II RT-PCR kit (Invitrogen, Carlsbad, CA, USA) and Unil2 primer and the PB1-F2 gene was amplified as mentioned above.

The total cellular protein was extracted using radio-immunoprecipitation assay (RIPA) buffer (Sigma, St. Louis, Missouri, USA). Protease inhibitor and Phosphatase inhibitor cocktails (Calbiochem, USA) were added to RIPA lysis buffer. After protein extraction the cellular debris was pelleted by centrifugation at 8000g for 10 min at 4°C. The protein concentration of the supernatant was determined by Nanodrop spectrophotometer (Nanodrop technologies, Wilmington, DE, USA) at 280 nm. Equal concentration of cell extract was fractioned on 12.5% Sodium dodecyl sulfate-polyacrylamide gel electrophoresis (SDS-PAGE) and the separated proteins were transferred by semi-dry blotting onto nitrocellulose membranes (Hybond C-Extra; GE Healthcare) for reaction with specific antibodies against the His-tagged proteins.

2.6. TUNEL Assay. For detection and relative quantification of apoptosis in A549 cells transfected with PB1-F2 construct from various influenza subtypes, the TUNEL Apoptosis Detection assay was performed (terminal deoxynucleotidyl transferase mediated dUTP nick end labeling), using a kit from APO-BrdU™ TUNEL Assay Kit (Invitrogen Life Technologies, Carlsbad, CA, USA). TUNEL assay detects the DNA fragmentation of apoptotic cells by exploiting the $3'$-hydroxyl ends of the DNA breaks. These hydroxyl groups are then modified by terminal deoxynucleotidyl transferase (TdT) enzyme, which adds deoxyribonucleotides in a template-independent fashion. Addition of the deoxythymidine analog 5-bromo-$2'$-deoxyuridine $5'$-triphosphate (BrdUTP) to the TdT reaction serves to label the break sites. Once incorporated into the DNA, BrdU can be detected by an anti-BrdU antibody using standard immunohistochemical techniques. The nuclear staining was performed with propidium iodide (PI). The assay was carried out according to the instructions of the manufacturer and slides were visualized using Olympus IX51 microscope. Apoptotic index (AI) was determined. AI = number of TUNEL-positive epithelial cells/total number of epithelial cells stained with PI from a total of 25 fields per sample.

2.7. Quantitative Real-Time PCR (qRT-PCR). Total cellular RNA from control and transfected A549 cells was extracted using TRIzol reagent (Invitrogen Life Technologies, Carlsbad, CA, USA) and purified using RNeasy mini kit (Qiagen, Carlsbad, CA). Comparison of the relative expression of 84 apoptosis-related genes between the samples was made with Human Apoptosis RT^2 Profiler PCR Array (SABiosciences, Frederick, MD). The total RNA was treated with DNase I (Roche Diagnostic, Germany) to eliminate genomic DNA contamination at 37°C for 20 minutes. One microgram of total RNA was converted to cDNA using RT^2 First Strand cDNA Synthesis kit (Qiagen, Carlsbad, CA) following the manufacturer's protocol. The cDNA was used for qPCR amplification in the Human Apoptosis RT^2 Profiler PCR Array (SABiosciences, Frederick, MD) using an ABI Prism 7300 detection system (Applied Biosystems, Foster City, CA) according to manufacturer's instructions. The PCR conditions and cycles were as follows: initial DNA denaturation 10 min at 95°C, followed by 40 cycles at 95°C for 15 sec, and final step at 60°C for 1 min. Melting curves were analyzed to determine the specificity of each reaction to the target. Reactions were conducted in duplicate and repeated at least three times for each experiment, and the mean values and standard deviations were calculated. To ensure that the primers produced a single and specific PCR amplification product, a dissociation curve was performed at the end of the PCR cycle. The data were presented as the fold

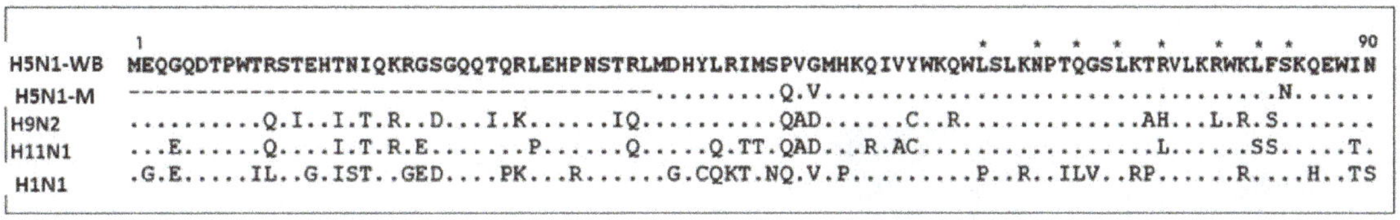

(a)

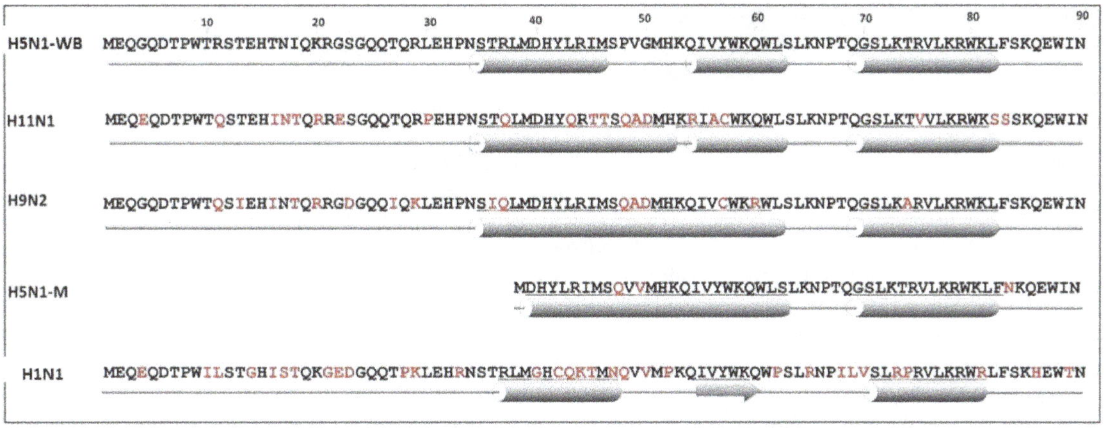

(b)

FIGURE 1: PB1-F2 sequence comparison and structural predictions. (a) Amino acid multialignment of the five PB1-F2 variants included in this study. PB1-F2 proteins analyzed *in silico* in this study from the influenza viruses are abbreviated as follows: A/chicken/India/WBNIV2653/2008 (H5N1-WB); A/chicken/India/NIV9743/2007 (H5N1-M); A/chicken/India/WB-NIV1057231/2010 (H9N2); A/aquatic bird/India/NIV-17095/2007 (H11N1); and A/WSN/1933 (H1N1). Eight amino acid residues which have been reported in literature to have significance in enhancing apoptosis and inflammation have been marked with asterisk. L62, R75, R79, and L82 are reported as inflammatory motifs, N66S as a pathogenic marker, L69 and L75 important for mitochondrial localization, K73 and R75 minimally required for apoptosis via mitochondria. (b) Secondary structure predictions were obtained with the software RaptorX program. Straight line structure represents the loop, cylinder represents the α Helix, and arrow head marks the β sheet structure.

change using the formula $2^{-\Delta\Delta CT}$ as recommended by the manufacturer.

2.8. ELISA Based Cytokine Quantification in Cell Super-natants. Cytokine levels (IL-1β, IL-6, IL-8, TNF-α, CCL2, CCL-5, CXCL10, and CXCL9) in supernatants of control and transfected cells were analyzed using human viral induced cytokines Multi-Analyte ELISArray kit (SABiosciences, Frederick, MD). Supernatants (1:2 diluted) were incubated in 96-well plates precoated with individual cytokine capture antibodies for 2 hours. After they were washed, biotin-conjugated cytokine detection antibodies were added for 30 minutes. Bound antibodies were detected with avidin-HRP and development solution. Color reactions were quantified at OD 450. Cytokine concentrations were determined by extrapolation from cytokine standard curves, according to the manufacturer's protocol. The experiments were repeated three times using supernatant from cells transfected with individual PB1-F2 constructs.

2.9. Statistical Analysis. Statistical analysis was carried out using GraphPad Prism 5.0 software (San Diego, CA, USA) by applying the Student t-test for 2 group comparisons. All the experiments were repeated at least 3 times for validation. The differences were considered significant at p < 0.05.

3. Results

In this study, we used PB1-F2 protein belonging to different subtypes of IAV and investigated its ability to induce apoptosis and modulate inflammatory response in transfected A549 cells.

3.1. Amino Acid Sequence Analysis and Secondary Structure Prediction of PB1-F2 Protein from Different Influenza A Viruses. The length and sequence of PB1-F2 protein greatly decide its contribution to viral pathogenicity. We compared the amino acid sequence of the PB1-F2 protein of the five IAVs, four of them were isolated from avian species while one of them from mammals. Since evolutionarily IAV moved from avian hosts to mammals [22], the amino acid sequence was compared with PB1-F2 protein of highly pathogenic avian influenza-H5N1 (H5N1-WB) virus. The amino acid alignment of PB1-F2 sequences showed variations at more than half of the amino acids positions (52/90), 19 of which were contributed by H1N1 (A/WSN/1933) alone. Avian influenza viruses in the study varied at 26 amino acid positions, hence showing 71.1% homology. All these variations were distributed in the entire length of the protein (Figure 1(a)).

Amino acid residues which are reported to be involved in modulating apoptosis and host immune responses are

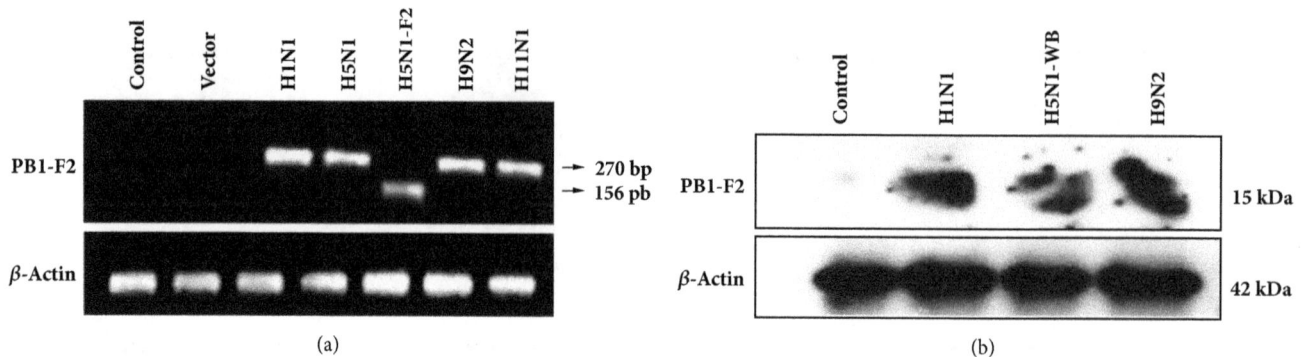

FIGURE 2: A549 cells grown in a 60 mm culture dish and transfected with different PB1-F2 constructs and empty pcDNA3.1 plasmid as vector (mock) control. Forty-eight-hour after transfection, the cells were lysed and PB1-F2 expression was detected by (a) reverse transcriptase PCR and (b) immunoblotting using a monoclonal mouse anti-HIS antibody. PB1-F2 protein expression analyzed in this study is abbreviated as follows: A/chicken/India/WBNIV2653/2008(H5N1-WB); A/chicken/India/NIV9743/2007 (H5N1-M); A/chicken/India/WB-NIV1057231/2010 (H9N2); A/aquatic bird/India/NIV-17095/2007 (H11N1); and A/WSN/1933 (H1N1).

depicted in Figure 1(a). The PB1-F2 proteins from H5N1-WB and H5N1-M have amino acids residues L62, R75, R79, and L82 which are inflammatory motifs of the protein [23]. Except for mammalian H1N1, PB1-F2 protein from all other subtypes isolated from avian species had Leucine at amino acid position 62. Double Leucine residue at positions 69 and 75 which is required for mitochondrial localization of PB1-F2 was not present in any of the strains. Amino acids K73 and R75 which are minimally required for apoptosis via mitochondria were present in the PB1-F2 sequence of both the H5N1-viruses. All the viruses had a full length 90 amino acid PB1-F2 protein except for H5N1-M which had an N-terminal truncation and was only 52 amino acid long.

The secondary structures of all the PB1-F2 proteins showed a typical helix-loop-helix structure except for H1N1 (WSN) which showed a small β sheet structure in between α-helices (Figure 1(b)). PB1-F2 structure of avian influenza viruses varied from each other in the length of their alpha-helices. Distinctive two α-helix breaks were seen for viruses isolated from avian species with a single break for H9N2 and H5N1-M but double breaks for H5N1-WB and H11N1 viruses. The α-helices towards the C-terminal end of PB1-F2 have been reported to be associated with viral pathogenicity [22]. Interestingly, PB1-F2 of H1N1-WSN displayed a unique β-sheet structure (54-59aa) in between the two α helices (Figure 1(b)).

3.2. Apoptosis in A549 Cells Transfected with PB1-F2 constructs of Different Influenza A Viruses. We expressed PB1-F2 protein of different influenza A viruses in A549 cells and investigated its ability to induce apoptosis in host cells using TUNEL assay and by studying the expression of apoptotic genes. The expression of PB1-F2 was confirmed at mRNA and protein level as shown in Figure 2. We could not able to detect expression of PB1-F2 protein in H5N1-M and H11N1 viruses in western blots. This could be due to the small size and shorter half-life of the protein in these subtypes (Figure 2(b)).

TUNEL assay revealed that only a small proportion of PB1-F2 transfected cells underwent apoptosis (Figure 3). This

could explain the relatively low proapoptotic effect of the PB1-F2 protein observed in the assay above. Number of apoptotic cells was counted from a population of more than 500 cells. At 24 h post-transfection, highest proportion (20%) of apoptotic cells were observed with H1N1 (WSN) PB1-F2 construct compared to 15% with H11N1-PB1-F2 and least with H5N1 (WB) PB1-F2 (5%).

To further explore the molecular mechanism of apoptosis triggered by transient expression of PB1-F2 protein belonging to various influenza subtypes in A549 cell line, we determined the expression of 84 key genes involved in programmed cell death by using the Human Apoptosis RT^2 Profiler PCR Array. This array includes TNF ligands and their receptors, members of the bcl-2, caspase, IAP, TRAF, CARD, death domain, death effector domain, and CIDE families, as well as genes involved in the p53 and DNA damage pathways.

The mRNA expression profile showed that 24 of the 84 genes analyzed in this study were upregulated with statistical significance ($p<0.05$) when compared with the mock control (empty vector). We grouped the gene transcripts in 4 groups based on function and signaling pathway. Group I included genes belonging to the TNF death receptors and ligand family, Group II included different caspases, Group III belonged to BCl-2 family member genes while Group IV belonged to different mitochondrial cell death inhibitor (IAP) genes involved in intrinsic regulation of apoptosis.

Analysis of group I genes revealed that there was statistically significant upregulation of expression of *TNFα* in A549 cells in response to PB1-F2 protein from H1N1 (WSN), H9N2, and H11N1 when compared to mock control. Most of the TNF family member genes showed significant higher expression (compared to mock) in response to PB1-F2 protein belonging to low pathogenic avian influenza viruses and WSN strain. The expression level of *TNF, TNFRSF9, TNFRSF10a,* and *TNFRSF21* was highest in H1N1-WSN of all the subtypes used for the study. However, expression of *TNFSF10* and *TNFRSF10a* genes was significantly higher in cells transfected with PB1-F2 of highly pathogenic avian influenza-H5N1-WB virus (Figure 4).

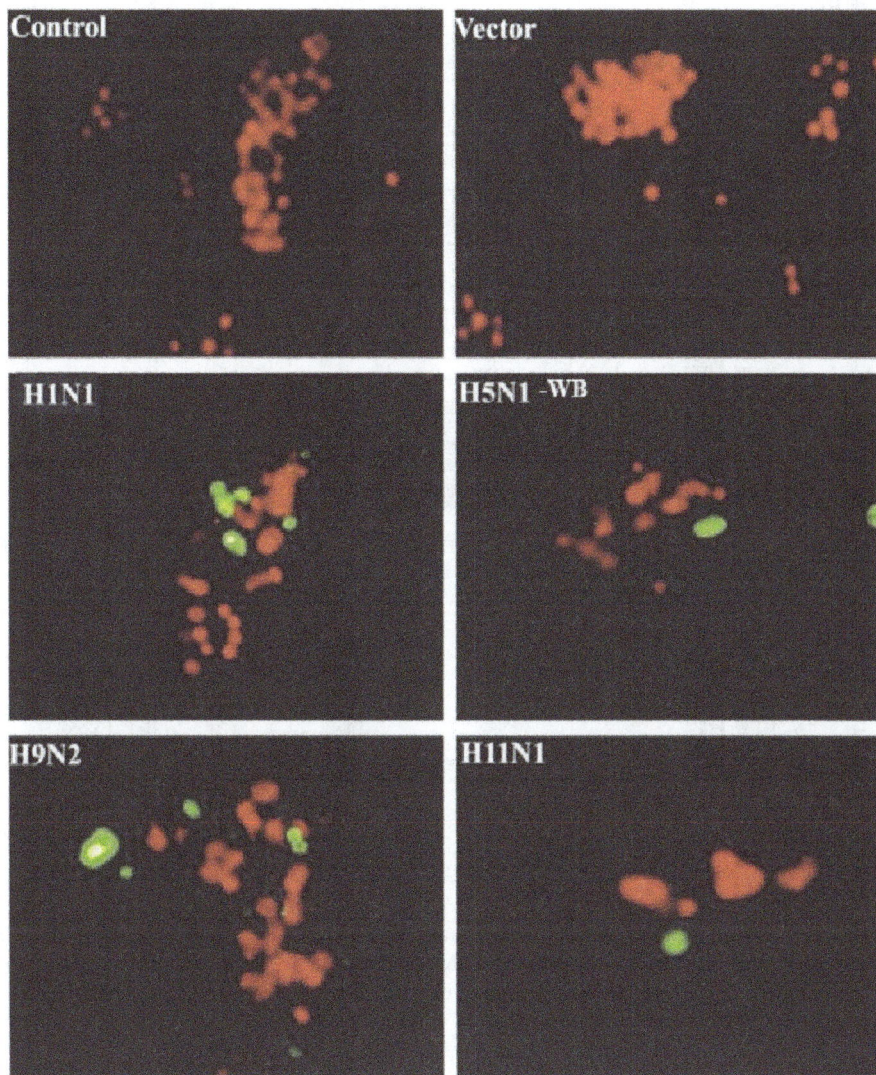

FIGURE 3: Ectopic expression of PB1-F2 protein predisposes cells to minimal apoptosis. A549 cells were transfected with PB1-F2 constructs and the apoptosis was observed by TUNEL staining with FITC-conjugated dUTP cells showing apple green fluorescence which were apoptotic. The overall proportion of death cells was measured by propidium iodide staining (seen as red in color).

Analysis of gene expression of caspase family (Group II) revealed that Caspase 8 (*CASP8*) was significantly upregulated in response to PB1-F2 of influenza A H1N1 and H11N1 strains. Caspase 10 (*CASP10*) and Caspase 14 (*CASP14*) which are in turn activated by *CASP8* were significantly upregulated in response to PB1-F2 of H9N2 strain (Figure 5). The gene expression of Caspase 1 was found considerably higher (>10 folds) in response to PB1-F2 of H1N1, H9N2, and H11N1 subtypes (Figure 5). Overall, *CASP1, CASP4, CASP5,* and *CASP14* showed highest expression in response to PB1-F2 protein belonging to H9N2 subtype.

Group III include genes belong to the BCL2 gene family which are both pro- and anti-apoptotic. BAK1 expression was significantly higher in response to PB1-F2 protein of H1N1 and H11N1 viruses (Figure 6(a)). The BCL2 ligands, *BCL2L1* and *BCL2L10*, showed highest expression in H11N1 and H9N2, respectively. *BID* was downregulated in response

to PB1-F2 of all the subtypes, although it was found to be not statistically significant (Figure 6(a)).

Group IV genes included IAPs (inhibitors of apoptotic protein: BIRC2 and BIRC3) and apoptosis protectors (BNIP1 and BNIP2) (Figure 6(b)). The gene expression of IAPs showed significant upregulation in A549 cells transfected with PB1-F2 constructs of H5N1. Interestingly, *BNIP1, BNIP2,* and *XIAP* genes were downregulated by PB1-F2 of all subtypes except for H5N1 (Figure 6(b)).

3.3. Cytokine Expression in A549 Cells Transfected with PB1-F2 Constructs of Different Influenza A Viruses . To understand and compare the proinflammatory effect of PB1-F2 protein from various influenza subtypes in A549 cell, we measured the expression of cytokines/chemokines in supernatants of PB1-F2 transfected cells using Multi-Analyte ELISArray Kit (Figure 7). The expression levels of inflammatory cytokines

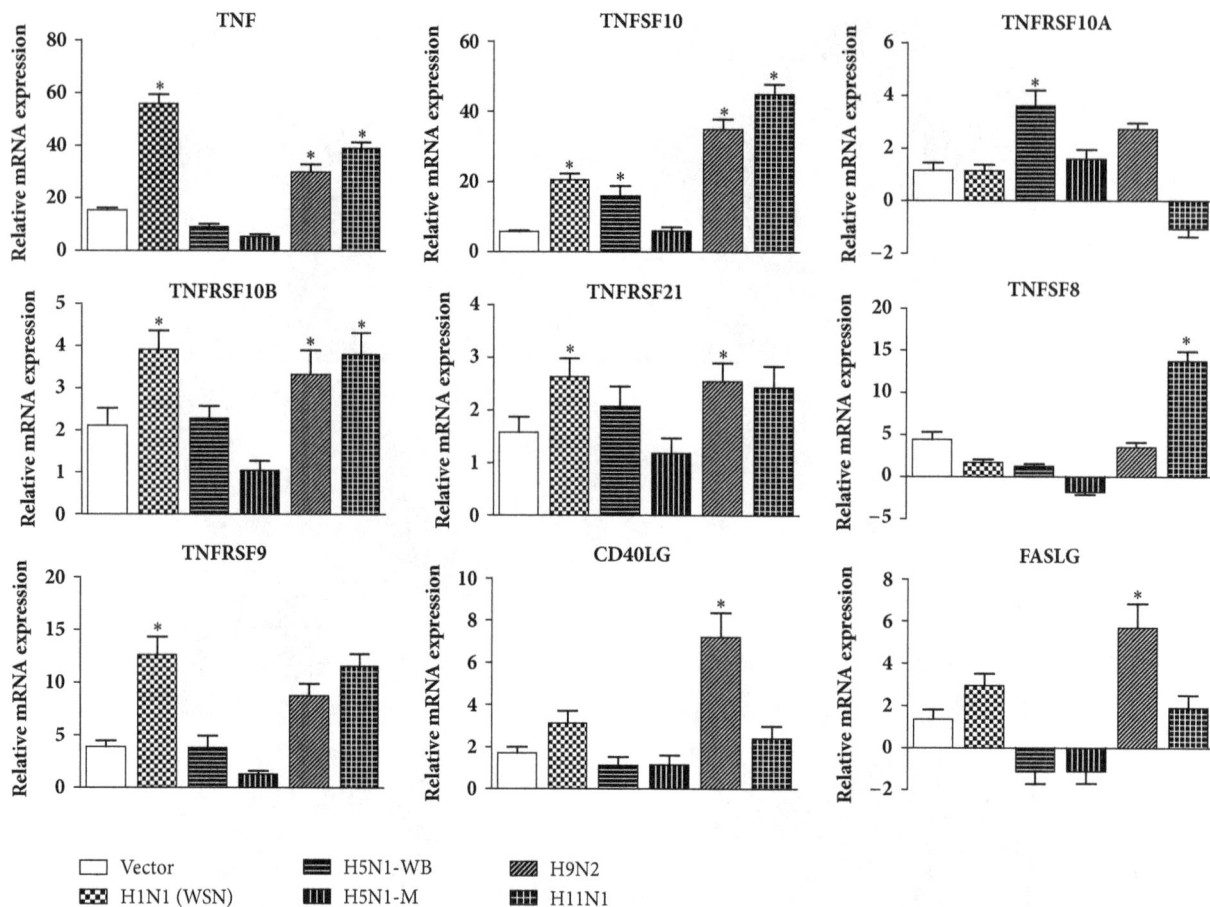

FIGURE 4: Impact of PB1-F2 expression on the mRNA expression levels of death receptor ligands and receptors. Gene expressions were normalized with the β-actin gene expression level and presented as fold increase relative to nontransfected cell controls. Data asterisks (*) indicate ($p* < 0.05$) and are calculated relative to the mock vector control which is also represented in the figure. This data is representative of three independent experiments. Error bars represent the ± SEM.

INFα, IL-6, MCP1 (CCL2), RANTES (CCL5), and IP-10 (CXCL10) were considerably and significantly (p<0.05) higher in H5N1-WB PB1-F2 transfected cells compared to mock control (empty vector) at 24 h post transfection. The levels of TNFα were significantly higher (p<0.05) for PB1-F2 protein from H1N1 (WSN), H9N2, and H11N1 viruses and this was in concurrence with the gene expression data. IFNα levels induced by H1N1 (WSN) PB1-F2 were lower than the level in control cells although it was not statistically significant. On the contrary the IFNα levels were significantly higher (p<0.05) in cells transfected with H5N1-WB PB1-F2 construct. Overall, highly pathogenic avian influenza H5N1-WB subtype was the strongest inducer for most cytokines/chemokines showing prominent induction for CXCL10, RANTES, INF-α, and IL-6 in transfected cells (Figure 7).

4. Discussion

PB1-F2 is a nonstructural viral protein and has been shown in most studies to be proapoptotic, inducing apoptosis in immune cells by targeting mitochondria [8, 24–26]. Basic

amphipathic helix present in the C-terminal region of PB1-F2 protein dissipates mitochondrial inner membrane potential and brings about apoptosis [27]. Regardless of these findings, the precise mechanism and function of PB1-F2 induced apoptosis and innate immune responses remain unclear. Most studies on PB1-F2 protein are with well-known laboratory strains of influenza viruses, thus, causing lack in knowledge on wide spectrum of IAV subtypes [24, 25]. Hence, to improve our understanding, we analyzed the amino acid sequences of PB1-F2 protein from various IAVs and compared their ability to induce apoptosis and inflammation. In this study, we have used a cell line derived from human lung epithelial cells as an *in vitro* model to study the host responses of influenza PB1-F2 proteins. This is because lung epithelial cells are the key target of influenza virus. Virions initially replicate in airway epithelial cells, which are later released from their apical side towards airspace where they encounter immune cells [28].

Reasons for IAVs to induce apoptosis have been debatable [29–31]. Apoptosis is a multifactorial event and in this study our first objective was to understand the role of PB1-F2 in inducing apoptosis and identify the pathway involved. We found that PB1-F2 protein was minimally apoptotic in

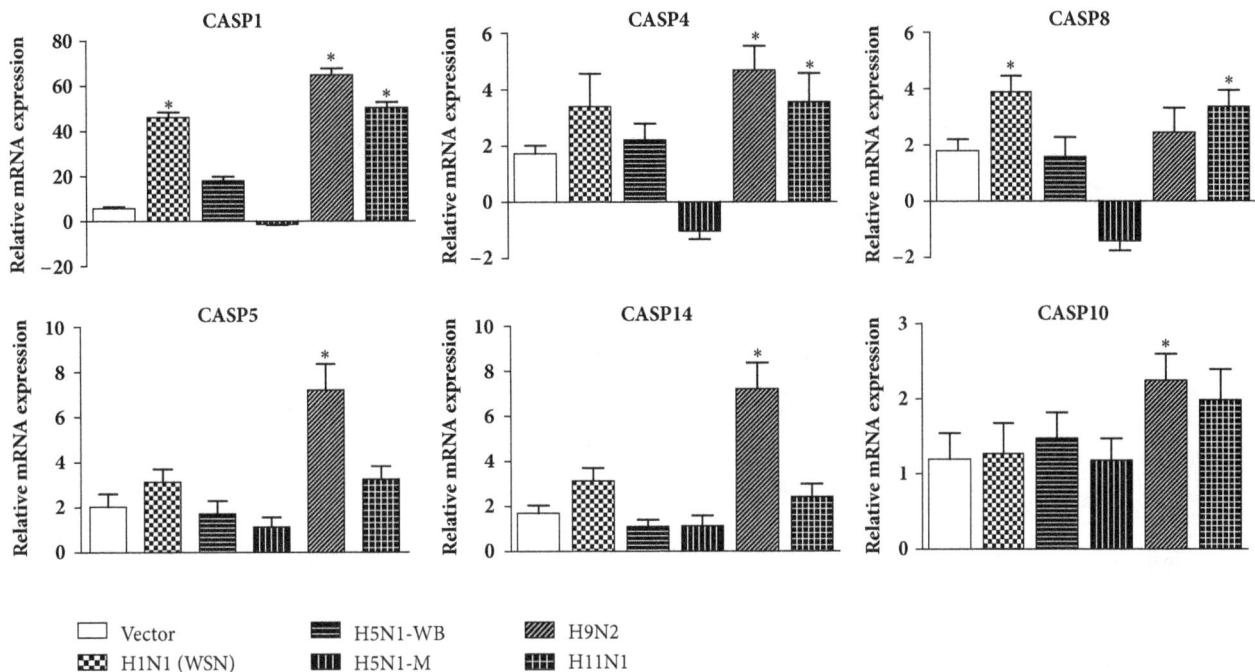

FIGURE 5: Impact of PB1-F2 expression of the mRNA expression levels of Caspases. Gene expressions were normalized with the β-actin gene expression level and presented as fold increase relative to nontransfected cell controls. Asterisks (*) indicate ($p* < 0.05$) and are calculated relative to the mock vector control which is also represented in the figure. This data is representative of three independent experiments. Error bars represent the ± SEM.

A549 cells. This is in congruence with previous studies, where PB1-F2 caused minimal apoptosis in epithelial cells [6, 24]. PB1-F2 constructs from H1N1, H9N2, and H11N1 viruses were more apoptotic than the two highly pathogenic H5N1 strains. Expression analysis of genes gave evidence that PB1-F2 protein brought about apoptosis of epithelial cells via death receptor signaling pathway. Among the death receptor ligands, expression of *TNFα* and *TRAIL* (*TNFSF10*) was significantly upregulated for the PB1-F2 constructs from H1N1, H9N2, and H11N1 (Figure 4). Expression of *FasL* was significantly upregulated in cells transfected with PB1-F2 construct from H9N2 subtype (Figure 4). Although death receptor ligands were significantly upregulated after transfection, there was feeble upregulation of Caspase 8 in H9N2 (Figure 5). This could be because TNFα and other death receptor ligands were induced during late stage of transfection. However, Caspase 8 showed statistically significant fold increase in cells transfected with H1N1 and H11N1 PB1-F2 constructs when compared to the mock (empty vector) control (Figure 5).

There was no apparent change in the levels of cytochrome c (data not shown) when measured by Human Cytochrome c Platinum ELISA kit (eBioscience; Affymetrix Co., USA). In contrast to the above, it has been hypothesized earlier that PB1-F2 along with apoptotic stimuli like TNFα sensitizes the cells to apoptosis via the proapoptotic effect of BID [24]. Cross-talk between the death receptor-mediated pathway and mitochondrial apoptotic pathway occurs when active Caspase 8 cleaves BID and the truncated BID translocates to the mitochondria causing release of cytochrome c and hence

apoptosis [31]. Interestingly, there was 20 fold increase in the expression of BIRC3/cIAP2 protein in cells transfected with H5N1-WB PB1-F2 construct (Figure 5(b)). BIRC3 is a member of the IAP family of proteins that inhibits apoptosis by binding to tumor necrosis factor receptor-associated factors TRAF1 and TRAF2, probably by interfering with activation of ICE-like proteases [32]. This could explain to some extent the least amount of apoptosis shown by A549 cells transfected with PB1-F2 from H5N1 subtype. We hypothesize that probably mitochondria are not permeabilized by PB1-F2 protein and the intrinsic pathway is not activated via BID in A549 cells. The minimal apoptosis brought about by PB1-F2 protein is via the extrinsic pathway wherein the Caspase 8 directly activates the executioner caspases 3, 6, and 7. The absence of amino acids essential for mitochondrial localization (L61, L75) in H5N1-PB1-F2 further explains its inability to induce apoptosis through intrinsic pathway (Figure 1).

The second objective of the study was to unravel the strain specific inflammatory response against PB1-F2 proteins. We observed that only H5N1-WB with complete PB1-F2 sequence showed a distinct pattern of cytokine/chemokine induction. H5N1-WB was the strongest inducer of five chemokines/cytokines: CXCL10, CCL5, CCL2, IFNα, and IL-6 (Figure 7). The mechanism that promotes high proinflammatory response in H5N1 influenza virus infection is not completely understood, but it is thought to be caused by specific amino acids in the PB1-F2 gene which facilitate this dysregulation of innate immune response [10, 14]. Presence of motifs L62, R75, R79, and L82 in the C-terminal of PB1-F2 gene in H3N2 viruses caused significant pathogenicity in

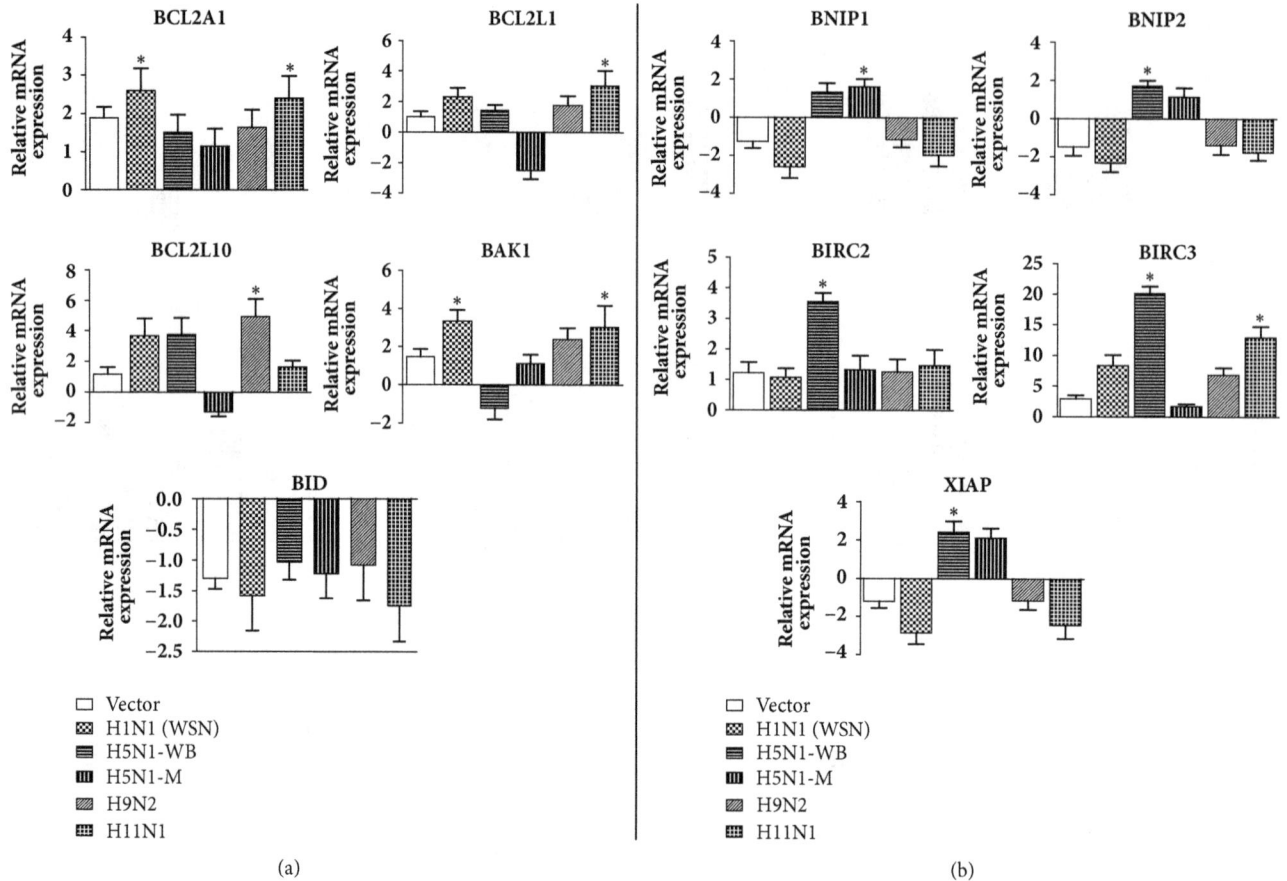

FIGURE 6: Impact of PB1-F2 expression on the mRNA expression levels of (a) BCl2 member family which govern the permeabilization of mitochondrial outer membrane. (b) IAP (Inhibitors of Apoptosis) and other antiapoptotic proteins. Gene expressions were normalized with the β-actin gene expression level and presented as fold increase relative to nontransfected cell controls. Data Asterisks ($*$) indicate ($p* < 0.05$) and are calculated relative to the mock vector control which is also represented in the figure. This data is representative of three independent experiments. Error bars represent the $\pm$ SEM.

mice [23, 33]. PB1-F2 of H5N1-WB carries this inflammatory motif, hence reiterating the significance of motif. However, it was worth noting that PB1-F2 of the other H5N1-M strain with a natural truncation in N-terminal end, carrying the inflammatory motif in the C-terminal end did not incite heightened levels of chemokines/cytokines [17, 18]. This highlighted the fact that N-terminal end of PB1-F2 has definite role in induction of inflammatory response and it needs to be explored in another study.

Early innate antiviral immune response primarily relies on the production of type I IFN [8]. PB1-F2 has been reported to modulate the induction of IFN in cell type or virus isolate specific manner. There are divergent findings suggesting ambiguities about the antagonistic or agonistic role of PB1-F2 in regulating the IFN response. One study, suggested PB1-F2 mediated exacerbation of IFN-β expression in human lung epithelial cells, while another study demonstrated prevention of transcriptional upregulation of type I IFN and other interferon stimulating genes which was associated with interaction of C-terminal domain of PB1-F2 with the mitochondrial antiviral signaling protein (MAVS) [8]. In our study, PB1-F2 construct from H5N1-WB strain caused upregulation of IFNα

and the levels were significantly higher than the cell and mock control. In contrast, PB1-F2 construct from H1N1-WSN strain prevented induction of IFNα in A549 cells. Thus, our result also proves that PB1-F2 brings about strain specific regulation of type1 IFN (Figure 7). It is very interesting to note that PB1-F2 construct from H5N1-WB initially exacerbated the IFNα levels which is detrimental to successful virus infection and then induced aberrant chemokine production thus supporting the hypothesis that H5N1 pathogenicity arises from the fatal effects of hypercytokinemia. We believe that PB1-F2 contributes significantly to this high pathogenicity. This is supported by our previous study, where we observed that approximately 96% of the H5N1 strains possessed complete PB1-F2 fragment. This suggests that PB1-F2 is positively selected in these strains and definitely necessary for the virus survival and proliferation [9]. *In silico* prediction of the secondary structure of the PB1-F2 from the strains used in this study revealed substantial differences in the secondary structure of the C-terminal regions of PB1-F2 proteins. We speculate that these differences in the conformations will definitely have bearing on the varying inflammatory response displayed by the proteins. However, these observations need

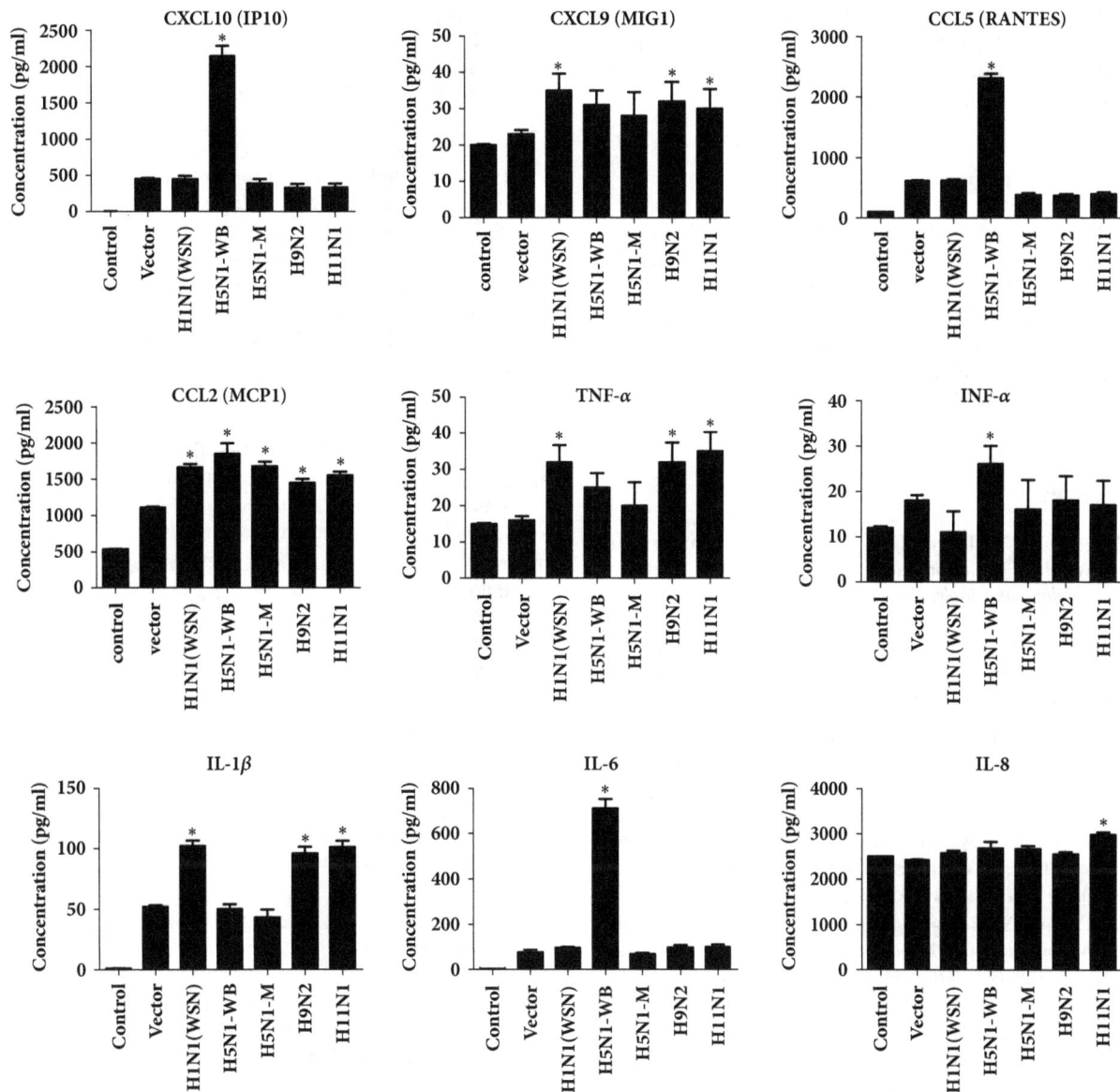

FIGURE 7: Levels of secreted cytokine/chemokine levels in response to PB1-F2 expression in A549 cells measured by ELISArrays. Data are presented as absorbance values at 450 nm. Concentration of the cytokine/chemokines is in pg/ml. Asterisks (∗) indicate ($p∗ < 0.05$) and are calculated relative to the mock vector control which is also represented in the figure. This data is representative of three independent experiments. Error bars represent the ± SEM.

to be confirmed by more sophisticated technology like NMR and CD spectroscopy.

Thus, our study suggests that ectopic expression of PB1-F2 induces minimal apoptosis in A549 lung epithelial cells. The induction of apoptosis is through death receptor mediated extrinsic pathway. PB1-F2 protein from H5N1 subtype induced least apoptosis but maximum proinflammatory response. This study emphasizes the strain specific agonist or antagonist effect of PB1-F2 on the expression of IFNα in A549 cells. Apart from C-terminal region of PB1-F2, N-terminal portion is also essential to induce exuberant proinflammatory response. This study was carried out using PB1-F2 constructs and not using any wild type or recombinant viruses created

by reverse genetics. We believe this is advantageous, mainly because of the overlapping nature of the viral ORFs and the interdependence in expression of PB1, PB1-F2, and PB1-N40, which makes it difficult to evaluate the contribution of individual protein [34]. This study is an attempt to fill in the gap in the literature regarding the effect of PB1-F2 from various IAV strains.

Authors' Contributions

Alok K. Chakrabarti conceived and designed the study. Gunisha Pasricha, Sanjay Mukherjee, and Alok K. Chakrabarti designed and performed the experiments. Gunisha Pasricha,

Sanjay Mukherjee, and Alok K. Chakrabarti performed data analysis and bioinformatics studies. Gunisha Pasricha, Sanjay Mukherjee, and Alok K. Chakrabarti wrote the paper. All authors read and approved the final manuscript.

Acknowledgments

The authors are grateful to Dr. D.T. Mourya, Director of ICMR-National Institute of Virology for his support. We are thankful to Dr. M. S. Chadha and Dr. S. D. Pawar for providing some viral strains used this study. We thank Mrs. V. C. Vipat and Mr. R. N. Khedkar for their technical support. This study was funded and supported by the Indian Council of Medical Research, Government of India.

References

[1] R. G. Webster, W. J. Bean, O. T. Gorman, T. M. Chambers, and Y. Kawaoka, "Evolution and ecology of influenza A viruses," *Microbiology and Molecular Biology Reviews*, vol. 56, no. 1, pp. 152–179, 1992.

[2] K. Shinya, A. Makino, and Y. Kawaoka, "Emerging and reemerging influenza virus infections," *Veterinary Pathology*, vol. 47, no. 1, pp. 53–57, 2010.

[3] S. Fukuyama and Y. Kawaoka, "The pathogenesis of influenza virus infections: the contributions of virus and host factors," *Current Opinion in Immunology*, vol. 23, no. 4, pp. 481–486, 2011.

[4] A. Vasin, O. Temkina, V. Egorov, S. Klotchenko, M. Plotnikova, and O. Kiselev, "Molecular mechanisms enhancing the proteome of influenza A viruses: an overview of recently discovered proteins," *Virus Research*, vol. 185, pp. 53–63, 2014.

[5] C. Klemm, Y. Boergeling, S. Ludwig, and C. Ehrhardt, "Immunomodulatory Nonstructural Proteins of Influenza A Viruses," *Trends in Microbiology*, vol. 26, no. 7, pp. 624–636, 2018.

[6] W. Chen, P. A. Calvo, D. Malide et al., "A novel influenza A virus mitochondrial protein that induces cell death," *Nature Medicine*, vol. 7, no. 12, pp. 1306–1312, 2001.

[7] H. M. Wise, A. Foeglein, J. Sun et al., "A complicated message: identification of a novel PB1-related protein translated from influenza A virus segment 2 mRNA," *Journal of Virology*, vol. 83, no. 16, pp. 8021–8031, 2009.

[8] Z. T. Varga, I. Ramos, R. Hai et al., "The influenza virus protein PB1-F2 inhibits the induction of type I interferon at the level of the MAVS adaptor protein," *PLoS Pathogens*, vol. 7, no. 6, Article ID e1002067, 2011.

[9] A. K. Chakrabarti and G. Pasricha, "An insight into the PB1F2 protein and its multifunctional role in enhancing the pathogenicity of the influenza A viruses," *Virology*, vol. 440, no. 2, pp. 97–104, 2013.

[10] G. Pasricha, A. C. Mishra, and A. K. Chakrabarti, "Comprehensive global amino acid sequence analysis of PB1F2 protein of influenza A H5N1 viruses and the influenza A virus subtypes responsible for the 20th-century pandemics," *Influenza and Other Respiratory Viruses*, vol. 7, no. 4, pp. 497–505, 2013.

[11] C.-J. Chen, G.-W. Chen, C.-H. Wang, C.-H. Huang, Y.-C. Wang, and S.-R. Shih, "Differential localization and function of PB1-F2 derived from different strains of influenza A virus," *Journal of Virology*, vol. 84, no. 19, pp. 10051–10062, 2010.

[12] H. Marjuki, C. Scholtissek, J. Franks et al., "Three amino acid changes in PB1-F2 of highly pathogenic H5N1 avian influenza virus affect pathogenicity in mallard ducks," *Archives of Virology*, vol. 155, no. 6, pp. 925–934, 2010.

[13] J. L. McAuley, J. E. Chipuk, K. L. Boyd, N. van de Velde, D. R. Green, and J. A. McCullers, "PB1-F2 proteins from H5N1 and 20th century pandemic influenza viruses cause immunopathology," *PLoS Pathogens*, vol. 6, no. 7, Article ID e1001014, 2010.

[14] M. Schmolke, B. Manicassamy, L. Pena et al., "Differential contribution of pb1-f2 to the virulence of highly pathogenic h5n1 influenza a virus in mammalian and avian species," *PLoS Pathogens*, vol. 7, no. 8, Article ID e1002186, 2011.

[15] G. M. Conenello, D. Zamarin, L. A. Perrone, T. Tumpey, and P. Palese, "A single mutation in the PB1-F2 of H5N1 (HK/97) and 1918 influenza A viruses contributes to increased virulence," *PLoS Pathogens*, vol. 3, no. 10, pp. 1414–1421, 2007.

[16] D. Zamarin, M. B. Ortigoza, and P. Palese, "Influenza A virus PB1-F2 protein contributes to viral pathogenesis in mice," *Journal of Virology*, vol. 80, no. 16, pp. 7976–7983, 2006.

[17] A. C. Mishra, S. S. Cherian, A. K. Chakrabarti et al., "A unique influenza A (H5N1) virus causing a focal poultry outbreak in 2007 in Manipur, India," *Virology Journal*, vol. 6, article no. 26, 2009.

[18] A. K. Chakrabarti, S. D. Pawar, S. S. Cherian et al., "Characterization of the influenza A H5N1 viruses of the 2008-09 outbreaks in India reveals a third introduction and possible endemicity," *PLoS ONE*, vol. 4, no. 11, Article ID e7846, 2009.

[19] Z. Xing, R. Harper, J. Anunciacion et al., "Host immune and apoptotic responses to avian influenza virus H9N2 in human tracheobronchial epithelial cells," *American Journal of Respiratory Cell and Molecular Biology*, vol. 44, no. 1, pp. 24–33, 2010.

[20] M. A. Larkin, G. Blackshields, N. P. Brown et al., "Clustal W and clustal X version 2.0," *Bioinformatics*, vol. 23, no. 21, pp. 2947-2948, 2007.

[21] E. Hoffmann, J. Stech, Y. Guan, R. G. Webster, and D. R. Perez, "Universal primer set for the full-length amplification of all influenza A viruses," *Archives of Virology*, vol. 146, no. 12, pp. 2275–2289, 2001.

[22] R. Kamal, I. Alymova, and I. York, "Evolution and Virulence of Influenza A Virus Protein PB1-F2," *International Journal of Molecular Sciences*, vol. 19, no. 1, Article ID 96, 2018.

[23] I. V. Alymova, A. M. Green, N. van de Velde et al., "Immunopathogenic and antibacterial effects of H3N2 influenza a virus PB1-F2 map to amino acid residues 62, 75, 79, and 82," *Journal of Virology*, vol. 85, no. 23, pp. 12324–12333, 2011.

[24] D. Zamarin, A. García-Sastre, X. Xiao, R. Wang, and P. Palese, "Influenza virus PB1-F2 protein induces cell death through mitochondrial ANT3 and VDAC1," *PLoS Pathogens*, vol. 1, no. 1, article e4, 2005.

[25] J. R. Coleman, "The PB1-F2 protein of Influenza A virus: increasing pathogenicity by disrupting alveolar macrophages," *Virology Journal*, vol. 4, article 9, 2007.

[26] J. L. McAuley, F. Hornung, K. L. Boyd et al., "Expression of the 1918 influenza A virus PB1-F2 enhances the pathogenesis of viral and secondary bacterial pneumonia," *Cell Host & Microbe*, vol. 2, no. 4, pp. 240–249, 2007.

[27] J. S. Gibbs, D. Malide, F. Hornung, J. R. Bennink, and J. W. Yewdell, "The influenza A virus PB1-F2 protein targets the inner

mitochondrial membrane via a predicted basic amphipathic helix that disrupts mitochondrial function," *Journal of Virology*, vol. 77, no. 13, pp. 7214–7224, 2003.

[28] M. Kurokawa, A. H. Koyama, S. Yasuoka, and A. Adachi, "Influenza virus overcomes apoptosis by rapid multiplication," *International Journal of Molecular Medicine*, vol. 3, no. 5, pp. 527–530, 1999.

[29] S. J. Stray and G. M. Air, "Apoptosis by influenza viruses correlates with efficiency of viral mRNA synthesis," *Virus Research*, vol. 77, no. 1, pp. 3–17, 2001.

[30] W. J. Wurzer, O. Planz, C. Ehrhardt et al., "Caspase 3 activation is essential for efficient influenza virus propagation," *EMBO Journal*, vol. 22, no. 11, pp. 2717–2728, 2003.

[31] S. Elmore, "Apoptosis: a review of programmed cell death," *Toxicologic Pathology*, vol. 35, no. 4, pp. 495–516, 2007.

[32] J. Silke and D. Vucic, "IAP family of cell death and signaling regulators," *Methods in Enzymology*, vol. 545, no. 2, pp. 35–65, 2014.

[33] J. McAuley, Y.-M. Deng, B. Gilbertson, C. Mackenzie-Kludas, I. Barr, and L. Brown, "Rapid evolution of the PB1-F2 virulence protein expressed by human seasonal H3N2 influenza viruses reduces inflammatory responses to infection," *Virology Journal*, vol. 14, no. 1, 2017.

[34] S. Tauber, Y. Ligertwood, M. Quigg-Nicol, B. M. Dutia, and R. M. Elliott, "Behaviour of influenza A viruses differentially expressing segment 2 gene products in vitro and in vivo," *Journal of General Virology*, vol. 93, no. 4, pp. 840–849, 2012.

20

Prevalence of Human Sapovirus in Low and Middle Income Countries

Mpho Magwalivha [ID],[1] Jean-Pierre Kabue,[1]
Afsatou Ndama Traore,[1] and Natasha Potgieter [ID][1,2]

[1]*Department of Microbiology, School of Mathematical and Natural Sciences, University of Venda, South Africa*
[2]*Dean of School of Mathematical and Natural Sciences, University of Venda, South Africa*

Correspondence should be addressed to Mpho Magwalivha; mpho.magwalivha@univen.ac.za

Academic Editor: Jay C. Brown

Background. Sapovirus (SV) infection is a public health concern which plays an important role in the burden of diarrhoeal diseases, causing acute gastroenteritis in people of all ages in both outbreaks and sporadic cases worldwide. *Objective/Study Design*. The purpose of this report is to summarise the available data on the detection of human SV in low and middle income countries. A systematic search on PubMed and ScienceDirect database for SV studies published between 2004 and 2017 in low and middle income countries was done. Studies of SV in stool and water samples were part of the inclusion criteria. *Results*. From 19 low and middle income countries, 45 published studies were identified. The prevalence rate for SV was 6.5%. A significant difference (*P=0*) in SV prevalent rate was observed between low income and middle income countries. Thirty-three (78.6%) of the studies reported on children and 8 (19%) studies reported on all age groups with diarrhoea. The majority (66.7%) of studies reported on hospitalised patients with acute gastroenteritis. Sapovirus GI was shown as the dominant genogroup, followed by SV-GII. *Conclusion*. The detection of human SV in low and middle income countries is evident; however the reports on its prevalence are limited. There is therefore a need for systematic surveillance of the circulation of SV, and their role in diarrhoeal disease and outbreaks, especially in low and middle income countries.

1. Introduction

An estimated number of 6.3 million deaths of children under the age of 5 years suffering from diarrhoea have been reported worldwide [1, 2]. In Africa, death due to diarrhoeal disease remains a major health concern, though it has decreased from 2.6 million to 1.3 million between 1990 and 2013 [3]. Diarrhoeal disease is the important cause of morbidity and mortality in low and middle income countries, also the third most frequent cause of death and greatest contributor to the burden of disease in children younger than 5 years of age [4]. The infection of human intestinal tract occurs through transmission at the household level due to different pathways such as ingestion of contaminated food and water, poor waste disposal, and person-to-person interactions in the households and community [4, 5]. Low and middle income countries still face challenges like inadequate human waste disposal, poor water quality, poor health status, and disease transmission through faecal-oral route [6].

Amongst diarrhoeal causing agents, Sapovirus (SV) is one of the enteric viruses that cause acute gastroenteritis in humans and animals. Sapoviruses were previously called "typical human Caliciviruses" or "Sapporo-like viruses" in the family Caliciviridae [7]. They are identified as nonenveloped, positive-sense, single-stranded ribonucleic acid (RNA) genome of approximately 7.1 to 7.7 kb in size with a poly(A) tail at the 3'-end [8–10]. Amongst the five designated genogroups (GI to GV), GIII infects porcine species [11–14], while GI, GII, GIV, and GV infect humans [15]. Currently, human SV genogroups are classified into 16 genotypes (comprising seven genotypes for GI and GII, respectively, and one genotype each for GIV and GV) through phylogenetic analysis of the complete capsid gene [15, 16]. Coinfections of SVs with other enteric viruses (such as noroviruses [NoVs], rotaviruses [RVs], astroviruses [AstVs], adenoviruses [AdVs], enteroviruses [EVs], and kobuviruses [KbVs]) have been noted in acute gastroenteritis outbreaks in humans [17–19].

This review summarises reports on SV detection and typing in low and middle income countries. In addition, it highlights the need to establish the relatedness of circulating SV strains in environmental (water) samples and clinical samples from communities in low and middle income countries (particularly rural settings). The time-frame chosen was 2004 to 2017 because of the availability of published data on human SV within the low and middle income countries.

2. Methodology

Two literature searches were carried out. The first literature search was performed using the terms: calicivirus, sapovirus, and developing countries, as listed by National Institutes of Health PUBMED library and ScienceDirect. A second literature search was independently done for each of the 139 "developing" countries accessed from the list published by the Society for the Study of Reproduction (http://www.ssr.org). Furthermore, the identified countries were then assessed according to the 2018 World Bank analytical classification report (http://datahelpdesk.worldbank.org/knowledgebase/articles/906519). For a successful search, each of the countries' names was combined with the following keywords: calicivirus, sapovirus, enteric viruses, and gastroenteritis. Studies identified by the search terms were selected for inclusion in the review based on the following inclusion criteria:

(a) Studies limited to human SV detected in clinical specimen and environmental water samples, reported in the 21st century.

(b) SV studies using laboratory molecular techniques including nested-PCR (nPCR), real time-PCR (RT-PCR), and RT-multiplex PCR.

Studies were excluded from the review if SV was detected in other mammalian species or animals or if the study was conducted in high income countries. In case of duplication of studies by authors, only one article was included.

Data was extracted from each selected study when provided: country name and its economic status (i.e., low income, lower, and upper middle income) as per the analytical classification report by World Bank, study setting (hospitalised, outpatient, and environment), study population (age group), population size, duration of the study, diagnostic method used, number of samples tested for SV (including their genogroups and genotypes), first author, and year of publication (Tables 1, 2 and 3).

The difference of SV data in middle and low income countries was analysed for statistical significance by Student's t-test using the simple interactive statistical analysis (SISA) at http:home.clara.net/sisa. Result with $P < 0.05$ was considered significant.

3. Results

A total of 138 articles published from 2004 to 2017 were identified from 19 low and middle income countries. After

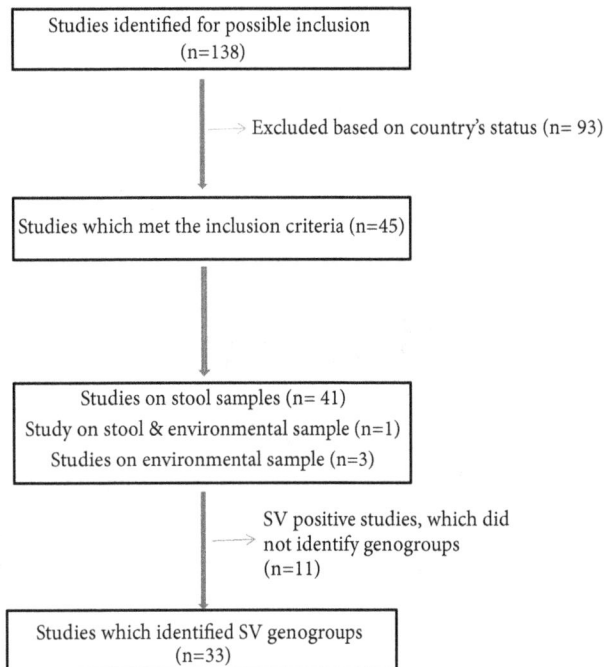

FIGURE 1: Schematic diagram showing search process for selection of studies reported.

selection based on the selection criteria (Figure 1), a total of 45 studies met the inclusion criteria. From 45 publications, 41 reported on clinical (stool) samples, 3 on environmental (water) samples, and 1 on both. Of the 42 studies conducted on clinical specimens, 66.7% (n=28) were done in hospitalised patients, 23.8% (n=10) in outpatients, and 9.5% (n=4) in both hospitalised and outpatient settings.

3.1. SV Age Distribution in Human Populations. The majority of studies (78.6%; 33/42) investigated SV in children less than 5 years of age and a further 19% (8/42) included all ages. However, only a single study investigated SV in adults with diarrhoea or acute gastroenteritis.

3.2. Seasonality. The detection of SV from clinical samples based on seasonality was reported in only 14.3% (6/42) of the studies. The majority (42.9%, 18/42) of the studies did not report on the time-frame of detection, 38% (16/42) of the studies showed inconsistent time-frame of detection, and 4.8% (2/42) of the studies showed detection throughout the year. Studies investigating SV in water sources in South Africa (SA) did not detect any seasonal peaks.

Five studies reported on samples collected within a period of 2 to 4 months, and these cases were not defined as outbreaks, while the duration period of sample collection for other 40 studies ranged over periods from 1 year to 5 years.

3.3. Sapovirus Detection and Genotyping. From the 42 included studies, 41 of these reported SV positive cases while only one study on adults reported negative results (Tables 1 and 2). Mixed infection of SV with bacteria and/or other enteric viruses was identified in 19.5% (8/41) of the studies, a

TABLE 1: Summary of human SV detection from 33 studies (stool samples) conducted in 14 non-African low and middle income countries.

Country	World Bank Classification as of year 2018	Study population	Study setup Population size	Study setup Study setting	Duration of study	Prevalence (seasons or defined period of incidence)	Method used	Rate of Detected Genotypes	Reference
Bangladesh	Lower middle income	Infants/ Children	917	HP with AGE	From 2004 to 2005	Oct 2004 – Jan 2005, Sept 2005	RT-PCR	2.7 % SV (*All in <3 yrs of age*) SV-GI.1, GI.2	Dey et al [20]
Brazil	Upper middle income	Children	305	HP severe GE	From March to September 2003	March, May - September	RT-PCR	15/305 (4.9%), mixed infection of SV and Astv in 1 sample SV-GII.1, SV-GI.1, SV-GI.2	Aragao et al [21]
		Children (0 - 10 yrs	159	OP (81 = diar; 78 = non-diar)	From April 2008 to July 2010	February, April	RT-PCR	2 of 81: 2.5% SV (GI.1, GII.2)	Aragao et al [22]
		Children (6-55 mn old)	539	Day Care (Healthy)	From October 2009 to October 2011	Not defined	RT- multiplex PCR	25/539 (4.6%) SV, SV-GI.1, GI.3	de Oliveira et al [23]
		Children, outpatients	212 129	HP OP With AGE	From 2012 to 2014	Not defined	Quantitative real-time PCR (qPCR)	12/341 (3.5%) **[9/12 – HP, 3/12 – OP].** SV-GI.1 dominant, GI.2, GI.6, GII.1, GV.1	Fioretti et al [24]
		Children < 10yrs	426 (156 of <3yrs tested)	HP with AGE	From January 2010 to October 2011	Aug & Sept	RT-PCR	6/156 (3.8%), SV-GI.1, GI.2, GII.2, GII.4	Reymao et al [25]
		Children	172	Community	From 1990 t0 1992	Not defined	Nested PCR	9/172 (5.2%) SV-GI.1, GI.7, GII.1, GV.2	Costa et al [26]
China	Upper Middle income	Children <5yrs old	500	OP with acute (477)/ persistent (23) diar	From August to November 2010	Aug – Nov 2010	RT-PCR	9/477: 1.89% SV (*<24 month children*), mixed infection of SV & AdV in 1 sample, SV-GI dominant, SV-GII & SV-GIV	Ren et al [27]
		Patients (1mn – 78yrs)	412	HP & OP with AGE	From August 2014 to September 2015	Not defined	RT-PCR	[9/412] 2.2% SV single infection, Co-infection: 2/412 ETEC with SV, 1/412 Salmonella sp with SV, 1/412 Salmonella sp with SV & AdV **Genogroups not defined**	Shen et al [28]

TABLE 1: Continued.

Country	World Bank Classification as of year 2018	Study population	Study setup Population size	Study setting	Duration of study	Prevalence (seasons or defined period of incidence)	Method used	Rate of Detected Genotypes	Reference
India (New Delhi)	Lower middle income	Children <10yrs	226	HP with AGE	From August 2000 to December 2001	Not defined	Multiplex two-step RT-PCR	23/226 (39%), mixed infection in 5 samples {NV-GII and SV-GI} SV-GI [22], GII [1]	Rachakonda et al [29]
Iran		Children	200	HP with AGE	From 2008 to 2009	Winter and in fall	RT-PCR	6/200 (3%), SV-GII	Parsa-Nahad et al [30]
	Upper middle income	Patients (3 mn - 69yrs; mean 15.3yrs	42	HP with AGE	From May to July 2009	May – July 2009	RT-PCR	11.9% SV **(patients with <5yrs of age)** SV-GI.2	Romani et al [31]
Mongolia	Lower middle income	Infants	36	households	From July to August 2003	Jul – Aug 2003	RT-PCR	1/36 (2.8%) pos for SV SV-GI	Hansman et al [11, 12]
Nicaragua	Lower middle income	Children <5yrs	330	(175 HP; 155 OP), with AGE /diar	From September 2009 to October 2010	Nov 2009- Feb/Mar 2010, May-Aug/Sept 2010	Real-time PCR	57/330 (17%): **HP = 15% [27/175], OP = 19% [30/155]**. SV-GI, GII, GIV {HP: GI1, GI.2; OP: GII.2, GII.3	Bucardo et al [32]
Pakistan	Lower middle income	Infants <6 to >35 mn	122 Pos: Enteric Viruses	HP with AGE	From 1990 to 1994	Mar, Aug - Oct	RT-PCR	13.9% SV detection (12.3% SV mono-infections, 1.6 mixed infection – AstV & SV), SV-GI	Phan et al [33]
		Infants & children <1 mn – 5yrs	517	HP with AGE	From 1990 to 1994	1990: Aug, Sept, Oct 1991: Jan, May, Jul, Oct 1992: Mar, Aug, Sep 1993: Sep 1994: Apr, July	RT-PCR	3.2 % SV SV-GI dominated, followed by GII, and GIV	Phan et al [34]
Papua New Guinea (Goroka)	Lower middle income	Children <5yrs	199	HP with AGE	From August 2009 to November 2010	Not defined	RT-PCR	4/199 (2%) SV, **Genogroups not defined**	Soli et al [35]

TABLE 1: Continued.

Country	World Bank Classification as of year 2018	Study population	Population size	Study setting	Duration of study	Prevalence (seasons or defined period of incidence)	Method used	Rate of Detected Genotypes	Reference
Peru	Upper middle income	Children <2yrs	599	300 non-diar, 299 diar	From 2007 to 2010	Four seasons	Quantitative reverse transcription-real-time PCR (qPCR)	9.0% overall: *12.4% [37/299] **diarrhoeal** – SV-GI1/2/4/5, GII.1/2/4/5, GIV, GV/1; *5.7% [17/300] **non-diarrhoeal** – SV-GII.5, GIV	Liu et al [36]
Philippines	Lower middle income	Children <5yrs	417	HP with AGE	From June 2012 to August 2013	Not defined	Real-time PCR	29/417 (7%) detection, (co-infection in 10/29: 6/10 with RV, 2/10 with NV, 2/10 with AstV). SV-GI1, GI.2, GII.1, GII.4 & GV	Liu et al [1, 2]
Thailand	Upper middle income	Infants	80 randomly selected	HP with AGE	From November 2002 to April 2003	Nov 2002 – April 2003	RT-PCR	15%: 11% single infection, 4% mixed infection – NoV & SV), SV-GI	Guntapong et al [37]
		Children <5yrs	248	HP with AGE	From 2002 to 2004	Not defined	RT-PCR	3/248 (1.2%) SV- single infections SV-GI	Khamrin et al [38]
		Children	296	HP with AGE	From May 2000 to March 2002	Jun-Jul, Jan-Mar, May-Jul, Mar.	RT-PCR	SV-GI [GI.1 &GI.2], GIV 25%, mixed infection I 1 sample (NV-GI and SV) SV-GI1, GI.4, GI.5, GII.1, GII.2	Malasao et al [39]
		All age groups	273	HP with AGE/diar	From January 2006 to February 2007	Early summer: March & April	RT-PCR	0.8% SV SV-GII/3	Kittigul et al [40]
		Children (Neonate to 5yrs old)	147	HP with AGE/watery	January to December 2005	Not defined	RT-PCR	5/147 (3.4%) SV SV-GI [GI.2, GI.1, GI.5] dominating, SV-GII.3	Khamrin et al [41]
		Pediatric patients	160	HP with AGE	January to December 2007	Throughout the year	RT-multiplex PCR	5/160 (3.1%) SV Genogroup not defined	Chaimongkol et al [42]
		Children <5yrs	567	HP with AGE	In 2007, and from 2010 to 2011	2007: Feb, Sept, Oct. & 2010: Dec	Semi-nested RT-PCR	7/567 (1.2%), SV-GI1	Chaimongkol et al [43]
		Adult (15yrs – 90yrs)	332	HP with diar	Year 2008	Not defined	RT- multiplex PCR	No SV detected	Saikruang et al [44]
		Patients	1141	HP with AGE	From 2006 to 2008	May - July	RT-PCR	1.1% SV, mixed infection of NoV-GII & SV in 2 samples Genogroup not defined	Pongsuwanna et al [45]

TABLE 1: Continued.

Country	World Bank Classification as of year 2018	Study setup		Study setting	Duration of study	Prevalence (seasons or defined period of incidence)	Method used	Rate of Detected Genotypes	Reference
		Study population	Population size						
		Children	448	HP with acute sporadic gastroenteritis	From December 1999 to November 2000	Not defined	RT-PCR	1/448 (0.2%) SV-GI	Hansman et al [46]
		Paediatric patients	1010	HP with viral AGE	From October 2002 to September 2003	Oct 2002 – Sep 2003, Rainy season (July)	RT-PCR	0.8% SV (0.4% monoinfection, 0.4% coinfection), **Genogroup not defined**	Nguyen et al [47]
Vietnam	Lower middle income	Pediatric	502	HP with AGE	From December 2005 to November 2006	Dry season	RT-PCR	1.2% SV	Nguyen et al [48]
		Children <5yrs	501	HP with AGE	From November 2007 to October 2008	Cooler months (Oct – Feb)	Real-time RT-PCR	1.4% SV SV-GI and SV-GII Co-infection of (NoV & SV) in 1 sample, of (NoV, SV, and RV) in 1 sample	Trang et al [49]
Independent States of the former Soviet Union	See information below describing the States	Children	495	HP with AGE	From January to December 2009	Jan - Mar, May – Aug	Real-time PCR	16/495 (3.2%) SV-GI.1 dominating	Chhabra et al [50]

HP = hospitalised patient; OP = outpatient; AGE = acute gastroenteritis; mn= month; yr(s) = year(s); diar = diarrhoea; SV = Sapovirus; G (I-IV) = genogroup (I-IV).
*Independent States of the former Soviet Union refers to **Armenia, Azerbaijan & Belarus** (upper middle income status), **and Georgia, Republic of Moldova & Ukraine** (lower middle income status).

TABLE 2: Summary of human SV detection from 9 studies (stool samples) conducted in 5 African countries.

Country	World Bank Classification as of year 2018	Study population	Population size	Study setting	Duration of study	Prevalence (seasons or defined period of incidence)	Method used	Rate of Detected Genotypes	Reference
Burkina Faso	Low income	Children	263 diarrhoeal, 50 non-diarrhoeal	Urban area (HP & OP)	From November 2011 to September 2012	Not defined	Real-time RT-PCR	9%: 27/263 (10.3%) {5/27 = *hospitalised*, 22/27 = *non-hospitalised*} & 3/50 (6%) SV-GII [GII.2, GII.1, GII.3], SV-GI.2	Ouedraogo et al [51]
		Children <5yrs	309 diarrhoeal	Not defined	From May 2009 to March 2010	Not defined	Real-time PCR	56/309 (18%) [mixed infection: with RV 25/56, with NV 5/56; single infection 20/56; Genogrouping {34/56}; SV-GI [GI.1, GI.4], GII [GII.1, GII.4, GII.6], GIV.I & GV.I	Matussek et al [16]
Ethiopia	Low income	All age groups	213 diarrheic samples	Government Health Care Centre	From June to September 2013	June-sept 2013	RT-PCR	9/213 (4.2%) One sequenced (SV-GII.1)	Sisay et al [52]
Kenya	Lower middle income	All age groups	334-Lwak & 524-Kibera.	Clinics with diar	From June 2007 to October 2008	Not defined	RT-PCR	5%: 13/334 (4%) and 31/524 (6%) SV **Genogroups not defined**	Shioda et al [3]
		Paediatric <13yrs	245	HP gastroenteritis	Year 2008	Not defined	Real-time RT-PCR	10/245 (4.1%) incl. one Mixed infection with NV **Genogroups not defined**	Mans et al [53]
		Patients 1mn to 87yrs mean 14yrs	190 [94 diar, 93 non-diar, 3 unknown]	Bio-wipes from rural households	From July 2007 to December 2008	Not defined	Real-time RT-PCR	16/190 (8.4%): (1 - 62yrs: mean 24yrs) **Genogroups not defined**	Mans et al [54]
South Africa	Upper middle income	Children	Selected) 296 of 477 SV-Pos (for characterisation)	HP with gastroenteritis	From April 2009 to December 2013	Not defined	Nested PCR	221 were characterised (genotyped) SV-GI [GI.1 - GI.3, GI.5, GI.6, GI.7], SV-GII [GII.1 - GII.7], SV-GIV	Murray et al [55]
		Children <5yrs	3103	HP diar	From 2009 to 2013	Higher in Summer & Autumn (Nov to Apr)	Real-time PCR	238/3103 (7.7%) SV **Genogroups not defined**	Page et al [56]
Tunisia	Lower middle income	Children	788 [408 HP, 380 OP]	Consulting for AGE	From January 2003 to April 2007	Not defined	RT-PCR Primer Noel, 1997	6/788 (0.8%) [Mixed infection: with RV 2/6; single infection 4/6]. Positive from OP samples SV-GI.1	Sdiri-Loulizi et al [57]

HP = hospitalised patient; OP = outpatient; AGE = acute gastroenteritis; mn= month; yr(s) = year(s); diar = diarrhoea; SV = Sapovirus; G (I-IV) = genogroup (I-IV).

TABLE 3: Summary of human SV detection from 4 studies (water samples) conducted in low and middle income countries.

Country	World Bank Classification as of year 2018	Samples		Duration	Prevalence (season)	Method used	Rate of detection	Reference
		Type	Size					
Brazil	Upper middle income	Wastewater	156	From 2012 to 2014	Summer and Autumn	Quantitative real-time PCR (qPCR)	51/156 (33%)	Fioretti et al [24]
South Africa	Upper middle income	River water	99	From 2009 to 2010	May, Aug, Nov (2009); Jan, April (2010)	RT-PCR	48/99 (48.5%)	Murray et al [58]
		Wastewater	51	From August 2010 to December 2011	August (2010), June, July (2011)	Real-Time qPCR	37/51 (72.5%)	Murray et al [59]
		Water (various source)	10	January and March 2012	January and March 2012	Real-Time PCR	8/10 (80%)	Murray and Taylor [60]

SV single strain was identified in 36.6% (15/41) of the studies, and mixed strains of SV were identified in 43.9% (18/41) of the studies. From the 41 studies, only 31 studies reported SV detection with identification of the genogroups/genotypes. Overall detection of SV strains showed SV-GI.1 and GI.2 as the most dominant [90% (28/31)] strain from different settings of studies, followed by SV-GII.1, GII.2, GII.3, and GII.4 with the least detection of SV-GIV strain and –GV (GV.2) strain. No study showed the occurrence of SV-GIV as a single detection but only in mixed infection cases.

The prevalence rate of SV from the 41 documented studies in low and middle countries was 6.19% with a range from 0.2% to 39%. Further breakdown showed significant difference ($P = 0$) in SV prevalence rate between low income (10.40%) and middle income (5.86%) countries. Although data on the prevalence of SV in African countries is limited, thus far, eight studies have been conducted in urban settings. Detection of SV from children in Africa is recorded with different incidence rates: in Tunisia [0.8%] [57], Burkina Faso [18%, 10.3%, respectively] [16, 51], and South Africa [4.1%, 7.7%, respectively] [53, 56]. The prevalence of SV in all ages was reported from South Africa [8.4%] [54], Ethiopia [4.2%] [52], and Kenya [4%] [3]. A predominance of SV-GIV (53/221, 24%) was noted in the South African study done on stool samples from hospitalised children with gastroenteritis [55].

Only 8.9% of studies reported SV in the environmental and waste water samples from low and middle income countries. The detection of SV-GI, SV-GII, and SV-GIV has been reported from polluted water sources by wastewaters and also on samples collected from treatment plants within selected areas of SA [58–60]. Sapovirus genogroups I and II were identified from river water samples, with detection rate of 48.5% (48/99) [58], while, in Brazil, SV-GI (genotypes 1 and 2) were detected (33%, 51/156) from the wastewaters [22], Table 3.

4. Discussion

This review provides a summary of studies conducted in developing countries on the detection of human SV. Only 45 (41 stool samples, 3 water samples, and 1 both stool and water sample) studies satisfied the inclusion criteria of this review highlighting the importance for systematic surveillance monitoring human SV circulating in developing countries (rural and urban communities). Very little is known about the contribution of human SV to diarrhoeal disease in developing countries; this is reflected in the fact that reported studies were only from 19 identified countries which include 5 African countries, namely, Burkina Faso, Ethiopia, Kenya, South Africa, and Tunisia (Table 2). A total of 78.6% (33/42) studies reported on children ≤5 years of age from the collected data, highlighting the role of SV in diarrhoeal disease amongst children in the developing countries. Hence, SV and other emerging enteric viruses, being underappreciated, can be an important cause of Norovirus negative outbreaks as reported by Lee and colleagues [61]. In addition, since it is difficult to culture human SV on cell lines [13], specialised molecular laboratories are needed for the investigation of such virus in the developing countries. Because of lack of funding and skills, the prevalence of enteric viruses is underreported in Africa and other developing countries [62]

Most of the studies (66.7%; 28/42) were done in hospitalised patients, and this might be due to the fact that SV infection sometimes leads to hospitalisation as illustrated from other studies [49, 63]. GEMS study reported SV amongst other enteric pathogens to have been associated with moderate to severe diarrhoea in developing countries [64]. The Millennium Development Goals (MDG) 2015 report shows disadvantaged settings being vulnerable as compared with the advantaged or developed settings, highlighting the effectiveness and affordability of treatments, and improved service delivery and political commitment playing a role in such settings. The statistical analysis of this review similarly showed a significant difference in the prevalence of SV in low income than in middle income countries ($P=0$).

The circulation of SV genogroups shows variability, with SV-GI and SV-GII detected frequently, while SV-GIV and SV-GV are rarely detected comparing to other genogroups [16]. An African study (Burkina Faso) reported SV-GII as the predominated strain, mostly in outpatients with diarrhoea (81.5%: 22/27), suggesting that this genogroup may be less virulent and require fewer hospital admissions. However, additional studies on outpatients will have to be conducted to confirm this observation. Although the detection of SV-GII is seen in diarrhoeal samples, it might be less virulent to cause severe symptoms leading to hospitalisation of patients, unlike SV-GI which is commonly known to be associated with severe symptoms and frequently detected in patients presenting with gastroenteritis [16, 32]. The detection of SV (GI, GII, GIV, and GV) in gastroenteritis outbreak cases has been reported in high income countries, however with less detection rate of SV-GII in both cases [14, 17, 61, 65].

Human SV infections cases relating to acute gastroenteritis in people of all ages have been identified worldwide [14]. Notwithstanding the potential selection biases present based on the studies available for inclusion, this review shows that the prevalence in children may be higher than in adults in low and middle income countries. In addition, the GEMS study in low and middle income countries highlights diarrheal disease in children as a leading cause of illness and death and also increasing the risk of delayed physical and intellectual development [66]. It has been reported that sporadic and outbreak cases caused by enteric viruses spread mainly by person-to-person contact, contaminated surfaces or objects, and contaminated water or food [67]. Therefore children are more vulnerable than adults within such exposed environment, probably because of immune system development. However, previous studies noted that gastroenteritis symptoms are usually self-limiting, and patients usually recover within a couple of days depending on the individual immune's response [49, 63]. Adults are likely to consider self-treatment by oral rehydration solution (ORS) which is the safe, effective, and low cost therapeutic option preventing dehydration [68], hence not consulting in healthcare facilities or likely due to self-respect.

Sapoviruses, like other enteric viruses, play an important role in the burden of disease worldwide. The GEMS conducted a three-year study in selected low and middle income countries, amongst children aged 0 to 59 months, and reported the detection of SV (3.5%) associated with diarrhoea [64]. However, there is no surveillance system on SV infection and prevalence in low and middle income countries, which means underreporting of sporadic cases of human SV and its epidemic are underestimated. Nevertheless, detection and comparison of the SV strains circulating in low and middle income countries (especially Africa) are currently underreported and this could be due to various techniques used for sampling and detection, including study site conditions.

Information on seasonality, patient history, area settings, and predicated pattern of transmission of viruses within the community provides knowledge needed to implement public health intervention strategies. Furthermore, detection of enteric viruses (such as SV) in environmental samples gives awareness of the circulation of infectious viral particles within the population and health-hazards which might be associated with the environment. The predictable effects of human waste disposal, water quality, and high rate of immunocompromised society have been a big concern in low and middle income countries, but there are still few documented reports on the detection of SV from environmental samples. This is highlighted by the finding of this study with high prevalence of SV in low income countries. The survival and development of children depend on good hygiene practices and use of clean drinking and domestic water on daily basis [4]. Monitoring of genetic diversity of the current circulating or emerging SV genogroups, possible water-borne transmission, and possible zoonotic infections amongst the communities is critical, and studies which can show the transmission of SV between the environment(s) (especially river water), domestic animals, and human should be considered, and the role that SV plays in diarrhoeal diseases [69].

5. Conclusion

This review found substantial evidence of SV proportion associated with diarrhoeal disease in low and middle income countries. However there is limited data reporting the detection of circulating SV strains. Therefore systematic surveillance of SV circulation within the communities in low and middle income countries is needed to assess sufficiently its role in diarrhoea disease.

References

[1] L. Liu, S. Oza, D. Hogan et al., "Global, regional, and national causes of child mortality in 2000–13, with projections to inform post-2015 priorities: an updated systematic analysis," *The Lancet*, vol. 385, no. 9966, pp. 430–440, 2015.

[2] X. Liu, D. Yamamoto, M. Saito et al., "Molecular detection and characterization of sapovirus in hospitalized children with acute gastroenteritis in the Philippines," *Journal of Clinical Virology*, vol. 68, pp. 83–88, 2015.

[3] K. Shioda, L. Cosmas, A. Audi et al., "Population-based incidence rates of diarrheal disease associated with norovirus, sapovirus, and astrovirus in Kenya," *PLoS ONE*, vol. 11, no. 4, Article ID e0145943, 2016.

[4] T. Govender, J. M. Barnes, and C. H. Pieper, "Contribution of water pollution from inadequate sanitation and housing quality to diarrheal disease in low-cost housing settlements of Cape Town, South Africa," *American Journal of Public Health*, vol. 101, no. 7, p. -e9, 2011.

[5] J. N. S. Eisenberg, J. C. Scott, and T. Porco, "Integrating disease control strategies: Balancing water sanitation and hygiene interventions to reduce diarrheal disease burden," *American Journal of Public Health*, vol. 97, no. 5, pp. 846–852, 2007.

[6] GBD Diarrhoeal Diseases Collaborators., "Estimates of global, regional, and national morbidity, mortality, and aetiologies of diarrhoeal diseases: a systematic analysis for the Global Burden of Disease Study 2015," *Lancet Infectious Diseases*, vol. 17, pp. 909–948, 2015.

[7] M. A. Mayo, "A summary of taxonomic changes recently approved by ICTV," *Archives of Virology*, vol. 147, no. 8, pp. 1655-1656, 2002.

[8] T. Oka, Z. Lu, T. Phan, E. L. Delwart, L. J. Saif, and Q. Wang, "Genetic characterization and classification of human and animal sapoviruses," *PLoS ONE*, vol. 11, no. 5, Article ID e0156373, 2016.

[9] KY. Green, "Caliciviridae: the noroviruses: specific virus families," in *Fields virology*, P. M. Howley, D. E. Griffin, R. A. Lamb, M. A. Martin, B. Roizman, and S. E. Straus, Eds., pp. 949–979, Lippincott Williams & Wilkins, Philadelphia, PA, USA, 5th edition, 2007.

[10] K.-O. Chang, S. S. Sosnovtsev, G. Belliot, Q. Wang, L. J. Saif, and K. Y. Green, "Reverse genetics system for porcine enteric calicivirus, a prototype sapovirus in the Caliciviridae," *Journal of Virology*, vol. 79, no. 3, pp. 1409–1416, 2005.

[11] G. S. Hansman, K. Natori, T. Oka et al., "Cross-reactivity among sapovirus recombinant capsid proteins," *Archives of Virology*, vol. 150, no. 1, pp. 21–36, 2005.

[12] G. S. Hansman, M. Kuramitsu, H. Yoshida et al., "Viral gastroenteritis in Mongolian infants [10]," *Emerging Infectious Diseases*, vol. 11, no. 1, pp. 180–182, 2005.

[13] G. S. Hansman, H. Saito, C. Shibata et al., "Outbreak of gastroenteritis due to sapovirus," *Journal of Clinical Microbiology*, vol. 45, no. 4, pp. 1347–1349, 2007.

[14] T. Oka, Q. Wang, K. Katayama, and L. J. Saif, "Comprehensive review of human sapoviruses," *Clinical Microbiology Reviews*, vol. 28, no. 1, pp. 32–53, 2015.

[15] T. Oka, K. Mori, N. Iritani et al., "Human sapovirus classification based on complete capsid nucleotide sequences," *Archives of Virology*, vol. 157, no. 2, pp. 349–352, 2012.

[16] A. Matussek, O. Dienus, O. Djeneba, J. Simpore, L. Nitiema, and J. Nordgren, "Molecular characterization and genetic susceptibility of sapovirus in children with diarrhea in Burkina Faso," *Infection, Genetics and Evolution*, vol. 32, pp. 396–400, 2015.

[17] N. Iritani, A. Kaida, N. Abe et al., "Detection and genetic characterization of human enteric viruses in oyster-associated gastroenteritis outbreaks between 2001 and 2012 in Osaka City, Japan," *Journal of Medical Virology*, vol. 86, no. 12, pp. 2019–2025, 2014.

[18] S. Iizuka, T. Oka, K. Tabara et al., "Detection of sapoviruses and noroviruses in an outbreak of gastroenteritis linked genetically to shellfish," *Journal of Medical Virology*, vol. 82, no. 7, pp. 1247–1254, 2010.

[19] S. Räsänen, S. Lappalainen, S. Kaikkonen, M. Hämäläinen, M. Salminen, and T. Vesikari, "Mixed viral infections causing acute gastroenteritis in children in a waterborne outbreak," *Epidemiology and Infection*, vol. 138, no. 9, pp. 1227–1234, 2010.

[20] S. K. Dey, T. G. Phan, T. A. Nguyen et al., "Prevalence of sapovirus infection among infants and children with acute gastroenteritis in Dhaka City, Bangladesh during 2004-2005," *Journal of Medical Virology*, vol. 79, no. 5, pp. 633–638, 2007.

[21] G. C. Aragão, D. d. Oliveira, M. C. Santos et al., "Molecular characterization of norovirus, sapovirus and astrovirus in children with acute gastroenteritis from Belém, Pará, Brazil," *Revista Pan-Amazônica de Saúde*, vol. 1, no. 1, 2010.

[22] G. C. Aragão, J. D. Mascarenhas, J. H. Kaiano et al., "Norovirus Diversity in Diarrheic Children from an African-Descendant Settlement in Belém, Northern Brazil," *PLoS ONE*, vol. 8, no. 2, p. e56608, 2013.

[23] D. Marques Mendanha de Oliveira, M. Souza, F. Souza Fiaccadori, H. César Pereira Santos, and D. das Dôres de Paula Cardoso, "Monitoring of calicivirus among day-care children: Evidence of asymptomatic viral excretion and first report of GI.7 norovirus and GI.3 sapovirus in Brazil," *Journal of Medical Virology*, vol. 86, no. 9, pp. 1569–1575, 2014.

[24] J. M. Fioretti, M. S. Rocha, T. M. Fumian et al., "Occurrence of human sapoviruses in wastewater and stool samples in Rio De Janeiro, Brazil," *Journal of Applied Microbiology*, vol. 121, no. 3, pp. 855–862, 2016.

[25] T. K. A. Reymão, J. D. M. Hernandez, S. T. P. D. Costa et al., "Sapoviruses in children with acute gastroenteritis from manaus , Amazon region, Brazil, 2010-2011," *Revista do Instituto de Medicina Tropical de São Paulo*, vol. 58, p. 81, 2016.

[26] L. C. P. D. N. Costa, J. A. M. Siqueira, T. M. Portal et al., "Detection and genotyping of human adenovirus and sapovirus in children with acute gastroenteritis in Belém, Pará, between 1990 and 1992: First detection of GI.7 and GV.2 sapoviruses in Brazil," *Journal of the Brazilian Society of Tropical Medicine*, vol. 50, no. 5, pp. 621–628, 2017.

[27] Z. Ren, Y. Kong, J. Wang, Q. Wang, A. Huang, and H. Xu, "Etiological study of enteric viruses and the genetic diversity of norovirus, sapovirus, adenovirus, and astrovirus in children with diarrhea in Chongqing, China," *BMC Infectious Diseases*, vol. 13, no. 1, article 412, 2013.

[28] H. Shen, J. Zhang, Y. Li et al., "The 12 gastrointestinal pathogens spectrum of acute infectious diarrhea in a sentinel hospital, Shenzhen, China," *Frontiers in Microbiology*, vol. 7, 2016.

[29] G. Rachakonda, A. Choudekar, S. Parveen, S. Bhatnagar, A. Patwari, and S. Broor, "Genetic diversity of noroviruses and sapoviruses in children with acute sporadic gastroenteritis in New Delhi, India," *Journal of Clinical Virology*, vol. 43, no. 1, pp. 42–48, 2008.

[30] M. Parsa-Nahad, A. R. Samarbaf-Zadeh, M. Makvandi et al., "Relative frequency of sapovirus among children under 5 years of age with acute gastroenteritis at the Aboozar Hospital, Ahvaz, Iran," *Jundishapur Journal of Microbiology*, vol. 5, no. 1, pp. 359–361, 2012.

[31] S. Romani, P. Azimzadeh, S. R. Mohebbi, S. M. Bozorgi, N. Zali, and F. Jadali, "Prevalence of sapovirus infection among infant and adult patients with acute gastroenteritis in Tehran, Iran," *Gastroenterology and Hepatology from Bed to Bench*, vol. 5, no. 1, pp. 43–48, 2012.

[32] F. Bucardo, Y. Reyes, L. Svensson, and J. Nordgren, "Predominance of norovirus and sapovirus in nicaragua after implementation of universal rotavirus vaccination," *PLoS ONE*, vol. 9, no. 5, Article ID e98201, 2014.

[33] T. G. Phan, M. Okame, T. A. Nguyen et al., "Human Astrovirus, Norovirus (GI, GII), and Sapovirus Infections in Pakistani Children with Diarrhea," *Journal of Medical Virology*, vol. 73, no. 2, pp. 256–261, 2004.

[34] T. G. Phan, M. Okame, T. A. Nguyen, O. Nishio, S. Okitsu, and H. Ushijima, "Genetic diversity of sapovirus in fecal specimens from infants and children with acute gastroenteritis in Pakistan," *Archives of Virology*, vol. 150, no. 2, pp. 371–377, 2005.

[35] K. W. Soli, T. Maure, M. P. Kas et al., "Detection of enteric viral and bacterial pathogens associated with paediatric diarrhoea in

Goroka, Papua New Guinea," *International Journal of Infectious Diseases*, vol. 27, pp. 54–58, 2014.

[36] X. Liu, H. Jahuira, R. H. Gilman et al., "Etiological role and repeated infections of sapovirus among children aged less than 2 years in a cohort study in a peri-urban community of Peru," *Journal of Clinical Microbiology*, vol. 54, no. 6, pp. 1598–1604, 2016.

[37] R. Guntapong, G. S. Hansman, T. Oka et al., "Norovirus and sapovirus infections in Thailand," *Japanese Journal of Infectious Diseases*, vol. 57, no. 6, pp. 276–278, 2004.

[38] P. Khamrin, N. Maneekarn, S. Peerakome et al., "Genetic diversity of noroviruses and sapoviruses in children hospitalized with acute gastroenteritis in Chiang Mai, Thailand," *Journal of Medical Virology*, vol. 79, no. 12, pp. 1921–1926, 2007.

[39] R. Malasao, N. Maneekarn, P. Khamrin et al., "Genetic diversity of norovirus, sapovirus, and astrovirus isolated from children hospitalized with acute gastroenteritis in Chiang Mai, Thailand," *Journal of Medical Virology*, vol. 80, no. 10, pp. 1749–1755, 2008.

[40] L. Kittigul, K. Pombubpa, Y. Taweekate, T. Yeephoo, P. Khamrin, and H. Ushijima, "Molecular characterization of rotaviruses, noroviruses, sapovirus, and adenoviruses in patients with acute gastroenteritis in Thailand," *Journal of Medical Virology*, vol. 81, no. 2, pp. 345–353, 2009.

[41] P. Khamrin, N. Maneekarn, A. Thongprachum, N. Chaimongkol, S. Okitsu, and H. Ushijima, "Emergence of new norovirus variants and genetic heterogeneity of noroviruses and sapoviruses in children admitted to hospital with diarrhea in Thailand," *Journal of Medical Virology*, vol. 82, no. 2, pp. 289–296, 2010.

[42] N. Chaimongkol, P. Khamrin, B. Suantai et al., "A wide variety of diarrhea viruses circulating in pediatric patients in Thailand," *Clinical Laboratory*, vol. 58, no. 1-2, pp. 117–123, 2012.

[43] N. Chaimongkol, P. Khamrin, R. Malasao et al., "Molecular characterization of norovirus variants and genetic diversity of noroviruses and sapoviruses in Thailand," *Journal of Medical Virology*, vol. 86, no. 7, pp. 1210–1218, 2014.

[44] W. Saikruang, P. Khamrin, B. Suantai et al., "Detection of diarrheal viruses circulating in adult patients in Thailand," *Archives of Virology*, vol. 159, no. 12, pp. 3371–3375, 2014.

[45] Y. Pongsuwanna, R. Tacharoenmuang, M. Prapanpoj et al., "Monthly distribution of norovirus and sapovirus causing viral gastroenteritis in Thailand," *Japanese Journal of Infectious Diseases*, vol. 70, no. 1, pp. 84–86, 2017.

[46] G. S. Hansman, L. T. P. Doan, T. A. Kguyen et al., "Detection of norovirus and sapovirus infection among children with gastroenteritis in Ho Chi Minh City, Vietnam," *Archives of Virology*, vol. 149, no. 9, pp. 1673–1688, 2004.

[47] T. A. Nguyen, F. Yagyu, M. Okame et al., "Diversity of viruses associated with acute gastroenteritis in children hospitalized with diarrhea in Ho Chi Minh City, Vietnam," *Journal of Medical Virology*, vol. 79, no. 5, pp. 582–590, 2007.

[48] T. A. Nguyen, L. Hoang, L. D. Pham et al., "Norovirus and sapovirus infections among children with acute gastroenteritis in Ho Chi Minh City during 2005-2006," *Journal of Tropical Pediatrics*, vol. 54, no. 2, pp. 102–113, 2008.

[49] N. V. Trang, L. T. Luan, L. T. Kim-Anh et al., "Detection and molecular characterization of noroviruses and sapoviruses in children admitted to hospital with acute gastroenteritis in Vietnam," *Journal of Medical Virology*, vol. 84, no. 2, pp. 290–297, 2012.

[50] P. Chhabra, E. Samoilovich, M. Yermalovich et al., "Viral gastroenteritis in rotavirus negative hospitalized children <5 years of age from the independent states of the former Soviet Union," *Infection, Genetics and Evolution*, vol. 28, pp. 283–288, 2014.

[51] N. Ouédraogo, J. Kaplon, I. J. O. Bonkoungou et al., "Prevalence and genetic diversity of enteric viruses in children with diarrhea in Ouagadougou, Burkina Faso," *PLoS ONE*, vol. 11, no. 4, Article ID e0153652, 2016.

[52] Z. Sisay, A. Djikeng, N. Berhe et al., "Prevalence and molecular characterization of human noroviruses and sapoviruses in Ethiopia," *Archives of Virology*, vol. 161, no. 8, pp. 2169–2182, 2016.

[53] J. Mans, J. C. de Villiers, N. M. Du Plessis, T. Avenant, and M. B. Taylor, "Emerging norovirus GII.4 2008 variant detected in hospitalised paediatric patients in South Africa," *Journal of Clinical Virology*, vol. 49, no. 4, pp. 258–264, 2008.

[54] J. Mans, W. B. van Zyl, M. B. Taylor et al., "Applicability of Bio-wipes for the collection of human faecal specimens for detection and characterisation of enteric viruses," *Tropical Medicine & International Health*, vol. 19, no. 3, pp. 293–300, 2014.

[55] T. Y. Murray, S. Nadan, N. A. Page, and M. B. Taylor, "Diverse sapovirus genotypes identified in children hospitalised with gastroenteritis in selected regions of South Africa," *Journal of Clinical Virology*, vol. 76, pp. 24–29, 2016.

[56] N. Page, M. J. Groome, T. Murray et al., "Sapovirus prevalence in children less than five years of age hospitalised for diarrhoeal disease in South Africa, 2009-2013," *Journal of Clinical Virology*, vol. 78, pp. 82–88, 2016.

[57] K. Sdiri-Loulizi, M. Hassine, H. Gharbi-Khelifi et al., "Molecular detection of genogroup I sapovirus in Tunisian children suffering from acute gastroenteritis," *Virus Genes*, vol. 43, no. 1, pp. 6–12, 2011.

[58] T. Y. Murray, J. Mans, and M. B. Taylor, "Human calicivirus diversity in wastewater in south africa," *Journal of Applied Microbiology*, vol. 114, no. 6, pp. 1843–1853, 2013.

[59] T. Y. Murray, J. Mans, W. B. van Zyl, and M. B. Taylor, "Application of a Competitive Internal Amplification Control for the Detection of Sapoviruses in Wastewater," *Food and Environmental Virology*, vol. 5, no. 1, pp. 61–68, 2013.

[60] T. Y. Murray and M. B. Taylor, "Quantification and molecular characterisation of human sapoviruses in water sources impacted by highly polluted discharged wastewater in South Africa," *Journal of Water and Health*, vol. 13, no. 4, pp. 1055–1059, 2015.

[61] L. E. Lee, E. A. Cebelinski, C. Fuller et al., "Sapovirus outbreaks in long-term care facilities, Oregon and Minnesota, USA, 2002-2009," *Emerging Infectious Diseases*, vol. 18, no. 5, pp. 873–876, 2012.

[62] J. P. Kabue, E. Meader, P. R. Hunter, and N. Potgieter, "Human Norovirus prevalence in Africa: A review of studies from 1990 to 2013," *Tropical Medicine & International Health*, vol. 21, no. 1, pp. 2–17, 2016.

[63] M. Lorrot, F. Bon, M. J. El Hajje et al., "Epidemiology and clinical features of gastroenteritis in hospitalised children: Prospective survey during a 2-year period in a Parisian hospital, France," *European Journal of Clinical Microbiology & Infectious Diseases*, vol. 30, no. 3, pp. 361–368, 2011.

[64] K. L. Kotloff, J. P. Nataro, W. C. Blackwelder et al., "Burden and aetiology of diarrhoeal disease in infants and young children in developing countries (the Global Enteric Multicenter Study,

GEMS): a prospective, case-control study," *The Lancet*, vol. 382, no. 9888, pp. 209–222, 2013.

[65] X. L. Pang, B. E. Lee, G. J. Tyrrell, and J. K. Preiksaitis, "Epidemiology and genotype analysis of sapovirus associated with gastroenteritis outbreaks in Alberta, Canada: 2004-2007," *The Journal of Infectious Diseases*, vol. 199, no. 4, pp. 547–551, 2009.

[66] K. L. Kotloff, W. C. Blackwelder, D. Nasrin et al., "The Global Enteric Multicenter Study (GEMS) of diarrheal disease in infants and young children in developing countries: epidemiologic and clinical methods of the case/control study," *Clinical Infectious Diseases*, vol. 55, supplement 4, pp. S232–S245, 2012.

[67] B. Lopman, P. Gastañaduy, G. W. Park, A. J. Hall, U. D. Parashar, and J. Vinjé, "Environmental transmission of norovirus gastroenteritis," *Current Opinion in Virology*, vol. 2, no. 1, pp. 96–102, 2012.

[68] H. J. Binder, I. Brown, B. S. Ramakrishna, and G. P. Young, "Oral rehydration therapy in the second decade of the twenty-first century," *Current Fungal Infection Reports*, vol. 16, no. 3, 2014.

[69] K. Kumthip, P. Khamrin, and N. Maneekarn, "Molecular epidemiology and genotype distributions of noroviruses and sapoviruses in Thailand 2000-2016: A review," *Journal of Medical Virology*, vol. 90, no. 4, pp. 617–624, 2018.

21

Cumulative Impact of HIV and Multiple Concurrent Human Papillomavirus Infections on the Risk of Cervical Dysplasia

David H. Adler,[1] Melissa Wallace,[2] Thola Bennie,[2] Beau Abar,[1] Tracy L. Meiring,[3] Anna-Lise Williamson,[3,4] and Linda-Gail Bekker[2]

[1]Department of Emergency Medicine, University of Rochester, Rochester, NY 14642, USA
[2]Desmond Tutu HIV Centre, Institute of Infectious Diseases & Molecular Medicine, Faculty of Health Sciences,
 University of Cape Town, Anzio Road, Observatory, Cape Town, South Africa
[3]Institute of Infectious Diseases & Molecular Medicine and Division of Medical Virology, Faculty of Health Sciences,
 University of Cape Town, Anzio Road, Observatory, Cape Town, South Africa
[4]National Health Laboratory Service, Groote Schuur Hospital, Cape Town, South Africa

Correspondence should be addressed to David H. Adler; david_adler@urmc.rochester.edu

Academic Editor: Finn S. Pedersen

Infection with HIV is known to increase the risk of cervical cancer. In addition, evidence suggests that concurrent infection with multiple human papillomavirus (HPV) genotypes increases the risk of cervical dysplasia more than infection with a single HPV genotype. However, the impact of the combination of HIV coinfection and presence of multiple concurrent HPV infections on the risk of cervical dysplasia is uncertain. We compared the results of HPV testing and Pap smears between HIV-infected and HIV-uninfected young women to assess the cumulative impact of these two conditions. We found that both HIV and the presence of multiple concurrent HPV infections are associated with increased risk of associated Pap smear abnormality and that the impact of these two risk factors may be additive.

1. Introduction

Coinfection with HIV has a significant impact on the natural history of high-risk human papillomavirus (HR-HPV) infections, the causative agents of cervical cancer. Women with HIV are more likely to be infected with HPV, and more likely to harbor multiple concurrent HPV infections (i.e., simultaneous infection with two or more HPV genotypes) [1]. While many HPV infections are transient, HIV-infected women are more likely to have persistent HPV infections [2, 3] which are associated with a greater incidence and progression of the precancerous lesions caused by HPV [4, 5]. Worsening immunodeficiency correlates with an increased risk of advanced cervical dysplasia [6]. Invasive cervical cancer, an AIDS defining illness, is 2 to 22 times more likely to develop in HIV-infected women [7].

Multiple concurrent HPV infections have been associated with an increased risk of persistent HPV infection [8] and of developing precancerous cervical lesions [9, 10] compared to infection with a single HPV genotype. Simultaneous infection with HIV and multiple concurrent HPVs may increase the risk of developing precancerous lesions more than either condition on its own.

Cervical cancer is the number one cancer cause of years of life lost in the developing world [11] where over eighty-five percent of global cases and deaths from cervical cancer occur, with the very highest rates in Sub-Saharan Africa [12, 13]. We compared the prevalence of multiple concurrent high-risk HPV infections between HIV-infected and HIV-uninfected South African young women and the association of these infections with cervical cytological abnormalities.

2. Materials and Methods

Between October 2013 and March 2015, we conducted a longitudinal study of 50 HIV-infected and 50 HIV-uninfected

TABLE 1: Participant demographics and behavioral variables.

| | HIV-infected | | | | HIV-uninfected | | | | |
	M (SD)	IQR	f	%	M (SD)	IQR	f	%	p
Age at first visit	19.6 (1.4)	19.0–21.0			18.4 (1.4)	17.0–19.3			<0.001
Smoker at any time									0.28
No			43	88%			47	94%	
Yes			6	12%			3	6%	
Lifetime sexual partners									0.05
1			11	22%			4	8%	
2–5			34	68%			44	88%	
>5			5	10%			2	4%	
Sexual partners in the last 6 months									0.65
1			48	96%			47	94%	
2–5			2	4%			3	6%	
Number of past pregnancies									0.84
0			33	66%			34	68%	
1			15	30%			15	30%	
2			2	4%			1	2%	

young, sexually active, South African women. Cohort enrollment occurred sequentially until the goal of 50 participants in each group was met. Study participants were recruited through the Youth Centre in Masiphumelele and the Hannan Crusaid Clinic in Gugulethu, townships in the Cape Town area of South Africa. The Youth Centre in Masiphumelele serves the general adolescent population of the township, while the Hannan Crusaid Clinic specifically serves the HIV-infected population of Gugulethu. Demographic and behavioral variables of our cohort are presented in Table 1. Informed consent (age 18 years or older) or parental consent along with signed adolescent assent (age 16-17 years) was obtained from all participants. The Institutional Review Boards of the University of Rochester and the University of Cape Town approved this study.

Serial HPV DNA testing, utilizing self-collected specimens, was conducted approximately every six months for all participants. In addition, all study participants underwent baseline Pap smear testing. Most study participants underwent additional Pap testing during the follow-up period. Subsequent Pap testing (not included in this analysis) was conducted at varying intervals depending on HIV status and previous Pap result as per the guidelines of the Western Cape Province. Only the enrollment HPV and Pap test results are used in this analysis. HIV status was confirmed upon enrollment and HIV-uninfected participants underwent HIV testing at every study visit. Women with a history of HPV vaccination or cervical surgery were excluded.

Roche Diagnostics Linear Array HPV Test was used for all HPV genotyping. This kit identifies 37 HPV genotypes including all oncogenic "high-risk" HPV genotypes (HR-HPV) as designated by the International Agency for Research on Cancer [14]. These 13 HR-HPV genotypes include types 16, 18, 31, 33, 35, 39, 45, 51, 52, 56, 58, 59, and 68. Specimens for HPV testing were obtained via self-sampling, performed in private, in which study participants were instructed to twirl a Dacron swab high in the vagina for 10 seconds. Specimens were then transported to the laboratory in Digene transport medium. The MagNA Pure Compact Nucleic Acid Isolation Kit (Roche) was used to extract DNA. All Pap smears were reported per the Bethesda system. Per laboratory protocol a pathologist reviewed all positive Pap smears. We defined any Pap not categorized as "negative," including atypical squamous cell of undetermined significance (ASCUS) to be abnormal.

Pearson χ^2 tests for independence were used to compare groups, and Pearson and Spearman correlations were used to examine bivariate associations. All analyses were performed using IBM SPSS 22.0.

3. Results

Preliminary descriptive cross-sectional data from the initial 85 participants in this cohort have been previously published [15]. Among our final and complete cohort, we found an overall HPV prevalence rate of 64% and HR-HPV prevalence rate of 38%. Multiple concurrent HPV infections were found among 41% of participants and multiple concurrent HR-HPV infections were found among 24%. Since a wealth of research has shown cytological abnormalities to be driven by the presence of an HR-HPV infection, we examined Pap test results only among women with at least one such infection. The results indicated that having multiple concurrent HPV infections was associated with twice the likelihood of an associated Pap smear abnormality compared to having a single HPV infection (34% versus 17%), although this difference did not reach statistical significance ($p = 0.15$).

Overall, HIV-infected women had a much greater burden of HPV infection. HIV-infected participants had greater proportions of multiple concurrent HPV infections (60% versus

TABLE 2: HPV infections, Pap test results, and HIV status.

	HIV-uninfected n (%)		HIV-infected n (%)	
	Normal Pap	Abnormal Pap	Normal Pap	Abnormal Pap
Number of HPV infections				
0	26 (59%)	0 (0%)	10 (26%)	0 (0%)
1	11 (25%)	2 (33%)	8 (21%)	2 (17%)
2	3 (7%)	1 (17%)	4 (11%)	2 (17%)
3	3 (7%)	2 (33%)	6 (16%)	1 (8%)
4	0 (0%)	1 (17%)	2 (5%)	2 (17%)
5	1 (2%)	0 (0%)	2 (5%)	1 (8%)
>5	0 (0%)	0 (0%)	6 (16%)	4 (34%)
Number HR-HPV infections				
0	35 (80%)	3 (50%)	22 (58%)	2 (17%)
1	6 (14%)	0 (0%)	5 (13%)	3 (25%)
2	3 (7%)	3 (50%)	6 (16%)	2 (17%)
3	0 (0%)	0 (0%)	3 (8%)	5 (42%)
4	0 (0%)	0 (0%)	1 (3%)	0 (0%)
5	0 (0%)	0 (0%)	0 (0%)	0 (0%)
>5	0 (0%)	0 (0%)	1 (3%)	0 (0%)

Note: number of HPV infections and number of HR-HPV infections were associated with age (Pearson $r = 0.23$, $p = 0.019$; $r = 0.20$, $p = 0.048$, resp.).

22%, $p < 0.001$) and multiple concurrent HR-HPV infections (36% versus 12%, $p = 0.004$). The average number of HPV infections was 2.92 for the HIV-infected group compared to 0.90 for the HIV-uninfected group ($p < 0.001$). Likewise, the average number of HR-HPV infections was greater among those with HIV (1.16 versus 0.36, $p = 0.001$). HIV-infected participants had a greater proportion of abnormal Pap smears at baseline (24% versus 12%), although this difference was not statistically significant ($p = 0.12$).

Among HIV-infected women only, those with two or more HPV infections were more likely to have an associated Pap abnormality compared to those with zero or one HPV infection (33% versus 10%, $p = 0.058$). When eliminating those women without an HPV infection and comparing HIV-infected women with multiple infections to those with a single HPV infection, the difference is still substantial (33% versus 20%) although not statistically significant ($p = 0.43$). Thus, there may be an additive effect in which the combination of HIV infection and the presence of multiple concurrent HPV infections increases the association with cervical dysplasia greater than either risk factor alone.

Table 2 compares and summarizes the cross-sectional epidemiological differences in HPV and cervical dysplasia between HIV-infected and HIV-uninfected study participants. When focusing on HR-HPV infections, we found that, among HIV-infected women, there was a similar proportion of abnormal Pap tests results for those with one HR-HPV infection (38%) and those with more than one (39%) ($p = 0.94$). Among HIV-uninfected women, the difference is greater (1 HR-HPV infection = 0%, 2 or more HR-HPV infections = 50%) though not statistically significant due to the small size of this subset ($p = 0.18$).

Overall, our HIV-infected participants were not severely immune-compromised. Their average CD4 count was 518/mm^3 (IQR = 366–598; CD4 counts were not available for 10 participants). Forty-four percent were receiving antiretroviral (ARV) medications at the time of testing. The average CD4 count for those on ARV medications was 482 and for those not on ARV medications was 547 ($p = 0.37$). CD4 count was not found to be significantly associated with abnormal Pap test ($p = 0.21$) or total HPV infections ($p = 0.07$), although higher CD4 counts were associated with fewer HR-HPV infections ($r = -0.36$, $p = 0.02$). No significant relationship between ARV medications and the total number of HPV infections ($p = 0.54$) or the number of high-risk HPV infections ($p = 0.77$) was identified.

Our bivariate analyses identified positive associations between age and number of HPV infections (Pearson $r = 0.23$, $p = 0.019$), number of HR-HPV infections (Pearson $r = 0.20$, $p = 0.048$), and baseline Pap test abnormality (Spearman $r = 0.20$, $p = 0.050$). There were no relationships observed between HPV infections or Pap abnormality and lifetime smoking, lifetime number of sexual partners, number of sexual partners in the last 6 months, and current use of condoms, birth control pills, or birth control injections (all p values > 0.10).

4. Discussion

Our results demonstrate a large burden of HPV among our study population with a notably high prevalence of multiple concurrent infections. This is consistent with previous data from South Africa [16]. We confirmed previous work that identified a strong positive association between HIV and both HPV infection and HPV-related cervical dysplasia. Moreover, we found that the increased risks for cervical dysplasia conferred by HIV infection and multiple concurrent HPV infection may be additive.

It is well established that HPV causes cervical cancer [17]. It is less clear, however, whether multiple concurrent HPV infections increase this risk. Some research has demonstrated an increased risk of carcinogenesis due to multiple concurrent infections compared to infection with a single HPV genotype [10, 18]. One study identified a linear association between the number of HR-HPV infections and the severity of associated cervical dysplasia [9]. In a study of HIV-infected women, however, multiple concurrent infections were not found to have an increased association with cervical dysplasia [19]. We found a significantly positive association (twice the probability) between multiple concurrent infections and the risk of associated Pap smear abnormality.

Likewise, HIV infection is known to increase the risk of cervical dysplasia [2]. Data from the present study demonstrate HIV-infected participants to have a twofold greater proportion of Pap abnormalities compared to their HIV-uninfected counterparts. Furthermore, when looking only at HIV-infected participants, a large but not statistically significant difference in association with Pap abnormality was found between those with a single HPV infection and those with multiple concurrent HPV infections, indicating a possible additive effect of HIV infections and infection with multiple concurrent HPVs in increasing the risk of cervical dysplasia.

The impact of multiple concurrent HPV infections on the duration of infection is uncertain. Some prior research has found no impact of multiple concurrent infections on HPV persistence [20, 21]. In contrast, in a study of genotype specific duration of HPV infection, investigators found that coinfection with multiple HPV types increased the duration of infection for all HR-HPVs as a group as well as for HPV16 specifically [8]. The persistence of HPV infection is of critical importance as it has been demonstrated to increase the risk for incident precancerous lesions of the cervix [4, 22]. Due to sample size limitations, this study did not assess the impact of multiple concurrent HPV infections on HPV persistence.

This study has several limitations. The generalizability of the findings presented may be limited due to the specific geographical area from which the study cohort was drawn. Despite the numerous statistically significant findings, the relatively small sample size may have restricted the identification of important differences between groups and subgroups. Likewise, multivariate analyses examining differences between HIV-infected and HIV-uninfected participants while controlling for behavioral and demographic variables were not feasible due to the sample size. Lastly, because of the relative immune competency of our HIV-infected group and sample size limitations, study findings cannot be correlated with the extent of immunosuppression among HIV-infected participants.

5. Conclusion

In our cohort, Pap smear abnormalities were identified twice as frequently among HIV-infected women as among their HIV-uninfected counterparts. In addition, among all women in the cohort, having multiple concurrent HPV infections

was associated with twice the likelihood of an associated Pap smear abnormality compared to having a single HPV infection. These risk factors may be additive given our finding that HIV-infected women with multiple concurrent HPV infections were more likely to have associated Pap smear abnormalities than HIV-infected women with a single HPV infection.

Acknowledgment

This work was supported by the National Institute of Allergy and Infectious Diseases at the National Institutes of Health (5 K23AI07759 to David Adler).

References

[1] G. M. Clifford, M. A. G. Gonçalves, and S. Franceschi, "Human papillomavirus types among women infected with HIV: a meta-analysis," *AIDS*, vol. 20, no. 18, pp. 2337–2344, 2006.

[2] L. S. Massad, L. Ahdieh, L. Benning et al., "Evolution of cervical abnormalities among women with HIV-1: evidence from surveillance cytology in the Women's Interagency HIV Study," *Journal of Acquired Immune Deficiency Syndromes*, vol. 27, no. 5, pp. 432–442, 2001.

[3] X.-W. Sun, T. V. Ellerbrock, O. Lungu, M. A. Chiasson, T. J. Bush, and T. C. Wright Jr., "Human papillomavirus infection in human immunodeficiency virus-seropositive women," *Obstetrics and Gynecology*, vol. 85, no. 5, part 1, pp. 680–686, 1995.

[4] N. F. Schlecht, S. Kulaga, J. Robitaille et al., "Persistent human papillomavirus infection as a predictor of cervical intraepithelial neoplasia," *The Journal of the American Medical Association*, vol. 286, no. 24, pp. 3106–3114, 2001.

[5] M. A. E. Nobbenhuis, J. M. M. Walboomers, T. J. M. Helmerhorst et al., "Relation of human papillomavirus status to cervical lesions and consequences for cervical-cancer screening: a prospective study," *The Lancet*, vol. 354, no. 9172, pp. 20–25, 1999.

[6] G. M. Clifford, S. Franceschi, O. Keiser et al., "Immunodeficiency and the risk of cervical intraepithelial neoplasia 2/3 and cervical cancer: a nested case-control study in the Swiss HIV cohort study," *International Journal of Cancer*, vol. 138, no. 7, pp. 1732–1740, 2016.

[7] L. A. Denny, S. Franceschi, S. de Sanjosé, I. Heard, A. B. Moscicki, and J. Palefsky, "Human papillomavirus, human immunodeficiency virus and immunosuppression," *Vaccine*, vol. 30, supplement 5, pp. F168–F174, 2012.

[8] H. Trottier, S. Mahmud, J. C. M. Prado et al., "Type-specific duration of human papillomavirus infection: implications for human papillomavirus screening and vaccination," *Journal of Infectious Diseases*, vol. 197, no. 10, pp. 1436–1447, 2008.

[9] B. D. Bello, A. Spinillo, P. Alberizzi et al., "Cervical infections by multiple human papillomavirus (HPV) genotypes: prevalence and impact on the risk of precancerous epithelial lesions," *Journal of Medical Virology*, vol. 81, no. 4, pp. 703–712, 2009.

[10] H. Trottier, S. Mahmud, M. C. Costa et al., "Human papillomavirus infections with multiple types and risk of cervical

neoplasia," *Cancer Epidemiology Biomarkers and Prevention*, vol. 15, no. 7, pp. 1274–1280, 2006.

[11] B. H. Yang, F. I. Bray, D. M. Parkin, J. W. Sellors, and Z.-F. Zhang, "Cervical cancer as a priority for prevention in different world regions: an evaluation using years of life lost," *International Journal of Cancer*, vol. 109, no. 3, pp. 418–424, 2004.

[12] F. X. Bosch, T. R. Broker, D. Forman et al., "Comprehensive control of human papillomavirus infections and related diseases," *Vaccine*, vol. 31, supplement 7, pp. H1–H31, 2013.

[13] K. S. Louie, S. De Sanjose, and P. Mayaud, "Epidemiology and prevention of human papillomavirus and cervical cancer in sub-Saharan Africa: a comprehensive review," *Tropical Medicine & International Health*, vol. 14, no. 10, pp. 1287–1302, 2009.

[14] V. Bouvard, R. Baan, K. Straif et al., "A review of human carcinogens—part B: biological agents," *The Lancet Oncology*, vol. 10, no. 4, pp. 321–322, 2009.

[15] D. H. Adler, M. Wallace, T. Bennie et al., "Cervical dysplasia and high-risk human papillomavirus infections among HIV-infected and HIV-uninfected adolescent females in South Africa," *Infectious Diseases in Obstetrics and Gynecology*, vol. 2014, Article ID 498048, 6 pages, 2014.

[16] A. C. McDonald, A. I. Tergas, L. Kuhn, L. Denny, and T. C. Wright Jr., "Distribution of human papillomavirus genotypes among HIV-positive and HIV-negative women in Cape Town, South Africa," *Frontiers in Oncology*, vol. 4, article 48, 2014.

[17] F. X. Bosch, A. Lorincz, N. Muñoz, C. J. L. M. Meijer, and K. V. Shah, "The causal relation between human papillomavirus and cervical cancer," *Journal of Clinical Pathology*, vol. 55, no. 4, pp. 244–265, 2002.

[18] S. J. Patel, N. R. Mugo, C. R. Cohen et al., "Multiple human papillomavirus infections and HIV seropositivity as risk factors for abnormal cervical cytology among female sex workers in Nairobi," *International Journal of STD and AIDS*, vol. 24, no. 3, pp. 221–225, 2013.

[19] J. E. Levi, S. Fernandes, A. F. Tateno et al., "Presence of multiple human papillomavirus types in cervical samples from HIV-infected women," *Gynecologic Oncology*, vol. 92, no. 1, pp. 225–231, 2004.

[20] N. G. Campos, A. C. Rodriguez, P. E. Castle et al., "Persistence of concurrent infections with multiple human papillomavirus types: a population-based Cohort Study," *Journal of Infectious Diseases*, vol. 203, no. 6, pp. 823–827, 2011.

[21] M.-C. Rousseau, J. S. Pereira, J. C. M. Prado, L. L. Villa, T. E. Rohan, and E. L. Franco, "Cervical coinfection with human papillomavirus (HPV) types as a predictor of acquisition and persistence of HPV infection," *Journal of Infectious Diseases*, vol. 184, no. 12, pp. 1508–1517, 2001.

[22] G. Y. F. Ho, R. D. Burk, S. Klein et al., "Persistent genital human papillomavirus infection as a risk factor for persistent cervical dysplasia," *Journal of the National Cancer Institute*, vol. 87, no. 18, pp. 1365–1371, 1995.

22

Patterns of Human Respiratory Viruses and Lack of MERS-Coronavirus in Patients with Acute Upper Respiratory Tract Infections in Southwestern Province of Saudi Arabia

Ahmed A. Abdulhaq,[1,2] Vinod Kumar Basode,[1] Anwar M. Hashem,[3,4]
Ahmed S. Alshrari,[5] Nassrin A. Badroon,[3] Ahmed M. Hassan,[3] Tagreed L. Alsubhi,[3]
Yahia Solan,[6] Saleh Ejeeli,[1] and Esam I. Azhar[3,7]

[1] *Unit of Medical Microbiology, Department of Medical Laboratory Technology, College of Applied Medical Science,*
Jazan University, Jazan, Saudi Arabia
[2] *Deanship of Scientific Affairs and Research, Jazan University, Jazan, Saudi Arabia*
[3] *Special Infectious Agents Unit, King Fahd Medical Research Center, King Abdulaziz University, Jeddah, Saudi Arabia*
[4] *Department of Medical Microbiology and Parasitology, Faculty of Medicine, King Abdulaziz University, Jeddah, Saudi Arabia*
[5] *Department of Basic Health Sciences, Faculty of Pharmacy, Northern Border University, Arar, Saudi Arabia*
[6] *Department of Public Health, Ministry of Health, Jazan, Saudi Arabia*
[7] *Department of Medical Laboratory Technology, Faculty of Applied Medical Sciences, King Abdulaziz University, Jeddah, Saudi Arabia*

Correspondence should be addressed to Ahmed A. Abdulhaq; alhaq444@gmail.com

Academic Editor: George N. Pavlakis

We undertook enhanced surveillance of those presenting with respiratory symptoms at five healthcare centers by testing all symptomatic outpatients between November 2013 and January 2014 (winter time). Nasal swabs were collected from 182 patients and screened for MERS-CoV as well as other respiratory viruses using RT-PCR and multiplex microarray. A total of 75 (41.2%) of these patients had positive viral infection. MERS-CoV was not detected in any of the samples. Human rhinovirus (hRV) was the most detected pathogen (40.9%) followed by non-MERS-CoV human coronaviruses (19.3%), influenza (Flu) viruses (15.9%), and human respiratory syncytial virus (hRSV) (13.6%). Viruses differed markedly depending on age in which hRV, Flu A, and hCoV-OC43 were more prevalent in adults and RSV, hCoV-HKU1, and hCoV-NL63 were mostly restricted to children under the age of 15. Moreover, coinfection was not uncommon in this study, in which 17.3% of the infected patients had dual infections due to several combinations of viruses. Dual infections decreased with age and completely disappeared in people older than 45 years. Our study confirms that MERS-CoV is not common in the southwestern region of Saudi Arabia and shows high diversity and prevalence of other common respiratory viruses. This study also highlights the importance and contribution of enhanced surveillance systems for better infection control.

1. Introduction

Acute respiratory tract infections (ARTIs) represent a major and global cause of morbidity and mortality amongst people of all ages. Millions of children under the age of 5 die annually due to respiratory infections [1, 2]. While respiratory viruses represent a major cause of ARTIs, their prevalence and transmission are often influenced by several geographic, demographic, and environmental factors [3]. Moreover, the lack of

easy and cheap diagnostic methods, the nonspecific symptoms, the large number of associated viruses, and the possibility of mixed infections usually hinder the timely and accurate identification of causative viral agents. Nonetheless, the most commonly detected human respiratory viruses are human respiratory syncytial virus (hRSV), influenza viruses (Flu), human rhinovirus (hRV), enterovirus (EV), human coronavirus (hCoV), human parainfluenza viruses (hPIV), human adenoviruses (hAdv), and human metapneumoviruses (hMPV) [4]. Several other emerging respiratory viruses such as the Middle East respiratory syndrome coronavirus (MERS-CoV) and avian influenza viruses (H5N1and H7N9) have also been reported in several other parts of the world.

In Saudi Arabia, more than 5 million cases of ARTIs are being reported annually [5] but their etiology remains largely uncharacterized due to several reasons. First, most previous studies have mainly focused on the annual Hajj pilgrimage [6–9] which might not accurately reflect locally circulating viral species. Second, while several other studies have investigated viral prevalence in ARTIs within Saudi Arabia, most of these reports have focused on specific viral pathogens [10–13]. Third, except for the city of Riyadh, the capital of Saudi Arabia [5, 11, 13–15], there is a very limited number of reports from other regions of the country [12, 16, 17].

Furthermore, the recent emergence of MERS-CoV in Saudi Arabia, its continued spread, and the associated high mortality rates (35–40%) clearly represent a serious public health and economic concern locally and globally [18]. MERS-CoV is a lineage C betacoronavirus (betaCoVs) [19] which causes symptoms ranging from asymptomatic or mild upper respiratory tract infection to severe infections associated with acute pneumonia and occasional systemic infection and multiorgan failure [20]. It was first reported in Saudi Arabia in 2012 [21] and later from 26 countries in the Arabian Peninsula, Africa, Europe, Asia, and North America [22]. Almost all reported MERS-CoV cases were linked to the Arabian Peninsula. Frequent sporadic cases of MERS-CoV and multiple hospital outbreaks have been reported in several cities in Saudi Arabia including Alahsa, Taif, Jeddah, and Riyadh [23, 24].

Jazan province is located in the southwestern part of Saudi Arabia just north of Yemen. It is the second smallest and the most densely populated province in Saudi Arabia with density of ~132/km^2 [25]. So far, there is barley any report on the prevalence of respiratory viruses in this region of the country. Furthermore, while several MERS-CoV surveillance studies have been conducted in several parts of Saudi Arabia, only few reports have examined the seroprevalence of MERS-CoV in Jazan province and found no serological evidence of MERS-CoV circulation in this region [26, 27]. These data are consistent with the overall prevalence pattern of MERS-CoV in Saudi Arabia especially that only one single MERS case has been reported from this province since 2012 [28]. Therefore, we undertook enhanced surveillance of those presenting with respiratory symptoms at five healthcare centers by testing all symptomatic outpatients between November 2013 and January 2014 (winter time) in southwestern region of Saudi Arabia.

2. Material and Methods

2.1. Samples. A total of 182 samples were collected from five healthcare centers in Jazan province, Saudi Arabia, between November 2013 and January 2014 (winter time). Nasopharyngeal swabs were collected from all symptomatic outpatients at all ages who presented with symptoms of ARTI to these five healthcare centers. Swabs were collected from 131 male and 51 female patients. All swabs were collected in virus transport media, stored at −80°C, and transported for processing at the Special Infectious Agents Unit, King Fahd Medical Research Center, King Abdulaziz University, Jeddah, Saudi Arabia. Ethical approval was obtained from Jazan University Ethical Committee (JUEC). All clinical information and laboratory results were collected and informed consent was obtained from all adult patients and parents of children.

2.2. RNA Extraction and cDNA Synthesis. In order to control for RNA extraction, samples were thawed on ice and 2 μl of armored RNA (aRNA) of hepatitis C virus (genotype 1a) was added to 200 μl of each sample as an internal control. RNA extraction was then carried out using QIAamp Viral RNA mini kit (Qiagen, Germany) according to manufacturer's instructions with final elution volume of 40 μl, and extracted RNA was stored at −80°C until use. Viral RNA was then reverse-transcribed using SuperScript III First-Strand Synthesis SuperMix kit (Invitrogen, USA). The reaction mixture consisted of 1 μl of random primers, 1 μl annealing buffer, and 6 μl of extracted RNA containing aRNA in a total volume of 8 μl. The mixture was incubated at 70°C for 5 min, and then the PCR block was cooled to 4°C. After that, 10 μl of 2x first-strand reaction buffer and 2 μl of enzyme mix were added to the mixture and incubated at 25°C for 5 min and then at 50°C for 50 min. Reaction mixture was then incubated at 85°C for 5 min to inactivate enzymes and cDNA was stored at −80°C until use. All samples were tested with Infiniti RVP plus assay on Infiniti Plus analyzer which identifies the following respiratory viruses: Influenza A virus (FluA), Influenza B virus (FluB), Influenza A-Swine H1N1 virus, hPIV (1, 2, 3, and 4), hRV (A and B), EV (A, B, C, and D), hCoV (HKU1, OC43, NL63, and 229E), hMPV (A and B), hRSV (A and B), and hAdv (A, B, C, and D) with 90% sensitivity and 100 specificity according to manufacturer (AutoGenomics, USA).

2.3. RVP Plus Infiniti Microarray Assay. Generated cDNA was PCR amplified in a multiplex reaction by adding 9.9 μl of ready to use amplification mixture containing dNTPs, multiplex primers mix, MgCl2 and reaction buffer (AutoGenomics, USA), 0.1 μl of platinum Taq polymerase (Invitrogen, USA), and 10 μl of cDNA. Cycling conditions were performed as follows; one cycle at 94°C for 2 min, followed by 39 cycles consisting of denaturation at 94°C for 30 sec, annealing at 55°C, and extension at 72°C for 1 min. Reaction was ended with a final extension at 72°C for 3 min followed by hold at 4°C. PCR products were then cleaned up from remaining dNTPs and PCR primers by adding 3 μl of Shrimp Alkaline Phosphatase (SAP), 0.75 μl of Exonuclease I (EXO) (GE Life Sciences, USA), and 0.5 μl of titanium Taq polymerase (Invitrogen, USA) to each PCR product, and the mixture was incubated for 30 min at 37°C, followed by 10 min at 94°C and a final hold at 4°C. All samples were then loaded in the Infiniti Plus analyzer for testing with RVP plus assay (AutoGenomics, USA) according to manufacturer's instructions. Briefly, samples were subjected to primer extension reaction using detection primers (AutoGenomics, USA) and fluorescent labeling with fluorescent nucleotides of the amplified product, followed by hybridization of the tagged amplified products to corresponding probes on the DNA microarray chips. Chips were then washed, scanned, and signals were recorded and analyzed. Specimen was considered positive when the ratio between virus signal and background signal was above threshold calculated by the manufacturer's software.

2.4. MERS-CoV RT-PCR. Extracted RNA from all samples was subjected to MERS-CoV upstream E-gene (UpE) detection using real-time RT-PCR on LightCycler 2.0 (Roche, Germany) as previously described [29] in a final volume of 20 μl. Negative (no-template control (NTC)) and positive controls were always included.

2.5. Statistical Methods. Data were statistically analyzed using the Statistical Package for Social Science software (SPSS v20.0; IBM Crop, Armonk, NY, USA). Descriptive statistics for continuous variables were compared using the nonparametric Mann–Whitney U test or the Kruskal-Wallis test. For categorical variables, the χ^2 test, the Fisher exact test, or the Z test was applied to evaluate the difference between proportions or to assess whether there were any associations between the proportions. A two-tailed probability value $p < 0.05$ was considered statistically significant. Values were expressed as mean or standard deviation and percentages wherever necessary.

3. Results

3.1. Epidemiological Data. Of the samples tested, 41.2% (75/182) were positive for one or more viruses (Table 1) including 72% (54/75) males and 28% (21/75) females. Saudis accounted for 84% (63/75) whereas non-Saudis represented 16% (12/75) of all the positive cases. Prevalence of positive samples was similar regardless of gender or ethnicity. Specifically, detection rates were 41.2% (54/131) and 41.1% (21/51) amongst all males and female patients, respectively, and 40.7% (63/155) and 44.44% (12/27) in Saudis and non-Saudis, respectively. Similarly, smoking seems to have no significant effect on the detection of respiratory viruses as rates were 39.4% (13/33) and 41.61% (62/149) in smokers and nonsmokers, respectively. As shown in Table 1, presentation to healthcare centers and detection rate of respiratory viruses gradually decreased with age. Respiratory viruses were more commonly detected in individuals under the age of 15 years who represented 34.7% (26/75) of the infected patients. Nonetheless, viral detection rate was 61.9% (13/21) in patients who aged 35–44 and presented to healthcare centers. This was followed by those who were older than 44 (45.45%, 5/11) and individuals under the age of 15 (41.94%, 26/62), indicating that presentation to healthcare facilities by older adults, compared to children under the age of 15 years, is usually associated with actual infections. Interestingly, marked differences in detection rates were observed between the different healthcare centers within the region. As shown in Table 1, viruses were more commonly detected in healthcare center 3 (48%, 36/75) and healthcare center number 4 (29.3%, 22/75) compared to the remaining 3 centers.

TABLE 1: Summary of demography characteristics of patients with upper respiratory tract infection in the community in Jazan province, Saudi Arabia.

Variable	Infected Number (%)	All subjects Number (%)
Total number	75 (41.2)	182 (100)
Gender		
Male	54 (72.0)	131 (72.0)
Female	21 (28.0)	51 (28.0)
Nationality		
Saudi	63 (84.0)	155 (85.2)
Non-Saudi	12 (16.0)	27 (14.8)
Smoking		
Yes	13 (17.3)	33 (18.1)
No	62 (82.7)	149 (81.9)
Age group (years)		
<15	26 (34.7)	62 (34.1)
15–24	16 (21.3)	48 (26.4)
25–34	15 (20.0)	40 (22.0)
35–44	13 (17.3)	21 (11.5)
>45	5 (6.7)	11 (6.0)
Healthcare center[¶]		
Center 1	2 (2.7)	2 (1.1)
Center 2	10 (13.3)	19 (10.4)
Center 3	36 (48.0)	68 (37.4)
Center 4	22 (29.3)	87 (47.8)
Center 5	5 (6.7)	6 (3.3)

[¶]Center 1 (Al Rowdah District, North), Center 2 (Al Rowdah District, South), Center 3 (Area 5), Center 4 (Al Safa District), and Center 5 (Al Shatea District).

3.2. Viral Prevalence and Clinical Profile. All the samples collected in this study were first tested for MERS-CoV by RT-PCR. Consistent with the current prevalence and circulation pattern of MERS-CoV in Saudi Arabia, all samples were deemed negative for MERS-CoV, suggesting that there is a lack of nasal carriage of MERS-CoV in the southwest region of Saudi Arabia. Next, all samples were tested for a variety of respiratory viruses using Infiniti RVP plus microarray assay. As shown in Table 2, out of the 75 infected patients, 62 (82.7%) were infected with a single respiratory virus. On the other hand, 17.3% (13/75) had coinfections with two respiratory viruses.

The most frequently detected viruses were hRV (40.9%, 36/88), hCoV-OC43 (15.9%, 14/88), FluA (13.6%, 12/88), hRSV-B (10.2%, 9/88), and hAdv (5.7%, 5/88). The most frequent coinfecting virus was hRV which was detected in 7 out of the 13 coinfected patients representing 26.9% (7/26) of the coinfected viruses (Table 2). This was followed by hCoV-OC43 (19.2%, 5/26), FluA (19.2%, 5/26), hAdv (15.4%, 4/26), hRSV-A (7.7%, 2/26), hRSV-B (7.7%, 2/26), and EV (3.9%, 1/26). Human RV was most commonly detected with hAdv (4/7) which was the most common coinfection (30.8%, 4/13) followed by hCoV-OC43 and FluA coinfection (23.1%, 3/13).

TABLE 2: Viruses identified in patients with upper respiratory tract infection.

Viruses	Single infections ($n = 62$) Number (%)	Coinfections ($n = 26$)[⁋] Number (%)	All infections ($n = 88$)[$] Number (%)
hRV	29 (46.8)	7 (26.9)	36 (40.9)
hCoV-OC43	9 (14.5)	5 (19.2)	14 (15.9)
hCoV-HKU1	2 (3.2)	0	2 (2.3)
hCoV-NL63	1 (1.6)	0	1 (1.1)
hRSV-A	1 (1.6)	2 (7.7)	3 (3.4)
hRSV-B	7 (11.3)	2 (7.7)	9 (10.2)
FluA	7 (11.3)	5 (19.2)	12 (13.6)
FluB	2 (3.2)	0 (0.0)	2 (2.3)
EV	2 (3.2)	1 (3.9)	3 (3.4)
hAdv	1 (1.6)	4 (15.4)	5 (5.7)
hMPVA	1 (1.6)	0	1 (1.1)

[⁋]Coinfections include all viruses detected from the 13 coinfected patients (2 × 13).

[$]Includes viruses detected in single infections (62) and in coinfections (26).

The remaining 6 coinfections were detected once and were due to unique combination of respiratory viruses (Table 3). Human RV was the only virus that was detected in all age groups and most commonly in the age group of 25–34 compared to younger or older individuals (Table 3). Interestingly, while male patients had more coinfections than females, detection of more than one virus decreased with age and completely disappeared in people older than 45 years regardless of gender (Table 3).

When comparing the overall clinical symptoms of infected and noninfected patients (Table 4), we found a significant association between the positive detection of viral infections and presentations of runny nose ($p = 0.014$), wheezing ($p = 0.007$), or lethargy ($p = 0.035$) to healthcare facilities. While the small number of patients limited our ability to examine the association between each viral infection and clinical symptoms, we observed a significant association between nasal congestion and single infections ($p = 0.0163$) compared to dual infections (Table 5).

4. Discussion

In the present study, we screened clinical samples collected from 182 symptomatic patients that presented with suspected ARTIs in the southwestern province of Saudi Arabia not only for MERS-CoV but also for other respiratory viruses. While several respiratory viruses were detected in 41.2% of the patients, we could not detect any evidence of MERS-CoV in the study population which is in accordance with previous reports from the region [26, 27]. Noteworthy, Alagaili and colleagues studied geographical circulation of MERS-CoV in dromedary camels in the Kingdom of Saudi Arabia in 2013 and did not find any evidence of MERS-CoV in Jazan province [30].

Interestingly, viral prevalence was almost similar in children under the age of 15 years and individuals older than 15

years. Specifically, 26 out of 62 children (41.9%) tested positive for one or more viruses. Similarly, 40.1% of all adults who are older than 15 years were infected. The rate of infection in children in our study is lower than previously reported rate in Riyadh (~61%) [13] or in Najran province in the southern region of Saudi Arabia (>74%) [17] (Figure 1). However, this difference is expected as both of these previous studies have mainly focused on hospitalized children ≤5 years of age. In contrast, we screened outpatients in primary healthcare settings and used a wider age range which is also consistent with the decrease in detection of respiratory viruses with increasing age (Table 1). Nonetheless, children in our study represented the largest proportion (34.7%) of all infected patients compared to other age groups.

Human RV accounted for more than 40.9% of the pathogens identified, followed by non-MERS-CoV human coronaviruses (19.3%), Flu viruses (15.9%), and hRSV (13.6%). On the other hand, detection of other respiratory viruses (hAdv, EV, and hMPVA) comprised 10.2% of all identified viruses. While hRV, Flu A, and hCoV-OC43 were more prevalent in adults older than 15 years, they were detectable in patients from most age groups. In contrast, RSV, hCoV-HKU1, and hCoV-NL63 were mostly restricted to children. Our data clearly suggest that hRV is an important cause of ARTIs in addition to Flu A and hCoV-OC43 during winter season. The overall prevalence of respiratory viruses in this study is in accordance with previous reports of respiratory viruses in children [5, 13, 17] and adults [9, 31] from Saudi Arabia. However, our data showed lower levels of RSV and hMPVA in children compared to studies from Yemen which is in very close proximity to Jazan province [32, 33]. While further studies are clearly required, these differences could be due to the additional risk factors in Yemen especially in children [32, 33].

Coinfection was not uncommon in this study, in which 17.3% of the infected patients had dual infection consistent with a recent study from Turkey [34]. However, it is higher than previously reported rates in children from Saudi Arabia [5, 17] most likely due to the methodological differences between the studies. The most common coinfecting virus was hRV (7/13) and it was detected with most viruses including hAdv, FluA, hRSV, and hCoV-OC43. Other frequently coinfecting viruses were FluA, hCoV-OC43, hRSV, and hAdv. Notably, hCoV-OC43 was the only coinfecting coronaviruses amongst the other non-MERS-CoV coronaviruses. Interestingly, hAdv was more frequently detected in coinfections (4 times) compared to single infections (one time), and all these coinfections were with hRV only. Furthermore, dual infections were more common in children and young adults between 15 and 24 years compared to other age groups. While viral coinfections are very frequent in hospitalized children with ARTIs, the impact of such infections in outpatients is not clear especially that several studies concluded that coinfections may or may not contribute to increased disease severity and the risk for hospitalization [35–38]. Nonetheless, more studies are required to better understand the clinical impact of coinfections.

Interestingly, the detection of respiratory viruses varied significantly between regions in our study. Two centers (3

TABLE 3: Demographic data of patients with upper respiratory tract infection by pathogen.

Infecting viruses	Number of patients	Age (years) Number (%)					Gender Number (%)	
		<15	15–24	25–34	35–44	>45	Male	Female
Total number	75	26 (34.7)	16 (21.3)	15 (20.0)	13 (17.3)	5 (6.7)	54 (72.0)	21 (28.0)
hRV	29	5 (19.2)	6 (37.5)	9 (60.0)	5 (38.5)	4 (80.0)	24 (82.8)	5 (17.2)
hCoV-OC43	9	3 (11.5)	1 (6.3)	1 (6.66)	4 (30.8)	0	7 (77.8)	2 (22.2)
hCoV-HKU1	2	1 (3.8)	0	0	0	1 (20.0)	0	2 (100)
hCoV-NL63	1	1 (3.8)	0	0	0	0	1 (100)	0
hRSV-A	1	0	1 (6.3)	0	0	0	1 (100)	0
hRSV-B	7	5 (19.2)	2 (12.5)	0	0	0	4 (57.1)	3 (42.9)
FluA	7	2 (7.7)	1 (6.3)	2 (13.33)	2 (15.4)	0	5 (71.4)	2 (28.6)
FluB	2	1 (3.8)	0	1 (6.66)	0	0	1 (50)	1 (50)
EV	2	2 (7.7)	0	0	0	0	1 (50)	1 (50)
hAdv	1	0	1 (6.3)	0	0	0	1 (100)	0
hMPVA	1	0	0	1 (6.66)	0	0	1 (100)	0
hRV + FluA	1	0	1 (6.3)	0	0	0	1 (100)	0
hAdv + hRV	4	2 (7.7)	1 (6.3)	1 (6.66)	0	0	2 (50)	2 (50)
hCoV-OC43 + FluA	3	1 (3.8)	1 (6.3)	0	1 (7.7)	0	3 (100)	0
hRV + hRSV-A	1	1 (3.8)	0	0	0	0	1 (100)	0
hRSV-A + hRSV-B	1	1 (3.8)	0	0	0	0	0	1 (100)
hCoV-OC43 + hRSV-B	1	0	1 (6.3)	0	0	0	1 (100)	0
FluA + EV	1	1 (3.8)	0	0	0	0	0	1
hCoV-OC43 + hRV	1	0	0	0	1 (7.7)	0	0	1

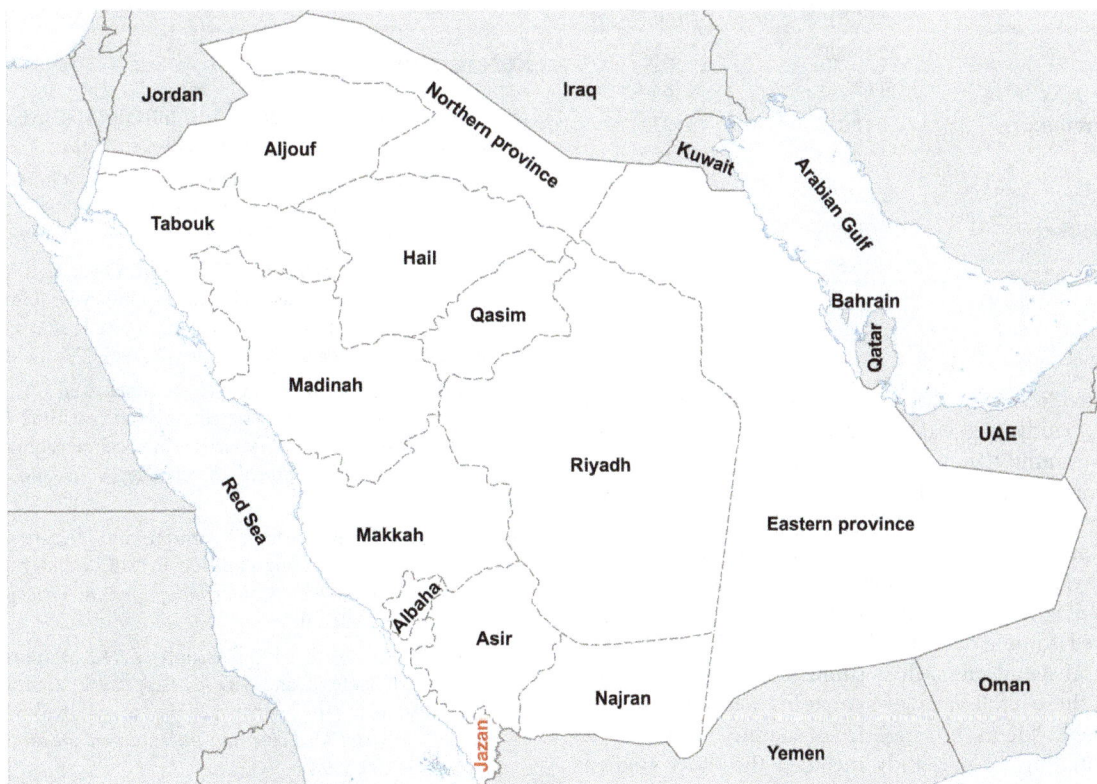

FIGURE 1: Map of Saudi Arabia showing the administrative provinces. Jazan province (red) is in the southwestern region of the country and north of Yemen.

TABLE 4: Clinical symptoms in infected and noninfected patients.

Clinical symptoms	Infected ($n = 75$)	Non-infected ($n = 107$)
	Number (%)	Number (%)
Fever	54 (72)	65 (60.7)
Cough	68 (90.7)	96 (89.7)
Runny nose	67 (89.3)*	80 (74.8)
Wheezing	17 (22.7)*	9 (8.4)
Headache	36 (48.0)	38 (35.5)
Sore throat	62 (82.7)	88 (82.2)
Difficult breathing	8 (10.7)	11 (10.3)
Lethargy	30 (40.0)*	27 (25.2)
Nausea	10 (13.3)	10 (9.3)
Nasal congestion	59 (78.6)	72 (67.3)
Earache	7 (9.3)	12 (11.2)

*Significant p value < 0.05.

TABLE 5: Clinical symptoms in patients with single infections and coinfections.

Clinical symptoms	Single infection ($n = 62$)	Coinfection ($n = 13$)
	Number (%)	Number (%)
Fever	43 (69.4)	11 (84.6)
Cough	55 (88.7)	13 (100)
Runny nose	56 (90.3)	11 (84.6)
Wheezing	13 (21)	4 (30.8)
Headache	29 (46.8)	7 (53.8)
Sore throat	51 (82.3)	11 (84.6)
Difficult breathing	5 (8.1)	3 (23.1)
Lethargy	25 (40.3)	5 (38.5)
Nausea	8 (12.9)	2 (15.4)
Nasal congestion	52 (83.9)*	7 (53.8)
Earache	4 (6.5)	3 (23.1)

*Significant p value < 0.05.

and 4) provided more than 77% of the positive cases in the current study, suggesting that these centers may serve a large community in the region compared to the other 3 centers and thus they could represent suitable sentinels for future surveillance studies. The similarity in clinical symptoms and manifestations of patients infected not only by respiratory viruses but also by some bacterial agents represent a hurdle in diagnosis based on clinical presentation. Although we observed some association between viral infection and some clinical symptoms such as running nose, lethargy, and wheezing, the overall array of symptoms were not specific most probably due to the small size of samples in this study. Other limitations in our study included the short studied period (winter season) and focusing of viral agents only. Nonetheless, the rapid advancements in molecular diagnostic methods such as microarray or multiplex PCR should aid in the epidemiological characterization of circulating viruses.

In conclusion, our data shows that circulating viruses in the southwestern province of Saudi Arabia are highly diverse and hRV represent a major pathogen in all age groups during winter season. Furthermore, it shows that MERS-CoV is infrequent in this region of Saudi Arabia compared to other regions most probably due to the limited number of dromedary camels, the reservoir host for MERS-CoV, and consequently their direct contact. Finally, use of multiplex assays such as the one used herein could help in the determination of the spectrum and diversity of respiratory viruses and in the implementation of effective control measures by public health authorities. This work shows the importance of enhanced surveillance in understanding the epidemiology of respiratory infections and therefore applying the appropriate control measures.

Competing Interests

The authors declare that they have no conflict of interests.

Acknowledgments

This study was funded Deanship of Scientific Affairs and Research, Jazan University, Jazan, Saudi Arabia (Grant 3475-36). The authors thank Dr. Mohammed Abdulhaleem Ahmed, Dr. Mohammed Saad Oraby, Dr. Maysa Saad Nadeem, Dr. Waad Hassaneen, and Dr. Shymaa Alsaid from different healthcare centers, Jazan Province, for helping in the collection of samples.

References

[1] A. M. Kesson, "Respiratory virus infections," Paediatric Respiratory Reviews, vol. 8, no. 3, pp. 240–248, 2007.

[2] R. E. Black, S. Cousens, H. L. Johnson et al., "Global, regional, and national causes of child mortality in 2008: a systematic analysis," The Lancet, vol. 375, no. 9730, pp. 1969–1987, 2010.

[3] C.-S. Khor, I.-C. Sam, P.-S. Hooi, K.-F. Quek, and Y.-F. Chan, "Epidemiology and seasonality of respiratory viral infections in hospitalized children in Kuala Lumpur, Malaysia: a retrospective study of 27 years," BMC Pediatrics, vol. 12, article 32, 2012.

[4] F. Raymond, J. Carbonneau, N. Boucher et al., "Comparison of automated microarray detection with real time PCR assay for detection of respiratory viruses in specimens obtained from children," Journal of Clinical Microbiology, vol. 47, no. 3, pp. 743–750, 2009.

[5] S. F. Fagbo, M. A. Garbati, R. Hasan et al., "Acute viral respiratory infections among children in MERS-endemic Riyadh, Saudi Arabia, 2012-2013," Journal of Medical Virology, vol. 89, no. 2, pp. 195–201, 2017.

[6] S. M. El-Sheikh, S. M. El-Assouli, K. A. Mohammed, and M. Albar, "Bacteria and viruses that cause respiratory tract infections during the pilgrimage (Haj) season in Makkah, Saudi Arabia," Tropical Medicine and International Health, vol. 3, no. 3, pp. 205–209, 1998.

[7] H. H. Balkhy, Z. A. Memish, S. Bafaqeer, and M. A. Almuneef, "Influenza a common viral infection among hajj pilgrims: time for routine surveillance and vaccination," Journal of Travel Medicine, vol. 11, no. 2, pp. 82–86, 2004.

[8] Y. Mandourah, A. Al-Radi, A. H. Ocheltree, S. R. Ocheltree, and R. A. Fowler, "Clinical and temporal patterns of severe pneumonia causing critical illness during Hajj," *BMC Infectious Diseases*, vol. 12, article 117, 2012.

[9] O. Barasheed, H. Rashid, M. Alfelali et al., "Viral respiratory infections among Hajj pilgrims in 2013," *Virologica Sinica*, vol. 29, no. 6, pp. 364–371, 2014.

[10] F. N. Al-Majhdi, A. Al-Jarallah, M. Elaeed, A. Latif, L. Gissmann, and H. M. Amer, "Prevalence of respiratory syncytial virus infection in Riyadh during the winter season 2007-2008 and different risk factors impact," *International Journal of Virology*, vol. 5, no. 4, pp. 154–163, 2009.

[11] S. Al Hajjar, S. Al Thawadi, A. Al Seraihi, S. Al Muhsen, and H. Imambaccus, "Human metapneumovirus and human coronavirus infection and pathogenicity in Saudi children hospitalized with acute respiratory illness," *Annals of Saudi Medicine*, vol. 31, no. 5, pp. 523–527, 2011.

[12] A. S. Abdel-Moneim, M. M. Kamel, A. S. Al-Ghamdi, and M. I. R. Al-Malky, "Detection of bocavirus in children suffering from acute respiratory tract infections in Saudi Arabia," *PLoS ONE*, vol. 8, no. 1, Article ID e55500, 2013.

[13] H. M. Amer, M. S. Alshaman, M. A. Farrag, M. E. Hamad, M. M. Alsaadi, and F. N. Almajhdi, "Epidemiology of 11 respiratory RNA viruses in a cohort of hospitalized children in Riyadh, Saudi Arabia," *Journal of Medical Virology*, vol. 88, no. 6, pp. 1086–1091, 2016.

[14] S. Al-Hajjar, J. Akhter, S. Al Jumaah, and S. M. Hussain Qadri, "Respiratory viruses in children attending a major referral centre in Saudi Arabia," *Annals of Tropical Paediatrics*, vol. 18, no. 2, pp. 87–92, 1998.

[15] S. S. Ghazal, M. Al Howasi, and D. Chowdhury, "Acute respiratory tract infections: epidemiological data, guided case management and outcome in a pediatric hospital in Riyadh," *Annals of Saudi Medicine*, vol. 18, no. 1, pp. 75–78, 1998.

[16] M. A. Al-Shehri, A. Sadeq, and K. Quli, "Bronchiolitis in Abha, Southwest Saudi Arabia: viral etiology and predictors for hospital admission," *West African Journal of Medicine*, vol. 24, no. 4, pp. 299–304, 2005.

[17] M. S. Al-Ayed, A. M. Asaad, M. A. Qureshi, and M. S. Ameen, "Viral etiology of respiratory infections in children in southwestern Saudi Arabia using multiplex reverse-transcriptase polymerase chain reaction," *Saudi Medical Journal*, vol. 35, no. 11, pp. 1348–1353, 2014.

[18] A. M. Al Shehri, "A lesson learned from Middle East respiratory syndrome (MERS) in Saudi Arabia," *Medical Teacher*, vol. 37, supplement 1, pp. S88–S93, 2015.

[19] D. Forni, R. Cagliani, A. Mozzi et al., "Extensive positive selection drives the evolution of nonstructural proteins in lineage C betacoronaviruses," *Journal of Virology*, vol. 90, no. 7, pp. 3627–3639, 2016.

[20] A. Bermingham, M. A. Chand, C. S. Brown et al., "Severe respiratory illness caused by a novel coronavirus, in a patient transferred to the United Kingdom from the Middle East, September 2012," *Euro Surveillance*, vol. 17, no. 40, Article ID 20290, 2012.

[21] A. M. Zaki, S. Van Boheemen, T. M. Bestebroer, A. D. M. E. Osterhaus, and R. A. M. Fouchier, "Isolation of a novel coronavirus from a man with pneumonia in Saudi Arabia," *New England Journal of Medicine*, vol. 367, no. 19, pp. 1814–1820, 2012.

[22] A. Zumla, D. S. Hui, and S. Perlman, "Middle East respiratory syndrome," *The Lancet*, vol. 386, no. 9997, pp. 995–1007, 2015.

[23] A. Assiri, G. R. Abedi, A. A. Bin Saeed et al., "Multifacility outbreak of middle east respiratory syndrome in Taif, saudi Arabia," *Emerging Infectious Diseases*, vol. 22, no. 1, pp. 32–40, 2016.

[24] H. M. Al-Dorzi, S. Alsolamy, and Y. M. Arabi, "Critically ill patients with Middle East respiratory syndrome coronavirus infection," *Critical Care*, vol. 20, no. 1, article 65, 2016.

[25] General Authority for Statistics, https://www.stats.gov.sa/sites/default/files/en-Census-Jizan-1425_1.pdf.

[26] Z. A. Memish, A. Alsahly, M. A. Masri et al., "Sparse evidence of MERS-CoV infection among animal workers living in Southern Saudi Arabia during 2012," *Influenza and other Respiratory Viruses*, vol. 9, no. 2, pp. 64–67, 2015.

[27] M. A. Müller, B. Meyer, V. M. Corman et al., "Presence of Middle East respiratory syndrome coronavirus antibodies in Saudi Arabia: A Nationwide, Cross-sectional, Serological Study," *The Lancet Infectious Diseases*, vol. 15, no. 5, pp. 559–564, 2015.

[28] WHO, "Middle East respiratory syndrome coronavirus (MERS-CoV)—Saudi Arabia," Disease Outbreak News, April 2016 http://www.who.int/csr/don/14-april-2016-mers-saudi-arabia/en/.

[29] E. I. Azhar, A. M. Hashem, S. A. El-Kafrawy et al., "Detection of the middle east respiratory syndrome coronavirus genome in an air sample originating from a camel barn owned by an infected patient," *mBio*, vol. 5, no. 4, 2014.

[30] A. N. Alagaili, T. Briese, N. Mishra et al., "Middle east respiratory syndrome coronavirus infection in dromedary camels in Saudi Arabia," *mBio*, vol. 5, no. 2, Article ID e00884-14, 2014.

[31] Z. A. Memish, M. Almasri, A. Turkestani, A. M. Al-Shangiti, and S. Yezli, "Etiology of severe community-acquired pneumonia during the 2013 Hajj-part of the MERS-CoV surveillance program," *International Journal of Infectious Diseases*, vol. 25, pp. 186–190, 2014.

[32] N. Al-Sonboli, C. A. Hart, A. Al-Aeryani et al., "Respiratory syncytial virus and human metapneumovirus in children with acute respiratory infections in Yemen," *Pediatric Infectious Disease Journal*, vol. 24, no. 8, pp. 734–736, 2005.

[33] N. Al-Sonboli, C. A. Hart, N. Al-Aghbari, A. Al-Ansi, O. Ashoor, and L. E. Cuevas, "Human metapneumovirus and respiratory syncytial virus disease in children, Yemen," *Emerging Infectious Diseases*, vol. 12, no. 9, pp. 1437–1439, 2006.

[34] C. Çiçek, A. Arslan, H. S. Karakuş et al., "Prevalence and seasonal distribution of respiratory viruses in patients with acute respiratory tract infections, 2002–2014," *Mikrobiyoloji Bulteni*, vol. 49, no. 2, pp. 188–200, 2015.

[35] M. Cebey-López, J. Herberg, J. Pardo-Seco et al., "Does viral co-infection influence the severity of acute respiratory infection in children?" *PLoS ONE*, vol. 11, no. 4, Article ID e0152481, 2016.

[36] M. Cebey-López, J. Herberg, J. Pardo-Seco et al., "Viral co-infections in pediatric patients hospitalized with lower tract acute respiratory infections," *PLoS ONE*, vol. 10, no. 9, Article ID e0136526, 2015.

[37] M. A. Marcos, S. Ramón, A. Antón et al., "Clinical relevance of mixed respiratory viral infections in adults with influenza A H1N1," *European Respiratory Journal*, vol. 38, no. 3, pp. 739–742, 2011.

[38] S. Kouni, P. Karakitsos, A. Chranioti, M. Theodoridou, G. Chrousos, and A. Michos, "Evaluation of viral co-infections in hospitalized and non-hospitalized children with respiratory infections using microarrays," *Clinical Microbiology and Infection*, vol. 19, no. 8, pp. 772–777, 2013.

23

Exosomes in Human Immunodeficiency Virus Type IPathogenesis: Threat or Opportunity?

Sin-Yeang Teow,[1] Alif Che Nordin,[2,3] Syed A. Ali,[3] and Alan Soo-Beng Khoo[1]

[1]*Molecular Pathology Unit, Cancer Research Centre (CaRC), Institute for Medical Research (IMR), 50588 Kuala Lumpur, Malaysia*
[2]*Faculty of Health Sciences, Universiti Teknologi MARA (UiTM), Bertam Campus, 13200 Kepala Batas, Pulau Pinang, Malaysia*
[3]*Oncological and Radiological Sciences, Advanced Medical and Dental Institute, Universiti Sains Malaysia, 13200 Kepala Batas, Pulau Pinang, Malaysia*

Correspondence should be addressed to Sin-Yeang Teow; ronaldsyeang@gmail.com

Academic Editor: Michael Bukrinsky

Nanometre-sized vesicles, also known as exosomes, are derived from endosomes of diverse cell types and present in multiple biological fluids. Depending on their cellular origins, the membrane-bound exosomes packed a variety of functional proteins and RNA species. These microvesicles are secreted into the extracellular space to facilitate intercellular communication. Collective findings demonstrated that exosomes from HIV-infected subjects share many commonalities with Human Immunodeficiency Virus Type I (HIV-1) particles in terms of proteomics and lipid profiles. These observations postulated that HIV-resembled exosomes may contribute to HIV pathogenesis. Interestingly, recent reports illustrated that exosomes from body fluids could inhibit HIV infection, which then bring up a new paradigm for HIV/AIDS therapy. Accumulative findings suggested that the cellular origin of exosomes may define their effects towards HIV-1. This review summarizes the two distinctive roles of exosomes in regulating HIV pathogenesis. We also highlighted several additional factors that govern the exosomal functions. Deeper understanding on how exosomes promote or abate HIV infection can significantly contribute to the development of new and potent antiviral therapeutic strategy and vaccine designs.

1. Introduction

The membrane-bound exosomes are present in a wide range of human fluids such as urine [1], plasma [2], saliva [3], ascites [4], breast milk [5], semen [6], bronchoalveolar lavage liquid [7], amniotic fluid [8], and cerebrospinal fluid [9]. These microvesicles are secreted from various types of immune cells such as dendritic cells (DCs) [10], macrophages [11], T cells [12], and B cells [13], as well as tumor cells from various cancers [14, 15]. Exosomes are mainly responsible for cell-cell communication processes such as cell proliferation [15], cell invasion [16], and immune and gene regulation [17, 18]. It is known that exosomes are derived from cellular endosomes, where the inward budding takes place on the endosomal multivesicular bodies (MVBs) to form the intraluminal vesicles (ILVs) [19]. The subsequent molecular mechanism then determines the fate of ILVs, entering the lysosomal degradation pathway or released extracellularly as exosomes upon fusion of MVB membrane with the plasma membrane [20].

Accumulative findings have demonstrated that exosomes highly resembled HIV particles in many aspects, from their physical properties to composition [21–24]. This has given rise to two models that explain these similarities [24]. First, the Trojan exosome hypothesis proposed that retroviruses are originated from exosomes following the evolution involving *gag* gene mutation [25]. This explained the ability of virus to exploit the preexisting exosome biogenesis pathway for viral dissemination and be able to infect cells in Env- and receptor-independent manner [26, 27]. The second model, however, is not in line with the evolutionary theory of the virus. Instead, the "crosstalk" or "hijacker" hypothesis suggested that the retroviruses have evolved to hijack the intercellular communication pathway of the host to promote HIV pathogenesis [28]. Although both models differ from each other, the similarity of the compositions (i.e., lipids, proteins, carbohydrates, and RNAs) between viral particles and exosomes suggests that exosomes may play an indispensable role in HIV pathogenesis.

Recently, several reports have demonstrated that exosomes contain internal cargoes that can inhibit HIV infection and replication [29–31]. These antiviral exosomes were mostly found in the body fluids such as semen and breast milk. However, the inhibitory action of exosomes is not well described compared to its viral infection enhancement effects. This may be due to the high abundance of HIV pathogenesis promoting molecules within the composition of exosomes, which may mask the existing antiviral effects, if any. By far, collective findings have shown that exosomes can either promote or inhibit HIV infection, with little understanding upon the critical factors and/or the exact mechanisms that determine the exosomal effects in viral infection. In general, the source (i.e., from different cell types and biological fluids) and the composition of exosomes may exert the decisive role in contribution to HIV/AIDS pathogenesis. More effort is required to thoroughly understand the exosomal function in HIV infection in order to benefit the development of new-era HIV/AIDS therapy and vaccine designs.

2. Morphological and Biological Properties of Exosomes and HIV Particles

Exosomes share several common structural and molecular properties with HIV. Physically, their size and density range from 50 to 150 nm in diameter [32] and 1.13 to 1.21 g/mL [33], respectively, and both are surrounded by a lipid bilayer. In addition to morphological similarities, they possess similar composition such as lipids (i.e., cholesterol and glycosphingolipids) [13], carbohydrates (i.e., high mannose and complex N-linked glycans) [34], proteins (i.e., tetraspanins, MHC molecules, actin, and TSG101) [35, 36], and RNA species [24]. Exosomes from HIV-infected cells are also enriched with viral proteins such as Nef [37] and viral RNAs [18, 38]. Due to these similarities, HIV-1 is believed to be generated by the same pathway of exosome biogenesis [24, 39]. Moreover, a substantial amount of the host component (e.g., MHC-II) can enter the viral particles [25]. This can be one of the mechanisms that is exploited by viruses to evade the host immune surveillance.

Despite sharing most of the biochemical features, HIV particles have a few principal differences in comparison with exosomes. First, HIV has more organized and uniform structures regardless of the type of infected cells while the structure of exosomal vesicles varies depending on the parental cell after the membrane budding [40]. Second, the exosomal contents are highly diverse from different sources while the biochemical content of HIV virions is steadily consistent. These differences allow the purification methods based on

iodixanol density gradients and immunoaffinity isolation to efficiently harvest exosome-free HIV virions [40, 41].

3. Distinct Functions of Exosomes from Different Sources

The composition of exosomes derived from biological fluids is highly variable from one to the other, which suggest the composition may define the distinct effects of exosomes (either promoting or inhibiting viral pathogenesis). Cumulative evidence suggested that the exosomal effect on HIV mainly depends on their cellular origins [42]. In most cases but not all, exosomes derived from HIV-infected cells are more virulent and enhance infection, while exosomes from uninfected cells have protective properties. In this section, we discuss the exosomal functions and their effects on HIV pathogenesis based on their sources or origins (summarized in Table 1).

3.1. Blood/Plasma/Serum. Human blood is where the HIV virions reside and is the main biofluid responsible for HIV transmission. The blood also contains various types of cells of both HIV-susceptible and uninfected cells that secrete exosomes. It has been postulated that HIV hijacks the exosome biogenesis pathway which carries various viral proteins and RNAs for viral dissemination process [24]. A general review on how exosomes enhance the spread of various infections has been recently reported [23, 42]. Exosomes secreted from HIV-infected cells had been found to contain chemokine receptors, CCR5 and CXCR4, that were delivered to recipient cells to facilitate HIV establishment and spreading [43, 44]. Exosomes from HIV-1 infected macrophages had also been known to facilitate viral transfer to uninfected cells [45]. Exosomes that contain HIV Nef protein have multiple pathogenic effects such as induction of T-cell apoptosis [37], inhibition of RNA interference [46], and downmodulation of cell surface molecules (i.e., MHC-I and CD4) for immune evasion [47]. Nef protein also induces exosomal secretion [37], thereby contributing to HIV/AIDS pathogenesis. Unlike the transfer of CCR5 and/or CXCR4 which primarily direct HIV infection, Nef proteins promote HIV infection by activating the uninfected cells.

Other viral components that are usually found in exosomes are HIV Gag [39], viral mRNA/miRNA [18], and pathogen-associated RNAs such as HIV trans-activation response (TAR) RNA [48] that could enhance the viral infection and replication in the recipient cells. Various host surface molecules (CD45, CD86, and MHC-II) have also been exported from HIV-infected cells via exosomes to silence the immune response [49]. Additionally, exosomes derived from infected dendritic cells (DCs) have a profound enhancing effect on CD4+ T cell infection [50]. More recently, Nef-induced exosome-associated ADAM17 (ADAM metallopeptidase domain 17) had rendered resting CD4+ T cells permissive to HIV-1 infection [51], whereas ADAM17 along with TNF-α has been known to synergistically activate the latent HIV-1 in primary CD4+ T lymphocytes and macrophages [52]. HIV particles are known to incorporate

TABLE 1: Dual effects of body fluids-derived exosomes against HIV infection.

Viral/antiviral effect	Exosomal source	Active component	Reference
Promote HIV infection	Blood/plasma/serum	CCR5 and CXCR4	[43, 44]
		Nef	[37, 46, 47]
		Gag	[39]
		Viral mRNA/miRNA	[18]
		TAR RNA	[48]
		CD45, CD86, and MHC-II	[49]
		ADAM17, TNF-α	[51, 52]
		Undefined (from DCs)	[50]
Inhibit HIV infection	Blood/plasma/serum	APOBEC3G	[29, 53, 54]
		CD4	[59]
		Interferon-alpha (IFN-α)	[63]
		Interferon-beta (IFN-β)	[64]
		Tumor necrosis factor (TNF-α)	[61, 65]
		Interleukins	[61, 62]
		Undefined (from CD8+ T cells)	[60]
	Breast milk	Lewis X	[66]
		Bile lipase	[67]
		IgA and IgG antibodies	[68]
		Mucin 1 (MUC-1)	[69]
		Oligosaccharides	[70]
		Undefined	[31]
	Semen	Mucin 6	[72]
		Undefined	[30, 71]
Promote/inhibit HIV infection (unexplored)	Urine Saliva Ascites Bronchoalveolar lavage liquid (BAL) Amniotic fluid Cerebrospinal fluid Vaginal fluid	Unexplored	Unexplored

various host cell components that enable the viruses to evade the immune system. The exchange of these components could be facilitated by the existing exosomes that are enriched with the host components. While extensive studies have been conducted on exosomes derived from HIV-infected cells, little is known of the role of exosomes derived from uninfected cells in viral pathogenesis. The increased exosomal secretion may have increased the susceptibility of uninfected cells to HIV; this however warrants future investigation.

Although exosomes are mainly enriched with viral components that promote HIV/AIDS pathogenesis, a few reports have shown that exosomes may potentially inhibit HIV infection. APOBEC3G, a host cellular protein, has been known to transfer from cell to cell through exosomes to protect the recipient cell from HIV infection [29, 53, 54]. APOBEC3G, the most prominent member of APOBEC3 (A3) proteins, is a cellular cytidine deaminase that restricts HIV replication by both DNA-editing and editing-independent activities [55, 56]. In order to function, the APOBEC3G must be incorporated into the virions [57]. The extensive

role of APOBEC3G in antiviral immunity has been recently reviewed [58]. The restrictive effect is more prominent in Vif-deficient HIV-1 than the wild-type strain as the Vif proteins target and counteract APOBEC3G for polyubiquitination and degradation by the 26S proteasome [29]. APOBEC3G proteins have been detected in human mammary tissues and were packaged into milk-borne virions and subsequently restricted HIV-1 infectivity [53]. Besides, it has also been shown that APOBEC3G inhibited viral replication by blocking the function of HIV-1 reverse transcriptase [54].

Although T lymphocytes are the primary reservoir of HIV infection, exosomes released from T lymphocytes had been found to have inhibitory effects against HIV. When compared to CD4-depleted exosomes from CD4+ T cells, CD4-containing exosomes efficiently inhibited HIV-1 infection [59]. This may be due to the masking of HIV-1 envelope proteins by the exosomal CD4 that has subsequently blocked HIV infection. Similarly, exosomes secreted from CD8+ T cells were able to suppress HIV transcription within the infected cells [60]. Other components from exosomes that inhibit viral infections are interleukins [61, 62], interferon-alpha [63], interferon-beta [64], and tumor necrosis factor (TNF-α) [61, 65].

3.2. Breast Milk. While exosome's dual-functions were seen in the blood-derived exosomes, current findings on milk-derived exosomes are skewed towards their antiviral effects, probably due to its role in providing natural passive immunity for infants. Exosomes were found to contain several components such as Lewis X [66], bile lipase [67], antibodies [68], mucin 1 (MUC-1) [69], and oligosaccharides [70] that inhibit the DC-mediated HIV transmission to CD4+ T lymphocytes. More recently, the antiviral effect of exosomes has been shown to be specifically derived from the milk since no HIV-1 inhibition was seen in plasma-derived exosomes when experimented in parallel [31]. In this study, the milk exosomes bound to monocyte-derived dendritic cells (MDDCs) and inhibited HIV-1 infection of MDDCs and the subsequent viral transfer to CD4+ T cells. Cumulative work showed that milk exosomes have a strong inhibitory effect; these protective effects may be transferred to the uninfected cells of newborns via breastfeeding as part of the passive antiviral immunity. This may also be a reason why the HIV-1 transmission via breastfeeding is rare. More efforts are currently in progress to evaluate the potential of utilizing milk exosomes in the antiviral therapy [42].

3.3. Semen. HIV-1 infection is notoriously known to be transmitted through sexual intercourse. Surprisingly, the semen-derived exosomes mainly possess antiviral effects compared to mediating HIV infection. It has been shown that exosomes purified from healthy individuals inhibit the HIV-1 replication in various cell types by blocking the postentry viral RNA reverse transcription [30]. The same group has also shown that exosomes in human semen were able to restrict the HIV transmission *in vivo* in LP-BM5-infected mice model [71]. Another group has also reported that the mucin-containing exosomes were able to prevent the HIV-1 transfer from DCs to CD4+ T cells [72]. Similar to the milk exosomes, these exosomes may presumably be transferred from one to another to exert the antiviral or protective effects. Yet, this seems to be inefficient as the number of HIV cases due to sexual transmission increases every year. Although several reports have highlighted the protective role of semen-derived exosomes, the exact mechanism that is involved is still unknown. This is important in order to control or prevent the HIV spreading, particularly through unprotected sexual intercourse.

3.4. Urine, Saliva, Ascites, and Other Biological Fluids. While the anti-HIV action of exosomes has been shown in human blood, semen, and breast milk, its antiviral potential in other sources such as saliva and ascites is yet to be determined. Since the composition of exosomes is heterogenous depending on the origins, it is of crucial importance to understand the factors that may eventually lead to its viral or antiviral effects. More efforts need to be done to reveal the activity of exosomes and the mechanisms in these biofluids as they may serve as an important source for antiviral therapies.

4. Decisive Factors for Exosome Functions in HIV Infection

It is interesting that the exosome nanovesicles packed with a very complex composition displayed two contradictory functions towards HIV infection (Table 1). However, little is known about the criteria that drive the ultimate role of exosomes in HIV/AIDS pathogenesis. Cumulative findings have demonstrated that the function of exosomes is mainly directed by their cellular origin and composition. Exosomes derived from HIV-infected T cells, monocyte/macrophage, and dendritic cells contain several components that abate viral infection [21, 24, 45]. These immune cells may release immune-regulatory factors that possess antiviral property such as interferon-alpha (IFN-α) [63], interferon-beta (IFN-β) [64], tumor necrosis factor-alpha (TNF-α) [65, 66], interleukins (ILs) [66, 67], and APOBEG3G [29, 53, 54], which are exported by exosomes from the cells. However, this antiviral implication might be masked due to the lytic replication of HIV that resulted in the pathogenic effects. To note, some of the viral molecules such as Nef and viral TAR RNA have been detected in exosomes which may further enhance the infection [37, 48]. Similarly, the antiviral activity of exosomes has been found in biological fluids (i.e., semen and breast milk) that are rich in immunological molecules [30, 31]. This indicates that the function of exosomes partly depends on the cellular origin. However, the exact components and their underlying mechanism that contributes to the antiviral action are left to be discovered.

In addition to the cellular origin of exosomes, the target or recipient cells in which exosomes are delivered also play a pivotal role in viral pathogenesis. Exosomes are known to transport from HIV-infected CD4+ T cells among themselves [51] as well as to other recipients cells such as dendritic cells and macrophages [45, 73]. The dendritic cells were also known to capture exosomes from the exosome-producer cells or infected CD4+ T cells and export to HIV-susceptible cells for the infection. These processes were termed as *trans*-infection and *trans*-dissemination, respectively [73, 74]. During the process, viral proteins such as chemokine receptors and Nef may be delivered to the recipient cells, thereby enhancing HIV infection and replication. Indeed, the ultimate functions of exosomes in the HIV-infected individuals largely rely on the composition of the nanosized exosomes. When both surface molecules and internal cargoes of exosome have net pathogenic effects, there is more likelihood that exosomes would contribute to viral pathogenesis and *vice versa*. The components of exosomes may also alter the intracellular signalling pathways [75, 76], thereby affecting HIV-1 infection.

It is well-known that the content of exosomes largely varies depending on their origins. However, it is intriguing that exosomes derived from the same source (i.e., blood-derived exosomes) have opposing effects in HIV pathogenesis. While individual components within the exosomes appear to have opposing effects on HIV pathogenesis, the net effect of the exosomes may depend on the relative strength of the effects. In addition to the net pathogenic effect of exosomal contents, it is noteworthy that variation in testing conditions such as culture conditions, exosome or virions preparation, cell infection status, and exosomal transfer or delivery status may also play role in determining the outcome of exosomes in viral infection. For example, the profile of exosomes generated *in vitro* from cultured cells may differ drastically from exosomes that are freshly harvested from biofluids. The variation of exosomes derived from tumor cells cultured *in vitro* and *in vivo* had been previously reported [77]. In this case, the experiments utilizing exosomes prepared *in vitro* may not represent the *in vivo* exosomal effects on viral infection. Second, the different methods used in exosome or virions preparation may also complicate the data interpretation. Exosomes prepared *in vitro* might be contaminated with HIV virions during the purification, thereby affecting the outcome of exosomes on viral pathogenesis. Third, the efficiency of exosomal transfer from the exosome producer cells to the recipient cells may also play a key role in viral pathogenesis. For instance, exosomes derived from cultured cells *in vitro* may not be as potent in the exchange of exosomal contents (i.e., virulence factors and/or antiviral components) as compared with *in vivo* transfer [42]. Moreover, the level of exosome secretion is dependent on the HIV infection status as Nef proteins mediate the exosome release [37]. All of these factors have to be taken into consideration by researchers before concluding the study outcome.

5. Development of a New-Era HIV Therapeutic Strategy

Cumulative findings demonstrate that the cell-encoded exosome pathways enhance HIV infection, supported by the Trojan exosome hypotheses [25] and the envelope protein- and receptor-independent viral dissemination without involving fusion events [26]. On the other hand, exosomes derived from the breast milk and semen exhibit modest antiviral effects [30, 31]. Both of these contradictory functions of exosomes pose a major impact to the existing strategy of anti-HIV drug development and urge the discovery of new-era therapeutics. For instance, the Env-targeting therapeutics may contribute to a low extent or none at all to the exosome-mediated infection that can take place without Env-receptor fusion. Hence, a more comprehensive strategy is needed to control the infection. The exosomal pathway in which HIV was hijacked for the viral dissemination can be targeted by newly designed inhibitors to target exosome biogenesis and exosome uptake in both Env-dependent and independent infections. Potential exosomal targets are cellular enzymes (RNases, proteases, and lipase), cytoplasmic proteins (TSG101, cyclophilins, MHC-II, and tetraspanins), and HIV-related proteins (e.g., Nef, CCR5, and CXCR4) [21, 23]. However, the specificity of the therapy must be monitored as the host-derived exosomes may trigger

deleterious side effects. Several reports have shown that exosomal inhibitors have reduced the overall Env-dependent infection [30, 31] and these suggest the applicability of inclusion of exosome inhibitors in the adjunctive HIV therapy. Exosomes that exhibit antiviral activity can be purified from the particular sources and developed into a potential therapy. Indeed, more promising action is envisioned when the exact components and mechanisms are revealed. It would also be interesting to determine whether the relative amount of HIV promoting versus HIV inhibiting exosomes could affect disease progression and thereby serve as potential biomarkers for prognosis. While this review focuses on the role of exosomes in HIV infection, these nanovesicles have also been shown to contribute to pathogenesis of other viral classes such as Hepatitis B and C virus [21, 78] and Herpes Simplex virus (HSV) [22]. Continuous efforts must be made to enhance the understanding of the exosomal function in these viral infections and their potential use in the development of antiviral therapies.

6. Conclusions

To summarize, the host-derived exosomes enhance HIV infection by several routes, including intercellular dissemination of viral components and immune evasion whereas only a limited number of reports have shown the antiviral function of exosomes. Numerous factors, such as the cellular origins, recipient cells, and the intracellular signalling that are affected by exosomes, are currently thought to contribute to the final outcome of exosome in the infection. Variation in exosome preparation and testing conditions may significantly affect the outcome of exosomes. Compared to other fields, the exosomal functions in HIV and other types of viral infections are apparently underexplored. More research is anticipated to improve the understanding of the association between exosomes and viral infections in order to reveal the potential of exosomes in the development of anti-HIV therapy.

Abbreviations

ADAM17:	ADAM metallopeptidase domain 17
AIDS:	Acquired immunodeficiency syndrome
CD:	Cluster of differentiation
DCs:	Dendritic cells
Env:	Envelope protein
HIV:	Human immunodeficiency virus
ILVs:	Intraluminal vesicles
MDDCs:	Monocyte-derived dendritic cells
MHC-II:	Major histocompatibility complex class II
MUC:	Mucin
MVBs:	Multivesicular bodies
TAR:	trans-Activation response.

Acknowledgments

The authors would like to thank the Director General of Health Malaysia for permission to publish this study and the Director of the Institute for Medical Research for his support.

References

[1] T. Pisitkun, R.-F. Shen, and M. A. Knepper, "Identification and proteomic profiling of exosomes in human urine," *Proceedings of the National Academy of Sciences of the United States of America*, vol. 101, no. 36, pp. 13368–13373, 2004.

[2] M.-P. Caby, D. Lankar, C. Vincendeau-Scherrer, G. Raposo, and C. Bonnerot, "Exosomal-like vesicles are present in human blood plasma," *International Immunology*, vol. 17, no. 7, pp. 879–887, 2005.

[3] A. Michael, S. D. Bajracharya, P. S. T. Yuen et al., "Exosomes from human saliva as a source of microRNA biomarkers," *Oral Diseases*, vol. 16, no. 1, pp. 34–38, 2010.

[4] F. Andre, N. E. C. Schartz, M. Movassagh et al., "Malignant effusions and immunogenic tumour-derived exosomes," *The Lancet*, vol. 360, no. 9329, pp. 295–305, 2002.

[5] C. Admyre, S. M. Johansson, K. R. Qazi et al., "Exosomes with immune modulatory features are present in human breast milk," *The Journal of Immunology*, vol. 179, no. 3, pp. 1969–1978, 2007.

[6] R. Sullivan, F. Saez, J. Girouard, and G. Frenette, "Role of exosomes in sperm maturation during the transit along the male reproductive tract," *Blood Cells, Molecules, and Diseases*, vol. 35, no. 1, pp. 1–10, 2005.

[7] C. Admyre, J. Grunewald, J. Thyberg et al., "Exosomes with major histocompatibility complex class II and co-stimulatory molecules are present in human BAL fluid," *European Respiratory Journal*, vol. 22, no. 4, pp. 578–583, 2003.

[8] S. Keller, C. Rupp, A. Stoeck et al., "CD24 is a marker of exosomes secreted into urine and amniotic fluid," *Kidney International*, vol. 72, no. 9, pp. 1095–1102, 2007.

[9] J. M. Street, P. E. Barran, C. L. Mackay et al., "Identification and proteomic profiling of exosomes in human cerebrospinal fluid," *Journal of Translational Medicine*, vol. 10, article 5, 2012.

[10] A. E. Morelli, A. T. Larregina, W. J. Shufesky et al., "Endocytosis, intracellular sorting, and processing of exosomes by dendritic cells," *Blood*, vol. 104, no. 10, pp. 3257–3266, 2004.

[11] D. G. Nguyen, A. Booth, S. J. Gould, and J. E. K. Hildreth, "Evidence that HIV budding in primary macrophages occurs through the exosome release pathway," *The Journal of Biological Chemistry*, vol. 278, no. 52, pp. 52347–52354, 2003.

[12] A. M. Booth, Y. Fang, J. K. Fallon et al., "Exosomes and HIV Gag bud from endosome-like domains of the T cell plasma membrane," *Journal of Cell Biology*, vol. 172, no. 6, pp. 923–935, 2006.

[13] R. Wubbolts, R. S. Leckie, P. T. M. Veenhuizen et al., "Proteomic and biochemical analyses of human B cell-derived exosomes: potential implications for their function and multivesicular body formation," *The Journal of Biological Chemistry*, vol. 278, no. 13, pp. 10963–10972, 2003.

[14] J. Wolfers, A. Lozier, G. Raposo et al., "Tumor-derived exosomes are a source of shared tumor rejection antigens for CTL cross-priming," *Nature Medicine*, vol. 7, no. 3, pp. 297–303, 2001.

[15] B. S. Hong, J.-H. Cho, H. Kim et al., "Colorectal cancer cell-derived microvesicles are enriched in cell cycle-related mRNAs that promote proliferation of endothelial cells," *BMC Genomics*, vol. 10, article 556, 2009.

[16] J. N. Higginbotham, M. Demory Beckler, J. D. Gephart et al., "Amphiregulin exosomes increase cancer cell invasion," *Current Biology*, vol. 21, no. 9, pp. 779–786, 2011.

[17] A. Clayton and M. D. Mason, "Exosomes in tumour immunity," *Current Oncology*, vol. 16, no. 3, pp. 46–49, 2009.

[18] H. Valadi, K. Ekström, A. Bossios, M. Sjöstrand, J. J. Lee, and J. O. Lötvall, "Exosome-mediated transfer of mRNAs and microRNAs is a novel mechanism of genetic exchange between cells," *Nature Cell Biology*, vol. 9, no. 6, pp. 654–659, 2007.

[19] W. Stoorvogel, M. J. Kleijmeer, H. J. Geuze, and G. Raposo, "The biogenesis and functions of exosomes," *Traffic*, vol. 3, no. 5, pp. 321–330, 2002.

[20] D. J. Katzmann, G. Odorizzi, and S. D. Emr, "Receptor down-regulation and multivesicular-body sorting," *Nature Reviews Molecular Cell Biology*, vol. 3, no. 12, pp. 893–905, 2002.

[21] D. G. Meckes Jr. and N. Raab-Traub, "Microvesicles and viral infection," *Journal of Virology*, vol. 85, no. 24, pp. 12844–12854, 2011.

[22] T. Wurdinger, N. N. Gatson, L. Balaj, B. Kaur, X. O. Breakefield, and D. M. Pegtel, "Extracellular vesicles and their convergence with viral pathways," *Advances in Virology*, vol. 2012, Article ID 767694, 12 pages, 2012.

[23] J. S. Schorey, Y. Cheng, P. P. Singh, and V. L. Smith, "Exosomes and other extracellular vesicles in host-pathogen interactions," *EMBO Reports*, vol. 16, pp. 24–43, 2015.

[24] N. Izquierdo-Useros, M. C. Puertas, F. E. Borràs, J. Blanco, and J. Martinez-Picado, "Exosomes and retroviruses: the chicken or the egg?" *Cellular Microbiology*, vol. 13, no. 1, pp. 10–17, 2011.

[25] S. J. Gould, A. M. Booth, and J. E. K. Hildreth, "The Trojan exosome hypothesis," *Proceedings of the National Academy of Sciences of the United States of America*, vol. 100, no. 19, pp. 10592–10597, 2003.

[26] Y.-H. Chow, D. Yu, J.-Y. Zhang et al., "gp120-independent infection of CD4- epithelial cells and CD4+ T-cells by HIV-1," *Journal of Acquired Immune Deficiency Syndromes*, vol. 30, no. 1, pp. 1–8, 2002.

[27] D. Marras, L. A. Bruggeman, F. Gao et al., "Replication and compartmentalization of HIV-1 in kidney epithelium of patients with HIV-associated nephropathy," *Nature Medicine*, vol. 8, no. 5, pp. 522–526, 2002.

[28] A. Pelchen-Matthews, G. Raposo, and M. Marsh, "Endosomes, exosomes and Trojan viruses," *Trends in Microbiology*, vol. 12, no. 7, pp. 310–316, 2004.

[29] A. K. Khatua, H. E. Taylor, J. E. K. Hildreth, and W. Popik, "Exosomes packaging APOBEC3G confer human immunodeficiency virus resistance to recipient cells," *Journal of Virology*, vol. 83, no. 2, pp. 512–521, 2009.

[30] M. N. Madison, R. J. Roller, and C. M. Okeoma, "Human semen contains exosomes with potent anti-HIV-1 activity," *Retrovirology*, vol. 11, article 102, 2014.

[31] T. I. Näslund, D. Paquin-Proulx, P. T. Paredes, H. Vallhov, J. K. Sandberg, and S. Gabrielsson, "Exosomes from breast milk inhibit HIV-1 infection of dendritic cells and subsequent viral transfer to CD4+ T cells," *AIDS*, vol. 28, no. 2, pp. 171–180, 2014.

[32] J. Conde-Vancells, E. Rodriguez-Suarez, N. Embade et al., "Characterization and comprehensive proteome profiling of exosomes secreted by hepatocytes," *Journal of Proteome Research*, vol. 7, no. 12, pp. 5157–5166, 2008.

[33] C. Théry, M. Boussac, P. Véron et al., "Proteomic analysis of dendritic cell-derived exosomes: a secreted subcellular compartment distinct from apoptotic vesicles," *The Journal of Immunology*, vol. 166, no. 12, pp. 7309–7318, 2001.

[34] L. Krishnamoorthy, J. W. Bess Jr., A. B. Preston, K. Nagashima, and L. K. Mahal, "HIV-1 and microvesicles from T cells share a common glycome, arguing for a common origin," *Nature Chemical Biology*, vol. 5, no. 4, pp. 244–250, 2009.

[35] C. Théry, L. Zitvogel, and S. Amigorena, "Exosomes: composition, biogenesis and function," *Nature Reviews Immunology*, vol. 2, no. 8, pp. 569–579, 2002.

[36] D. E. Ott, "Cellular proteins detected in HIV-1," *Reviews in Medical Virology*, vol. 18, no. 3, pp. 159–175, 2008.

[37] M. Lenassi, G. Cagney, M. Liao et al., "HIV Nef is secreted in exosomes and triggers apoptosis in bystander CD4$^+$ T cells," *Traffic*, vol. 11, no. 1, pp. 110–122, 2010.

[38] D. M. Pegtel, K. Cosmopoulos, D. A. Thorley-Lawson et al., "Functional delivery of viral miRNAs via exosomes," *Proceedings of the National Academy of Sciences of the United States of America*, vol. 107, no. 14, pp. 6328–6333, 2010.

[39] Y. Fang, N. Wu, X. Gan, W. Yan, J. C. Morrell, and S. J. Gould, "Higher-order oligomerization targets plasma membrane proteins and HIV gag to exosomes," *PLoS Biology*, vol. 5, no. 6, article e158, 2007.

[40] R. Cantin, J. Diou, D. Bélanger, A. M. Tremblay, and C. Gilbert, "Discrimination between exosomes and HIV-1: purification of both vesicles from cell-free supernatants," *Journal of Immunological Methods*, vol. 338, no. 1-2, pp. 21–30, 2008.

[41] E. Chertova, O. Chertov, L. V. Coren et al., "Proteomic and biochemical analysis of purified human immunodeficiency virus type 1 produced from infected monocyte-derived macrophages," *Journal of Virology*, vol. 80, no. 18, pp. 9039–9052, 2006.

[42] M. Madison and C. Okeoma, "Exosomes: implications in HIV-1 pathogenesis," *Viruses*, vol. 7, no. 7, pp. 4093–4118, 2015.

[43] M. Mack, A. Kleinschmidt, H. Brühl et al., "Transfer of the chemokine receptor CCR5 between cells by membrane-derived microparticles: a mechanism for cellular human immunodeficiency virus 1 infection," *Nature Medicine*, vol. 6, no. 7, pp. 769–775, 2000.

[44] T. Rozmyslowicz, M. Majka, J. Kijowski et al., "Platelet- and megakaryocyte-derived microparticles transfer CXCR4 receptor to CXCR4-null cells and make them susceptible to infection by X4-HIV," *AIDS*, vol. 17, no. 1, pp. 33–42, 2003.

[45] I. Kadiu, P. Narayanasamy, P. K. Dash, W. Zhang, and H. E. Gendelman, "Biochemical and biologic characterization of exosomes and microvesicles as facilitators of HIV-1 infection in macrophages," *Journal of Immunology*, vol. 189, no. 2, pp. 744–754, 2012.

[46] M. Aqil, A. R. Naqvi, A. S. Bano, and S. Jameel, "The HIV-1 Nef protein binds argonaute-2 and functions as a viral suppressor of RNA interference," *PLoS ONE*, vol. 8, no. 9, Article ID e74472, 2013.

[47] L. R. Gray, D. Gabuzda, D. Cowley et al., "CD4 and MHC class 1 down-modulation activities of nef alleles from brain- and lymphoid tissue-derived primary HIV-1 isolates," *Journal of NeuroVirology*, vol. 17, no. 1, pp. 82–91, 2011.

[48] A. Narayanan, S. Iordanskiy, R. Das et al., "Exosomes derived from HIV-1-infected cells contain trans-activation response element RNA," *The Journal of Biological Chemistry*, vol. 288, no. 27, pp. 20014–20033, 2013.

[49] M. T. Esser, D. R. Graham, L. V. Coren et al., "Differential incorporation of CD45, CD80 (B7-1), CD86 (B7-2), and major histocompatibility complex class I and II molecules into human immunodeficiency virus type 1 virions and microvesicles: implications for viral pathogenesis and immune regulation," *Journal of Virology*, vol. 75, no. 13, pp. 6173–6182, 2001.

[50] R. D. Wiley and S. Gummuluru, "Immature dendritic cell-derived exosomes can mediate HIV-1 trans infection," *Proceedings of the National Academy of Sciences of the United States of America*, vol. 103, no. 3, pp. 738–743, 2006.

[51] C. Arenaccio, C. Chiozzini, S. Columba-Cabezas et al., "Exosomes from human immunodeficiency virus type 1 (HIV-1)-infected cells license quiescent CD4+ T lymphocytes to replicate HIV-1 through a Nef- and ADAM17-dependent mechanism," *Journal of Virology*, vol. 88, no. 19, pp. 11529–11539, 2014.

[52] C. Arenaccio, S. Anticoli, F. Manfredi, C. Chiozzini, E. Olivetta, and M. Federico, "Latent HIV-1 is activated by exosomes from cells infected with either replication-competent or defective HIV-1," *Retrovirology*, vol. 12, article 87, 2015.

[53] C. M. Okeoma, A. L. Huegel, J. Lingappa, M. D. Feldman, and S. R. Ross, "APOBEC3 proteins expressed in mammary epithelial cells are packaged into retroviruses and can restrict transmission of milk-borne virions," *Cell Host & Microbe*, vol. 8, no. 6, pp. 534–543, 2010.

[54] X. Wang, Z. Ao, L. Chen, G. Kobinger, J. Peng, and X. Yao, "The cellular antiviral protein APOBEC3G interacts with HIV-1 reverse transcriptase and inhibits its function during viral replication," *Journal of Virology*, vol. 86, no. 7, pp. 3777–3786, 2012.

[55] R. K. Holmes, F. A. Koning, K. N. Bishop, and M. H. Malim, "APOBEC3F can inhibit the accumulation of HIV-1 reverse transcription products in the absence of hypermutation. Comparisons with APOBEC3G," *Journal of Biological Chemistry*, vol. 282, no. 4, pp. 2587–2595, 2007.

[56] A. J. Schumacher, G. Haché, D. A. MacDuff, W. L. Brown, and R. S. Harris, "The DNA deaminase activity of human APOBEC3G is required for Ty1, MusD, and human immunodeficiency virus type 1 restriction," *Journal of Virology*, vol. 82, no. 6, pp. 2652–2660, 2008.

[57] T. M. Alce and W. Popik, "APOBEC3G is incorporated into virus-like particles by a direct interaction with HIV-1 gag nucleocapsid protein," *The Journal of Biological Chemistry*, vol. 279, no. 33, pp. 34083–34086, 2004.

[58] S. Stavrou and S. R. Ross, "APOBEC3 proteins in viral immunity," *Journal of Immunology*, vol. 195, no. 10, pp. 4565–4570, 2015.

[59] J. V. De Carvalho, R. O. De Castro, E. Z. M. Da Silva et al., "Nef neutralizes the ability of exosomes from CD4+ T cells to act as decoys during HIV-1 infection," *PLoS ONE*, vol. 9, no. 11, Article ID e113691, 2014.

[60] A. Tumne, V. S. Prasad, Y. Chen et al., "Noncytotoxic suppression of human immunodeficiency virus type 1 transcription by exosomes secreted from CD8$^+$ T cells," *Journal of Virology*, vol. 83, no. 9, pp. 4354–4364, 2009.

[61] R. T. Bailer, B. Lee, and L. J. Montaner, "IL-13 and TNF-α inhibit dual-tropic HIV-1 in primary macrophages by reduction of surface expression of CD4, chemokine receptors CCR5, CXCR4 and post-entry viral gene expression," *European Journal of Immunology*, vol. 30, no. 5, pp. 1340–1349, 2000.

[62] D. Creery, W. Weiss, G. Graziani-Bowering et al., "Differential regulation of CXCR4 and CCR5 expression by interleukin (IL)-4 and IL-13 is associated with inhibition of chemotaxis and human immunodeficiency virus (HIV) type 1 replication but not HIV entry into human monocytes," *Viral Immunology*, vol. 19, no. 3, pp. 409–423, 2006.

[63] J. Li, K. Liu, Y. Liu et al., "Exosomes mediate the cell-to-cell transmission of IFN-α-induced antiviral activity," *Nature Immunology*, vol. 14, no. 8, pp. 793–803, 2013.

[64] S. D. Barr, J. R. Smiley, and F. D. Bushman, "The interferon response inhibits HIV particle production by induction of TRIM22," *PLoS Pathogens*, vol. 4, no. 2, Article ID e1000007, 2008.

[65] B. R. Lane, D. M. Markovitz, N. L. Woodford, R. Rochford, R. M. Strieter, and M. J. Coffey, "TNF-α inhibits HIV-1 replication in peripheral blood monocytes and alveolar macrophages by inducing the production of RANTES and decreasing C-C chemokine receptor 5 (CCR5) expression," *The Journal of Immunology*, vol. 163, no. 7, pp. 3653–3661, 1999.

[66] M. A. Naarding, I. S. Ludwig, F. Groot et al., "Lewis X component in human milk binds DC-SIGN and inhibits HIV-1 transfer to CD4$^+$ T lymphocytes," *Journal of Clinical Investigation*, vol. 115, no. 11, pp. 3256–3264, 2005.

[67] M. A. Naarding, A. M. Dirac, I. S. Ludwig et al., "Bile salt-stimulated lipase from human milk binds DC-SIGN and inhibits human immunodeficiency virus type 1 transfer to CD4$^+$ T cells," *Antimicrobial Agents and Chemotherapy*, vol. 50, no. 10, pp. 3367–3374, 2006.

[68] M. Requena, H. Bouhlal, N. Nasreddine et al., "Inhibition of HIV-1 transmission in trans from dendritic cells to CD4$^+$ T lymphocytes by natural antibodies to the CRD domain of DC-SIGN purified from breast milk and intravenous immunoglobulins," *Immunology*, vol. 123, no. 4, pp. 508–518, 2008.

[69] E. Saeland, M. A. W. P. de Jong, A. A. Nabatov, H. Kalay, T. B. H. Geijtenbeek, and Y. van Kooyk, "MUC1 in human milk blocks transmission of human immunodeficiency virus from dendritic cells to T cells," *Molecular Immunology*, vol. 46, no. 11-12, pp. 2309–2316, 2009.

[70] P. Hong, M. R. Ninonuevo, B. Lee, C. Lebrilla, and L. Bode, "Human milk oligosaccharides reduce HIV-1-gp120 binding to dendritic cell-specific ICAM3-grabbing non-integrin (DC-SIGN)," *British Journal of Nutrition*, vol. 101, no. 4, pp. 482–486, 2009.

[71] M. N. Madison, P. H. Jones, and C. M. Okeoma, "Exosomes in human semen restrict HIV-1 transmission by vaginal cells and block intravaginal replication of LP-BM5 murine AIDS virus complex," *Virology*, vol. 482, pp. 189–201, 2015.

[72] M. J. Stax, T. van Montfort, R. R. Sprenger et al., "Mucin 6 in seminal plasma binds DC-SIGN and potently blocks dendritic cell mediated transfer of HIV-1 to CD4$^+$ T-lymphocytes," *Virology*, vol. 391, no. 2, pp. 203–211, 2009.

[73] N. Izquierdo-Useros, M. Naranjo-Gómez, J. Archer et al., "Capture and transfer of HIV-1 particles by mature dendritic cells converges with the exosome-dissemination pathway," *Blood*, vol. 113, no. 12, pp. 2732–2741, 2009.

[74] N. Izquierdo-Useros, M. Naranjo-Gómez, I. Erkizia et al., "HIV and mature dendritic cells: Trojan exosomes riding the Trojan horse?" *PLoS Pathogens*, vol. 6, no. 3, Article ID e1000740, 2010.

[75] L. Abraham and O. T. Fackler, "HIV-1 Nef: a multifaceted modulator of T cell receptor signaling," *Cell Communication and Signaling*, vol. 10, no. 1, article 39, 2012.

[76] L. Urbanelli, A. Magini, S. Buratta et al., "Signaling pathways in exosomes biogenesis, secretion and fate," *Genes*, vol. 4, no. 2, pp. 152–170, 2013.

[77] X. Xiang, Y. Liu, X. Zhuang et al., "TLR2-mediated expansion of MDSCs is dependent on the source of tumor exosomes," *The American Journal of Pathology*, vol. 177, no. 4, pp. 1606–1610, 2010.

[78] T. N. Bukong, F. Momen-Heravi, K. Kodys, S. Bala, and G. Szabo, "Exosomes from hepatitis C infected patients transmit HCV infection and contain replication competent viral RNA in complex with Ago2-miR122-HSP90," *PLoS Pathogens*, vol. 10, no. 10, Article ID e1004424, 2014.

PERMISSIONS

LIST OF CONTRIBUTORS

D. Ajith Roni, A. Sathish Kumar, Lalit Nihal and K. Sridhar
Medical Gastroenterology, Narayana Medical College Hospital, Nellore, Andhra Pradesh 524002, India

Rama Mohan Pathapati and Sujith Tumkur Rajashekar
Clinical Pharmacology, Narayana Medical College Hospital, Nellore, Andhra Pradesh 524002, India

B. R. Alkali, A. I. Daneji and A. A. Magaji
Faculty of Veterinary Medicine, Usmanu Danfodiyo University, PMB 2346, Sokoto, Sokoto State, Nigeria

L. S. Bilbis
Faculty of Science, Usmanu Danfodiyo University, PMB 2346, Sokoto, Sokoto State, Nigeria

Swapnil Subhash Bawage, Pooja Munnilal Tiwari, Shreekumar Pillai, Vida Dennis and Shree Ram Singh
Center for NanoBiotechnology Research, Alabama State University, Montgomery, AL 36104, USA

Hasan Kweder, Michelle Ainouze, Joanna Brunel, Denis Gerlier, EvelyneManet and Robin Buckland
CIRI, International Center for Infectiology Research, Universit´e de Lyon, 69007 Lyon, France
Inserm, U1111, 69007 Lyon, France
Ecole Normale Sup´erieure de Lyon, 69007 Lyon, France
Centre International de Recherche en Infectiologie, Universit´e Lyon 1, 69007 Lyon, France
CNRS, UMR 5308, Lyon, France

Shashi Khare, Inderjeet Gandhoke and Arvind Rai
Division of Microbiology, National Centre for Disease Control, 22 Sham Nath Marg, Delhi 110054, India

Sachin Kumar, Supriya Singh and L. S. Chauhan
Division of Microbiology, National Centre for Disease Control, 22 Sham Nath Marg, Delhi 110054, India
Division of Biotechnology, National Centre for Disease Control, 22 Sham Nath Marg, Delhi 110054, India

Hanu Ram
Division of Biotechnology, National Centre for Disease Control, 22 Sham Nath Marg, Delhi 110054, India

Bano Saidullah
Discipline of Life Science, School of Science, Indira Gandhi National Open University, Delhi 110068, India

Judith M. Ball
Department of Pathobiology, Texas A&M University, College Station, TX 77843, USA

Rebecca D. Parr
Department of Pathobiology, Texas A&M University, College Station, TX 77843, USA
Department of Biological Sciences & Arkansas Biosciences Institute, Arkansas State University, Jonesboro, AR 72401, USA
Department of Biology, Stephen F. Austin StateUniversity, Nacogdoches, TX 75962, USA

Fabricio Medina-Bolivar, Lingling Fang, Tianhong Yang, Luis Nopo-Olazabal, Richard L. Atwill and Pooja Ghai
Department of Biological Sciences & Arkansas Biosciences Institute, Arkansas State University, Jonesboro, AR 72401, USA

Katelyn Defrates, Emily Hambleton and Megan E. Hurlburt
Department of Biology, Stephen F. Austin StateUniversity, Nacogdoches, TX 75962, USA

John F. Arboleda and Silvio Urcuqui-Inchima
Grupo Inmunovirología, Facultad de Medicina, Universidad de Antioquia (UdeA), Calle 70 No. 52-51, Medell´ın, Colombia

Malihe Moradzadeh and Elnaz Naderi
Department of Modern Sciences and Technologies, School of Medicine, Mashhad University of Medical Sciences, Mashhad, Iran

Sirous Tayebi, Hossein Poustchi, Ghodratollah Montazeri and Ashraf Mohamadkhani
Liver and Pancreatobiliary Diseases Research Center, Digestive Diseases Research Institute, Tehran University of Medical Sciences, Tehran, Iran

Kourosh Sayehmiri
Psychosocial Injuries Research Centre, Ilam University of Medical Sciences, Ilam, Iran

Parisa Shahnazari
Monoclonal Antibody Research Centre, Avicenna Research Institute, ACECR, Tehran, Iran

Moses Olubusuyi Adewumi
Department of Virology, College of Medicine, University of Ibadan, Ibadan, Oyo State, Nigeria

Temitope Oluwasegun Cephas Faleye
Department of Virology, College of Medicine, University of Ibadan, Ibadan, Oyo State, Nigeria
Department of Microbiology, Faculty of Science, Ekiti State University, Ado Ekiti, Ekiti, Nigeria

Johnson Adekunle Adeniji
Department of Virology, College of Medicine, University of Ibadan, Ibadan, Oyo State, Nigeria
WHO National Polio Laboratory, University of Ibadan, Ibadan, Oyo State, Nigeria

Bamidele Atinuke Coker and Felix Yasha Nudamajo
Department of Microbiology, Faculty of Science, University of Ibadan, Ibadan, Oyo State, Nigeria

Shahla Shahsavandi, Mohammad Majid Ebrahimi, Shahin Masoudi and Hasan Izadi
Razi Vaccine & Serum Research Institute, Karaj 31976 19751, Iran

Hasan Kweder, Michelle Ainouze, Camille Lévy, Els Verhoeyen, François-Loïc Cosset, EvelyneManet and Robin Buckland
INSERM-U1111, 69007 Lyon, France
ENS-Lyon, 69007 Lyon, France
University of Lyon, UCB-Lyon1, 69007 Lyon, France
LabEx Ecofect, University of Lyon, 69007 Lyon, France

Sara Louise Cosby
School of Medicine, Dentistry and Biomedical Sciences, Queen's University, BT7 1NN Belfast, UK

Claude P. Muller
Institute of Immunology, Public Research Center for Health/LNS, 20A rue Auguste Lumi`ere, L-1950 Luxemburg, Grand-Duchy of Luxembourg, Luxembourg

Mario Quijada and Ricardo Lleonart
Center of Cellular and Molecular Biology of Diseases, Instituto de Investigaciones Cient´ificas y Servicios de Alta Tecnolog´ıa (INDICASAT AIP), Building 219, Ciudad del Saber, Apartado 0843-01103, Panam´a, Panama

Carolina de la Guardia
Center of Cellular and Molecular Biology of Diseases, Instituto de Investigaciones Científicas y Servicios de Alta Tecnología (INDICASAT AIP), Building 219, Ciudad del Saber, Apartado 0843-01103, Panamá, Panama
Department of Biotechnology, Acharya Nagarjuna University, Guntur, India

Sahar Essa and Widad Al-Nakib
Department of Microbiology, Faculty of Medicine, Kuwait University, 24923 Safat, Kuwait

Abdullah Owayed
Department of Pediatrics, Faculty of Medicine, Kuwait University, 24923 Safat, Kuwait

Haya Altawalah
Virology Unit, Mubarak Hospital, Ministry of Health, 24923 Safat, Kuwait

Mousa Khadadah and Nasser Behbehani
Department of Medicine, Faculty of Medicine, Kuwait University, 24923 Safat, Kuwait

Soumyabrata Nag
Department of Microbiology, IIMSAR & BCRH, Haldia,West Bengal, India

Soma Sarkar and Manideepa SenGupta
Department of Microbiology, Medical College Kolkata, 88 College Street, Kolkata,West Bengal, India

Debprasad Chattopadhyay
ICMR Virus Unit, I.D. and B.G. Hospital, GB-4, 1st Floor, 57Dr. S. C. Banerjee Road, Beliaghata, Kolkata, India

Sanjoy Bhattacharya
Department of Medicine, Medical College Kolkata, 88 College Street, Kolkata,West Bengal, India

Rahul Biswas
Department of Community Medicine, A.I.I.H. & P.H., Kolkata,West Bengal, India

Joao Leandro Paula Ferreira, Rosangela Rodrigues, Andre Minhoto Lança and Luis Fernando de Macedo Brigido
Laboratório de Retrovírus, Centro de Virologia, Instituto Adolfo Lutz, Avenue Dr. Arnaldo 355, 01246-902 São Paulo, SP, Brazil

Valeria Correia de Almeida and Taisa Grotta Ragazzo
Centro de Referência em DST/Aids, 13013-051 Campinas, SP, Brazil

Simone Queiroz Rocha and Denise Lotufo Estevam
Centro de Referência e Treinamento em DST/Aids, 04121-000 São Paulo, SP, Brazil

Ivan Sanz, Silvia Rojo and Raúl Ortiz de Lejarazu
Valladolid National Influenza Centre, Avenida Ramón y Cajal No. 7, 47005 Valladolid, Spain

Microbiology and Immunology Service, University Clinic Hospital of Valladolid, Avenida Ram´on y Cajal s/n, 47005 Valladolid, Spain

Mar Justel
Microbiology and Immunology Service, University Clinic Hospital of Valladolid, Avenida Ramón y Cajal s/n, 47005 Valladolid, Spain

Sonia Tamames, José Eugenio Lozano and Tomás Vega
Consejería de Sanidad, Junta de Castilla y León, Paseo de Zorrilla No. 1, 47007 Valladolid, Spain

Carlos Disdier
Pulmonology Service, University Clinic Hospital of Valladolid, Avenida Ram´on y Cajal s/n, 47005 Valladolid, Spain

Luiz Gustavo Gardinassi
Department of Biochemistry and Immunology, Ribeirão Preto Medical School, University of São Paulo, 14049-900 Ribeirão Preto, SP, Brazil

Samuel I. Anyanwu, Akins Doherty, Michael D. Powell, Ming B. Huang, Claudette Mitchell and Gale W. Newman
Department of Microbiology, Biochemistry and Immunology, Morehouse School of Medicine, Atlanta, GA, USA

Chamberlain Obialo and Khalid Bashir
Department of Medicine, Morehouse School of Medicine, Atlanta, GA, USA

Alexander Quarshie
Clinical Research Center, Morehouse School of Medicine, Atlanta, GA, USA

Gunisha Pasricha, Sanjay Mukherjee and Alok K. Chakrabarti
Microbial Containment Complex, National Institute of Virology, Sus Road, Pashan, Pune 411021, India

Mpho Magwalivha, Jean-Pierre Kabue and Afsatou Ndama Traore
Department of Microbiology, School of Mathematical and Natural Sciences, University of Venda, South Africa

Natasha Potgieter
Department of Microbiology, School of Mathematical and Natural Sciences, University of Venda, South Africa
Dean of School of Mathematical and Natural Sciences, University of Venda, South Africa

David H. Adler and Beau Abar
Department of Emergency Medicine, University of Rochester, Rochester, NY 14642, USA

Melissa Wallace, Thola Bennie and Linda-Gail Bekker
Desmond Tutu HIV Centre, Institute of Infectious Diseases & Molecular Medicine, Faculty of Health Sciences, University of Cape Town, Anzio Road, Observatory, Cape Town, South Africa

Tracy L. Meiring
Institute of Infectious Diseases & Molecular Medicine and Division of Medical Virology, Faculty of Health Sciences, University of Cape Town, Anzio Road, Observatory, Cape Town, South Africa

Anna-Lise Williamson
Institute of Infectious Diseases & Molecular Medicine and Division of Medical Virology, Faculty of Health Sciences, University of Cape Town, Anzio Road, Observatory, Cape Town, South Africa
National Health Laboratory Service, Groote Schuur Hospital, Cape Town, South Africa

Vinod Kumar Basode and Saleh Ejeeli
Unit of Medical Microbiology, Department of Medical Laboratory Technology, College of Applied Medical Science, Jazan University, Jazan, Saudi Arabia

Ahmed A. Abdulhaq
Unit of Medical Microbiology, Department of Medical Laboratory Technology, College of Applied Medical Science, Jazan University, Jazan, Saudi Arabia
Deanship of Scientific Affairs and Research, Jazan University, Jazan, Saudi Arabia

Nassrin A. Badroon, Ahmed M. Hassan and Tagreed L. Alsubhi
Special Infectious Agents Unit, King Fahd Medical Research Center, King Abdulaziz University, Jeddah, Saudi Arabia

Anwar M. Hashem
Special Infectious Agents Unit, King Fahd Medical Research Center, King Abdulaziz University, Jeddah, Saudi Arabia
Department of Medical Microbiology and Parasitology, Faculty of Medicine, King Abdulaziz University, Jeddah, Saudi Arabia

Esam I. Azhar
Special Infectious Agents Unit, King Fahd Medical Research Center, King Abdulaziz University, Jeddah, Saudi Arabia

Department of Medical Laboratory Technology, Faculty of AppliedMedical Sciences, King Abdulaziz University, Jeddah, Saudi Arabia

Ahmed S. Alshrari
Department of Basic Health Sciences, Faculty of Pharmacy, Northern Border University, Arar, Saudi Arabia

Yahia Solan
Department of Public Health, Ministry of Health, Jazan, Saudi Arabia

Sin-Yeang Teow and Alan Soo-Beng Khoo
Molecular Pathology Unit, Cancer Research Centre (CaRC), Institute for Medical Research (IMR), 50588 Kuala Lumpur, Malaysia

Alif Che Nordin
Faculty of Health Sciences, Universiti Teknologi MARA (UiTM), Bertam Campus, 13200 Kepala Batas, Pulau Pinang, Malaysia
Oncological and Radiological Sciences, Advanced Medical and Dental Institute, Universiti Sains Malaysia, 13200 Kepala Batas, Pulau Pinang, Malaysia

Syed A. Ali
Oncological and Radiological Sciences, Advanced Medical and Dental Institute, Universiti Sains Malaysia, 13200 Kepala Batas, Pulau Pinang, Malaysia

Index